LIBRARY
College of St. Scholastica
Duluth, Minnesota 55811

TREATMENT AND DISPOSAL OF LIQUID AND SOLID INDUSTRIAL WASTES

OTHER RELATED PERGAMON TITLES OF INTEREST

Books

ALBAIGES:
Analytical Techniques in Environmental Chemistry

HUTZINGER *et al*:
Aquatic Pollutants — Transformation and Biological Effects

MOO-YOUNG and FARQUHAR:
Waste Treatment and Utilization — Theory and Practice of Waste Management
(Proceedings of the 1978 Symposium)

MOO-YOUNG:
Waste Treatment and Utilization — Theory and Practice of Waste Management
(Proceedings of the 1980 Symposium)

ZOETEMAN:
Sensory Assessment of Water Quality

Journals

PROGRESS IN WATER TECHNOLOGY

WATER RESEARCH

WATER SUPPLY AND MANAGEMENT

Full details of all the above titles and a free specimen copy of any Pergamon journal available on request from your nearest Pergamon office.

TREATMENT AND DISPOSAL OF LIQUID AND SOLID INDUSTRIAL WASTES

PROCEEDINGS OF THE THIRD TURKISH-GERMAN
ENVIRONMENTAL ENGINEERING SYMPOSIUM, ISTANBUL, JULY 1979

Edited by
KRITON CURI
Associate Professor of Civil Engineering,
Boğaziçi University, Bebek, Istanbul, Turkey

PERGAMON PRESS
OXFORD · NEW YORK · TORONTO · SYDNEY · PARIS · FRANKFURT

U.K.	Pergamon Press Ltd., Headington Hill Hall, Oxford OX3 0BW, England
U.S.A.	Pergamon Press Inc., Maxwell House, Fairview Park, Elmsford, New York 10523, U.S.A.
CANADA	Pergamon of Canada, Suite 104, 150 Consumers Road, Willowdale, Ontario M2J 1P9, Canada
AUSTRALIA	Pergamon Press (Aust.) Pty. Ltd., P.O. Box 544, Potts Point, N.S.W. 2011, Australia
FRANCE	Pergamon Press SARL, 24 rue des Ecoles, 75240 Paris, Cedex 05, France
FEDERAL REPUBLIC OF GERMANY	Pergamon Press GmbH, 6242 Kronberg-Taunus, Hammerweg 6, Federal Republic of Germany

Copyright © 1980 Pergamon Press Ltd.

All Rights Reserved. No part of this publication may be reproduced, stored in a retrieval system or transmitted in any form or by any means: electronic, electrostatic, magnetic tape, mechanical, photocopying, recording or otherwise, without permission in writing from the publishers.

First edition 1980

British Library Cataloguing in Publication Data
Turkish-German Environmental Engineering Symposium, *3rd, Bogazici University, 1979*
Treatment and disposal of liquid and solid industrial wastes.
1. Factory and trade waste - Congresses
I. Curi, Kriton
628.5'4   TD897   80-40993
ISBN 0-08-023999-4

In order to make this volume available as economically and as rapidly as possible the author's typescript has been reproduced in its original form. This method has its typographical limitations but it is hoped that they in no way distract the reader.

Printed in Great Britain by A. Wheaton & Co. Ltd., Exeter

CONTENTS

Foreword ix

B. Hanisch
    Basic Aspects of Industrial Wastewater
    Disposal in the Federal Republic of
    Germany 1

E. Erenler
    Measures in Industrial Wastewater Treatment
    with Respect to its Reuse 9

J. W. Hernandez
    Criteria for Irrigation with Industrial Wastes 15

J. K. Bewtra, R. L. Droste, H. I. Ali
    The Significance of Power Input in the
    Testing and Biological Treatment of
    Industrial Wastewater 23

P. Harremoës
    Advanced Treatment by Manipulation of
    Microbiological Processes 49

W. W. Eckenfelder, Jr.
    Technology for the Control of Toxic
    Pollutants from Industrial
    Wastewater Discharges 65

S. Uzman, S. Yongacoḡlu, B. Walker, I. J. Dunn
    Dynamics of Phenol and Oxygen Uptake
    in a Biological Wastewater Treatment
    Reactor 73

H. Urün
    The Investigation of the Equation Describing the
    Substrate Removal Mechanism on an Inclined-Plane
    Biological Filter 89

R. Y. Tokuz
    The Response of Activated Sludge
    To High Salinity 101

F. Sengül, A. Müezzinoğlu
    Biodegradability of Aqueous Solutions of
    Ecotoxic Organic Chemical Compounds 115

G. K. Anderson, T. Donelly, D. J. Letten
    Anaerobic Treatment of High-Strength
    Industrial Wastewaters 131

H. Z. Sarikaya
    Interactions Between Ferrous
    Iron Oxidation and Phosphate 143

D. Orhon, O. Tünay
    Effect of Recycle Ratio on the Activated
    Sludge Treatment of Metal-Containing Wastes 161

I. Sekoulov
    Laboratory Scale Studies as a Design Aid for
    Industrial Wastewater Treatment Plants 169

U. Sestini
    Assistance to Small Industries in
    the Treatment of Their Wastewater 181

K. Curi, S. G. Velioğlu, V. Diyamandoğlu
    Treatment of Olive-Oil Production Wastes 189

Ö. Velicangil, J. A. Howell
    In Situ Preparation of Self-Cleaning
    Ultrafiltration Membranes 207

M. Suerth
    A Case Study of Food Production
    Wastewater Difficulties 215

H. M. Rüffer
    Treatment of Yeast Factory Waste 221

S. Muttamara, N. C. Thanh
    A Study of Brewery Wastewater
    Characteristics and Treatment 227

A. C. Saatci
    Environment-Protecting Installation
    for the Cellulose Industry 241

A. Hamza
    Wastewater Renovation in Paper Reprocessing 249

M. Z. A. Khan
    Conditioning and Disposal of Industrial
    Sludges by Direct Slurry Freezing Process 261

H. Fleckseder
    Possibilities in Reducing the Effluent
    from Chemical Pulping Operations 269

B. Baysal, F. Sengül, A. Müezzinoğlu, A. Samsunlu
    A Case Study of the Treatment of
    Wastewater from Paper Mills 283

Kh. Krauth, K. F. Staab
    The Biological Treatment of Sewage
    from Hide Glue Factories 297

J. Leentvaar, H. M. M. Koppers, W. G. Buning
    Coagulation-Flocculation of Beet-Sugar Wastewater 303

N. Taygun, N. Sendökmen, G. Ülkü
    Purification of Sugar Factory Wastewater
    in the RT-Lefrancois System 317

S. G. Saad
    Case Study of Rice Starch Waste Treatment 329

T. Okubo, J. E. Ishihara, J. Matsumoto
    Fundamental Studies on the Treatment of
    Fish-Processing Wastewater by the
    Activated Sludge Process 345

A. Samsunlu
    Evaluation of the Treatability of
    Industrial Wastewaters in Izmir by
    Bacterial Respiration Measurements 357

E. Al-Khatib, I. I. Esen
    Pollution Control in the Shuaiba
    Industrial Area in Kuwait 369

H. Mitwally
    Problems Caused by Industrial Waste
    in Alexandria and Their Remedial Action 383

M. Onuma, T. Omura
    Evaluation of Water Quality in
    the Kitakami River 391

O. Tabasaran
    Practicable Models for Solid
    Waste Recycling 403

H. Mooss
    Experiences with a Two-Stage Refuse
    Composting System at the Refuse
    Composting Plant in Salzburg, Austria 413

L. Karlsson
    Investigation and Reuse of Industrial
    Solid Waste From Stockholm 427

E. Erdin
    A Study of Selected Industrial
    Solid Waste in Izmir, Turkey 437

W. Schenkel
    Waste Management and Raw Materials Policy 445

N. L. Nemerow
    Costs of and Innovative Solutions
    for Industrial Waste Treatment 469

S. A. S. Almeida, R. G. Ludwig
    Economic Design of Industrial
    Wastewater Treatment Systems in Brazil 475

S. Muthuswamy
    Wastewater Treatment by Using
    Bamboo-Bladed Rotors 485

M. Karpuzcu
    A Mathematical Model for Nutrient
    Cycle in Reservoirs 495

Author Index 511

Subject Index 513

FOREWORD

In view of the increasing importance of the effect of industrial wastes on the pollution of the environment, it was decided that the topic of the third Turkish-German Environmental Engineering Symposium should be "Treatment and Disposal of Liquid and Solid Industrial Wastes". It was believed that, in this way exchange of information, discussion of experiences, and in more general terms dissemination of knowledge on this subject will be achieved, resulting in the development of technology related to the protection of the environment.

This Symposium was the third in the series of the symposia known as the "Turkish-German Environmental Engineering Symposia" which started in 1975 and have been repeated every two years. The first two were of a bi-national character. In this one, however, taking into consideration the international character of the subject, it was kept open to all scientists and engineers, and thus it gained an international character.

In the present volume forty-three papers contributed by scientists from eighteen countries of different degree of development are included. Thus, information about methods used or recommended to be used to abate industrial pollution under different conditions are presented.

The people who should be thanked for the organization of this Symposium are so numerous that it is practically impossible to name all here. However, I cannot avoid expressing my thanks to the administration of Boğaziçi University as well as to my colleaques of the Civil Engineering Department among which special thanks are expressed to Dr. *Giray Velioğlu*.

Thanks are also expressed to all the lecturers. This volume is the result of their contribution.

Appreciation is also extended to *Miss Marion Leith* for her assistance in improving the wording of the text as well as to *Miss Meral Akyol* for directing the typing of this manuscript.

Boğaziçi University　　　　　　　　　　　　　　　　　　　　　　　Kriton CURI
　April, 1980

Basic Aspects of Industrial Wastewater Disposal in the Federal Republic of Germany

B. HANISCH

Institut für Siedlungswasserbau, University of Stuttgart

ABSTRACT

The protection of surface waters against harmful wastewater disposals is important with respect to ecological reasons from a general water quality management point of view. Among the effective measures in order to avoid damage of waterbodies by industrial wastewater discharges, the following are the most important:

- the internal changes of water saving cycles
- the recovery, capturing and reuse of several substances
- avoiding the internal production of toxic substances - particularly those with a long time human physiological effect - by the utilization of less hazardous substitutes
- the advanced wastewater treatment and decontamination
- the separate collection of different wastewaters inside the factories, when a separate treatment seems advantageous

The most biologically degradable organic industrial effluents can be treated without any difficulties by the well known biological processes, whereby certainly prolonged aeration times or two staged biological systems have to be chosen. For the removal of the difficult biodegradeable or resistent matter and toxic non-iron metal compounds physico-chemical processes can be utilized. Most of these processes have been well established in the chemical engineering and the drinking water supply technology, but these processes have to be further developed in order to be applicable in the wastewater treatment technology. By the effluent charge act proclaimed in 1976 an incentive is given to carry out the wastewater treatment measures in a relatively short amount of time.

RELATIONSHIPS to WATER RESOURCES MANAGEMENT

The highly developed industry plays a dominating part in German economy. It is credited for 56% of the gross national product and is supplying 51% of the places for the working people. Its productivity is clearly pretty important in determining the living standard of the inhabitants.

Due to this heavy industrialization and its centralization, problems in water resources management have to be expected. The annual water demand of the industry is close to 12.2 billion m^3, while domestic households, small businesses, public utilities and agricultural plants only need 7.6 billion m^3 per year. In a water demand prognosis for the year 2000 the figures are 22 and 11 billions m^3/year, respectively. The supply of such substantial amounts of drinking and process water of adequate quality causes difficulties. At most 16 billion m^3 of the technically accessible ground water can be captured in the whole country. Additionally, the ground water aquifers are mostly situated away from the main demand areas. Therefore, the water demand is mainly met by surface water reservoirs. In 1973, 10% of the domestic drinking water and 62% of the process water for the industry came from surface water resources. Due to the disposal of domestic and industrial wastewaters, receiving water bodies are in many cases so strongly polluted that it is no longer possible to use them as a source of drinking or process water. In many cases the reuse of these waters becomes possible only after they are subject to a very expensive water treatment.

Over the years in economically important industrial areas, the disproportion of natural resources and water demand, required special measures in order to guarantee the necessary water supply. In the Ruhr-river valley, for example, dams were constructed at the beginning of this century in order to store the flood water of the clean rivers. Approximately at the same time the first long distance pipeline between Stuttgart and the Donau-Valley near Ulm was constructed. This pipeline had a length of about 100 km and contained high quality groundwater. Nowadays, this part of the country, receives the main part of its drinking - and process water from Lake Constance. Two long distance pipes with an overall length of 246 km and with a flow capacity as high as 7.5 m^3/s are used for this purpose.

In the absence of sufficient water resources in the neighbourhood of the demand area, the protection of the surface water against pollution and contamination by the disposal of domestic and industrial wastewaters, is in fact very important with respect to their future use as resources for the drinking and process water supply. Besides the economical utilization of clean water bodies along with their "social values" and their importance for the total ecological system is generally underestimated. The environmental program of the federal government after many lists the aims of pollution control as follows:

- to maintain the quality of water bodies
- to improve the quality of water bodies already strongly polluted in order to obtain at least the second best of four quality classifications (corresponding to the β-mesasaprobic state within the saprobic system).

Water bodies that are situated within recreational areas and currently unpolluted are to be kept at the highest quality state.

CURRENT STATUS

Table 1 shows the quantity of sewage disposed by the German industries. The figures are taken of a water statistics from 1973. Within the industry groups those branches are mentioned which have to be considered the most important as to their pollution load. In single cases there might also be other branches, not given in Table 1, but decisively influencing the local pollution load on water bodies.

Table 1: Sewage Disposal from German Industries -1973-

INDUSTRY	Sewage Quantity million m^3/yr	%
Mining	1346	12.2
coal mining	1234	11.2
Raw Material&producer's goods	8198	74.6
iron	1744	15.9
oil refining	569	5.2
chemical	4237	38.5
paper mills & pulp	839	7.6
Investment goods	582	5.3
Commodity goods	416	3.8
Food & Luxury goods	448	4.1
Total	10990	100.0

As can be observed from Table 1, in 1973, 10990 million m^3 of sewage have been discharged by industrial firms. The amount of cooling water was about 76% of the total discharges. Actually the cooling water need not be treated before discharging it into the sewer or into the receiving waterbody, but nevertheless due to its high temperature it is not harmless. The other 24%, about 2621 million m^3, is sewage which need treatment. About 736 million m^3 are discharged into sewers to be treated together with the domestic sewage in a common treatment plant.

The residual 1885 million m^3 are flowing directly into the receiving waterbody (Fig. 1).

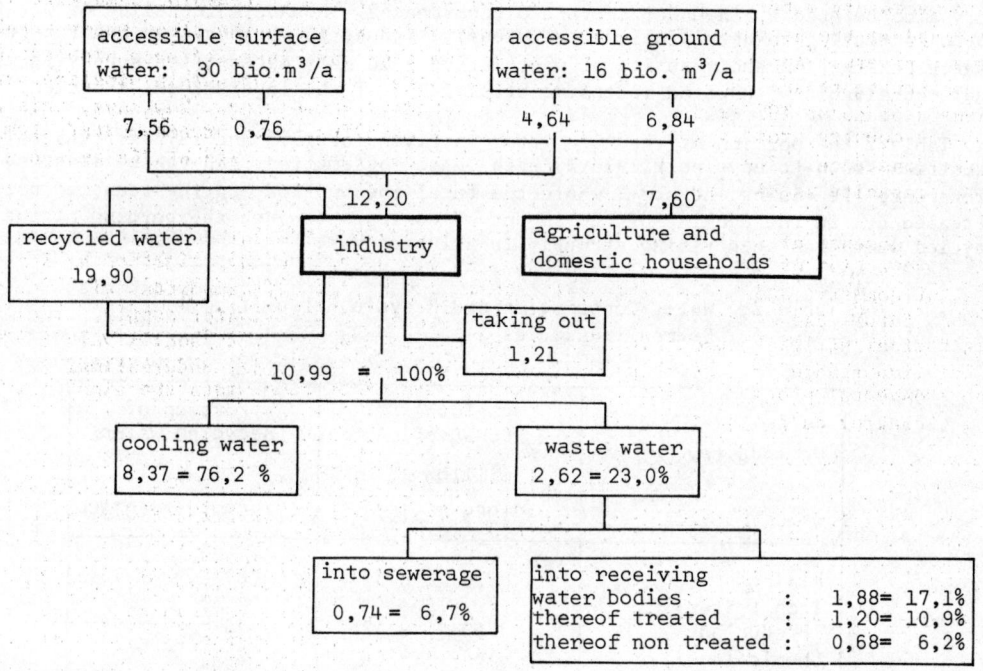

Fig. 1: Water demand and quantity of sewage from industries in Germany -1973-

The joint treatment of organic wastewaters from small and medium firms, and domestic sewage is often simpler and cheaper for the firms than the treatment in individual small plants. By the inter-mixing with domestic or other organic industrial wastes, the wastewater from an industry frequently becomes biologically degradable even if this would not be possible by a separate biological treatment. In addition, the required effluent standards are met with a higher reliability than in separate small treatment plants. For the rivers and seas being used as receiving water bodies the direct disposal of industrial wastewater is more hazardous. Such direct discharge wastewaters mainly originate from large scale manufacturing plants located near large streams, and from medium or small firms whose wastewaters cannot be connected to a domestic sewage treatment plant due to their isolated location or to their specific characteristic of wastewater. Thus, in some cases small rivers may also be loaded by direct industrial wastewater discharges. From recorded information it is observed that 1203 million m^3 are treated (although not always adequately) while 682 million m^3 are discharged without any treatment.

Comparing the total quantity of sewage from domestic households and small businesses of 3100 million m^3 with that of the industry, it should not be forgotten that the industrial wastewaters have much higher pollution levels and due to their properties of its solids they are often more hazardous than the domestic sewage being treated in most cases by mechanical and biological means.

MEASURES TO REDUCE THE QUANTITY of SEWAGE and ITS SOLIDS

Over the years a lot has been said about "recycling", and the idea of the reuse of valuable wastewater is not so new. In many ore processing, coal washing and paper manufacturing plants the recycling of water is an old practice where the water is clarified by sedimentation or flotation in tanks or ponds to be reused. Besides water reuse the captured solids - in paper mills, for example, the fiber materials - may also be used again under certain circumstances. Nowadays recycling of water has been increasing in nearly all water intensive industrial branches in order to reduce water consumption and quantity of wastewaters. In many cases water supply is no longer conceivable without an intensive utilization of the internal water cycles.

In Table 2 the portion of recycled water is given for 1959 and 1973 for several industries known to have particularly high water consumption. As can be observed, with regard to the mining industry, where the total consumption and the recycled part decreased due to the decline of the amount of coal processing, the portion of the recycled water has been rising strongly in all branches.

Table 2: Water Consumption and Recycle Ratios from German Industries

Industry	Total Consumption (million m^3/yr) 1959	1973	Portion of Recycled water % 1959	1973
Mining	8970	6757	87	78
Iron & Steel Producing	3769	9162	25	52
Chemical	3164	9162	25	52
Oil Refining	918	3510	62	82
Paper Mills & Pulp	1268	2211	33	60
Food	617	1653	28	69

For example, the chemical industry with its huge water consumption succeeded to rise the portion of recycled water from 25% in 1959 to 52% in 1973 by internal precautions. All industrial activities in the Federal Republic of Germany consumed 36,980 million m^3 water in 1973 which corresponds to 1688 liters per capita per day. Of this amount about 68% could be recovered by internal water cycles. Clearly, it is not possible to increase the utilization of the water cycles arbitrarily; furthermore, it can be assumed that several firms have approximately reached the limits given by the loss of water and the concentration of undesirable substances.

Besides the recycling of water, further processing of solids during the manufacturing processes is achieving more and more importance from an economical profit point of view in addition to pollution prevention. Thus, for example, the phenols and cresols of the coking plant wastes can be reused as raw material which are becoming rare and more expensive. Also, copper and nickel from pickling plant wastewater and chromium from galvanic baths or tanning auxiliaries have been particularly of economical interest. It is also important, in the manufacturing plants where water is used as a transport medium for solid and fluid wastes, which are produced separately, to keep these out of the water whenever it can be achieved without any difficulty. This can also be economically advantageous if the waste consists of reusable substances: for example, milk, sugar and albumin can be recovered from the whey originating in diaries. Albumin, fibrin and similar substances all being valuable raw materials for the food processing, the pharmaceutic and the chemical industry can also be recovered from slaughter house wastes. All considerations on productivity should always take into account the recovery and further processing of such substances and the fact that a lot of money can be saved at wastewater treatment.

TECHNOLOGICAL ASPECTS

Without regard to the relatively less polluted cooling waters the industrial effluents being directly discharged into the receiving water bodies have to be treated in accordance with the regulations. The extraordinary diversity and the associated technological problems of industrial wastewater treatment are documented by about 25000 to 30000 international publications. Of course in this essay it is only possible to briefly present some of the actual developments in the Federal Republic of Germany.

For the treatment of biologically degradable industrial effluents the activated sludge and trickling filter plants are known to be well matured technical processes. Certainly, in some cases an equalization of strongly fluctuating quantities and concentrations in adequately constructed basins is necessary in order to avoid shock loads to the biological units. Degradeable wastes causing rapid acid fermentation has to be fed into the biological units as fresh as possible. Wastewater from diaries for example is aerated in an equalization basin to prevent acid formation. Presence of high concentrations or a relatively persistent organic matter, such as the effluents of the chemical industry, often require two staged systems or extended aeration times. From earlier experiences it can be concluded that aeration by pure oxygen may also be advantageous.

Among the high concentrated non-biodegradable effluents that are of concern to water pollution control authorities is the pulp production wastes. In spite of the extended evaporation of the sulphite liquor and the biological treatment of the wastewater, inadmissable high loadings of refractory lignin compounds could not always be avoided. Recently, a considerable progress in treating such wastewaters has been made by a new process developed in the catchment area of Lake Constance. In the process the wastewater is first treated in an adsorption reactor, containing fine granulated aluminium oxide (Al_2O_3) and is fed into a two staged activated sludge plant. The spent aluminium oxide is regenerated during operation in a rotating drum. Due to this treatment the $KMnO_4$ - demand of the wastewater is reduced by about 95% with effluent values between 100 and 159 mg/ℓ. The final BOD_5 only comes to some few mg/ℓ.

Not only with respect to the economics but also because of harmfulness of its effluents the chemical industry is of particular importance. Due to the diversity of its intermediate and final products, of the manufacturing processes and also of the properties of the wastewater the chemical industry can be considered as the most heterogenous industrial domain. This variety is also valid for the structure of the branch: while about one half of the approximately 4500 chemical firms has less than 10 employees, the three IG-Farben-successors are found -as to their turnover- at the top of the big companies in the Federal Republic of Germany. Following production areas have to be mentioned:

- dye-staff
- pharmaceutics
- plastic fibers
- detergents and means for personal hygiene

Only in small firms - according to the nature of their chemical product - the composition of the wastewater is relatively uniform. However, most chemical industries produce wastewaters of distinct nature and with toxic influence, while processing raw materials into semimanufactured goods and into final products. These effluents can be divided into the following groups

- organic, biodegradeable
- organic, non-biodegradeable
- inorganic, strongly saline
- inorganic, metal containing

In order to permit a "regular" treatment the separate collection of these wastewaters in the factory has to be presumed. Especially in old works, which grew up during the last decades from small firms into their present size, it is often very troublesome and costly to convert the old combined wastewater collection systems into separate systems.

The biodegradable chemical effluents are treated in most cases by activated sludge plants which frequently need, with respect to the concentration and the nature of the organic matter, relatively long residence times. In the past years large biological treatment plants for chemical industries have been set into operation. Among these there is, for example, the treatment plant of the BASF in Ludwigshafen - the largest in the Federal Republic of Germany treating 7.2 m^3/s of wastewater with an aeration time of 12 hours.

In a modern, chemical manufacturing plant the wastewater usually contains toxic substances. Due to their ecological influence, and especially effects on the drinking water supplies certain non-iron metal compounds and biologically persistent organic compounds, that are concentrated in the food chains or hazardous due to their physiological long time effects, cause a lot of concern. Among the non-iron metals particularly the mercury, cadmium and lead compounds have to be mentioned, while among the biologically persistent organic substances the chlorinated hydrocarbons are most hazardous: more than up till now, the scientists will have to be faced with the question "how to replace substances resulting in pollution by less hazardous matter without an essential deterioration of the manufacturing conditions and the products themselves?" For scientists and engineers, the main task is the further development of physico-chemical processes suitable for the removal of such toxic matter, e.g., chemical precipitation, activated carbon adsorption, ion exchange and reverse osmosis, so that they may be installed with a satisfying performance and acceptable cost.

The scope of the toxic of this essay would be expanded too much, if additional industrial branches were presented. The substantial tasks and trends to be found in the chemical industry, however, will also come across in the other water intensive industrial branches, although the technology of wastewater treatment may differ depending on the nature of the wastewater.

FINANCIAL ASPECTS - EFFLUENT CHARGE ACT

The treatment and decontamination of the industrial wastewaters is not only a technological but also a financial problem for the industrial firms. By water pollution control acts and -orders of the Federal States in Germany - being very similar - the disposal of wastewater into a public waterbody is only permitted, if the effluent quality meets the standards according to the necessities of water pollution control. In the past it has been shown that both in the domestic and in the industrial field the legal orders can only be put partly through by regulations. The water pollution control boards fear to employ these regulations strictly, when the technological feasibility is doubted or serious economical consequences, often connected with the argument that working places will be lost, are pointed out by the industrial firms. In negotiations with the authorities a large company often has a better position due to having fundamental information on their status. So it can be stated, that the necessary wastewater treatment measures, in many manufacturing plants, have been carried out very slowly, and thus pollution of the receiving water bodies were unavoidable.

This situation shall be improved by the new "effluent charge act" passed through the Federal Parliaments in July, 1976. It is based on the "principle of the originator" meaning the producer of a wastewater is responsible for its treatment and disposal. By this law for the disposal of a domestic or industrial wastewater a charge has to be payed the amount of which is determined by its "hazardousness". The "hazardousness" is computed from a table and specified by the so called "Schadeinheit" (SE). The following factors are taken into consideration in the computation:

- the quantity of wastewater
- the concentration of settleable matter
- the concentration of chemically oxidizable matter
- the concentration of mercury and cadmium
- the fish toxicity

The charges will have to be paid for the first time on January 1, 1981. The amount to be paid will be increasing year by year till 1986 (starting at 12 DM/SE up to 40 DM/SE). The higher the efficiency of the sewage treatment and decontamination will obviously reduce the amount to be paid.

The following three advantages are expected by the authorities:

1. The wastewater discharging firms and communities are economically encouraged to reduce the hazardousness of their effluent, for example, by the introduction of water saving cycles, by capturing and reusing several substances, or by constructing wastewater treatment and decontamination plants.

2. The largest and capital plants (from a water pollution point of view) are constructed first, so an economical water system is promoted.

3. By these charges money will be available for measures to maintain and to improve the quality of water bodies. They may also be used as an aid in cases of economical rigour or to achieve a particularly high efficiency of wastewater treatment at a certain location, which would not be realizable alone by the incentive effect of the taxes.

The amount of time passed since the proclamation of the effluent charge act is too short to make a reliable statement on the effects observed up-to today. Nevertheless, large cities and companies, the most important pollutors of the water bodies, are ready to install further treatment stages in addition to mechanical and biological units in order to achieve a high quality effluent, more than ever as a consequence of the law.

DISCUSSIONS

Hernandez, J.W. (U.S.) : Do you have the same problem as we in the U.S. of having millions of barrels of concentrated hazardous liquid industrial waste disposed of in open pits and ponds throughout the country?

Hanisch, B. (Germany) : The use of pits and ponds requires space, which you have in the U.S. and which we lack in Germany. For this reason we tend to discourage this method of disposal and strongly urge that other methods be found of getting rid of such hazardous waste.

Measures in Industrial Wastewater Treatment with Respect to its Reuse

E. ERENLER

*At Present University of Stuttgart,
Ege University, Izmir*

INTRODUCTION

All measures aiming at the recovery of solids from sewage or wastes, solids which can again be utilized in new production processes in order to obtain economically valuable goods, are called recycling. Water like air is an indispensable basis of life for men, animals and plants.

Water plays, also an important part as a production factor for the industry. As it is not possible to increase the quantity of water within the natural cycle, we have to take care of it in order to preserve its quality. In the last decades this logical principle has unfortunately been violated more or less in a thoughtless way.

The protection of waters against an overload by sewage is a national and international aim in order to preserve it for the different utilizations of men.

Industry is one of the biggest water uses. It is particularly involved in the protection of the water reservoirs, because it needs water in huge quantities for the processing of goods as a means for chemical reactions and for cooling purposes. As these conversions due to the natural laws, never can be carried out totally to the final products and on the other hand the quantity of matter is maintained during these production processes, inevitably organic or inorganic impurities are collected in the wastewater.

So, in order to protect the water bodies, modern industry is forced to lower the amount of the disposed wastewater and the concentration of impurities as well as to watch carefully the residual influence on the receiving waters.

The most important means of sewage purification in an industrial plant are the following:

a) The construction of sewage purification plants.
b) Utilization or reuse of certain sewage-borne matter.
c) In-plant measures, like the intensification of recycling by the reuse of water, reduction of water consumption by changing production processes, etc.

RECYCLING AND SPIRALING OF THE WATER

In many industries, like the processing of coal, the production of steel, pulp and paper, the required water volume is often much higher than the volume of the produced goods. Because of that for a long time methods for deminishing the costs by reusing the water as often as possible are investigated. Cooling water, being polluted only very slightly during operation, its recycling through cooling towers is very reasonable. By cooling the water inside the production plant the disadvantageous heating of the receiving water is avoided.

The water which is polluted during processing has to be purified before being discharged into the receiving water body in order to meet certain standarts established by law . Due to the high investment and operation costs, it can be often more economical to purify the process water only partly in order to reuse it again internally for production purposes. This system has already been introduced succesfully in several industry branches. Two kinds of reuse have to be distinguished : recycling and spiraling of the water. Recycling means, that the water is passing always through the same points between production and sewage treatment. The water losses due to evaporation or sludge disposal are substituted by fresh water. Spiraling, on the other hands means the reuse of water for subordinate purposes. Spiraling is applied if recycling would be too expensive.

TREATMENT IN ORDER TO OBTAIN DRINKING WATER QUALITY

The question whether the wastewater should be treated intensively in order to obtain an effluent of drinking quality can only be answered after considering all conditions connected to each particular case. Therefore, difficulties in finding appropriate water supply, high water prices or the fact that water of a particularly high quality is required may be the most decisive factors. If for example the surface refinement of metal pieces with water being completely free of salts is adyantageous, this has to be considered at sewage treatment.

In large industrial plants it is often reasonable to aim at a separate treatment for the effluents of different processes if their degree of pollution varies greatly. Thereby it is often possible to clean only a part of the wastewater with low costs so that it can be recycled. An interesting example is the production water recycling facility of an automobile manufacturing plant constructed in 1972. After an expansion of the productivity it was no longer possible, to cover the increased water demand by the existing supply facilities. So they investigated the possibility of cleaning part of the required water (max. 600 m^3/h) and to recycle it. After a profound examination of the properties of the sewage from the different manufacturing departments they decided finally to separate 300 m^3/h and to clean this portion so far, that its reuse in a cycle becomes possible.

The selected sewage contained besides solids (dirt - and lacquer particles). From this waste, a relatively high portion of dissolved salts. Above all, the phosphate- and sulphate-compounds should have been reduced so that the water again obtains the quality of well water. The developed water treatment process contained the following steps:

a) Collection of the water and flotation of light solids inside the production.
b) Solid separation in a contact reactor with a simultaneous precipitation of phosphates and reduction of the carbonate hardness.
c) Sandfiltration.
d) Partial desalination.
e) Neutralization.
f) Pumping in order to feed the pure water into the production water net.
g) Sludge treatment.

The residual quantity of water (max. 300 m^3/h) is cleaned and discharged into the receiying water body. By the described measures the water demand of the works could be reduced to one half, so that they got independent of the supply with external water.

ION EXCHANGE PLANT WITH WASH-WATER-RECYCLING

The recycling of wash water from electroplating shops through an ion exchanger is not a real sewage treatment process. During the periodical regeneration of the ion exchangers, concentrated wastewaters are produced, which have to be treated carefully before being discharged into the sewerage.

ADVANTAGES OF THE WASH-WATER-RECYCLING

If the recycling of wash water from electroplating shops through ion exchangers is preferred to the traditional treatment decontamination and neutralization, this may have economical and technical reasons. Economical reasons which lead to a recycling of water may be an urgent water deficit of high costs. Among the technical advantages, the most important is that the ion exchange resins remove very securely and completely the wash water born ions if the concentration is below a certain fix maximum given during construction. Certain fluctuations of the concentration of the impurities are equalized. A break-through of hazardous matter into the receiving water body is not possible, because the ion exchanger is not connected to the channel. The devices for the measurement of the conductivity which are necessary in desalinization plants are secure and reliable. The completed load of the resins is automatically indicated by the increasing of the conductivity. An electrical contact, caused by an optical or acoustic signal, may take out of line the operated filter or change over to a unit in parallel if the limiting value is reached. So it is an important advantage of the ion exchangers that there is always salt-free water available for the washing of the pieces. There-fore the formation of spots or films on the pieces due to crystallized salts is impossible. The quality of the wash water is always good and equal.

RECYCLING IN THE TEXTILE INDUSTRY

The textile industry is a big water consumer. Due to the continious increase in charges for the fresh - and wastewater, this branch is also forced to recycle its wastewater. Due to the great deal of problems originated in the kind of manufacturing as well as the size of the works, no generally applicable processes can be proposed. The necessary treatment steps and their sequence have to be confirmed for each case by lab scale or pilot scale investigations in order to guarantee an optimum performance and economy of the recycling system.

A WASTEWATER RECYCLING - EXAMPLE FROM THE INDUSTRY

When in 1936 the place for the new works of the Volkswagen production was determined, a very important factor for choosing the neighbourhood of the Mittelland Channel was that this waterway appeared to be particularly suitable for the delivery of fuel to the power station.

Unfortunately by this channel it was not possible to supply enough drinking water for such a large industrial plant. Also the discharge of sewage was limited. The amount of water which they were for the cooling of the turbines in the power station, for the compressors and welding machines in the production as well as for the melting furnaces in the foundry was in the order of several thousand m^3/day. Water was also needed in the painting electroplating and hardening shops. Shortly water was needed in all parts of the production and also for the sanitary facilities which have the size of a town. As there was not enough water available, a storm water tank with a length of 700 m and a width of 350 m was constructed. This tank was afterwards expanded to 700 x 500 m with a volume of 1.5×10^6 m^3.

When the plans were finished they decided to use the storm water for production purposes. After this, the second step was only a very short one: If the discharge of the sewage and also the supply of fresh water is so difficult, why should they not clean their own wastewater, feed it into the storm water tank and reuse it in a cycle?

Today this is not only an economical solution but also a good example for an applied environmental pollution control. Now the factory covers its water demand by two dams in the Harz, by two own water drawing areas and by recycling the cooling water and the effluent of the sanitary installations. Compared to the daily water demand of 700 000 m^3 the daily fresh water supply of 35 000 m^3 is very small. This is only 5%. Nearly the same quantity is excess water, flowing into the river Aller while 95% of the water demand (665 000 m^3/d) is flowing in a continuous cycle.

In order to be able to reuse sewage, extensive purification and clarification plants of different kinds are necessary; it is not possible to treat aggressive acids in the same way like the residues of kitchens. Therefore the wastewaters are already treated at the place of their origin, afterwards separated and fed into two different sewerage systems.

In the above described works, three types of sewage are originating: storm water, industrial wastes and domestic sewage. The storm water is captured on the roofs, collected on the street or on the large parking areas. This storm water is certainly not clean but can be used as production water without any treatment.

The mechanically loaded industrial sewage in this plant is also easily cleaned. It originates, for example in the air cleaning plants of the grinding departments. It contains grinding dust and small metal scops. These solids can be removed by filtration and the water can be reused. This water, together with the storm water, is fed into a sewerage and afterwards lifted into the large reservoirs by a pumping station.

The chemically loaded industrial sewage has to be treated most carefully, because it contains salts, acids, alkaline solutions or emulsions. It is often extremely toxic. In the metal industry it is not possible to work without toxic substances, for example cyanogen - and chromium compounds are needed, in the hardening shops, in the foundry and in the electroplating department. These toxic substances have to be removed by adding other chemicals. Cyanogen compounds for example are detoxicated by iron sulphate in order to obtain harmless substances. As a result of this the toxic matter disappears but on the other hand strong and aggressive acids are formed, which should not remain in the water. They have to be neutralized by adding further chemicals. In a final treatment plant the salts and flocs formed by chemical precipitation are removed. The chemical sludge is not toxic and neutral. It is dewatered in a filter press and deposited finally. The water having passed these complicated purification steps, can be introduced again into the production water cycle.

The domestic wastewater, which consists of the water originating in the kitchens, toilets and showers is collected in a suitable sewerage and fed into a biological treatment plant. The technique of purification and recycling of this waste is an imitation of nature. One principle of water pollution control is the treatment of mechanically or biologically polluted water directly at the place of its origin, for example the sewage from a grinding shop is filtered immediately, also the concentrates from hardening - or electroplating shops are decontaminated and neutralized at once.

Several industrial firms treat their waste in another way. They bring together all biologically, mechanically and chemically polluted waters as well as the storm water in a so called combined sewerage system and feed it into a central sewage purification plant. There the whole water is cleaned in a combined process by chemical and biological methods. This system is not demanding for a treatment plant at each sewage production place and because of that at first sight, it seems to be reasonable and at least cheaper. But by experience it was shown that it is easier to decontaminate the chemically loaded sewage where the hazardous substances are most concentrated. If this hazardous wastewater is dilluted by another sewage and the toxic substances are found only in traces (which have to be removed nevertheless), the removal of these matter is more difficult, because a much higher quantity of sewage has to be treated. On the other hand if a separate system like in the VW-works is applied a variety of different treatment plants spread over the whole work area, is necessary. Moreover these chemical plants, each of which can be considered as a small chemical manufacturing system, are not cheap, too. But in most cases this system is economical, and in any case it is the most secure one, because each impurity like salts, acids, alcaline solutions or toxic substances can be removed directly by the best suitable chemicals. Furthermore if ever a leakage in the sewerage takes place only decontaminated water would penetrate into the ground water, but in no case toxic substances.

A good example for a separate system is the sewage treatment of the VW-works which was already mentioned above. The cleaned water arriving from the different treatment plants is collected in the large retention tank. In this tank an additional natural purification process takes place. After a detention time of about one week a big pumping station lifts the water back into the manufacturing plant, where it is reused.

The good quality of the water in the retention tank is demonstrated by the goldfish, swimming in a test basin filled with water of the large retention tank.

THE COOLING WATER CYCLE

In parallel to this production water cycle there are additional water cycles for the turbines in the power station, for the foundry furnaces and similar devices. Since the water itself is not polluted in these cycles - it is only heated and cooled again - filters, or other purification units are not necessary. Only a part of this cooling water - about 40% - is taken out and after utilization discharged into the Mitteland-Channel.

The heating of the water which is induced by the disposal of these cooling waters is cancelled again by surface cooling in the harbour.

Criteria for Irrigation with Industrial Wastes

J. W. HERNANDEZ

The College of Engineering, New Mexico State University

ABSTRACT

The safe disposal of industrial wastes has become a public issue in the United States; this has led to increased regulatory emphasis on alternative methods of waste management, particularly land disposal. The advantages and disadvantages of using industrial wastewaters for irrigation are outlined, as are the different operational techniques that can be employed. The mechanisms for pollutant stabilization are briefly discussed. The paper provides a review of many of the design considerations involved in an acceptable waste management system that utilizes irrigation as a means of disposal.

INTRODUCTION

Land disposal has once again become an acceptable means of wastewater management in the United States; I mean acceptable in the sense that governmental agencies are encouraging this method as a favored alternative to further treatment prior to discharge into a receiving stream. Certainly, various forms of irrigation were among the earliest practices in American sanitary engineering.

The renewed interest in well-designed land-disposal operations comes about because of national legislation (the 1972 amendments to the Federal Water-Pollution Control Act, the 1976 Resources Conservation and Recovery Act, and the 1977 Clean Water Act) that has encouraged their use. Regulation of the generation, transportation, treatment and disposal of hazardous wastes been proposed. A long list of materials defined as hazardous has been published.

Perhaps the strongest motivation for the design of adequate land-disposal sites for industrial wastes is the "horror-stroy" treatment that the country's press has accorded to the indiscrimate dumping of threatening materials into uncontrolled pits, ponds, lagoons, and landfills. The statistics are alarming to the general public: a quarter of a million United States industrial plants produce over 35 million metric tons of contaminated materials a year; thousands of drums and barrels of waste solvents and toxic materials are added annually to those already in storage (Costle, 1979). Public concern has led industry to respond with active programs to design and operate safe and secure land-disposal facilities. Irrigation by means of industrial wastes will continue to be one of the major methods of waste management to be utilized.

ADVANTAGES and DISADVANTAGES OF IRRIGATION WITH WASTEWATER

The use of irrigation as a means of ultimately disposing of industrial wastes must be subjected to critical analysis and compared to other approaches that are feasible for a given water system. There are some general advantages and disadvantages associated with this method, some of which may be quantified. A summary of the most common follows.

ADVANTAGES	DISADVANTAGES
1. Potentially lower energy costs.	1. Potentially greater population exposure to hazardous materials.
2. Elimination of surface discharge of nutrients.	2. Potential concentration of nutrients in storm-water run-off.
3. Potential means of meeting effluent water quality standards while providing only minimal pre-irrigation treatment.	3. Potential concentration of nitrates in the shallow groundwater.
4. Potential conservation of water resources and supplement of existing irrigation supplies.	4. Potential odor problems.
5. Potential increase in crop yields.	5. Possible concentration of toxic substances in harvested crops and food chain.
6. Potential improvement in control of waste materials by limiting dispersion and eliminating the need for diluting water in receiving streams.	6. Potential concentration of waste materials in the groundwater and their uncontrolled migration.
7. Possible creation of bodies of recreational water and associated aquaculture.	7. Uncertainty of the risks involved in the use of surface and groundwaters and in crops grown in wastewaters.
8. Enlargement of greenbelt areas and lowdensity open space.	8. Requirement for very long-term commitments to be made on site use.
9. Fewer work-skills required for operating staff and lower degree of treatment prior to use for irrigation.	9. Inclusion of unusual or unexpected toxic wastes in the waste stream, possibly undetected until damage is done.
10. Ability to handle shock loadings of pollutants and hydraulic overloads.	10. Inability to accept wastewaters for crop irrigation during periods of intense cold or heavy rain.
11. Possibly lower initial capital investment needed.	11. Controlled access to site may be demanded for many years after operations cease.

MECHANISMS OF REDUCTION OF POLLUTANT HAZARDS

Problems and health hazards associated with manufacturing wastes can be reduced or eliminated in many cases by the use of these effluents for irrigation. Some of the several possible mechanisms involved in pollution reduction are:

1. Biological. Crop and sod microorganisms will both act to take up and occasionally accumulate certain materials, particularly growth nutrients. Organic-waste

Criteria for Irrigation with Industrial Wastes

materials will be decomposed by soil organisms-adding nutrition for plant growth in the available soil system. Both anaerobic and aerobic conditions can occur in irrigated fields; each process could act to stabilize wastes. Conversion to gaseous forms is one possibility. The growth of soil fungi and bacteria will tend to limit reproduction of pathogenic organisms through competition. Particulate matter in a waste stream will be held in the shallow surface soils by the increase in soil microorganisms; resistance to soil erosion and pollutant movement is provided by roots and other plant fibers.

2. Physical. Sedimentation and soil filtration tend to remove particulate matter from wastewaters and to make them susceptible to other forms of stabilization. Capillary action helps maintain moisture available in the shallow-surface region and also acts to facilitate gas transfer both into and out of this region.

3. Chemical. Many diverse, complex chemical interactions take place in the top soil in an irrigated field. Precipitation of heavy metals may occur; precipitation of phosphates in calcium and aluminum-oxide complexes will likely occur. The relative concentrations of various cations in soils will be determining factors in what precipitates are formed. Adsorption an clay particles is another strong possibility for cations and for some anions. Substitution and ion exchange can also be dominant processes. Neutralization can and will take place. A shift in the pH may happen, making the formation of volatile forms of some pollutants possible, resulting in their loss to the atmosphere.

Oxidation and/or reduction reactions may occur; some by-products may be converted to soluble forms and carried off with a water flow. Similarly, particular cations may be oxidized to form insoluble complexes. On the surface, photochemical reactions are possible; this mechanism will probably be most important when organic wastes are involved. Factors that influence the rates and nature of these reactions are soil, moisture and temperature; soil characteristics and texture; particle size-distribution and permeability; cation species and their relative concentrations; presence of organic matter, availability of nutrients, and the pH of soils.

SYSTEM DESIGN CONSIDERATIONS

Several factors must be taken into consideration in the selection of a land disposal site-and in the design of the irrigation system. Among these are the following.

1. Treatment prior to irrigation. A waste that can be characterized as "hazardous" should not be used for irrigation without treatment. Guidelines on the constitution of such dangerous effluents include wastes.

   a) That are combustible with a flash point at or below 60^0C;

   b) That have vapor pressure lower than 78 mm of mercury at 25^0C;

   c) That are corrosive if the pH is less than 3 or greater than 12, or will corrode steel at a rate of 1/4 inch or more per year at 130^0F;

   d) That are reactive by being thermally unstable, tending toward autopolymerization, explosion, or reactive by generating a toxic gas;

   e) That contain constituents that exceed the limits for safe drinking water by a factor of ten;

   f) That are toxic to aquatic life or plants;

   g) That tend to accumulate or concentrate in the food chain to levels that are toxic to human beings; and,

h) That contain pathogenic organisms; i.e., total coliform counts in excess of 2,000 per liter would be unacceptable for the irrigation of recreational areas, while counts greater than 22 per liter would be unacceptable for food crops (Dawson, 1979).

2. Site-selection criteria. Public-health considerations should be foremost in choosing disposal areas. A primary concern should be the potential pollution of a public water supply through surface run-off of storm waters or percolation of toxic materials into the groundwater table. Areas that offer a high degree of isolation and long-term security should be selected. Dikes, diversions, and retaining dams can be used to improve the integrity of a site. Locations with impermeable soil systems are important where pollutants are involved that can concentrate to toxic levels in the food chain. Clay soils that are subject to swelling and clogging will be appropriate for irrigation operations with some wastes, but not others. In general, level areas with good, deep soils that can be or have been farmed will provide the best sites for irrigation with wastewater. Soils shallower than twenty inches are not satisfactory, nor is land with a slope greater than twenty percent (Powell, 1975). Areas with a high groundwater table should be avoided, and zones with rock outcrops are not acceptable.

A storage basin, possibly requiring a sealed bottom and designed to retain one to three months of effluent, is a good idea. This alternative offers operators some flexibilty and ample protection in the event of unexpected problems and wheather conditions.

3. Irrigation prectices. A number of different methods of application and operational techniques can be followed in the land disposal of industrial wastes. The two principal factors that will determine the selection of processes are the nature of the effluent and of the disposal site.

   a) Methods of application. The traditional methods of applying water are from fixed sprinkler heads, noving sprinklers, or by flooding and ridge-and-furrow systems. Spray systems have definite advantages in that the land need not be leveled nor must crops be planted in lines. The year-round application of water may be practised, even in very cold regions. On the other hand, the three disadvantages are high initial costs, continued energy costs for pumping, and problems with suspended solids in the sprinkler nozzles. Flooding requires intensive site preparation and the selection of crops that can withstand inundation. The ridge-and-furrow method is well-suited to row crops, but extensive preparation is necessary on relatively flat sites. Winter watering by flood or ridge-and-furrow methods may be limited in colder months. Geographic considerations, soils, crops, and weather will combine to determine the method of selection in most cases.

   b) Operational techniques. Overland flow might be selected in cases where large volumes of wastewater with low concentrations of dissolved pollutants are to be managed. The effluent reaching a drainage course must meet stream standards. Where the nature of a wastewater is such that pollution of the shallow groundwater will not occur, and where a very permeable soil system exists, large volumes of water can be disposed of through infiltration and percolation. To be satisfactory, soils must be well drained, and the depth to the shallow-water table should be greater than the height of capillary rise, plus a safety factor.

   In warmer regions where wheather conditions permit, a disposal system could be operated to produce optimum crop yields. Here to volume of water per acre will be lower than in systems using overland flow or ground-water recharge. High-rate irrigation can also be employed; this represents a combination of the previous three operational techniques. The relative volumes that may be disposed of are in acre-feet per acre per year: optimal crop irrigation, one to five; high-rate irrigation one to ten; overland flow, five to twenty-five; and infiltration-percolation, eleven to five hundred (Powell, 1975).

4. Crop Selection. The choice of planting should be based on the regional growing season; the volume of water available for disposal; herbal tolerance to toxic materials and salt levels; need for and availability of nutrients; ease of harvest; role of the vegetation in the food chain if toxic materials are involved; and the market for the product.

5. Hydraulic and pollutant loading. The hydraulic application rate will depend upon the type of irrigation system employed. Watering of three to four inches per week can be sustained for long periods of time-in a summer season, for example; however, lower rates should be anticipated during colder months. Pollutant loading depends on many factors, but some general guidelines may be useful.

   a) The sodium-calcium ratio is calculated by the equation below when concentrations are in milliequivalents per liter:

$$SAR = \frac{Na^+}{\sqrt{\frac{Ca^{2+} \; Mg^{+2}}{2}}}$$

   SAR should not exceed a value of 9; problems of soil texture, soil permeability, and plant uptake of water and nutrients begin to occur at SAR values greater than 3 (EPA, 1975).

   b) Salinity (total dissolved solids) can result in decreased yields and, in some cases of high salt concentration, no growth of certain crops. Problems begin when electrical conductivity values exceed one micromho per cm, (EPA, 1975).

   c) Chlorides in irrigation water can cause leaf burn when concentration exceeds three milliequivalents per liter (EPA, 1975).

   d) Nitrogen loadings may be limited, by regulation, to what crops are able to incorporate into new plant growth (75 to 300 pounds of N per acre per year is typical), but an analysis may show that other mechanisms are involved in stabilizing nitrogen compounds in the soil system or in converting them to molecular nitrogen.

   e) Phosphorus compounds will not cause serious problems if overland flow into a stream can be prevented; phosphates are adsorbed on soil colloids and can form insoluble complexes with calcium, iron, and aluminum compounds.

   f) Certain ions, such as cadmium, may cause some difficulties because of potential concentration in the food chain and others because of their toxicity to plants. The table below provides limitations developed for the Environmental Protection Agency for maximum concentrations of spesific ions (Powell, 1975).

RECOMMENDATIONS AND CONCLUSION

This paper offers general suggestions for irrigation with industrial effluents. Each site, waste, crop, climate, etc. creates a somewhat unique situation. As a consequence, the formulation of absolute design rules is impractical. Each installation must be designed to meet the specific exigencies. Irrigation with industrial wastes will typically provide a means of meeting rigorous effluent standards at a relatively low cost.

Table 1 Recommended and Estimated Maximum Concentrations
of Specific Ions In Irrigation Waters (Mg/l)[1]

Element	Removal Mechanism[2]	For Waters Used up to 20 Years on Fine-Textured Soil, pH 6.0 to 8.5	
		3 ft/yr Application Recommended Limit	8 ft/yr Application Estimated Limit
Aluminum	PR, S	20.0	8.0
Arsenic	AD, S	2.0	0.8
Beryllium	PR	0.50	0.2
Boron	AD, W	2.0-10.0	2.0
Cadmium	AD, CE, S	0.050	0.02
Chromium	AD, CE, S	1.0	0.4
Cobalt	AD, CE, S	5.0	2.0
Copper	AD, CE, S	5.0	2.0
Fluoride	AD, S	15.0	6.0
Iron	PR, CE, S	20.0	8.0
Lead	AD, CE, S	10.0	4.0
Manganese	PR, CE, S	10.0	4.0
Nickel	AD, CE, S	2.0	0.8
Selenium	AE, W	0.020	0.02
Zinc	AD, CE, S	10.0	4.0

[1] These levels will normally not adversely affect plants or soils.
[2] AD = Adsorption with iron or aluminum hydroxide, pH dependent; AE = Anion exchange; CE = Cation exchange; PR = Precipitate, pH dependent-iron and manganese are also subject to changes by oxidation reduction reaction; S = Strong strength of removal; W = Weak strength of removal.

REFERENCES

Costle, D.M. (1979). Spending Some Popularity: The Politics of Hazardous Waste Management. *Industrial Water Engineering,* March/April 1979, 8.

Dawson, (1979). Hazardous Sludge Management Rules. *Sludge Magazine,* January/February 1979, 12-14.

EPA (1975). *Evaluation of Land Application Systems: Evaluation Checklist and Supporting Commentary.* U.S. Environmental Protection Agency, Technical Bulletin EPA-430/9-75-001, March 1975.

Powell, G.M. (1975). *Land Treatment of Municipal Wastewater Effluents, Design Factors, Part II.* CH2M HILL, Denver, Colorado, USA, September 1975.

DISCUSSIONS

Hanisch, B. (Germany) : This lecture is of great interest, not only in agricultural regions but also in industrial countries. In areas with a comparatively wet climate, do you have problems with surface irrigation?

Hernandez, J.W. (U.S.) : Very wet climates will limit irrigation as a disposal practice. However, with special plants - for example, food-processing and canning factories which operate seasonally, or plants such as those in northern climates like Indiana, where the ice forming in the winter later melts and seeps gradually into the soil - limited irrigation can be practiced if the surface run-off does not cause serious problems of pollution.

Leentvar, J. (Holland) : One of the important aspects of irrigation is its usefulnes not only in crop-feeding but also in the disposal of industrial wastewater. You have already mentioned the disadvantage of hazardous effects. But when an area is flooded with industrial wastewater, there is also the possibility of a biochemical reaction, which could lead to objectionable odours. This problem makes irrigation unsuitable as a disposal practice in densely populated countries, especially in Europe.

Hernandez, J.W. (U.S.) : This is a very good point. I agree that objectionable odours are a problem and should be listed among the disadvantages of using industrial wastes in irrigation.

Günay, J. (Turkey) : Which industrial wastes are best suited to irrigation?

Hernandez, J. (U.S.) : This depends on the climate and available space. In general, any kind of organic wastewater is suitable, for example, the wastes from sugar mills, from canning, dairy and most of the other food industries, where the stream of water is fairly dilute.

The Significance of Power Input in the Testing and Biological Treatment of Industrial Wastewater

J. K. BEWTRA*, R. L. DROSTE** and H. I. ALI***

*Department of Civil Engineering, University of Windsor, Windsor, Ontario, Canada
**Department of Civil Engineering, University of Ottawa, Ottawa, Ontario, Canada
***Department of Civil Engineering, Ain-Shams University, Abbasia, Cairo, Egypt

ABSTRACT

The significance of mechanical power input into a biological system during testing and treatment is discussed. Laboratory studies on the influence of power input on biological activity were conducted in the following three parts:

  i) Evaluation of the biochemical oxygen demand of polluted waters under turbulent conditions;

  ii) Effect of mechanical blending of mixed liquor in an activated sludge system;

  iii) Effect of mechanical blending on the aerobic digestion of waste activated sludge.

It is shown that both 5-day BOD rate constant increase significantly if the contents of the BOD bottles are stirred. The blending of mixed liquor/return sludge significantly increase the overall substrate utilization rate and the endogenous decay rate. An improvement in the auto-oxidation rate means a reduced volume of sludge for disposal.

INTRODUCTION

The growing importance of effective water pollution control has created a need to improve methods for the analysis of polluted waters, treatment of wastewaters and treatment of sludges.

Biological treatment is the most common form of secondary treatment of both domestic and industrial wastewaters. Reasons for this include the ability of microorganisms to flocculate dissolved and colloidal biodegradable organic matter and to oxidize thereby reducing its quantity in the system. The activated sludge process is the most popular method of biological treatment. In this process, flocculated biological growth is continuously circulated in an aeration tank and brought in contact with organic matter in the presence of oxygen. The organic pollutants get adsorbed, absorbed, oxidized and finally synthesized into biomass. This biomass is subsequently separated by sedimentation. It is becoming a common practice to digest aerobically the excess waste sludge produced in the activated sludge plant.

The oxygen demand for the bio-degradation of organic matter in a waste is quantitatively determined by the biochemical oxygen demand (BOD) test. In the standard laboratory BOD test (Standard Methods, 1975), the wastewater sample is diluted with BOD dilution water and is incubated under quiescent conditions at a fixed temperature. The values of the BOD rate constant, obtained in the laboratory, have been used frequently in predicting the rate of biological stabilization of putrescible matter when discharged into the streams. These values have also been used in the design of aerobic biological systems in which the concentrations of both the substrate and the microorganisms are low.

It has been shown (Hartman and Launberger, 1968; Fair and co-workers, 1965) that the input of external power to a biological flocculation-oxidation system resuts in a better mixing and agitation and therefore more efficient performance. Similarly, it has been reported in the literature (Lordi and Heukelekian, 1964; Gannon, 1966) that the presence of turbulence in flowing streams significantly influences the rate of biochemical oxidation. This paper discusses the results obtained in Laboratory studies with varying rates of power input, on various aspects of the testing and biological treatment of wastewaters.

THEORY AND LITERATURE REVIEW

Substrate uptake in a biological oxidation system depends upon the following:

　　　i) The metabolic rates of the microbial cells;

　　　ii) The mass transfer of nutrients, including oxygen and substrate, to the cell surface and removal of waste products therefrom; and

　　　iii) The dissolution of oxygen in the aeration basin.

A qualitative representation of this phenomena in an activated sludge mixed liquor shown in Fig.1. Fig. 2 shows the various rate-controlling processes occurring in and around the bio-floc for substrate and oxygen utilization during biochemical oxidation (Ali, 1972). In this illustration, R is the rate of conversion of solid substrate into solution, R_2 and R_2' are the rates of transport of dissolved substrate and oxygen to the vicinity of a metabolizing organism, R_3 and R_3' are the rates of substrate and oxygen transport through the stagnant liquid film surrounding the cell, R_4 and R_4' are the rates of substrate and oxygen transport through the cell membrane, and R_b is the rate of biochemical reaction within the cell.

The calculations of Wuhrmann (1963) indicated that is much more likely for substrate, rather than oxygen, to be diffusion limited through the floc matrix in slowing rate of reaction. Mueller and his co-workers (1966) found that the oxygen transfer to or through the flocs was generally not limiting under normal operating conditions, but that a low oxygen concentration could be limiting if floc sizes were too large. Baillod and Boyle (1969a, 1969b) have demonstrated that the glucose uptake rates of blended flocs were higher than those of a nonblended control al lower concentrations of glucose; furthermore, that the control deviated from zero-order uptake kinetics at a higher glucose concentration than did the blended flocs. From the oxygen uptake data they concluded that glucose uptake was impaired due to an intra-particle resistance to mass transfer.

Many authors have noted beneficial increases in substrate uptake rates with increased power input, but have not quantified them. Baars (1965) has mentioned that increased treatment efficiency may be expected only when bacteria near the interior of the floc are supplied with the required quantities of food and oxygen. Kalinske (1971) has argued that the most effective means of improving efficiency would be achieved through better mass transfer rates.

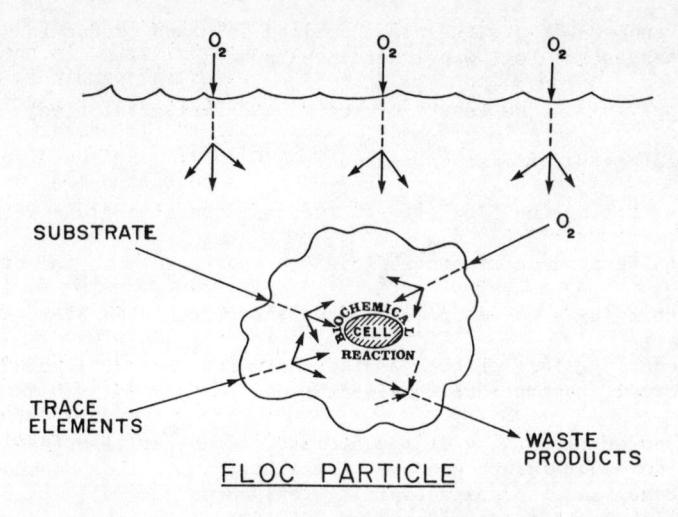

Fig. 1. Mass Transfer Processes in Activated Sludge Mixed Liquor

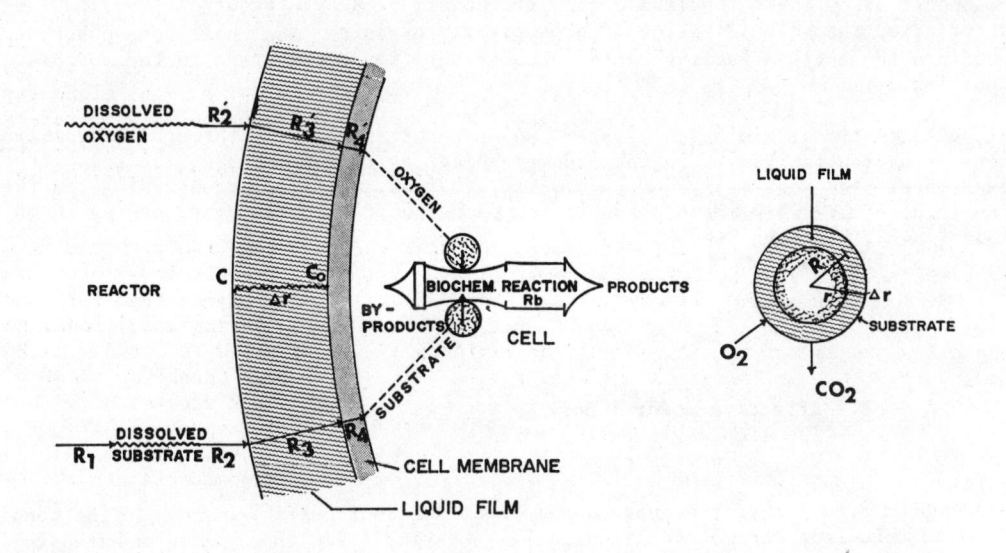

Fig. 2. Schematic Representation of Rate-Controlling Processes involved in Substrate and Oxygen Utilization During Biochemical Reactions

According to Fair and co-workers the power applied to mixed liquor in a biological treatment system serves the following useful purposes:

    i) Supplying the necessary oxygen to the biological floc;

    ii) Holding the biomass in suspension;

    iii) Restricting the floc size to increase the floc-fluid interface

    iv) Transferring nutrients efficiently and rapidly to the biomass;

    v) Washing away waste products from the flocculated biomass;

    vi) Producing flocs of such a size and weight that they can seperate from the effluent by normal gravitational sedimentation.

The flocculation and settleability of the biomass is another important parameter. Good flocculation does not insure good settleability; the solids concentration, the density of the sludge, and its compactibility influence overall settleability Heukelekian and Weisberg (1956); Ford and Eckenfelder (1967) observed that the sludge volume index (SVI) increased with the bound water content of the sludge. No consistent relationship between floc diameter and SVI was found by Finstein and Heukelekian(1967), but the more filamentous sludge samples had the higher SVI's.

Aerobic stabilization of sludge essentially involves the endogenous metabolism of the microorganisms present in the sludge when no fresh exogenous substrate is available to them. Under such conditions, bacteria utilize food stored within their cells and the synthesis of additional bacterial protoplasm becomes minimal. After the stored food is metabolized, the active microorganisms in the sludge obtain energy through the degradation of the sludge mass either by the metabolism of their own intracellular components or by the metabolism of the protoplasm of dead microorganisms. In the latter case, the solubilization of protoplasmic organics and their subsequent diffusion into the cells of active bacteria constitute important steps in the aerobic sludge digestion.

In the conventional biological system, power input is provided through the combinations of hydraulic flow, aeration and mechanical agitation. To describe the effective power input, the root mean square velocity gradient, G, s^{-1}, corresponding to the mean value of the work input per unit volume per unit time, as developed by Camp and Stein (1943), has been used.

$$G = \sqrt{\frac{P}{\mu V}} \qquad (1)$$

Where

    P = Effective power input, w
    μ = The liquid viscosity, Ns/m^2
    V = The basin volume, m^3

It has been argued that the above development does not sufficiently describe conditions in turbulent regimes (Kalinske, 1971). Parker (1970) has given an extensive theoretical development and experimental study of floc breakup in turbulent regimes. He found that floc breakup was proportional to G^2, which was related to the total effects in the microscale range. There are many factors besides floc breakup and reflocculation, such as diffusion through and to the remaining range of sizes of floc aggregates, partial inactivation of enzymes, etc., which are affected by shear. It would seem most logical to relate the sum total of these effects to G and utilize the best correlation between them and G.

Testing and Biological Treatment of Industrial Wastewater

The other parameter needed to describe power input is the time, θ, in s/d, for which energy is expended in the process. The quantity $G\theta$ then gives an index for the total energy expenditure. It appears that microorganisms can withstand a high amount of power input. Allen (1944) had to pass sludge five to six times through a high-speed blender to achieve maximum plate count.

BOD AND ACTIVATED SLUDGE MODELS

In 1925, Streeter and Phelps (1925) proposed that the rate of biochemical oxidation of carbonaceous organic matter, dL/dt, is proportional to the remaining concentration of unoxidized substances, L. Therefore, the oxygen demand (BOD), y, for time t is given by the following equation:

$$y = L_a - L = L_a \{1 - e^{-k'(t-t_0)}\} = L_a \{1 - 10^{-k(t-t_0)}\} \qquad (2)$$

where L_a is the ultimate oxygen demand of carbonaceous matter, mg/ℓ; k is the rate coefficient of BOD removal per day; and t_0 is the lag period in days (Ali, 1972).

The model developed by Lawrence and McCarty (1970) for an activated sludge system was utilized in this study. A completely mixed reactor with no influent solids is assumed. The net growth of biological solids is given by

$$\frac{dM_a}{dt} = Y \frac{dS_e}{dt} - k_e M_a \qquad (3)$$

where M_a is the active mass concentration, mg/ℓ'; t is the time in days; Y is the yield coefficient, mg/mg; S_e is the substrate concentration in the effluent, mg/ℓ; and k_e is the endogenous decay rate per day.

Substrate removal is described either by the Monod (1949) equation,

$$\frac{dS_e}{dt} = \frac{k_m S_e M_a}{K_s + S_e} \qquad (4a)$$

or by the average rate formulation proposed by Goodman and Englande (1974),

$$\frac{dS_e}{dt} = k S_e M_a \qquad (4b)$$

Two important process control parameters are the sludge age, t_s in days, and the process loading factor (F:M ratio), U.

$$t_s = M_a / (dM_a/dt) \qquad (5)$$
$$U = (dS_e/dt/M_a) \qquad (6)$$

Dividing Eq. 3 by M_a, one obtains the following relation between t_s and U:

$$1/t_s = YU - k_e \qquad (7)$$

Due to difficulties in measuring the active mass concentration, the volatile suspended solids, M_v, is usually substituted for M_a in the above equations.

If diffusional resistance through the floc particles is significant, the effective substrate uptake rate of the microorganisms is less than that predicted by considering their metabolic efficient alone. Atkinson and co-workers (1974) have developed a complex equation for substrate uptake that depends on the volume to area ratio of the microbial flocs. The relation can be simply described by multiplying k_m or k by E, an efficiency factor which increases as the floc volume to area decreases. It is not feasible at this time to measure this ratio in an aeration basin.

MATERIAL AND METHODS

The study on the influence of power input on biological activity in wastewater testing and treatment is divided into the following three phases:

i) Evaluation of the biochemical oxygen demand of polluted waters under turbulent conditions;

ii) Effect of mechanical blending of mixed liquor in an activated sludge system;

iii) Effect of mechanical blending on the aerobic digestion of waste activated sludge.

All tests in these studies were carried out according to the *Standard Methods* (1975), unless otherwise stated.

Several raw, primary settled, biologically treated wastewater and synthetic sewage samples were tested to determine the influence of turbulence on the progression of their biochemical oxygen demand. Two sets of BOD bottles were inoculated for each experiment. The first set was incubated under the standard quiescent condition. In the second set, the contents were kept continuously agitated with magnetic stirrers. The stirrer's speed could be varied between 0 and 1400 rpm. It was maintained at 680 rpm in all experiments except when studying the effect of stirring speed on BOD progression All BOD bottles were incubated at 20 ± 1^0C, except while studying the temperature effect. The procedure was standardized after a considerable number of experiments conducted under various conditions (Ali, 1972). When nitrification was expected to occur simultaneously with oxidation of carbonaceous matter, nitrifying bacteria were suppressed either by maintaining 3M ammonium chloride concentration in the BOD bottles or by pasteurizing the sample and subsequently reseeding with domestic wastewater. The standard plate count method was adopted for the bacteriological examination of the BOD bottle contents. The 'Slope Method', the 'Moment Method', and the 'Simplified Graphical Method' were modified and then adopted in analyzing BOD data. Ali (1972) prepared a computer programme for each of these modified methods and showed that the best values for BOD-progression parameters were obtained by taking an average of the individual values obtained by each method.

The study on the effect of mechanical blending of mixed liquor in an activated sludge system was divided into two phases: batch and continuous flow operations. The more easily controlled and operated batch setup experiments provided general information on synthetic sewage and its treatment. This basic information was utilized to establish operational parameters in continuous flow studies. The schematics of the batch and continuous operations are shown in Figs. 3 and 4. For comparison purposes, control units were usually run parallel with experimental units. The intense shear was provided by blending with a ten-speed Osterizer Model Galaxie Ten Blender. The duration and frequency of blending was controlled by an interval timer.

The feed was made up of glucose and beef extract, at a COD ratio of 1:2, respectively,

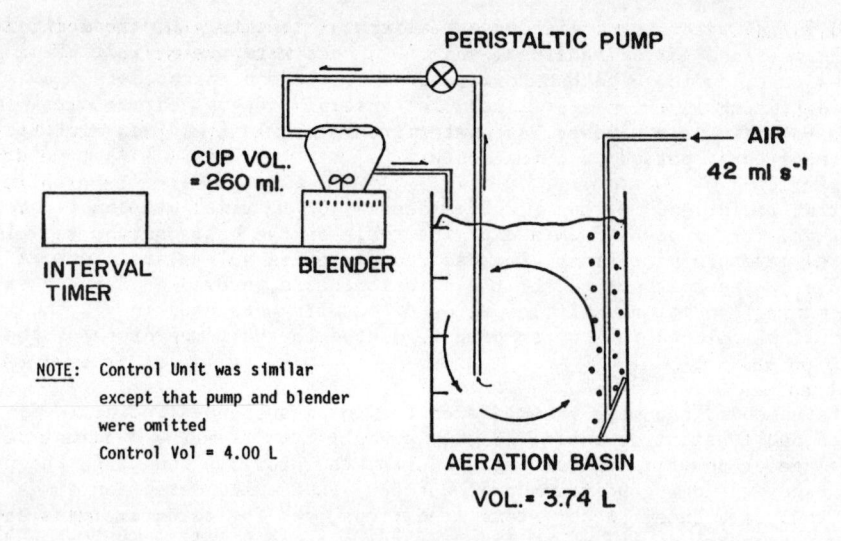

Fig. 3. Schematic of Batch Operation

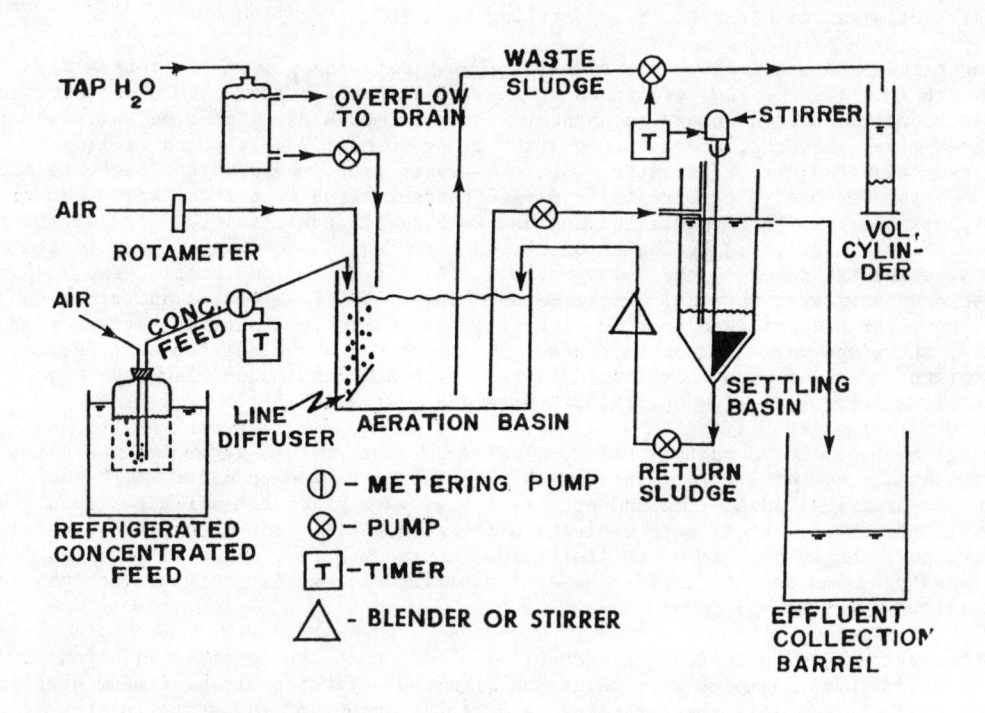

Fig. 4. Schematic of Continuous Flow Model

dissolved in tap water to provide essential trace elements. It was refrigerated at 0 - 4°C between and during feedings, and COD checks were run periodically. For the continuous runs, sodium bicarbonate was added to the concentrated feed to insure adequate buffering in the mixed liquor. The initial sewage seed was obtained from the aeration basin of a Windsor Wastewater Treatment Plant and was acclimatized to the substrate for a period of three months.

The aeration basin, 450 mm long x 260 mm deep x 100 mm wide, was constructed of plexiglass There were grooves machined at intervals on the basin bottom and sides to accommodate moveable partitions. The desired operating volume was 4.0 ℓ. A perforated tee-diffuser, placed at one end of the aeration basin, provided adequate oxygen and complete mixing conditions. Airflow rates of 42 ml/s were used in all the runs. The control unit was placed in a water bath regulated to the temperature of the blended unit.

Steady-state conditions were assumed when the total suspended solids (TSS) for both units remained constant or exhibited only a small drop from the previous day's value. If sludge age or power input were changed from the previous run, then the units were operated under the new conditions at least for six to nine detention times before starting analyses. Under steady-state conditions, the following analyses and measurements were carried out: mixed liquor total suspended solids (MLTSS) and volatile suspended solids (MLVSS), soluble effluent COD and BOD, mixed liquor COD, mixed liquor pH, mixed liquor dissolved oxygen (DO), endogenous and respiring oxygen uptake rates (OUR), mixed liquor viscosity, and flow rate through a blender cup. At the end of each run, the waste mixed liquor from each unit was settled for 1/2 hour and the supernatant was examined for TSS, VSS and total COD. Also, the sludge volume index (SVI) was measured after 1/2 hour settling in a 100 ml or 1000 ml cylinder.

Flow rates were measured volumetrically or with metering pumps. For this study, soluble effluent COD or BOD is defined as any filtrate which passes through Whatman GF/C filter paper. Endogenous oxygen uptake rates were measured by placing a volume of wasted mixed liquor of known VSS in a 300 ml or 60 ml BOD bottle and filling it with oxygen-saturated distilled water. Then the oxygen probe measured the decrease in oxygen probe measured the decrease in oxygen concentration with time, which was recorded continuously. The respiring oxygen uptake rate procedure was the same except that a quantity of feed, proportional to the quantity fed to the mixed liquor in the aeration basin, was added to the volume of mixed liquor in the BOD bottle. The respiring oxygen uptake rates were calculated both in the initial 2 - 4 min. and after the rate of change in oxygen concentration with time had become constant. In the endogenous case, the slope remained constant after one minute of probe adjustment. Further details of operating procedures used in the batch and continuous flow systems are explained elsewhere (Droste, 1978; Droste and Bewtra, 1978).

Sludge age was chosen as the primary control parameter. This parameter was varied through the range of 2 to 6 days in each phase of operation to determine kinetic constants. Organic loading rate and detention time were fixed for all batch runs, but these control parameters were varied considerably in the continuous operation in order to push the systems to their limits. One continuous flow run utilizing real sewage as substrate was used to verify the applicability of results obtained with the synthetic sewage runs.

In order to study the effect of mechanical blending on the aerobic digestion of waste activated sludge, samples were collected from the activated sludge treatment plant at Windsor, Ontario. This plant receives a variable amount of industrial wastes and during a party of this study, alum and lime were added to raw sewage to reduce phosphates in the effluent. Mixed liquor samples were collected near the outlet end of the aeration tank and were concentrated. Eight to twelve liters of thickened sludge were continuously aerated in jars for a period of 9 to 11 days at a room temperature ranging from 21 to 23°C. The sludge in one or more jars was blended periodically in an Osterizer Blender either in a batch operation or in a continuous operation. In the batch operation the entire contents of the aeration jar were removed periodically and blended; whereas in the continuous operation, the jar contents were circulated continuosly through the blender. The frequency of the blender operation was controlled by a timer. Parallel tests were conducted on the unblended sludge to act as controls.

The parameters varied with the intensity and frequency of blending. The effect of washing the sludge either with 0.05 M phosphate buffer of pH 7.45 or with distilled water was also determined. The digested sludge characteristics were determined in terms of the progressive biodegradation of volatile suspended solids, VSS, and its settleability in 30 minutes. Oxygen uptake rates at 20 ± 1^0C were measured polarografhically with YSI Biological Oxygen Monitor, Model 53. Generally, the sample for oxygen uptake measurement was withdrawn from the jar about 30 minutes after the blending had been completed and the sludge had been aerated. The sludge was assumed to behave like Bingham plastic, and its viscosity was determined with a Brookfield Synchrolectric Viscometer. The dissolved oxygen concentration in both jars was always maintained above 2 mg/ℓ. It was observed that a change in DO concentration above this level had no significant effect on the rate of autooxidation. Further details are given elsewhere (Bokil, 1972).

RESULTS AND DISCUSSION

Experiments were conducted on primary settled sewage samples to study the effect of varying stirring speed between 0 and 1320 rpm on BOD progression. It was observed that the BOD values for any incubation period and at any stirring speed, except at a very high speed, were higher than the corresponding BOD value under quiescent conditions. A statistical analysis of L_a values showed no significant influence of stirring speed on the ultimate carbonaceous oxygen demand, However, the lag period values, which were always negative, had decreased with an increase in stirring speed.

The variation in the rate constant, k in day $^{-1}$, with a change in the stirrer's speed, N in revolutions per minute, is given by

$$k_{stirred} = k_{quiescent} \{1.035 + 3.404 \times 10^{-4} N\} \quad (8)$$

with r = 0.866

In order to obtain an appropriate value of the rate constant for applying to the actual conditions in a stream or in a treatment process, a proper value of N to represent field conditions should be obtained and used in Eq. 8. Ali (1972) has suggested a method for obtaining the equivalent value of N from the power dissipated per unit volume of fluid under actual conditions.

Eight sets of experiment were conducted on raw sewage samples and k, L_a and 5-day BOD values were obtained for both stirred and quiescent conditions. It was observed that when the samples were stirred continuously, k values for raw sewage increased significantly with less than 0.10 % probability of drawing an erroneous conclusion. The total average increase in k value was 25%, with maximum and minimum increases of 47% and 13%, respectively. There was a significant increase in 5-day BOD values, ranging from 5 to 29%, with an average of 18% for the tested samples. Values of L_a for continuously stirred samples showed no significant increase as compared to standard values at a 90% level of confidence. In all the twelve samples of primary settled sewage, 5-day BOD values under stirring conditions were significantly higher than those under quiescent conditions, even at a 99% confidence level. The percentage increase in the 5-day BOD ranged between 8% and 37%, with an overall average of 20%. Here again, the k values showed an increase due to stirring, but the variation in L_a values was insignificant. The increase in k values ranged between 4% and 170%, with an overall average increase of 70% for the tested samples. For nine final effluent samples, it was observed that both the 5 - day BOD and L_a values increased when the incubated samples were stirred continuously. The increase in 5-day BOD values ranged between 13% and 70%, with an average increase of 44%. The L_a values showed a significant increase in the stirred bottles, with an overall increase of 15%. Here again, the stirring resulted in a considerable increase in the k values. The percentage increases ranged between 18% and 134%, with an average of 75%. It was demonstrated that ammonium chloride as well as pasteurization provided only a short term suppression of nitrification. Nine sets of BOD progression experiments were carried out on a synthetic medium containing 250 mg/ℓ of dextrose. The BOD dilution water used in these experiments was seeded with primary settled sewage. In some experiments, ammonium chloride was added to suppress nitrification. Typical BOD progressions under stirred and quiescent conditions are shown in Fig. 5. Both k and 5-day BOD values showed a significant increase due to stirring, but L_a values were not affected significantly. The increase in k values ranged between 9% and 157%, with an average of 86%. Likewise, the increase in the 5-day BOD ranged between 2% and 14%. All raw sewage, primary settled sewage, and final effluent samples showed a negative lag period exhibiting an immediate oxygen demand. Lag period values were substantially reduced when the BOD bottle contents were contantly stirred. However, the synthetic medium exhibited a positive lag period and this was considerably diminished when the BOD bottle contents were stirred continuously.

Nine experiments were conducted on a synthetic medium to study the effect to temperature and turbulence on rate constant values. It was observed that the turbulence affects the rate of substrate utilization by microorganisms when incubated at different temperatures. The turbulent conditions yielded a higher rate of substrate utilization at all incubation temperatures between 10 and 35°C. On the other hand, no significant changes were observed in the ultimate carbonaceus oxygen demand, L_a, due to change in temperature. When the (k_T/k_{20}) values of stirred samples were compared statistically with those obtained under quiescent conditions, no significant difference was found at 90% confidence limits. Therefore, the following equation, originally obtained for k values at different temperatures in the absence of turbulence can also be applied in the presence of turbulence:

$$k_T = k_{20} \, \theta^{(T-20)} \tag{9}$$

where θ is the thermal coefficient for the given wastewater in the absence of turbulence. (Ali, 1972)

The viable cell count in BOD bottles under stirred and unstirred conditions, are also plotted along with the corresponding BOD values in Fig. 5. The results obtained with raw sewage, primary sewage, and final effluent showed bacterial growth similar to that obtained with a synthetic medium (Ali, 1972). These results illustrate the following

      i) The number of viable cells in the stirred bottles was always higher than the corresponding number obtained under quiescent conditions;

      ii) Stirring increased the rate of logarithmic growth and reduced the lag period, but there was no apparent change in the rate of logarithmic death;

      iii) The stationary phase lasted for a few hours under both testing conditions;

      iv) Although the time to reach the maximum bacterial population, starting from the time of incubation, was approximately the same, the net growth period under stirring conditions was longer.

In the case of the synthetic medium (Fig. 5), it was also noticed that the appearance of maximum viable population corresponded to the plateu in the BOD curve. All experiments on the synthetic medium showed a pronounced occurence of a plateau in oxygen uptake lasting for a longer duration. The BOD exertion corresponding to a dying-off in

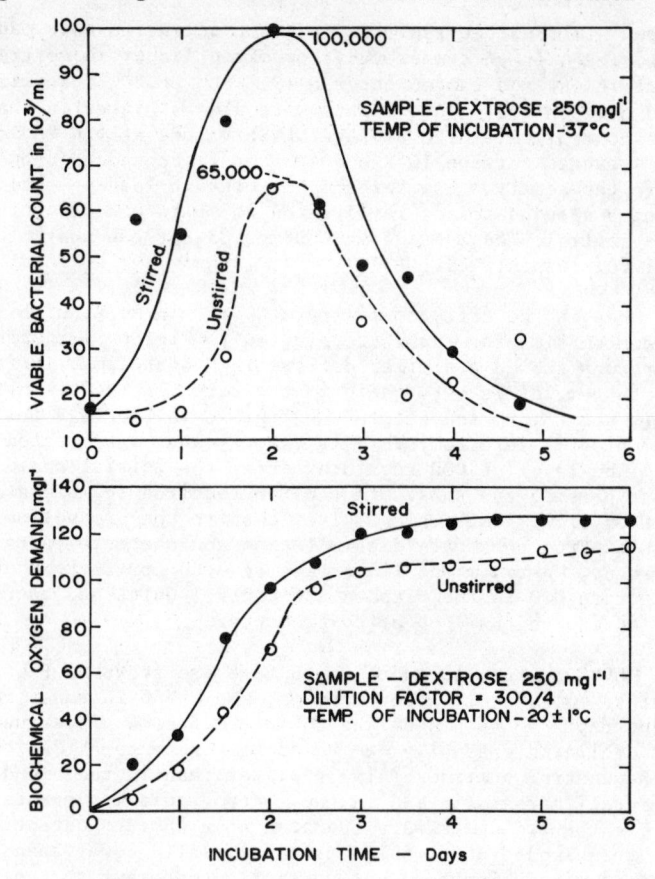

Fig. 5. Effect of Stirring on Bacterial Growth and BOD Progression Using Synthetic Medium

viable population, did not show any pronounced rise in the BOD values.

Thus, it is shown that by continuous stirring of the BOD bottle contents the progression of biochemical oxygen demand is accelerated as compared to that under quiescent conditions in all types of samples tested. Also, bacteriological examination of the BOD bottle contents has revealed that, irrespective of the type of substrate, the number of viable bacterial cells in stirred bottles is always higher than that obtained under standard quiescent conditions. Therefore, the differences observed in BOD and rate constant values between stirred and quiescent conditions are due to the increased rate of bacterial metabolic activity with an increase in turbulence. Ali (1972) has proposed a model to explain such a behaviour of bacteria under turbulent conditions. According to him, the influence of turbulence on the BOD rate constant is due to the following:

    i) An increase in the net rate of material transport into the cell,

    ii) An increase in the rate of removal of by-products accumulating around the cell membrane, thereby improving the environmental conditions surrounding the cell.

The next phase of the study was the mechanical blending of mixed liquor in an activated sludge system. All fourteen runs in the batch operation were made with a detention time of 8 hours. The average temperatures of mixed liquor in control and experimental units were fairly close and ranged between 19.1 and 26.2°C for different runs. The pH values remained between 7.37 and 8.28 and were always higher in the blended unit as compared to the control unit. The average MLVSS values at the end of the detention period (Table 1) ranged between 1042 and 4039 mg/ℓ, depending upon the sludge age to be maintained in the aeration chamber and the organic loading rate. The average volatile content of suspended solids in aeration chambers was 80% for the blended unit and 81% for the control. The average mg COD/mg VSS for the control and blended units were 1.52 and 1.56, respectively.

The typical changes in the effluent filtrate COD, S in mg/ℓ, with time, t, in days, during one detention time for Run 7 are plotted in Fig. 6 for both control and blended units. Other runs showed a similar decline in S with time. It is apparent that the rate of substrate removal is very rapid in the first 15 to 30 min, when the substrate concentration is very high, and becomes negligible in the last few hours when the substrate concentration is very low. This is exactly what is expected from the kinetic Eqs. 4a and 4b. The level of COD remaining after the initial rapid substrate removal was necessary to provide the substrate tension required to maintain the residual at a nearly constant or slowly decreasing value, despite the production of difficult-to-degrade substances from substrate dissimilation and bacterial auto-oxidation. In a seperate experiment, wasted mixed liquor was aerated for forty hours. It was observed that the filtrate COD in the blended and control units had increased due to the destruction of solids and buildup of by-products.

The changes in MLVSS in Runs 13 and 14 with time are shown in Fig. 7, both for control and blended units. Immediately after feeding, the MLVSS increase repidly in both the units due to the rapid synthesis of the substrate, reach a peak value in about an hour and then start declining gradually due to dominating auto-oxidation. In these two runs, the MLVSS in the control unit were always higher than in the blended unit. Also, the average incremental increase in VSS in the control unit was greater than in the blend-unit. Thus, if the two units were operated at a shorter detention time, excess sludge produced in the blended unit would be substantially lower than in the control. Actually, the VSS in mixed liquor do not correctly represent the concentration of active biomass. Total viable counts per mg of VSS in control sludge samples were several times the corresponding values for blended sludge samples in both Runs 13 and 14. This clearly indicates that a much smaller amount of active biomass in the blended unit was able to achieve the same substrate removal efficiency as that obtained in the control unit.

The values for the residual filtrate COD, S_e in mg/ℓ, for different runs were always considerably less in the blended unit as compared to the control unit. These values increased with a decrease in sludge age. The major part of this residual filtrate COD is considered to be non-degradable or very difficult to degrade. The fraction of feed which is non-removable was calculated for each run (Droste, 1978), and the non-removable COD in substrate, S_n, was found to range between 5% and 15%, of the feed COD. It was always higher in the control than in the blended unit and increased with a decrease in sludge age. It is believed that the blending energy had not only reduced the floc size but had also made it more effective in attacking certain organic matter. Further evidence of a non-degradable fraction of substrate is the low values for the residual filtrate BOD, generally less than 10% of the corresponding residual filtrate COD for each run, whereas the BOD of the feed was approximately 62% of the COD.

The non-removable COD, S_n, is considered to be composed of two components, organic matter in the feed, which cannot be degraded by bacteria, and the minimum concentration of degradable organic matter in mixed-liquor required for proper diffusion through the biomass film. The first part is assumed to be dependent on the feed and the characteristics of organic matter in the biomass. The second part depends on the nature of the biomass and is expected to be smaller for the blended unit because of smaller film thickness.

Table 1 Average Operating Conditions for Different Runs-Batch Operation

	Control			Blended				
Run No.*	U day^{-1}	t_s day	MLVSS* mg/ℓ	U day^{-1}	t_s day	MLVSS* mg/ℓ	G s^{-1}	θ s/d
1	0.44	5.68	2825	0.42	6.02	3083	6537	178
2	0.45	4.28	2670	0.40	5.30	3087	21014	166
3	0.38	6.35	3341	0.37	6.35	3384	6540	558
4	0.41	6.35	2648	0.46	6.35	2846	21200	554
5	0.78	1.90	1383	0.92	1.90	1374	7270	545
6	1.15	1.90	1042	1.14	1.90	1119	24700	544
7	1.16	1.90	1044	1.11	1.90	1152	24800	169
8	0.36	6.35	3288	0.41	6.35	3284	18100	167
9	0.58	3.98	2119	0.69	3.98	1887	22700	166
10	0.59	3.98	1996	0.58	3.98	2184	7150	535
11	0.60	3.98	2032	0.64	3.98	2032	22200	511
12*	0.61	3.98	4039	0.66	3.98	3869	18000	509
13	0.81	2.47	1421	0.86	2.47	1413	7230	509
14	0.47	5.13	2738	0.51	5.13	2617	6600	509

*Amount of feed for Run 12 was twice that for other runs.

To achieve the same unit rate of substrate utilization, a much higher substrate concentration is required in the control unit as compared to the blended unit, due to the slower diffusion rate in the control floc. Using Eq. 4a, it is estimated (Droste, 1978) that the affinity constant, K_s, in the blended unit will be at least 10% lower than in the control when both units receive the same amount of substrate. The data for each run are plotted in Fig. 8 and 9 to determine the growth yield coefficient, Y, and the endogenous rate constant, k_e, in Eq. 7, and the following correlations were obtained statistically:

$$\text{Control}: \quad t_s^{-1} = 0.479\ U - 0.023 \text{ with } r = 0.988 \tag{10}$$

$$\text{Blended}: \quad t_s^{-1} = 0.500\ U - 0.049 \text{ with } r = 0.984 \tag{11}$$

It is evident that the growth yield cofficient was not affected significantly due to blending. Similar values for Y have been reported by other investigators also (Lawrence and McCarty, 1970). On the other hand, k_e values increased significantly, clearly indicating that the auto-oxidation of the biomass is improved considerably by blending the mixed liquor. Eqs. 10 and 11 show that for the same U and Y values an improvement in the endogenous rate constant indicates more rapid auto-oxidation of the

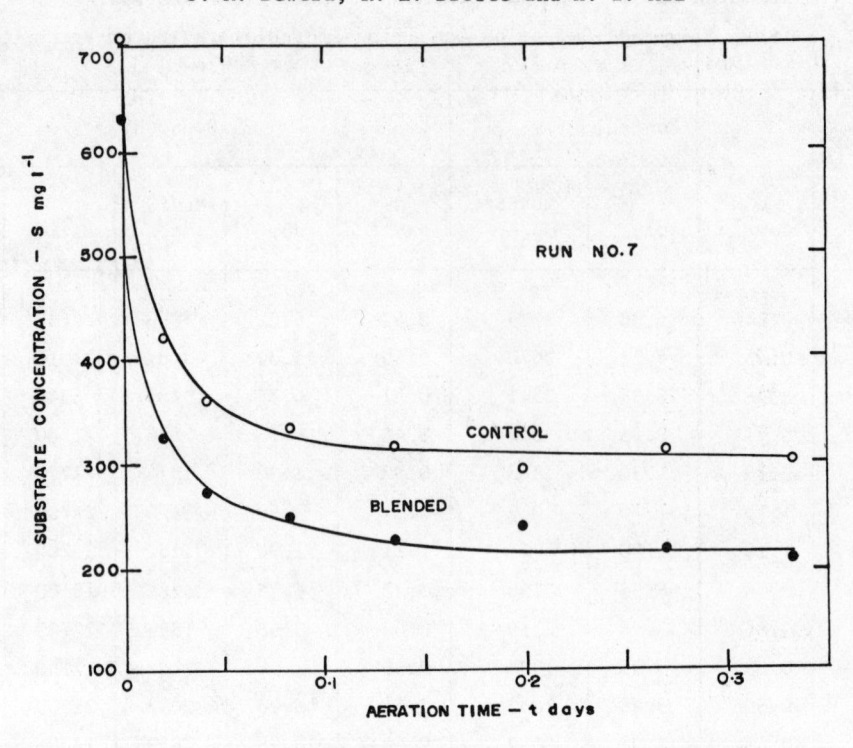

Fig. 6. Typical Substrate Utilization with Time

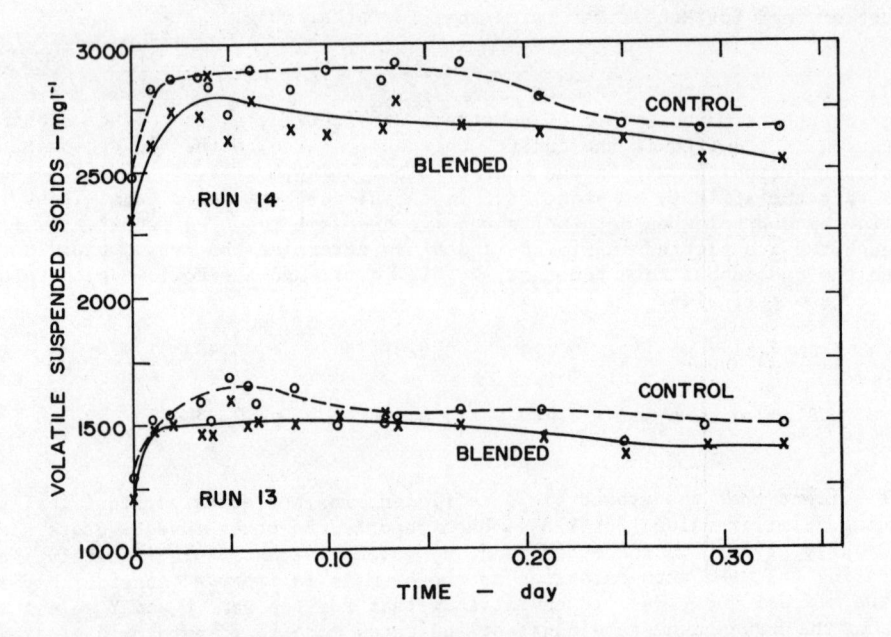

Fig. 7. Typical Changes in Mixed Liquor VSS with Time

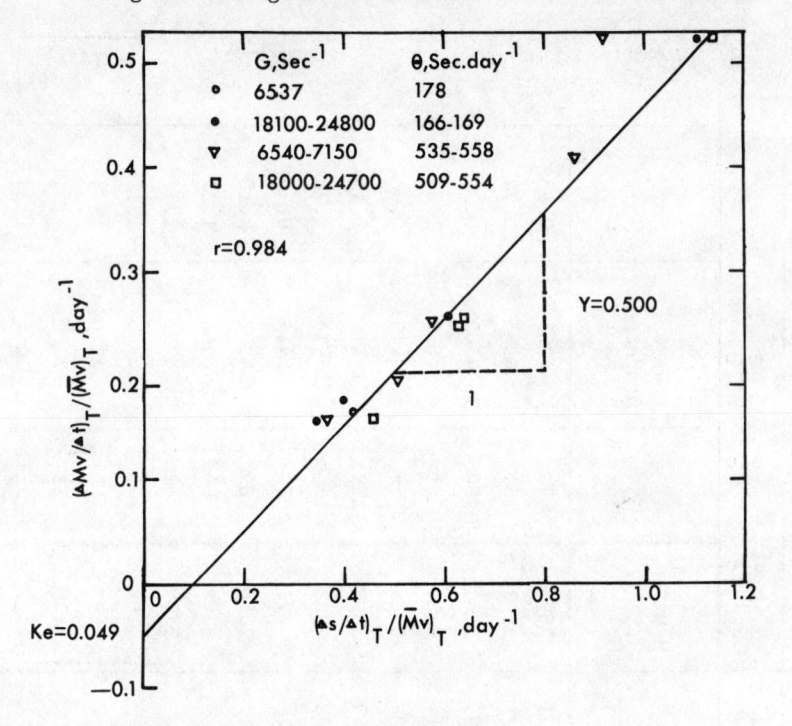

Fig. 8. Growth Kinetics-Blended

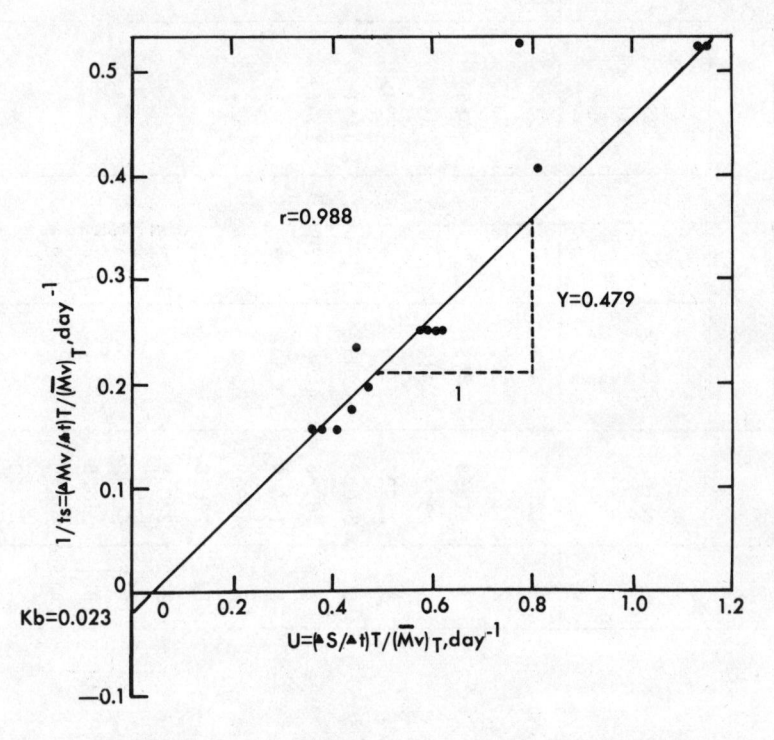

Fig. 9. Growth Kinetics-Control

Table 2 Average Operating Conditions for Different Runs - Continuous System

Run	Feed rate mgd	Control MLVSS mg/ℓ	Control S_e mg/ℓ	Control VSS produced mgd	Control U d^{-1}	Experimental MLVSS mg/ℓ	Experimental S_e mg/ℓ	Experimental VSS produced mgd	Experimental U d^{-1}
1	12875	2480	42	5578	1.21	3695	28	5288	0.83
2	12571	3968	45	5573	0.73	3927	44	4885	0.74
3	11963	2853	35	4650	0.99	2834	36	4365	0.99
4	23927	4934	71	8596	1.14	4569	71	7812	1.23
5	23927	5309	45	7285	1.06	5789	56	7419	0.96
6	36088	4903	131	11511	1.69	5694	139	10119	1.45
7	36088	4404	126	15479	1.88	4228	134	13916	1.96
8	11029	3780	31	2877	0.70	3776	33	2444	0.69
9	33089	3562	109	15343	2.17	3650	109	14202	2.10
10B1	22058					3699	86	8344	1.37
10B2	22058					3708	82	8092	1.37
11B1	21688					4164	75	7773	1.21
11B2	21688					3694	83	7510	1.35
12B1	20420					3610	73	7650	1.31
12B2	20420					3522	75	7047	1.34

biomass, thereby resulting in a reduction in the volume of excess sludge produced.

It was observed that at the end of a given detention time both the COD and VSS in the control supernatant were always higher than in the blended supernatant, on the average by a factor of about 1.69 for COD and 4.46 for VSS. Thus the blending had actually decreased the wash-out solids in the treated waste. The average ratios between the supernatant COD and VSS in the wash-out solids were 1.88 for control and 2.75 for blended units.

The SVI in each case was within the acceptable limits. Although SVI values were generally higher for the blended unit than for the control unit, the visual observations showed that the supernatant was usually clearer for the blended unit. It was observed from microscopic examination that many more fine particles were present in the supernatant from the control unit. Microscopic examination revealed fewer or no ciliated motile protozoans in the blended unit as compared to the control unit. The blended flocs were always much more uniform than the granular, irregular flocs of the control unit. Viscosity observations further supported the floc behaviour. The blended ML viscosity was always higher than the control due to the stability of the blended floc matrix.

The endogenous oxygen uptake rate, OUR, in mg O_2/day/mg VSS, was usually lower for the blended mixed liquor. On the other hand, the respiring OUR was mostly higher for the blended mixed liquor, decreasing with an increase in sludge age in both units. It was also observed that the specific OUR in the blended unit exhibited its highest percentage increase over that in the control unit during the initial rapid removal phase. However the total oxygen uptake in one detention time for the two units was about the same because of a higher proportion of solids in the control unit. A portion of mixed liquor from both units was washed with 0.05 M phosphate buffer several times. The difference in endogenous OUR valus for washed and unwashed portions was negligible.

The average operating conditions for the twelve runs with the continuous operation are summarized in Table 2. In this table B1 and B2 refer to the two blended systems in the last three runs. The B2 unit always had more power input than the B1 unit. The shear input values, GT, where T is the total time in sd^{-1} for which mechanical power was applied, were near 9×10^6 d^{-1} during the first nine runs. The average energy expenditure in the experimental unit was six times that in the control unit in these nine runs. Detention times were approximately 4 h. There was no significant variation in pH, temperature or DO values between the units under the various operating conditions. Average DO values were kept above 1.7 mg/ℓ. The BOD of the effluents was usually less than 20 mg/ℓ.

Employing Equations 4a and 4b, the following substrate uptake constants were obtained for the first nine runs.

For the Monod formulation:

$$\text{Control} : \quad \frac{1}{M_v} \frac{dS_e}{dt} = \frac{3.81 \, S_e}{132 + S_e} \tag{12}$$

$$\text{Experimental} : \quad \frac{1}{M_v} \frac{dS_e}{dt} = \frac{2.49 \, S_e}{72 + S_e} \tag{13}$$

For the average rate formulation:

$$\text{Control} : \frac{1}{M_v} \frac{dS_e}{dt} = 0.0112\, S_e + 0.495 \quad (14)$$

$$\text{Experimental} : \frac{1}{M_v} \frac{dS_e}{dt} = 0.0101\, S_e + 0.482 \quad (15)$$

The correlation coefficient, r, was over 0.85 for all of the above. With each type of formulation the data were found to correlate well.

Utilizing Eq. 7, the following yield and endogenous decay coefficients were obtained:

$$\text{Control} : Y = 0.520 \,; \quad k_e = 0.131\, d^{-1} \quad (16)$$

$$\text{Experimental} : Y = 0.483 \,; \quad k_e = 0.120\, d^{-1} \quad (17)$$

The net yield of solids was 10% lower in the experimental unit. Formulation of the oxygen utilization in a similar manner as Eq. 3 provided the same information as the yield coefficient and endogenous decay results. The experimental unit was utilizing more oxygen for oxidation than the control unit and less oxygen for endogenous decay.

The COD balances around the system also showed that 16% more COD being satisfied in the experimental unit. By taking the average difference in solids production between the two units and the average COD content of the solids, this value was computed to be 14.4%. The average COD content of the MLVSS was 1.44 and 1.45 for the experimental and control units, respectively. No significant relative deviation between the COD oxidized and solids produced was observed at different values of sludge age. Plate counts utilizing nutrient agar were higher in the control unit. On comparing the experimental to the control unit, it was found that the overall COD oxidation efficiency was larger than 1.16.

The high power inputs did not produce a corresponding improvement in effluent quality. However, solids production had shown a decrease. Seperation of the microbial flocs by blending resulted in an overall increase in the metabolic efficiency. Before a final assessment can be made, the overall cost-benefits with regard to sludge disposal must be considered.

According to Eq. 3, the kinetic expression for aerobic biodegradation in the endogenous phase becomes.

$$\frac{dM_a}{dt} = -k_e M_a \quad (18)$$

or

$$\frac{M_a}{M_{a_o}} = e^{-k_e t} = 10^{-kt} \quad (19)$$

where M_a and M_{a_o} are the concentrations of active biomass, measured as VSS, in the system at digestion times t and t = 0, respectively; k_e and k are the fractions of VSS digested per day by endogenous respiration. A typical plot of the remaining VSS, as a percent fraction of the initial VSS, versus time is shown in Fig. 10. The value of G is the average value for the entire duration of the experiment. The slope of the plots, k, in Fig. 10 gives the fractional rate of decrease in biodegradable VSS during auto-oxidation. After a digestion period of 8 to 9 days, the slopes of both the lines start decreasing. Similar linear plots were obtained for other sets of observations also; however, the digestion period after which the slope started changing varied. Also, different sludge samples gave different rates of auto-oxidation (Table 3). It has been reported that the auto-oxidation rate of any sludge depends upon the type of wastewater, microbial content of the sludge, age, and temperature (Bokil, 1972). When the organic matter in the sludge becomes difficult to biodegrade, the value of the slope, k, starts decreasing. The increased rate of biodegradation in the case of blended sludge is attributed to the following:

      i) The breaking of the floc matrix and reduction in the size of the resulting floc particles, with a consequent increase in the surface area;

      ii) The decrease in the diffusional resistance to the movement of soluble substances of bioparticles smaller in size than the larger parent floc;

      iii) The solubilization of certain organics present in the sludge as a result of endogenous respiration;

      iv) The more frequent renewal of contacts between the surfaces of active microorganisms and the surrounding medium.

From Fig. 10, it can be seen that the settleability of sludge is also improved as a result of blending. Visual observations on the supernatant of the blended sludge also showed that generally it did not become turbid during the first 3 to 4 days. Later on the turbidity increased gradually, but did not become so great as to mean a substantial loss of fine particles in the effluent. The improved settleability of the sludge due to blending can be attributed to the more rapid mineralization of the blended sludge, as well as some decrease in the quantity of bound water due to the reduction in particle size. Heukelekian and Weisberg (1956) have shown that in a normal zoogleal sludge poor settleability is associated with high bound water content.

Table 3 shows that the rate of auto-oxidation increased with an increase in blending intensity, G, when all the sludges were blended for the same duration. In the light of arguments stated earlier, it is reasonable to expect an increase in the auto-oxidation rate due to an increase in the intensity of blending. Similarly, the settleability of the sludge improved due to the increase in the intensity of blending.

When the sludge was washed with 0.05 M phosphate buffer and blended, there was an improvement in the auto-oxidation rate and the settleability, as the aerobic digestion progressed. The washing of sludge samples with pH 7.45 buffer solution at the start of the experiment maintained a fairly constant pH in the aeration jar. However, in the case of sludge samples washed with tap water and also in the case of the unwashed samples, the pH went down to an acidic range between 5 and 6.5. This drop in pH was due to the interaction of the CO_2 produced and alkalinity in the sludge. The pH values obtained for the samples washed with buffer solution were considered to be the favorable pH range for the bacterial auto-oxidation. It was thought that the washing operation might have removed the particles which were hard to degrade, as well as some toxic by-products of the bacterial auto-oxidation. It was also noticed that the effects of these factors became significant only when accompanied by blending.

The oxygen uptake rates for the blended sludge were higher than those for the control sludge during the first few days of aerobic digestion. The increase in the oxygen uptake rate corresponded to the increased rate of auto-oxidation. The day-to-day varia-

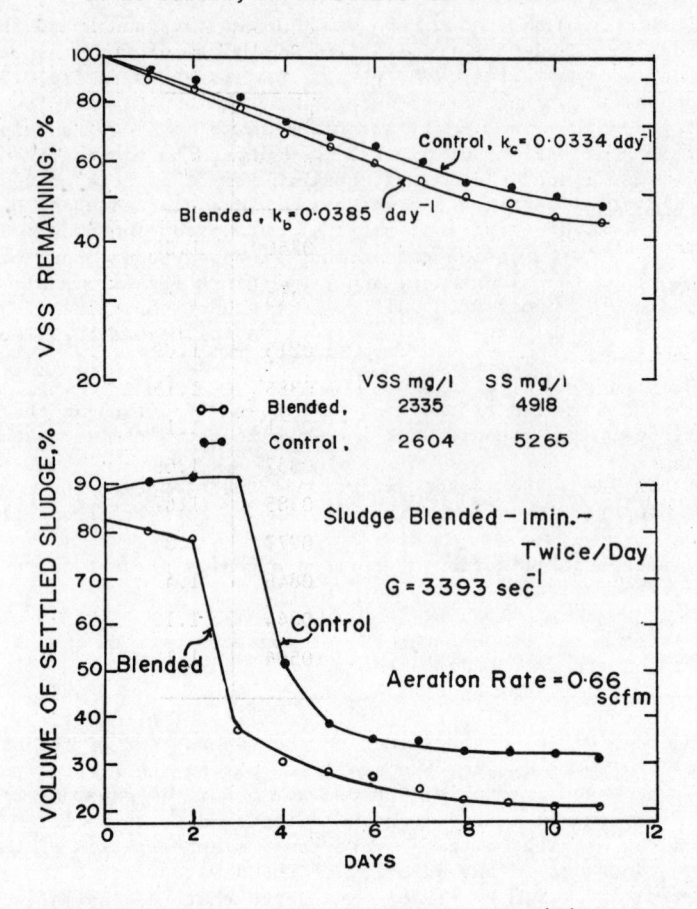

Fig. 10. Typical Auto-Oxidation and Settleability of Sludge

tions in the oxygen uptake rate were attributed to the predominance of different species of bacteria at different times. The environment in the aerobic digestion jar changed from day-to-day with respect to substrate, pH, and chemical components that were formed or degraded, and this resulted in a succession of dominant species. Plate counts on blended and control sludges showed such a shifting predominance. The bacterial count in the supernatant of the blended sample was higher than in the supernatant of the control sludge. Therefore, this supports the previous statement that the blending of sludge is not likely to increase the loss of fine particles in the supernatant.

Figure 11 shows the relationship between the blending energy input to the sludge and the corresponding auto-oxidation rate. The energy input is plotted as an average energy-duration function, $G\theta$, over the duration of the experiment, where θ is the total time of blending the sludge each day. The auto-oxidation rate is plotted relative to that of control k_b/k_c. All these data were obtained with continuous circulation of sludge through the blender. The frequency of blending ranged from 1 min in 5 min to 1min in 15 min and the sludge circulation rate was regulated so as to completely displace the blended sludge from the blender cup before the next blending started. This plot shows that with an increase in the blending energy input, expressed by function $G\theta$, there is a corresponding increase in the rate of auto-oxidation. The relationship between $G\theta$ and k_b/k_c is linear. However, the upper limit for this linear relationship could not be established with the commercial blenders used in this study.

Table 3 Auto - Oxidation Rate and Settling Characteristics of
Sludge Under Different Conditions of Blending

No.	Frequency of Blending	G sec^{-1}	k_c Control day^{-1}	k_b Blended day^{-1}	k_b/k_c	Final Settled Sludge Volume, % Control	Final Settled Sludge Volume, % Blended
1	Once/run	3043	0.0250	0.0250	1.00	22	20
2	Alternate days	3995	0.0192	0.0215	1.12	16	11
3	Twice/day	2000	0.0211	0.0215	1.09	98	35
4	Twice/day	3393	0.0334	0.0385	1.15	32	20
5	Once/day	3885	0.0410	0.0471	1.15	26	20
6	Twice/day	3361	0.0308	0.0337	1.09	40	32
7	Twice/day	2679	0.0230	0.0385	1.67	59	38
8	Once/day	4728	0.0596	0.0777	1.30	--	--
9	Once/day	4302	0.0591	0.0848	1.43	98	55
10	Twice/day	4082	0.0726	0.0844	1.16	--	--
11	Twice/day	2377	0.0451	0.0506	1.12	95	83

CONCLUSIONS

Based on the results discussed in this paper, the following conclusions can be drawn about the influence of mechanical power input during the testing and biological treatment of industrial wastewaters:

1. Irrespective of the source of the sample, the 5-day BOD values obtained in stirred bottles are consistently higher than those obtained under standard conditions. The rate constant values also increase significantly in all types of samples when the bottle contents are stirred continuously. However, the increase in the ultimate oxygen demand due to the stirring of BOD bottle contents is pronounced only in some cases of the final effluent. The bacteriological examination of the BOD bottle contents reveals that the bacterial growth and the metabolic activity are improved because of turbulence. The increase in 5-day BOD and k values is attributed to the increased population of viable microorganisms which, in turn, is due to an improvement in the environmental conditions surrounding the cells as well as a decrease in the stagnant liquid film around each cell, which would tend to increase the rate of substrate transfer inside the cells.

2. In the batch operation, the blending significantly reduces the COD remaining in the effluent after one detention time. This nonremovable fraction of substrate increases with a decrease in sludge age. Blending the total mixed liquor even at very low shear rate values for only 3 min/day lowers the half velocity constant, K_s, by at least 10%, thereby increasing the overall rate constant. This also results in considerably reducing the non-removable COD. The growth yield coefficient is not altered significantly due to blending, but the endogenous decay rate increases appreciably. An improvement in auto-oxidation rate means a reduced volume of excess sludge produced. The settleability and clarification of blended mixed liquor floc is good. The blending of mixed liquor reduces both the COD and VSS in supernatant, thereby indicating that

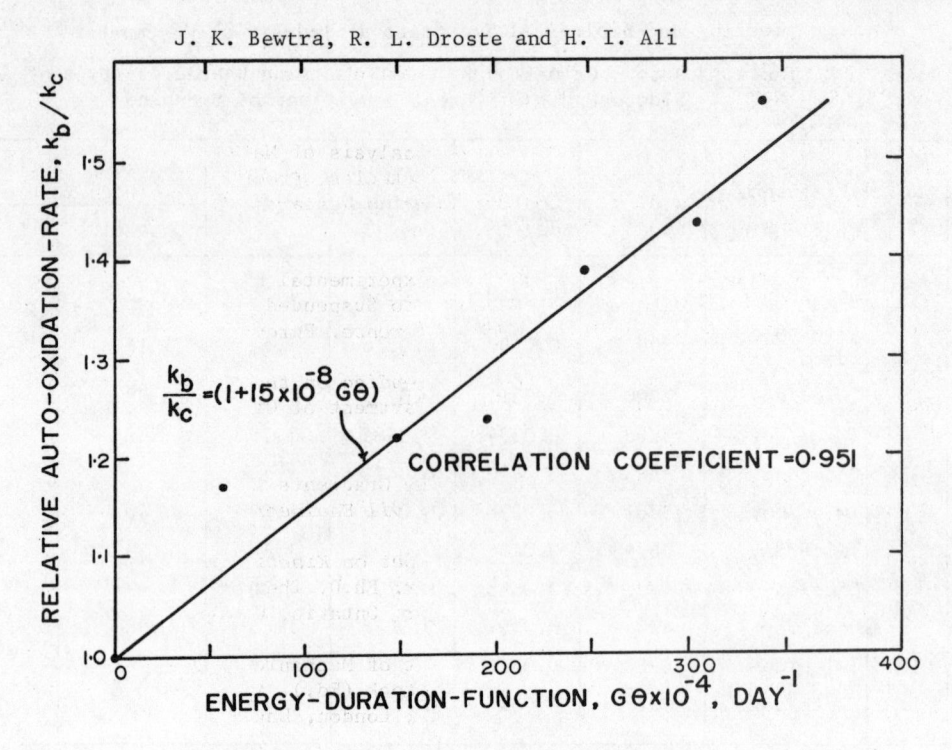

Fig. 11. Correlation Between Energy-Duration-Function and Relative Auto-Oxidation-Rate

the wash-out of solids is decreased.

The continuous operation of the activated sludge system confirmed the above conclusions. It was further demonstrated that the oxidation efficiency of the active microorganisms in the blended unit increased.

    3. When the waste activated sludge is mechanically blended in an aerobic digestion process, the rate of auto-oxidation as well as settleability of the sludge are improved significantly. The rate of auto-oxidation increases linearly with an increase in the value of $G\theta$, indicating that the higher the input blending energy, the more rapid the auto-oxidation. Washing the sludge with 0.05 M phosphate buffer together with blending further improves its auto-oxidation rate and settleability.

ACKNOWLEDGEMENT

These research projects were supported by Grant No. A5642 from the National Research Council of Canada.

REFERENCES

Ali, H.I. (1972) *Influence of Turbulence on BOD Progression*. Ph.D. Thesis. Department of Civil Engineering, University of Windsor, Windsor, Ontario, Canada.

Allen, L.A. (1944). The Bacteriology of Activated Sludge. *Journal of Hygiene*, 43, 424.

Atkinson, B. (1974). *Biochemical Reactors*. Pion Ltd., London.

Baars, J.K. (1965). Bacterial Activity in Pollution Abatement. *Journal of the Institution of Sewage Purification, 36*.

Baillod, C.R. and W.C. Boyle (1969a). An Analysis of Mass Transfer Limitations in Substarte Removal by Biological Floc Particles. *Proceedings of the ASCE Second National Symposium on Sanitary Engineering Research, Development and Design*. Cornell University, Ithaca, N.Y.

Baillod, C.R. and W.C. Boyle (1969b). An experimental Evaluation of the Intra - Particle Resistance in Substrate Transfer to Suspended Zoogloeal Particles. *Proceedings of the 24th Industrial Waste Conference*. Purdue University.

Bokil, S.D. (1972) *Effect of Mechanical Blending on the Aerobic Digestion of the Waste Activated Sludge*. Ph. D. Thesis. Department of Civil Engineering, University of Windsor, Windsor, Ontario, Canada.

Camp, T.R. and P.C. Stein, (1943). Velocity Gradients and Internal Work in Fluid Motion. *Journal of the Boston Society of Civil Engineers, 30*, 219.

Droste, R.L. (1978). *Influence of Power Input on Kinetics and Process Variables in Activated Sludge Treatment of Wastewater*. Ph.D. Thesis. Department of Civil Engineering, University of Windsor, Windsor, Ontario, Canada.

Droste, R.L. and J.K. Bewtra (1978). Effect of Mechanical Blending of Mixed Liquor in Batch Activated Sludge System. In G.Mattock (Ed.), *New Processes of Wastewater Treatment and Recovery*. Ellis Howood, Ltd., London, England. p.159.

Fair, G.M., Gemmel, R.S. and M.N. Nugent (1965). Power Dissipation in Biological Flocculation. In J.K. Baars (Ed.), *Advances in Water Pollution Research*, Vol. 2. Pergamon Press, New York. p.201.

Finstein, M.S. and H.Heukelekian, (1967). Gross Dimensions of Activated Sludge Flocs with Reference to Bulking. *Journal of the Water Pollution Control Federation, 39*, 33.

Ford, D.L. and W.W. Eckenfelder, Jr. (1967) Effect of Process Variables on Sludge Floc Formation and Settling Characteristics. *Journal of the Water Pollution Control Federation, 39*, 1850.

Gannon, J.J. (1966). River and Laboratory BOD Rate Considerations. *Proceedings of the American Society of Civil Engineers, 135*.

Goodman, B.L. and A.J. Englande, Jr. (1974). A Unified Model of the Activated Sludge Process. *Journal of the Water Pollution Control Federation, 46*, 312.

Hartmann, L. and G. Laubenberger (1968). Influence of Turbulence on the Activity of Activated Sludge Floc. *Journal Water Pollution Control Federation, 40*, 670.

Heukelekian, H. and E. Weisberg (1956). Bound Water and Activated Sludge Bulking. *Sewage and Industrial Wastes, 28*, 558.

Kalinske, A.A. (1971). Effect of Dissolved Oxygen and Substrate Concentration on the Uptake Rate of Microbial Suspension. *Journal of the Water Pollution Control Federation, 43*, 73.

Lawrence, A.W. and P.L. McCarty (1970). Unified Basis for Biological Treatment Design and Operation. *Journal of Sanitary Engineering Division, Proceedings of the American Society of Civil Engineers. 96*, SA3, 757.

Lordi, D. and H. Heukelekian (1964). The Effect of Rate of Mixing on the Deoxygenation of Polluted Water. *Proceedings of the 19th Industrial Waste Conference, 530*. Purdue University.

Monod, J. (1949). The Growth of Bacterial Cultures. *Annual Reviews of Microbiology, 3*, 371.

Mueller, J.A., Boyle, W.C. and E.N. Lightfoot (1966). Oxygen Diffusion Through a Pure Culture Floc of Zooglea Remigera. *Proceedings of the 21st Industrial Waste Conference*, Part 2, *964*. Purdue University.

Parker, D.S. (1970). *Characteristics of Biological Flocs in Turbulent Regimes*. Ph. D. Thesis. University of California, Berkeley, California.

APHA-AWWA-WPCF (1975). *Standard Methods for the Examination of Water and Wastewater*. 14th ed. American Public Health Association, New York.

Streeter, H.W. and E.B. Phelps (1925). A study of the Pollution and Natural Purification of the Ohio River. *USPHS Public Health Bulletin, 146*. Washington, D.C.

Wuhrmann, K. (1963). Effect of Oxygen Tension on Biochemical Reactions in Sewage Purification Plants. In W.W. Eckenfelder, Jr. and B.J. McCabe (Eds.), *Advances in Biological Waste Treatment*, Vol. 27. Pergamon Press, New York.

DISCUSSIONS

Şentürk, H. (Turkey) : Have you ever conducted experiments related to actual industrial wastewater treatment plants?

Bewtra, K. (Canada) : In the first phase of our study, we used synthetic sewage in addition to using domestic sewage or wastewater. During the second phase, synthetic sewage was used after initial acclimatization of the system with domestic wastewater. No two industrial wastewaters are identical. So when a synthetic sewage with known characteristics is used, the applicability of the experimental results will be more extensive. Results from using a specific industrial waste would have only very limited applicability.

Samsunlu, A. (Turkey) : You have worked with the Lawrence and McCarty equation, as I have done. Do you know of any treatment plant designed on the basis of that model and have you ever compared its actual treatment efficiency with that of the model?

Bewtra, K. (Canada) : I do not believe that we design our actual treatment plants on the basis of any model. We tend to use rules of thumb and draw from practical experience. In running experiments with a 14-liter jar, it is better to use a completely mixed system and we found that the data obtained in the laboratory fitted well with the Lawrence and McCarty model. No tests were conducted in actual treatment plants. In any case the emphasis here is not on the model but on showing the improved effect of blending on effluent BOD values.

Hanisch, B. (Germany) : I know of some cases where specific industrial wastewaters, similar to pure solutions, have been tested in pilot plant studies over a comparatively long period, and as a result, parameters have been found which would allow wider application of such models.

Leentvar, J. (Holland) : How did you accurately measure the power input, the G-value, in your experiments, using a blender?

Bewtra, K. (Canada) : The information on the actual power input when the blender was operated at different speeds was obtained from the manufacturers. That power input value was used. In the experiments conducted a correlation was found between viscosity and suspended solids concentration. So for different suspended solids concentrations the corresponding value of viscosity was found. The G-value was then calculated using Eq; 1 in the paper.

Leentvar, J. (Holland) : Why didn't you use a stirred vessel?

Bewtra, K. (Canada) : Our experiments were conducted at a low range of θ, only part of the sludge being blended at a time. We were blending at very high power, which a normal stirring system could not have provided. We used a heavy-duty domestic blender. A much larger blender unit would have been required for the entire contents of the aeration tank to be blended at the same time.

Leentvar, J. (Holland) : I think you are interested not so much in the G-value or power input, but more in the average floc diameter as a result of agitation. If you knew that, you could calculate diffusion coefficients inside the flocs, wouldn't that be a suitable direction for your next study?

Bewtra, K. (Canada) : It is true that information on floc diameters would be very important in calculating diffusion coefficients etc., but in practice accurate determination of floc diameters has not been possible.

Advanced Treatment by Manipulation of Microbiological Processes

P. HARREMOËS

Department of Sanitary Engineering, Technical University of Denmark

ABSTRACT

Throughout the last twenty years, significant insight into the mechanisms of biological treatment has been gained by the use of mass balance and kinetic approaches. In spite of remaining uncertainties, it is manifest that the purification result is governed predominantly by the organic load (kg BOD/kg VSS·d) -irrespectively of the particular plant configuration. Economic data show-loaded plants are both superior in performance and economically competitive. Nitrification is also feasible in sufficiently low-loaded plants. All arguments are in favour of choosing low-loaded plants up to a certain plant size (approximately 40,000 m^3/d). Nitrogen removal by denitrification can be achieved at little extra cost either by exposing the bacteria to alternating aerobic and anaerobic conditions or by mixing the liquor at a high recirculation rate. There are indications that phosphorus removal can be achieved by similar manipulation of the bacteria, but until the actual mechanism of P-removal is understood it is difficult to reach reliable design criteria.

INTRODUCTION

Microbiological processes have been used for sewage treatment since the turn of the century. For decades technological development was based on trial and error; by such empirical methods, suitable, techniques were developed for the removal of organic matter. A lack of intimate knowledge of the mechanisms involved brought into play a certain amount of myth about the suitability of certain process configurations, and traces of this can still be found in the water industry.

However, during the sixties and seventies a significant development of our basic understanding of the processes has taken place and is still continuing. As usual, improved understanding removes the myth and leads to the more rational design and operation of biological treatment systems. Furthermore, a set of new possibilities emerges, e.g.nitrification, denitrification, phosphorus removal-either biologically, by biologically mediated precipitation or by sheer chemical precipitation, with or without chemical addition. Potentially, advanced treatment can be achieved by mere manipulation of the processes involved. This requires an intimate understanding of the mechanisms an understanding, to the development of which numerous researchers all over the world strive to contribute their share. Through international communication these bits and pieces of new knowledge merge into a framework of understanding that will provide the basis for the rational design of biological treatment schemes.

THE MYTH ABOUT BIOLOGICAL TREATMENT

For the municipality or industry that produces the waste, sewage treatment should provide purification to meet effluent standards set by regulatory agencies with the receiving

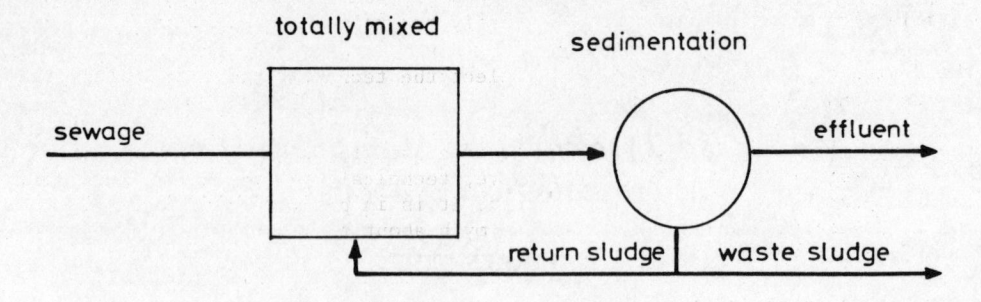

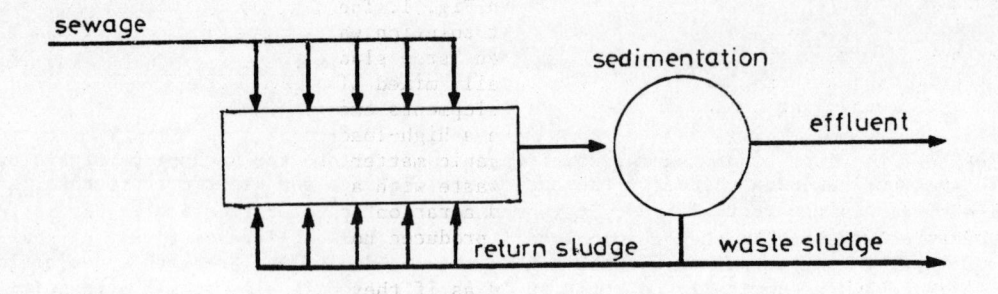

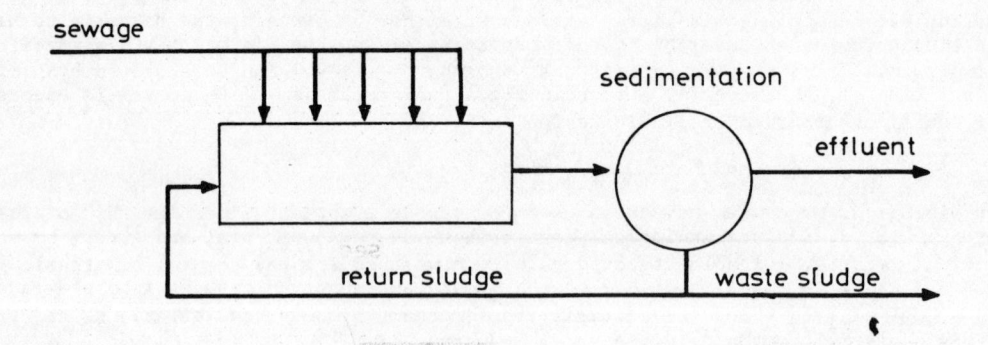

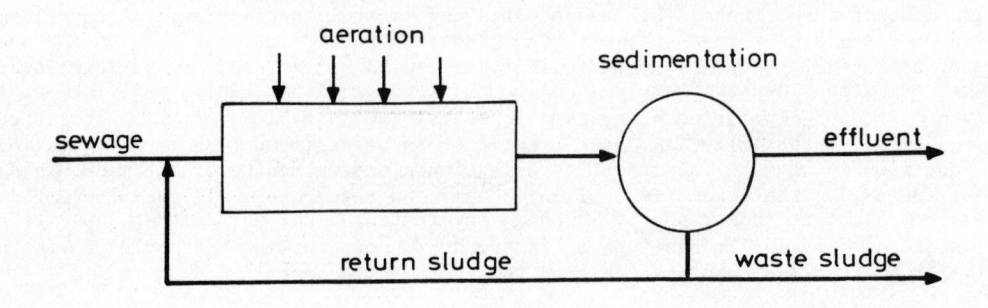

Fig. 1. Modifications of the Original Longitudinal Aeration Tank

capacity of the receiving waters in view. This goal should be achieved as economically as possible with facilities that can be easily and reliably operated and maintained.

For the design engineer, the goal is to select the technology that will meet these requirements and that is no easy task.

For the manufacturer of the technology, it is a question of producing facilities that can meet the requirements and be competitive, technical and economical. Furthermore, the manufacturer has to promote his product. It is in particular in these sales drives, that we find significant traces of the old myth about the business. A number of examples will illustrate this.

At the very beginning, activated sludge plants were made with longitudinal aeration tanks. At an early stage it was recognized that the oxygen demand at the head of the tank was much bigger than at the end. The simple idea was to make an equivalent uneven distribution of the air supply, as shown in Fig. 1. The magic term for this simple modification was "tapered aeration". The next solution was to distribute the loading of the tank. The term was "step aeration". Even "step sludge return" was suggested. Then in the fifties it was claimed that the totally mixed plant was superior (Fig. 1). At each end of the loading scale, two new developments emerged. The "contact stabilization" process (Fig. 2) is generally used in a high-loaded plant which relies on the adsorbtion of particulate and colloidal organic matter on the biological flocs-a process that is totally unsuitable for industrial waste with a high degree of soluble organic matter. The other extreme was the "extended aeration" process, which for a period was advocated with the erroneous claim that it produced no waste sludge.

Particular technologies have been advocated as if they would have a magic effect on the biological process. The "rapid block system" was just a very simple way of returning the sludge. It suffered greatly from the fact that it could not be controlled adequately and has faded into oblivion where it belongs. The "counter current" plant simply has an efficient aeration system, but no special effects on the biological process can be

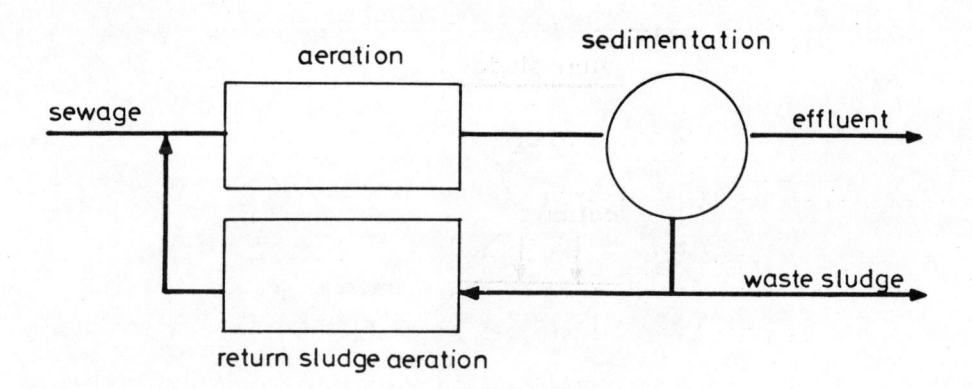

Fig. 2. Contact-Stabilization

claimed. The latest myth of the business is the "pure oxygen" plant, which is also just another way of providing oxygen for the aerobic bacteria. The only effect on the biological processes that can reasonably be claimed is a higher oxygen concentration in the mixed liquor, which can economically be achieved using pure oxygen, because of the increased driving force in the aeration process.

As long as actual understanding of the mechanisms by which competitive performance is achieved does not exist, a healthy amount of scepticism should be exercised when evaluating the pragmatic results of the postulated magic of BOD-removal.

BOD-REMOVAL

The processes of concern in biological treatment were until a decade ago only those related to the removal of organic matter. The fact is that nitrification was suppressed because it could cause rising sludge in the final clarifier due to nitrogen bubbles from denitrification in the anaerobic settling zone of the clarifier.

The breakthrough in the understanding of the process kinetics of biological processes came with the application of Michaelis-Menten kinetics in combination with the process hydraulics of the biological treatment plant. This is a suprising fact, because Michaelis-Menten kinetics is valid for soluble substrates only, which is contrary to the fact that only 20-40 percent of organic matter carried to the aeration tank is in a soluble form. Figure 3 shows the consecutive processes that are expected to take place when particulate and colloidal matter is adsorbed, hydrolysed and degraded. Figure 4 shows the result of a batch-test and illustrates that the BOD of the liquid reaches a minimum, before which adsorbtion and after which hydrolysis predominately occurs, while significant degradation has not yet taken place. This is what happens in contact stabilization, but the kinetics of adsorbtion in particular has never been very well described. The process can work well only on sewage with little soluble BOD. At present it cannot be termed a reliable process-though the literature contains several examples of its successfull performance.

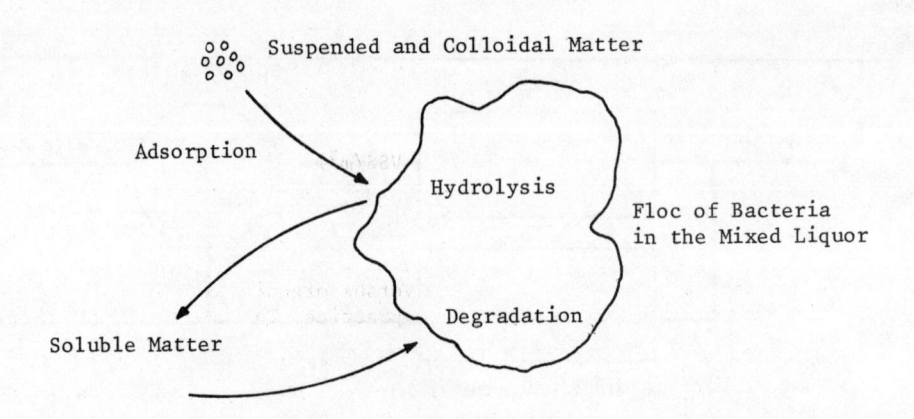

Fig. 3. Sequence of Degradation of Suspended and Colloidal Organic Matter

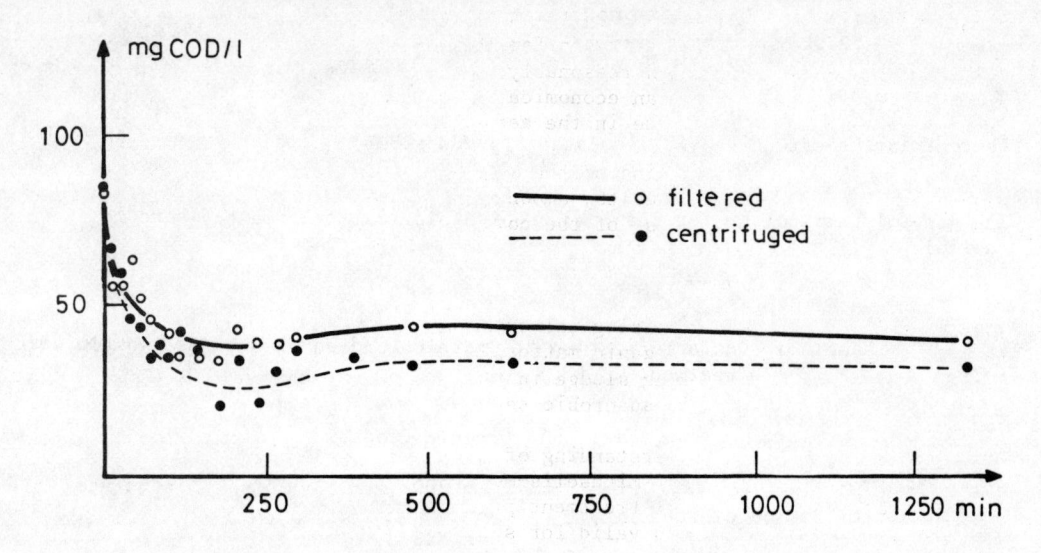

Fig. 4. The Result of COD Measurements in the Bulk Liquid in a Batch Test with Raw Sewage and Activated Sludge

Through the process kinetics it is easy to show that the organic loading (also called the food to mass ratio), OL, is the key factor.

$$OL = \frac{QS_i}{XV} \quad \left(\frac{g\ BOD}{g\ VSS \cdot d}\right) \tag{1}$$

where

Q = flow (m^3/d)

S_i = influent BOD (g/m^3)

X = bacterial concentration (g VSS/m^3)

V = volume of aeration tank (m^3)

Figure 5 shows the degree of purification versus organic load theoretically for selected kinetic data, compared with the results in practice. The sludge age, SA, is closely related to the organic load:

$$SA = \frac{VX}{SP} \tag{2}$$

Where SP is waste sludge production (g VSS/d).

$$SA \cdot OL = \frac{QS_i}{SP} = \frac{1}{Y} \qquad (3)$$

Where Y is the effective yield constant (g VSS/g BOD).

The general terminology for the loading of biological treatment plants is shown in Table 1.

Table 1 Characterization of Organic Load and Sludge Age in Biological Treatment

Characterization of Load	Organic Load $\frac{g\ BOD}{g\ VSS \cdot d}$	Sludge Age d
Low	OL < 0.3	10 < SA
Medium	0.3 < OL < 0.6	3 < SA < 10
High	0.6 < OL	SA < 3

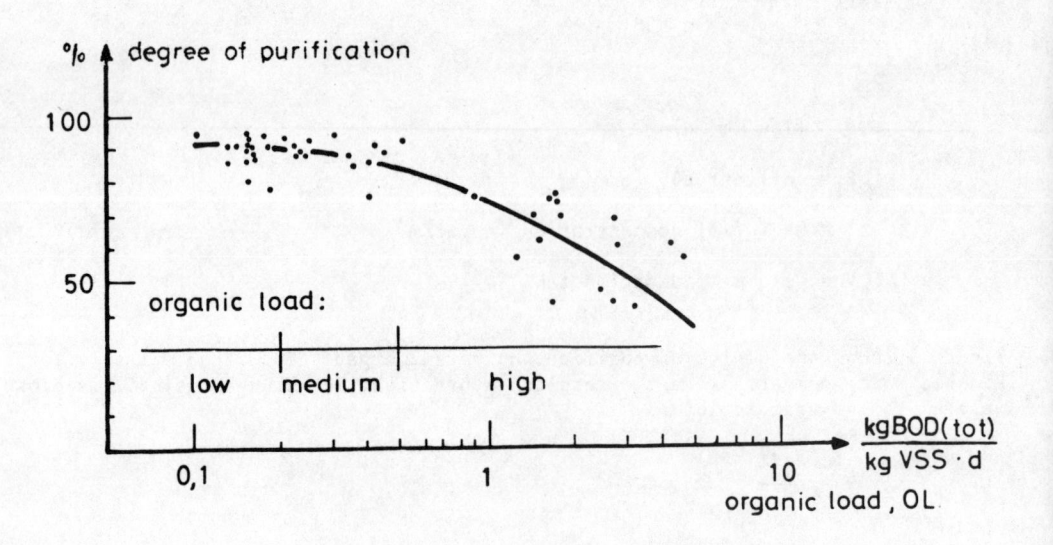

Fig. 5. Degree of Purification Versus Organic Load for Treatment of Domestic Sewage

As shown in Fig. 5, the purification improves with decreased loading and increased sludge age. However, there are a number of reservations to be made. From a certain loading downwards, the BOD-removal is governed more by the performance of the clarifier than by the biological processes. Due to the above-mentioned adsorption phenomena, good performance can be experienced at high loadings but good performance is less reliable. This is shown in Fig. 6.

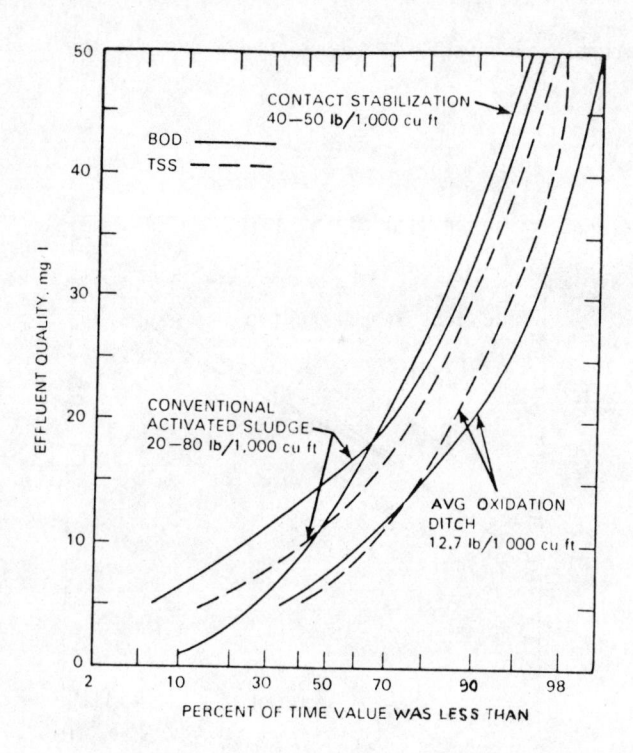

Fig. 6. Comparison of Effluent Quality and Reliability between Low-Loaded and High-Loaded Treatment Plants (Bender, 1979).

The concentration of sludge in the aeration tank varies significantly. The tradition in the USA and Europe is different. In the USA the mixed liquor suspended solids level is often found in the region 2000-3000 g VSS/m^3, while 5000-6000 g VSS/m^3 (even 8000) is often found in Europe. It is difficult to find an explanation for this, it appears to be a question of tradation. The effect of this on volume for an equal load can easily be seen from the formula. The smaller X is, the larger the V required.

Equally it can be seen that for a constant X, the maximum of which is governed predominantly by the efficiency of the clarifier, the volume must increase for the same performance to be reached. One would expect that the cost of construction would increase significantly, but this may well be offset by the lower operation and maintenance cost of the low-loaded plant.

Figure 7 shows the cost of four process configurations, of which extended aeration (loading in the region 0.02-0.05 g BOD/g VSS.d) comes out superior for small plants both with respect to performance and cost. For larger plants the oxidation ditch comes out superior too. Most oxidation ditch plants are low loaded.

The important conclusion is that low loading and a greater sludge age are superior in all respects for small and medium-size plants. It is an important feature that they are easy to maintain, reliable even with mediocre maintenance and that the waste sludge needs no further stabilization at low loadings and only aerobic stabilization at higher loadings.

In a country with little experience in sewage treatment the low-loaded plants are the best solution.

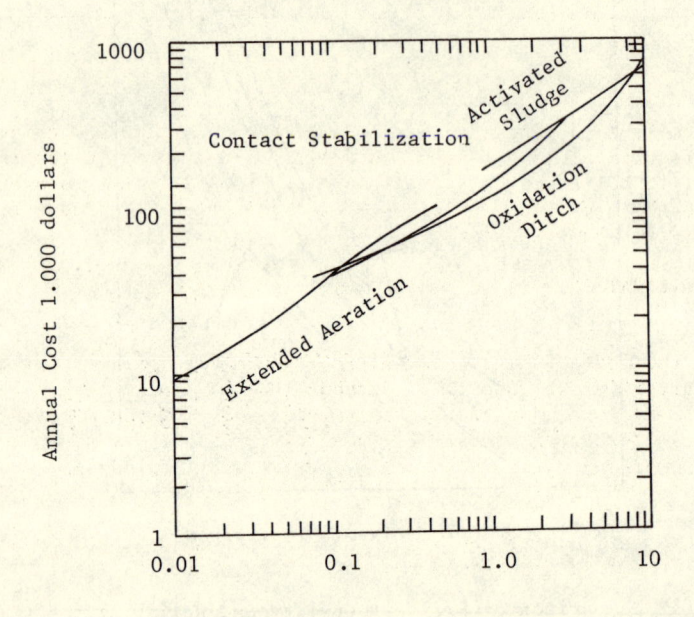

Fig. 7. Economical Comparison of Four Different Types of Treatment Plants (1 mgd ~ 3800 m^3/d) (Bender, 1979)

NITRIFICATION

By now it has been proved by many that discharge of ammonia to sensitive, receiving waters (small rivers in particular) can give rise to serious oxygen problems. Thus, ammonia has to be oxidized in the treatment plant before discharge. Before the discovery in the early sixties of the kinetics of the nitrifying bacteria, *Nitrosomonas* and *Nitrobacter* nitrification in treatment plants belonged to the traditional magic of the business. Today there is no particular problem in obtaining nitrification. The requirement is that the relatively slow-growing nitrifiers are not washed out of the plant. The sludge age, which is equivalent to the retention time of bacteria, must be greater than the reciprocal of the growth rate of the nitrifiers. The growth rate of nitrifiers is sensitive to temperature. The required sludge age for achieving nitrification is given in Table 2 for different temperatures.

Table 2 Required Age in Order to Achieve Nitrification

Temperature ^{0}C	Sludge Age days
5	25
10	10
15	5
20	3

It can be seen from Table 4 that nitrification can be difficult to achieve at low temperatures. It requires a long sludge age (SA>25) and a very low loading (OL<0.1). On the other hand, in hot climates nitrification can be difficult to avoid unless the plant receives a high load. Thus, precautions to avoid floating sludge from denitrification in the clarifier have to be taken under all circumstances.

DENITRIFICATION

Denitrification is a process performed by facultative bacteria which are capable of using nitrogen in the form of nitrite or nitrate as oxidizing agent for the degradation of organic matter in the absence of oxygen. The oxidized nitrogen is reduced to molecular nitrogen. Water in contact with air is already 80% saturated with molecular nitrogen and only a low rate of denitrification will cause supersaturation and bubble formation. These bubbles are formed in or on the bacterial flocs, which are then lifted to the surface; this is disastrous for a clarifier based on sedimentation.

Denitrification may occur even in an apparently aerobic mixed liquor. The reason is that the bacteria are flocculated into aggregates of considerable size. The oxygen in the bulk water can penetrate only a small distance into large flocs, the interior of which then becomes anaerobic and available for denitrification.

This phenomena is illustrated in Fig. 8. Thus, denitrification will occur to some degree in any nitrifying plant.

The original reason for introducing nitrogen removal through denitrification as a deliberate measure was due to protection against eutrofication caused by either of two nutrients, nitrogen or phosphorus. However, reasons for the introduction of denitrification have multiplied with development:

1. Nitrogen removal in order to protect against eutrofication.

2. Savings of energy by the use of the nitrate already produced, rather than oxygen, oxidizing agent for the removal of organic matter. Where nitrification is required, it may actually pay to introduce denitrification rather than waste the oxidizing by discharging the nitrate.

3. Denitrification may be used for the control of pH. Nitrification is an acidity-producing process, while denitrification is an alkalinity-producing process. In water of low alkalinity, nitrification may cause decreases in pH to such an extent that the process is inhibited. This

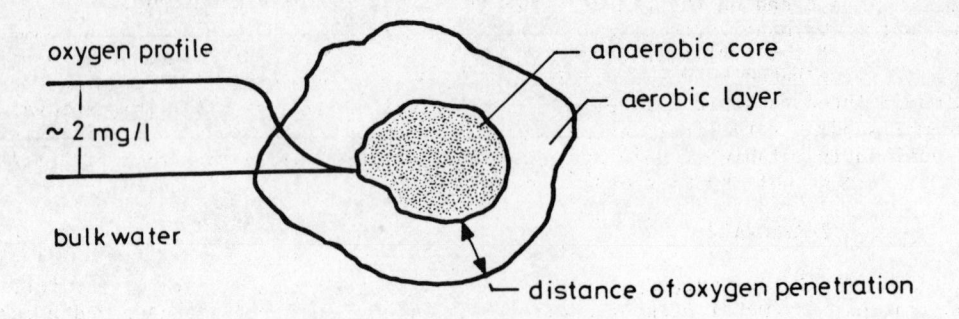

Fig. 8. Oxygen Penetration into Flocs

can be overcome by simultaneous alkalinity production through denitrification.

4. Nitrate in drinking water can be poisonous above certain levels. Where treated sewage is discharged to rivers that provide water for a drinking-water supply, downstream removal at the treatment plant may be required.

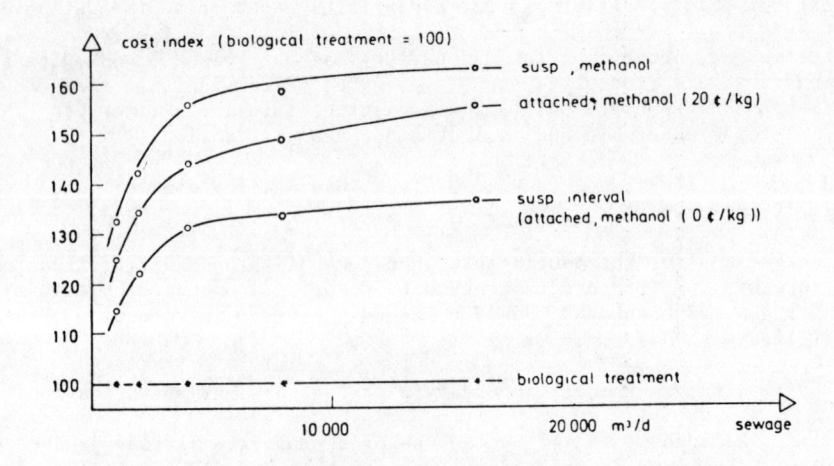

Fig. 9. Cost of denitrification Using Methanol as a Carbon and Energy Source (Christensen and Harremoës, 1978).

Denitrification can be achieved in a number of ways. The most popular one at the beginning was based on the use of methanol as a carbon and energy source. However, the cost of methanol turned out not to be competitive, as seen in Fig. 9, compared to the use of the available carbon in the raw sewage as a carbon and energy source. This is illustrated in Fig. 10. Figure 11 shows how the biological treatment plant can be manipulated to achieve a very high degree of nitrogen removal. This removal can be accomplished at very little extra cost, as shown in Fig. 10. The remaining nitrogen is predominantly soluble organic nitrogen (abbreviated SON) in the order of 2 mg/ℓ, which can be removed only by some other means.

P-REMOVAL

While N-removal has reached a stage of development which makes it fit for full-scale application, P-removal without chemical addition still lacks adequate understanding for reliable full-scale design. However, three different schemes have been employed with success in a full-scale plant. These plant configurations are shown in Fig. 12. Equal results can be obtained by the alternative scheme shown in Fig. 11(a).

There is still a debate as to the basic excess removal as compared to conventional treatment plants. Until further clarification is reached it is impossible to predict a reliable performance, because it may not be possible to secure the conditions that are required, e.g. those conditions that will cause selection of bacteria capable of excess removal by one theory or conditions, such as CA and/or Fe concentrations and adequate floc size, that will cause precipitation in the bulk or the floc phase.

This is an interesting subject for further research. Potentially, the removal of BOD, N and P can be achieved simply by manipulation of the biological treatment process.

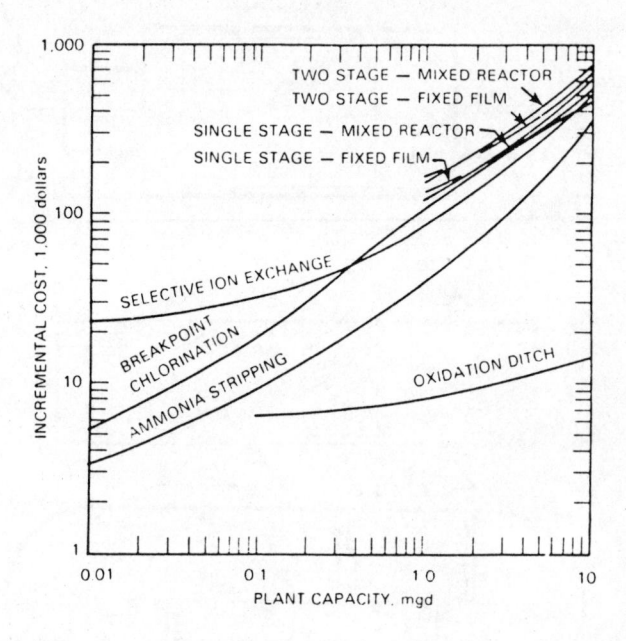

Fig. 10. Incremental Total Annual Cost for Nitrogen Removal (Bender, 1979)

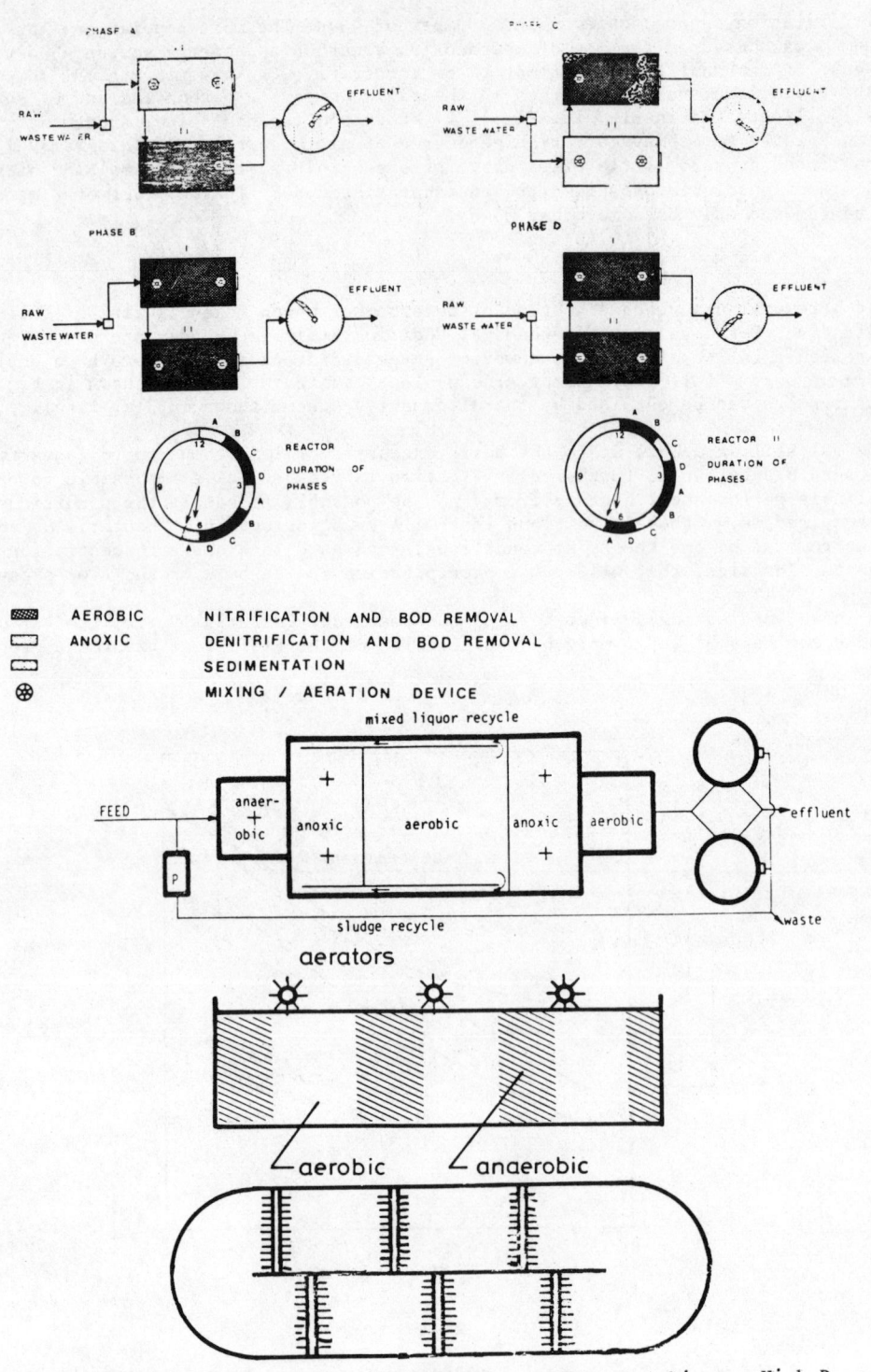

Fig. 11. Treatment Plant Configurations that can Achieve a High Degree of Nitrogen Removal (Akvadan-Harvey; Barnard, 1976)

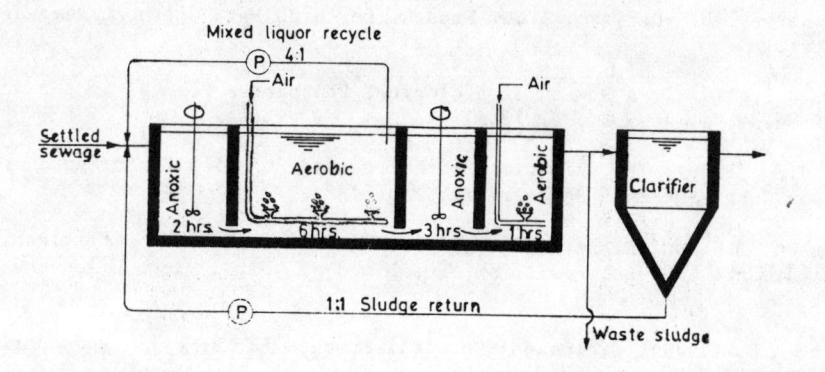

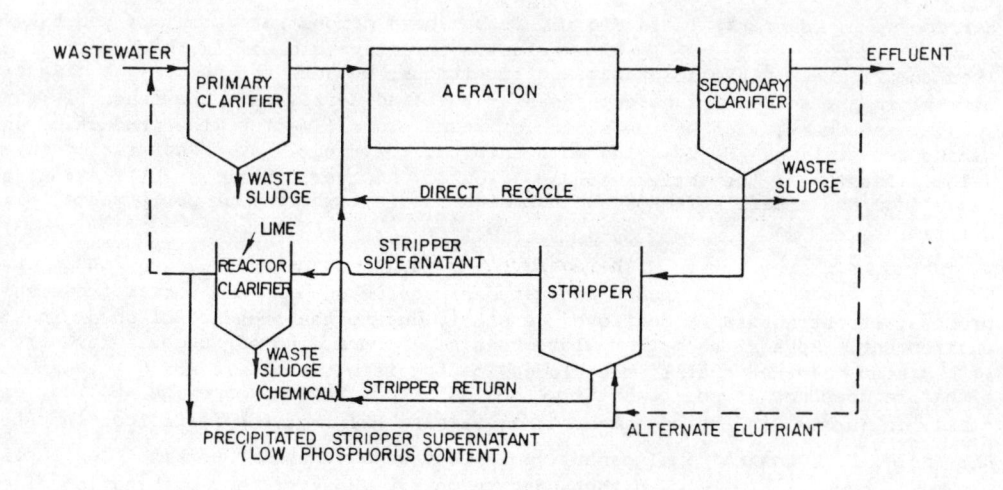

Fig. 12. Treatment Plant Configurations which have Achieved P-removal.
(Barnard, 1976; Drnevich, 1979)

REFERENCES

Akvadan-Harvey. Bio-Denitro, A New Process for High Rate Nitrogen Removal from Wastewater.

Barnard, J.L. (1976). A Review of Biological Phosphorus Removal in the Activated Sludge Process. *Water S.A.*, 2, 136-144.

Bender, J.H. (1979). The Oxidation Ditch Process: Superior Performance and Reliability at Low Cost. *EPA Env. Res. Brief.*, Jan. 1979.

Christensen, M.H. and P. Harremoës (1978) Nitrification and Denitrification in Wastewater Treatment. *In Water Pollution Microbiology*, Vol. 2, John Wiley and Sons, Inc., New York.

Drnevich, R.F. (1979). Biological-Chemical Process for Removing Phosphorus at Reno/Sparks, NY. EPA-600/2-79-007.

DISCUSSIONS

Taygun, N. (Turkey) : In low-loaded biological treatment systems activated sludge is said to stabilize. What quantity of activated sludge is produced in this system?

Harremoës, P. (Denmark) : In Fig. 5, it has been demonstrated that if you talk of loading relative to active biomass in the treatment plant, then the figures in the low range are fictitious, because the percentage of active biomass in a treatment plant decreases as the load decreases. The reason is that you get an accumulation in the sludge of inert organic matter, the product of endogenous respiration. So by going down into the low range, you do not really obtain a lower loading of the active biomass. You do, however, obtain stabilization of the sludge and a safety factor against high loading, where your performance will deteriorate.

Taygun, N. (Turkey) : With regard to the sugar industry, we have found that when activated sludge is used in the normal treatment process, its structure is destroyed generally due to the presence of sugar in the environment. Sphaeritus natans microorganisms form a kind of fungus. In Germany it has been observed that if the sludge loading is between 0.12 and 0.16, such a situation does not arise. Can the method described in your paper be applied, especially in Turkey, in the treatment of wastes containing NH_4 from the fertilizer industry?

Harremoës, P. (Denmark) : I cannot claim that we have used the principles outlined in the sugar industry. However, any combination of biodegradable organic matter and nutrients can be handled by these methods. Of course, different concentrations will call for different configurations which could be worked out on the basis of these principles, but the result might be plants which look different from those employed in the treatment of sewage, as opposed to the treatment of industrial wastewater.

Saad, S.G. (Egypt) : How often would you transfer from the aerobic to the anaerobic treatment process?

Harremoës, P. (Denmark) : The optimum period of alternating the mode of operation varies in practice from between 2 and 4 hours. That is, we change from tank to tank every hour or every second hour. This develops a population of facultative bacteria which can readily change from aerobic to anaerobic conditions and back on a continuous basis. The nitrifiers, which are strictly aerobic, are not harmed by the exposure to anaerobic conditions.

Rüffer, H.R. (Germany) : The Dutch and the Austrians have recently published findings showing that the sludge volume index in totally mixed activated sludge basins is higher than that in other basins. What effect does the alternating aerobic-anaerobic process have on the sludge volume index?

Harremoës, P. (Denmark) : Generally speaking, it is my impression from the literature that you can, if you look hard enough, find an operational parameter related to the sludge volume index - for example, the rate of circulation, well-mixed "tapered aeration", organic sludge load or F/M ratio. I do not think that anyone has yet found any meaningful correlation with the sludge volume index. In those nitrogen-removal plants we have knowledge of, i.e. 5 or 6 plants, including our own pilot plants which are up to full-scale, we have had sludge volume indexes in the range of 50 to 80. I stress that this applies only to these treatment plants. I do not claim that this will be universal. With regard to many other statements found in the literature concerning the relationship between operation, loading etc. and sludge volume index, we have no idea whether they are universally applicable.

Technology for the Control of Toxic Pollutants from Industrial Wastewater Discharges

W. W. ECKENFELDER, JR.

Department of Environmental and Water Resources Engineering,
Vanderbilt University, Nashville, Tennessee

This paper will consider the removal of priority pollutants in the context of existing wastewater treatment technology. Several cases may be considered:

a) Removal of the pollutant to acceptable levels through existing BPT technology.

b) Pretreatment of selected wastewater streams for the removal of priority pollutants prior to discharge to existing BPT technology.

c) Separate treatment facilities for priority pollutant removal or modifications to existing technology.

The primary alternatives available today are shown in Fig.1 Three factors need to be evaluated for discharge of priority pollutants to existing BPT plants: firstly, whether the concentration of the pollutant discharged will be inhibitory or toxic to the plant performance; secondly, whether the concentration of the pollutant will be reduced to acceptable discharge levels; and thirdly, whether accumulations in the sludge will create problems for ultimate sludge disposal, such as the accumulation of heavy metals in a biological sludge, which will mitigate against land disposal.

A number of the priority pollutants are biogedradable in a conventional activated sludge plant. Table 1 lists several of these.

In order for degradation to occur, the biological system must be acclimated to the organics in question. This is particularly true of the more complex compounds. An acclimation procedure has been described by Pitter (1976) and Adams, Ford, and Eckenfelder (1979). A laboratory acclimation study should be conducted to determine (a) the rate of degradation of the pollutant and the terminal effluent concentrations achievable in the existing or proposed plant and (b) maximum concentrations of the pollutant to avoid inhibition in the biological process, if any. In order to ensure continuous operation of the biological facility and avoid upsets, the priority pollutant should be retained in a storage tank and fed to the treatment plant at a controlled rate as defined by (a) and (b) above.

The controlled rate should be such that the concentration in the aeration basin does not exceed the defined inhibition level as defined by (b) above. This problem would normally not exist in a completely mixed basin since the concentration in the basin would be the same as the effluent, or the discharge limitation level. The kinetics of removal in a mixed wastewater can be expressed by the relationship:

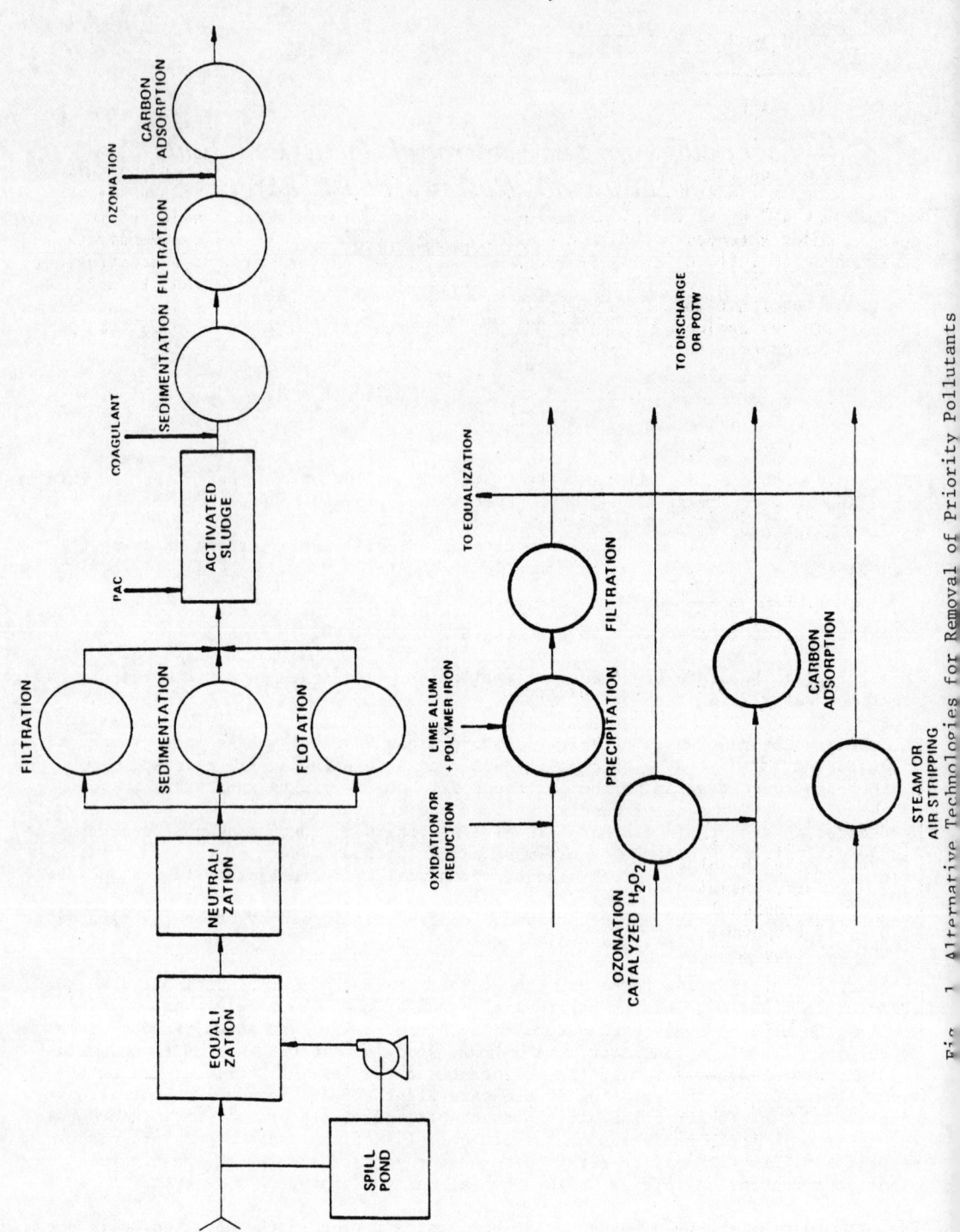

Fig. 1 Alternative Technologies for Removal of Priority Pollutants

Technology for the Control of Toxic Pollutants

$$\frac{S_o - S_e}{X_v t} = k \frac{S_e}{S_o} \qquad (1)$$

Depending on the influent and effluent desired and the reaction rate of the pollutant in question, $X_v t$ must be adjusted accordingly. This is illustrated by the data reported by Tabak and Barth (1978) on the aerobic oxidation of benzidine added to municipal sewage. Under extended aeration conditions (F/M = 0.05), benzidine was completely degraded up to influent concentrations of 1 mg/l. At higher influent concentrations the effluent increased linearly as shown in Fig. 2. It is significant to note that six weeks' acclimation was required to achieve maximum degradation rates. This is in accord with the projected kinetics which show that effluent quality is related to F/M in accordance with the relationship:

$$\frac{S_e}{S_o} = \frac{1}{K/F/M + 1} \qquad (2)$$

Equation (2) is obtained by combining Equation (1) with the expression for the organic loading F/M, in which

$$F/M = \frac{S_o}{X_v t}$$

Table 1 Biological Degradability of Priority Pollutants

Compound	Percent Removed (Based Upon Cod)	Rate of Biodegradation (mg COD/g VSS-hr)
2,4 - Dichlorophenol	98.0	10.5
2,4 - Dimethylphenol	94.5	28.2
3,4 - Dimethylphenol	97.5	13.4
3,5 - Dimethylphenol	89.3	11.1
2,5 - Dinitrophenol	*	
2,6 - Dinitrophenol	*	
2,4 - Dinitrophenol	85.0	6.0
Nitrobenzene	98.0	14.0
m-Nitrotoluene	98.5	21.0
p-Nitrotoluene	98.0	32.5
Phenol	98.5	80.0

*2,5 and 2,6 - Dinitrophenol were at higher concentrations not degraded. 2,6 - Dinitrophenol was at lower concentrations decomposed with long adapted activated sludge (40 days). 2,5 - Dinitrophenol was biochemically stable.

In order to maintain low effluent quality at increased influent concentrations, the F/M would have to be reduced. With substances with very low degradability such as benzidine, this is not practical for higher influent concentrations and the influent feed rate would have to be controlled to a level that would ensure the desired

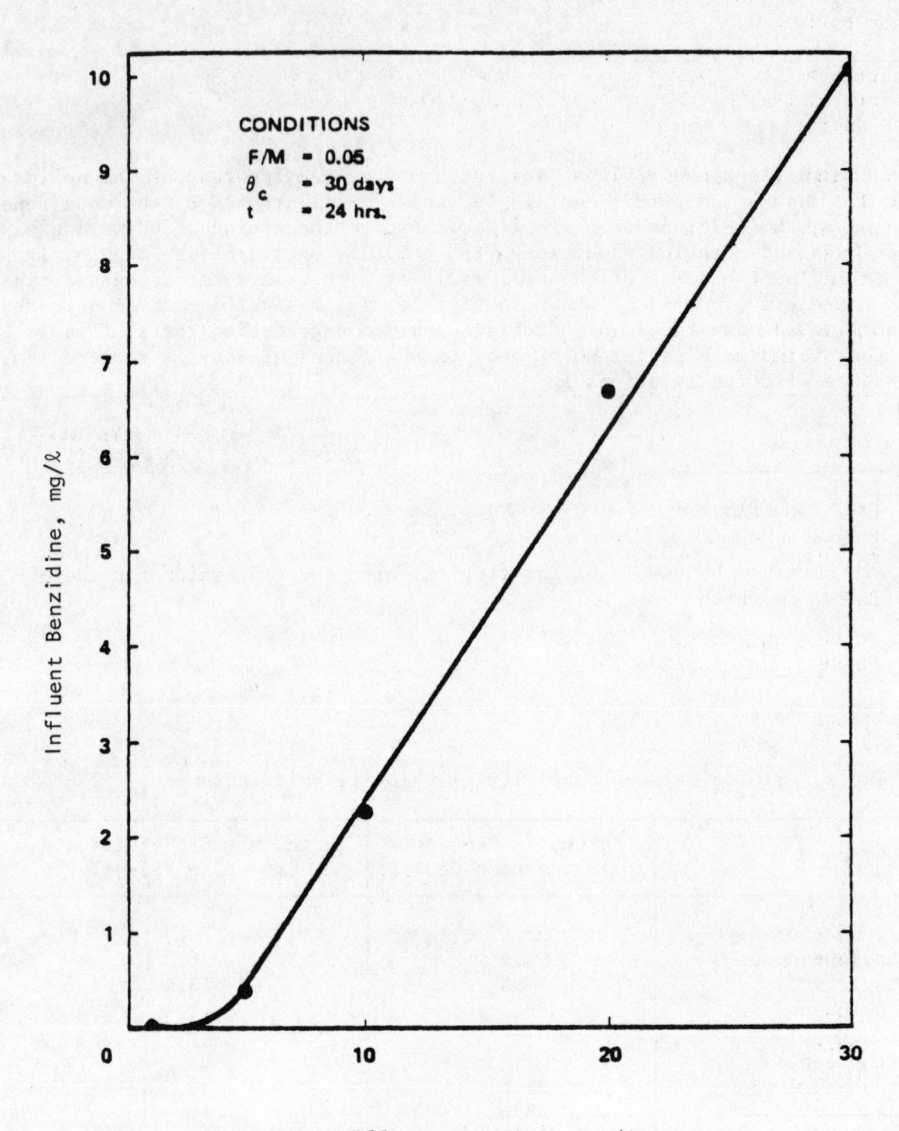

Fig. 2. Aerobic Degradation of Benzidine

effluent quality. Investigations should also be made to ensure that oxidation by-products are not toxic.

Winter operation needs to be considered since the degradation rate is a function of basin temperature. It has also been shown in a number of cases that inhibition is increased at decreased operating temperatures.

In many cases, the most cost-effective solution will be pretreatment of the priority pollutant at the source. Several alternatives exist to achieve this, as shown in Fig. 1. Heavy metals may be removed by precipitation. (Alternative technologies include ion exchange, reverse osmosis and evaporation.)

Table 2 Removal of Heavy Metals by Precipitation (Patterson, 1977)

Metal	Process	Effluent Level	Constraint
Arsenic	Prec. with $S^=$ at ph 6-7 Carbon adsorbtion (low levels) $Fe(OH)_3$ co-precipitation	0.05 mg/l 0.06 mg/l 0.05 mg/l	
Barium	$BaSO_4$	0.5 mg/l	
Cadmium	$Cd(OH)_2$ ppt (ph 10.0) $Fe(OH_3)$ co-precipitation at pH 8.5 H_2O_2 oxidation	0.1 mg/l none -	complexing ions e.g. CN^- require pretreatment oxidizes CN^- and Cd to oxide
Copper	$Cu(OH)_2$ $Fe(OH)_3$ co-precipitation	0.2 mg/l 0.3 mg/l	pH 9.0 - 10.3 pH 8.5
Lead	$Pb(OH)_2$ $Pb(OH)_3$ Pb S	0.5 mg/l 0.001 mg/l -	pH 10.0 pH 8.0 - 9.0 pH 7.5 - 8.5
Mercury	$Fe(OH)_3$; $Al(OH)_3$ co-precipitation	0.1 mg/l	Na_2S added
Nickel	$Ni(OH)_2$	0.15 mg/l	pH 10.0
Selenium	Se S	0.05 mg/l	pH 6.5
Zinc	$Zn(OH)_2$	-	pH 8.5

Reported efficiencies are shown in Table 2. In some cases pretreatment by oxidation or reduction is required to effect optimal removal by precipitation. For example, Cr^{+6} must be reduced to Cr^{+3} in order to precipitate $Cr(OH)_3$. Complexed metals will require pretreatment. Cyanide complexes require oxidation of the cyanide. If high concentrations of ammonia are present, ammonia removal is required prior to precipitation of those metals which complex with ammonia. If high removal of the total metal is required, filtration, following precipitation and sedimentation, may be required. If the waste volume and the quantity of precipitate are relatively low, the precipitated mixture may be clarified directly through a diatomite filter. This offers the advantage of producing a dry cake for disposal. Depending on the content of other residuals, the effluent may be directly discharged or sent to the equalization basin for further treatment.

Chemical oxidation of priority pollutants may be achieved by ozonation or, in some cases, catalyzed H_2O_2. Under practical conditions, complete degradation of fairly unreactive compounds, such as saturated hydrocarbons and halogenated aliphatic compounds, does not occur with ozone alone; but current research has shown that ozone with an additional energy source, i.e. sonification, ultra-violet radiation or a catalyst, will readily decompose many refractory compounds.

In many cases, the initial by-products of ozonation are fairly degradable in conventional biological treatment. For example Yocum, Mayes and Myers (1977) showed that the BOD_5/TOC ratio increased from 0.11 to 0.98 following ozonation in polyol wastewater. The change in biodegradability of organics following ozonation can be defined by the procedures described by Adams, Ford and Eckenfelder (1979). It has recently been shown that preozonation prior to carbon adsorption enhances removal in a granular carbon column (Ford, 1976). Results from a municipal-industrial wastewater are shown in Fig. 3.

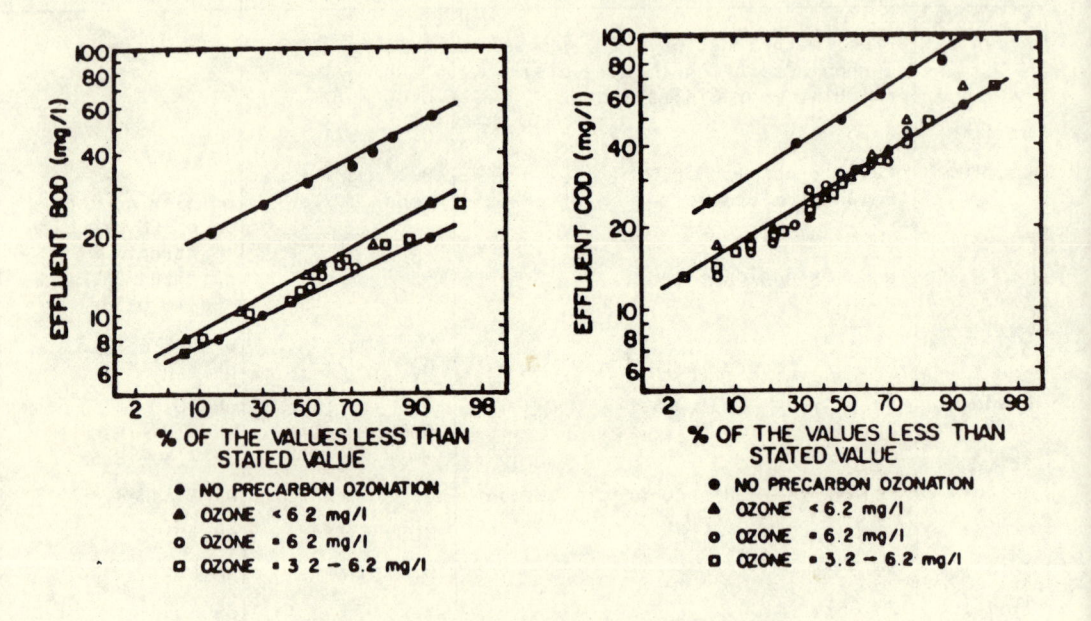

Fig. 3. Effect of Ozonation on BOD and COD Removal Through Activated Carbon Columns.

Carbon adsorption can be applied for the removal of many of the priority pollutants. A comprehensive study by Dobbs, Middledorf and Cohen (1978) has provided data for a variety of organic compounds including priority pollutants.

Selected pretreatment by carbon adsorption may markedly enhance the performance of a biological treatment plant. Table 3 shows the results obtained by pretreating a pesticide wastewater stream through granular carbon adsorption prior to discharge to an activated sludge plant. Volatile organics may be removed by air or stream stripping.

There are several modifications to existing technology that may be employed to remove priority pollutants to acceptable discharge levels. The most promising is the appli-

Table 3 Reaction Rate Coefficients with and
Without Carbon Treatment of a Pesticide Wastewater

	Non - carbon treated K (d^{-1})	Carbon treated K (d^{-1})
28°C	2.25	23.1
8°C	0.81	6.5

cation of powdered activated carbon (PACT) to the activated sludge process. The activated carbon removed by adsorption pollutants that are otherwise non-degradable. This may have the effect of enhancing the oxidation rate by adsorption of inhibitory substances as well as of removing priority pollutants to an acceptable level. Results reported by Hutton and Robertaccio (1978) are shown in Table 4.

Table 4 Comparison of Pact and Activated Sludge for Organics Removal
(Hutton and Robertaccio, 1978)

Organic	Effluent Concentration, mg/l	
	Pact	Activated Sludge
Chlorinated Pesticides	0.017	0.35
Organo-sulfur pesticides	0	15.0
Organo-phosphate pesticides	1.23	3.03
PCB's	0.008	0.13

The question arises as to what procedures might best be applied to provide the most cost-effective solution to a given problem. Assuming for this case that a biological treatment plant is presently treating the major wastewater flows, several procedures may be followed, relative to the acceptable removal of priority pollutants. (This discussion does not consider in-plant process change resulting in the removal of the priority pollutant from the wastewater.) In most cases, removal of the priority pollutant through existing technology would be the most cost-effective. Assuming the existing technology is an activated sludge plant, it would be necessary to define:

a) The biodegradation rate of the pollutant in the existing wastewater mixture (Fig.2).

b) The maximum concentration in the influent to avoid inhibition. (This may be the case in a multi-stage or plug flow system, since in a single completely mixed basin the basin contents are equal to the effluent concentration).

c) Whether the operational modes of the plant would enable sufficient degradation to meet effluent criteria.

If degradation is not sufficient to achieve the required effluent limitation, PACT addition to the aeration basin can be considered. This could be evaluated in laboratory scale reactors. Source pretreatment could consider stripping, ozonation or carbon adsorption.

NOMENCLATURE

S_o - Influent organic content, mg/ℓ
S_e - Effluent organic content, mg/ℓ
X_v - Volatile suspended solids, mg/ℓ
t - Aeration detention time
K - Biodegradation rate coefficient, days^{-1}
F/M - Organic loading, lbs organic applied/day/lb MLVSS

REFERENCES

Adams, C., D.L. Ford and W.W. Eckenfelder (1979). *Development of Design Criteria for Wastewater Treatment Processes*. Enviropress, Inc., Nashville, TN.

Dobbs, R.A., R.J. Middledorf and J.M. Cohen (1978). Carbon Adsorption Isotherms for Toxic Organics. EPA Report, Cincinnati, Ohio, May 1978.

Ford, D.L. (1976). Current State of the Art of Activated Carbon Treatment. Open Forum on Management of Petroleum Refinery Wastewaters, Tulsa, Okla.

Hutton, D.G. and F.L. Robertaccio (1978). Use of Powdered Activated Carbon for Textile Wastewater Pollution Control. Symposium on Textile Industry Technology, EPA.

Patterson, J. (1977). *Wastewater Treatment Technology*. Ann Arbor Press, Michigan.

Pitter, P. (1976). Determination of Biological Degradability of Organic Substances. *Water Research, 10,* 231.

Tabak, H.H. and E.F. Barth (1978). Biodegradability of Benzidine in Aerobic Suspended Growth Reactors. *Jour. Wat. Poll. Cont. Fed., 50,* 552.

Yocum, F.H., J.H. Mayes and W.A. Myers (1977). Pretreatment of Industrial Wastes with Ozone. *AIChE Symposium Series, Water,* p. 217.

Dynamics of Phenol and Oxygen Uptake in a Biological Wastewater Treatment Reactor

S. UZMAN, S. YONGACOĞLU, B. WALKER and I. J. DUNN

Technisch-Chemisches Laboratorium ETHZ, CH-8092 Zurich, Switzerland

ABSTRACT

The objective of this work was to study the dynamic responses of an activated sludge reactor which is treating phenolic wastewater. Various types of dynamic experiments were made to measure short and long term phenol adaptation of the sludge, phenol toxicity, and the relationship between oxygen uptake rates and phenol uptake rates. Comparison of the phenol and oxygen uptake rates showed that the latter was not a simple function of phenol concentration. The substrate inhibition rate kinetics which has been reported for phenol in the literature was not found applicable to the dynamic experiments performed. Some success was, however, achieved using differential mass balance and kinetics principles for the dynamic simulations. The main significance of the work is the establishment of oxygen uptake rate as a parameter which correlates qualitatively with phenol uptake. The sensitivity and accessibility of the dynamic electrode measurements of the oxygen uptake rate suggest its use as a basis for the control of phenol feed rates in wastetreatment processes.

INTRODUCTION

Phenol is highly toxic and is present in wastewater from oil refineries, coking plants and plastics processing. Not only is it damaging to fish at concentrations above 1 mg/l (Zdybiewska, 1968), but in conbination with chlorine it imparts a bad taste to drinking water at low concentrations of 0.005 mg/l. Many methods are available for treating phenolic wastewater (Lanovette, 1977), but for low concentrations the activated sludge process is common. Special precautions must, however, be taken for the successful treatment of toxic organics. It is generally suggested that equalization tanks are necessary for protection against shock loading and that a 24 h residence time is recommended (Paulson, 1976).

A number of workers have investigated the dynamic and steady-state operation of pure and mixed cultures for the bio-oxidation of phenol (Pawlowsky, Howell and Chi, 1973; Yang and Humphrey, 1975; Holladay and others, 1978). It is known that phenol is oxidized rapidly at low concentrations, but that at high concentrations (> 300 mg/l) it is toxic to all organisms.

The present work is directed at measuring the dynamic response of a laboratory activated sludge system to changing phenol load conditions.

EXPERIMENTAL EQUIPMENT AND PROCEDURES

A schematic drawing of the reactor and companion equipment is shown in Fig. 1. The 5 l reactor (A) was stirred and aerated. It was continuously fed (Pawlowsky and Howell, 1973), and the reactor effluent was passed into a 6.6 sedimentation tank (B).

The sludge was recycled to the reactor continuously. The reactor could be subjected to variations in inlet concentration and flow rate using two reservoirs and pumps.

Samples were withdrawn into an automatic sample collector every 0.3 - 1 h using a membrane filter (C) and a pumping system. These samples were analyzed for TOC (TOCsin-II Analyzer, Phase Separations Ltd., Queensferry, Great Britain), and for phenol using the modified 4-AAP method (Yang and Humphrey 1975). Oxygen uptake rates were determined by turning the aeration on and off automatically between 20 and 50 % saturation and measuring the dissolved oxygen concentration with an electrode (EIL) located in the reactor (D). The oxygen uptake rate ($Q_{O_2} X$) was obtained from the slope of the recorded curve during the period without aeration. The dry weight biomass concentration (X), which changed only slowly, was obtained from hourly samples. A Q_{O_2} value was thus obtained every 5 min. These methods are similar to those used previously in a study of nontoxic substrate shocks (Mona, Dunn and Bourne, 1979).

A synthetic wastewater, termed "substrate" here, was fed together with phenol in all experiments. Its composition ensured carbon-limited growth: casein peptone, 0.182 g; meat extract, 0.13 g; yeast extract, 0.13 g; glucose, 0.15 g; $MgSO_4$, 0.012 g; NaCl, 0.04 g; $CaCl_2$, 0.03 g; $MnSO_4$, 0.01 g; K_2HPO_4, 0.20 g; and KH_2PO_4, 0.06 g. The above proportions were mixed with water to achieve the desired TOC substrate feed concentration. Phenol was fed separately, dissolved in water at the desired concentration. The reactor was seeded with activated sludge from the aeration tank of a Zurich treatment plant.

The experiments described here as "batch" were in reality obtained by a stop-flow continuous mode of operation in which the feed, outlet and recycle to the reactor were closed and a pulse of substrate and phenol was injected. "Semi-batch" systems were batch experiments with a continuous feed of phenol. Continuous experiments involved varying the feed flow rate of phenol in a square wave fashion.

The basic feed conditions for all continuous experiments were a substrate feed rate of 250 ml/h at a concentration of 250 mg C/l. Unless otherwise noted, the feeding was carried out before and after all experiments. During the semi-batch and continuous experiments, phenol was fed separately at a concentration of 10 mg/ml at the volumetric rates given in the figures for each experiment.

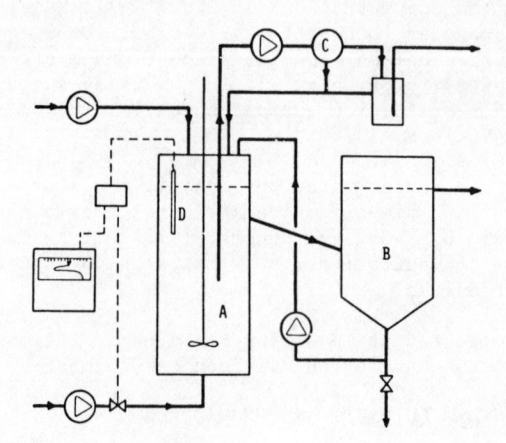

Fig. 1. Schematic Drawing of Reactor Equipment as Described in the Text

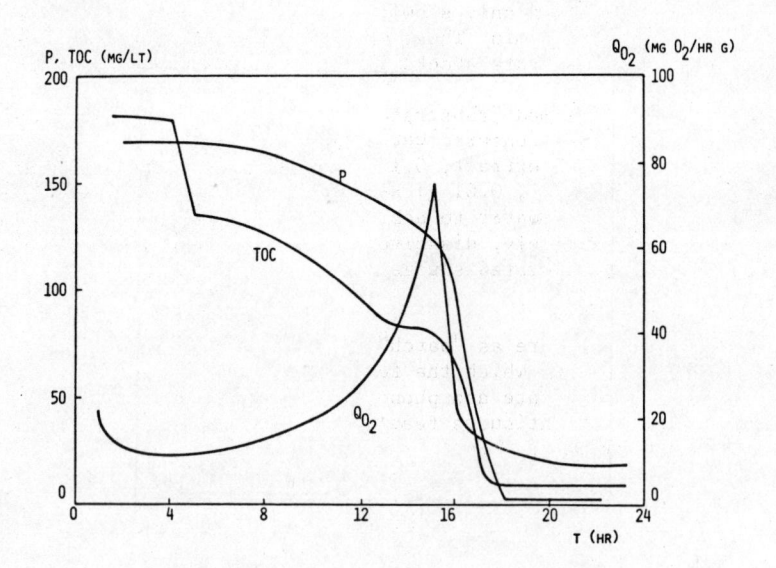

Fig. 2. Batch Experiment on 14 March, 1978. S_I = 50 mg/l; P_o = 200 mg/l ; V =5.5 l ; X = 1.144 g/l

DISCUSSION OF RESULTS

The results are presented in Figs. 2-18. The values for the initial concentrations (S_I and P_I), the feed times and mass feed rates (S_o and P_o), and the biomass concentration (X) are given in the captions to the figures. The dates on which the experiments were performed are also given.

Figs. 2,3 and 4 show the results from 3 batch kinetics experiments which were performed under identical conditions, starting with the substrate under steady-state conditions and adding with phenol. The times required to degrade the phenol completely

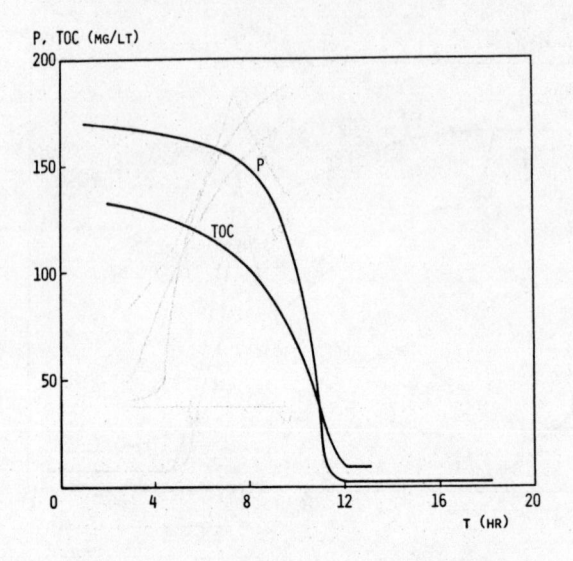

Fig. 3. Batch Experiment on 20 March, 1978. S_I = 50 mg/l; P_I = 200 mg/l; V = 6 l; X = 1.495 g/l.

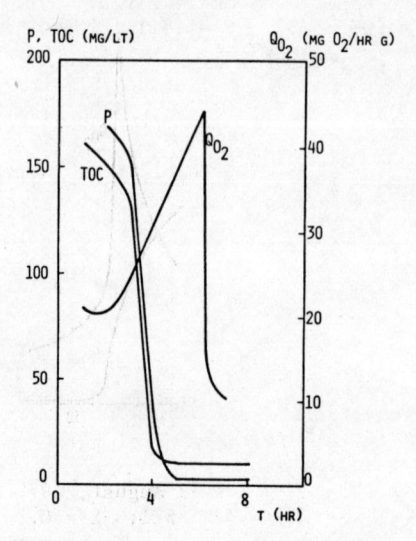

Fig. 4. Batch Experiment on 22 March, 1978. S_I = 50 mg/l; P_I = 200 mg/l; V = 6 l; X = 1.900 g/l

were 18, 12 and 5 h, respectively. This can be attributed to an adaption of the population to phenol during this period. The effect is particularly noticeable between experiments 3 and 4, performed two days apart, in which the biomass increased by 30% though the degradation time decreased by more than 50% . Fig. 2 shows a rapid decrease in TOC after 4 h, which can be attributed to oxidation of the substrate. The phenol concentration begins falling rapidly only after 14 h, at which point the oxygen uptake rate rises rapidly and falls when the phenol is exhausted.

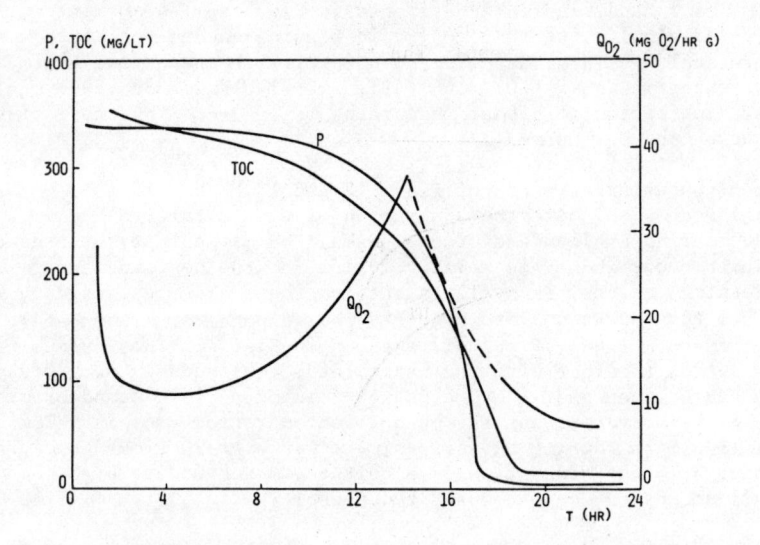

Fig. 5. Batch Experiment on 11 April, 1978. S_I = 50 mg/l; P_T = 500 mg/l; V = 5.5 l; X = 1.611 g/l

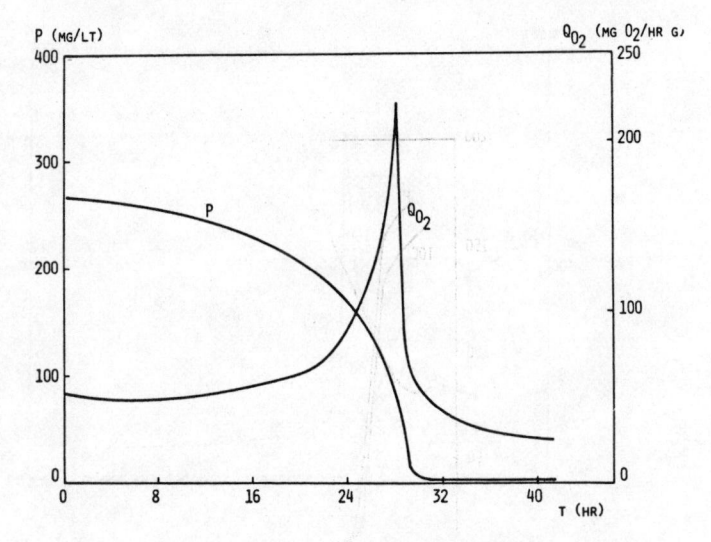

Fig. 6. Batch Experiment on 9 August, 1978. S_I = 50 mg/l; P_I = 250 mg/l; V = 5.5 l; X = 0.392 g/l

Fig. 5 shows a definite initial inhibition effect which causes the oxygen uptake rate first to decrease and later to increase as the oxidation process becomes more rapid after 10 h.

The experiments represented in Figs. 6 and 7 were performed with different initial P_I and show clearly that as P_I is increased, the initial lag time for bio-oxidation is increased.

The results of 1 h semi-batch experiments in Figs. 8-11 demonstrate that phenol is oxidized if fed slowly, but that too rapid feeding causes inhibition. A comparison of the theoretical P values, assuming no uptake, with the actual maximum P value gives for these experiments 91 (20), 166 (49), 250 (170) and 334 (296) mg/l, respectively. Increasing inhibition at higher feed rates is reflected by these figures and this is also shown by the Q_{O_2} curves.

In the continuous experiments of Figs. 12 and 13, phenol was fed for 3 h each day on 4 days in succession. Here the adaptation of the culture to phenol is clearly shown by the decreasing amplitude of the p peaks. The oxygen uptake rates exhibit double peaks, which occur when P is relatively low, at the begining and immediately after the phenol-feeding periods. The valleys between the peaks are caused by inhibition effects when phenol concentrations are highest. The response of a completely adapted system to much higher shock loads of phenol is shown in Fig. 14. This system was fed 10 mg / ml phenol solution, together with the basic feed, continuously for 20 days at a rate of 7 ml/h. It can be seen that the 5 h phenol-feeding periods cause prominent increases in the oxygen uptake rates, but the phenol concentrations remain below 2 mg/l. The phenol shock of 1500 mg/h during this experiment was over 20 times greater than the normal rate. Apparently the adaptation of a culture to accomodate high phenol loads requires only small amounts of continuously fed phenol.

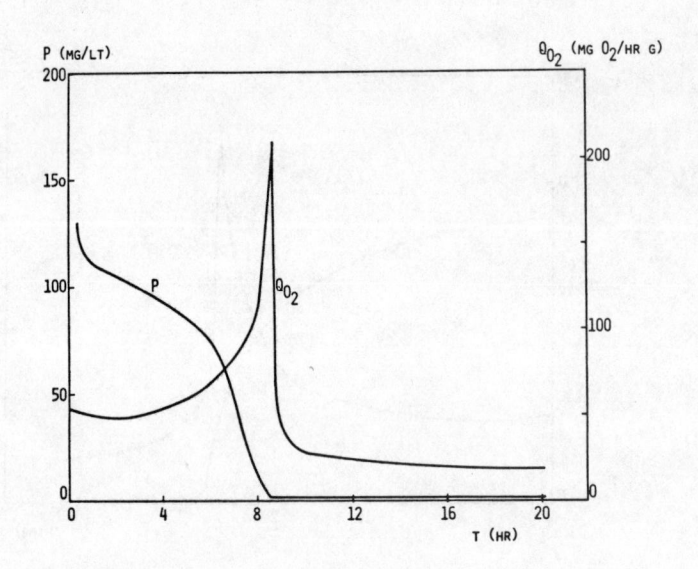

Fig. 7. Batch Experiment on 18 August, 1978. S_I = 50 mg/l; P_I = 150 mg/l; V = 5.5 l; X = 0.434 g/l

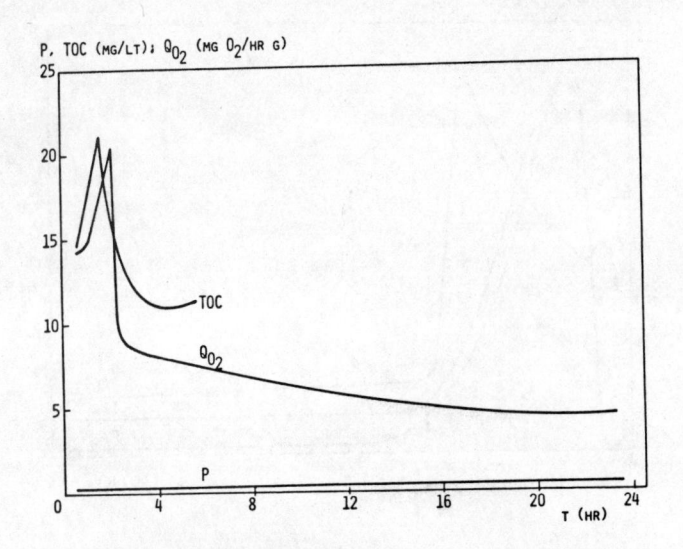

Fig. 8. Semi-Batch Experiment on 12 October, 1978. Fp = 50 ml/h; P_o = 10 mg/ml Added for 1 h; V = 5.5 l; X = 2.396 g/l

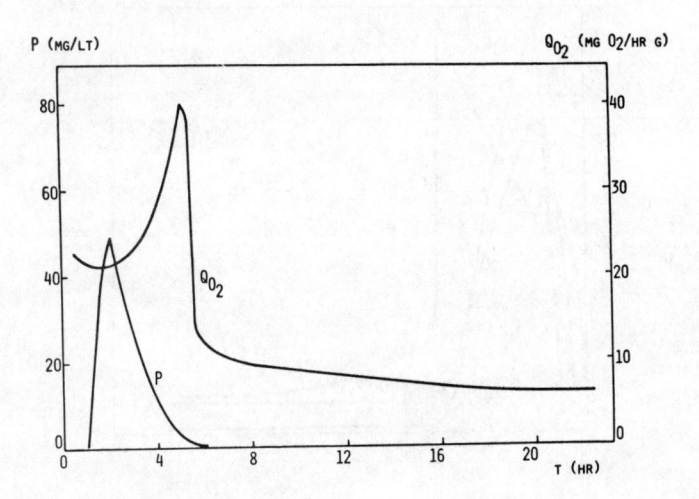

Fig. 9. Semi-Batch Experiment on 17 October, 1978. Fp = 100 ml/h; P_o = 10 mg/ml Added for 1 h; V = 6 l; X = 1.589 g/l

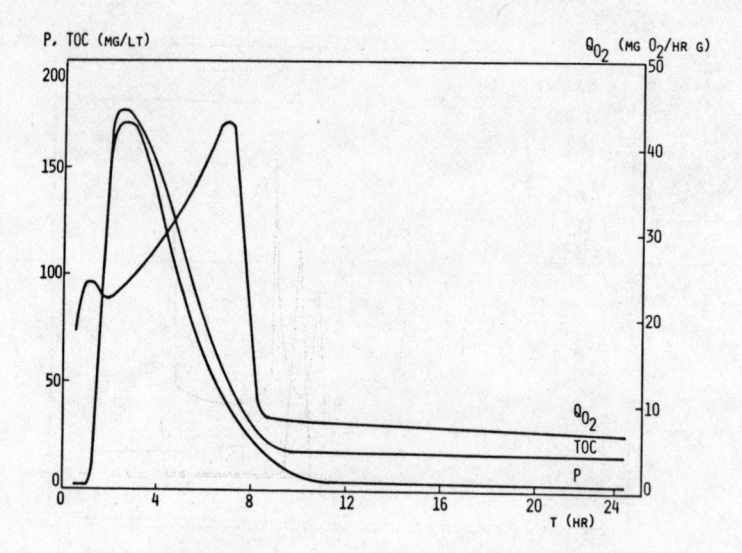

Fig. 10. Semi-Batch Experiment on 18 October, 1978. Fp = 150 ml/h; P_o = 10 mg/ml Added for 1 h; V = 6 l; X = 1.643 g/l

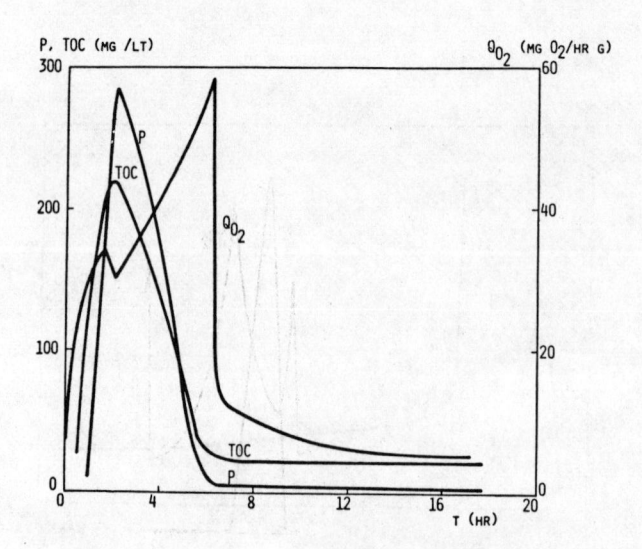

Fig. 11. Semi-Batch experiment on 19 October, 1978. Fp = 200 ml/h; P_o = 10 mg/ml Added for 1 h; V = 6 l; X = 1.704 g/l

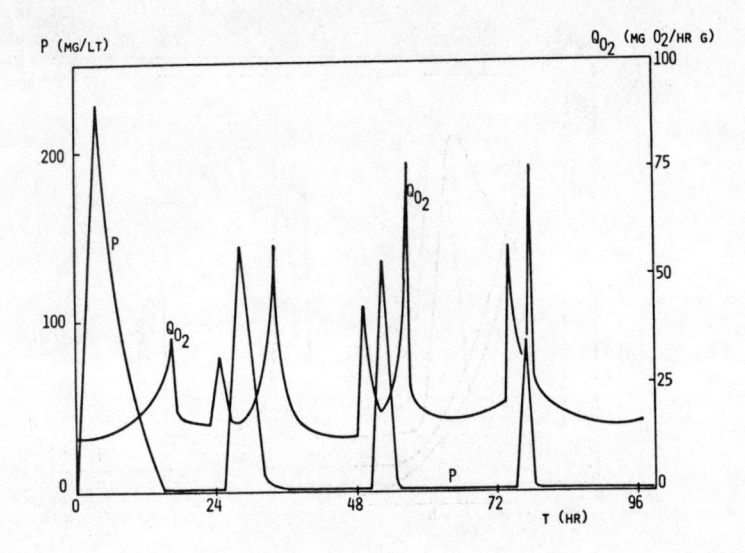

Fig. 12. Continuous Experiment on 19-22 September, 1978. Fp = 50 ml/h; P_o = 10 mg/ml Added for 3 h on 4 Days in Succession.

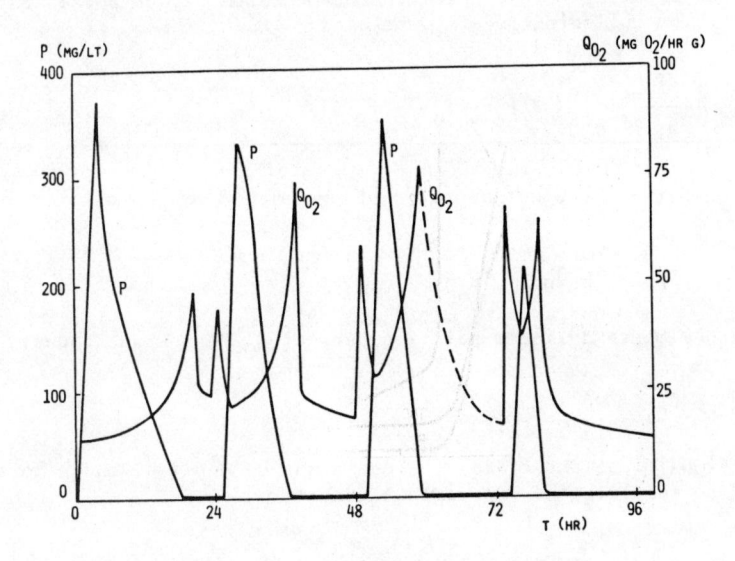

Fig. 13. Continuous Experiments on 3-6 October, 1978. Fp = 100 ml/h; P_o = 10 mg/ml Added for 3 h on 4 Days in Succession.

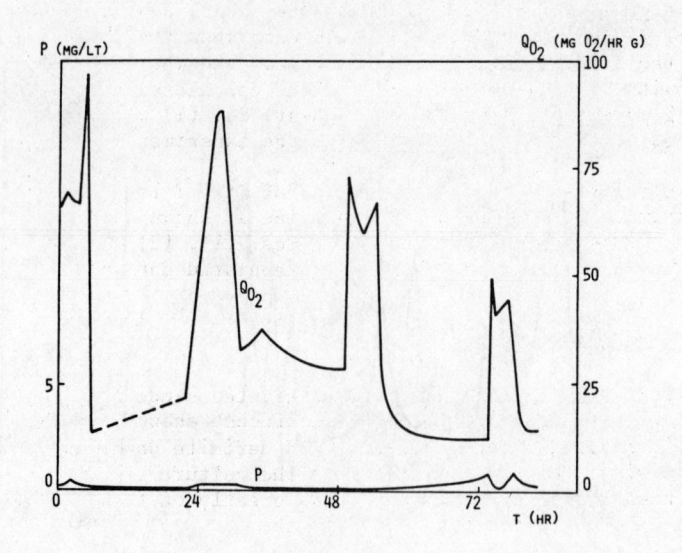

Fig. 14. Continuous Experiments on 20-23 June, 1978. Fp 150 ml/h;
P_o 10 mg/ml Added for 5 h on 4 Days in Succession

DYNAMIC MODELLING OF THE EXPERIMENTS

As found with previous dynamic experiments with a laboratory activated sludge reactor (Mona, Dunn and Bourne, 1979), changes in the biomass concentrations were so small that they could not be measured. An attempt was therefore made, using a dynamic phenol balance for the aeration tank, to describe the time variant phenol concentration. The phenol uptake kinetics were coupled to the measured oxygen uptake rates with a yield coefficient. The following models were used:

$$\frac{dP}{dt} = (F_P P_o - F_1 P + r_P V) / V \tag{1}$$

The relation between the uptake rates of phenol and oxygen was:

$$r_P = Q_{O_2} X Y_{P/O} \tag{2}$$

For some experiments a linear relation between Q_{O_2} and P was found:

$$Q_{O_2} = C + kS \tag{3}$$

Often the simplifications of Eq. (3) could not be applied, and a more general approach was needed to simulate the experimental data:

$$Q_{O_2} X = f_1(P) \qquad r_P = f_2(Q_{O_2} X) \tag{4}$$

where the functions were fitted to the experimental data.

The data of Fig. 15 were calculated from the phenol and Q_{O_2} changes which took place between hours 24 and 36 in Fig. 13. These data show that hysteresis occurred; this was ignored and the simplification of Eq. (3) was used where C = 55 mg/h-g and k = -0.1 1/h-g. Employing Eq. (2) with $Y_{P/O}$ = 0.52 as calculated from steady-state results and the operating parameters from the experiment, Eq. (1) was integrated numerically to yield the results that were compared with the experimental data in Fig. 16.

Attempts to simulate the batch experiments of Fig. 7 revealed that the Q_{O_2} = f(P) could not be approximated at all by a straight line as shown in Fig. 17, nor was $Y_{P/O}$ constant. Some success was obtained by using Eqs. (1), (2) and (4), where the functions were expressed in a simulation program as tabulated data calculated from Figs. 7 and 17.

CONCLUDING DISCUSSION

Experience with the adaption of various activated sludge seed cultures has shown that phenol adaptation is relatively easy. Significant changes in the phenol uptake rates during dynamic loading experiments produced definite day-to-day changes in the phenol oxidation activity. The time required for the culture to recover from sudden increases in phenol concentration appeared to depend greatly on the culture's past exposure to phenol and on the concentration of the shock.

Oxygen uptake rates, as measured dynamically with an electrode, showed that the culture was extremely sensitive to changes in feed rate and concentration. Oxygen uptake rates changed within minutes of any change in operating conditions. The direction of change depended on the previous phenol concentration level. For example, a decrease in Q_{O_2} could indicate a decrease in P at low concentrations, or toxicity effects when P was in a higher range. No simple relationship between phenol and oxygen uptake rates existed for the dynamic experiments. This work has, however, succeeded in establishing oxygen uptake rate as a measurable quantity which correlates qualitatively with phenol uptake rates and toxic concentration levels.

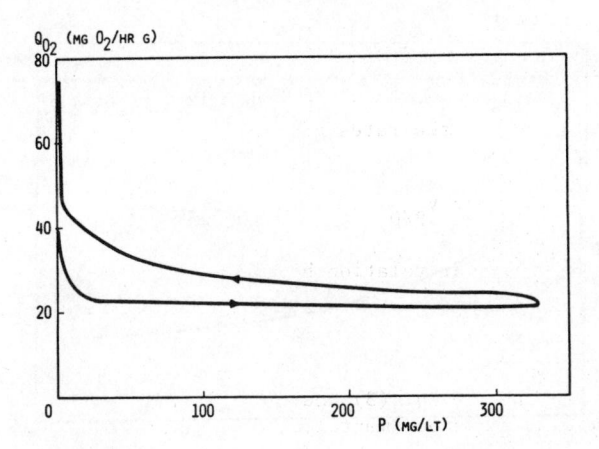

Fig. 15. Data Replotted from Fig. 13 Showing the Q_{O_2} Variation with P Between t = 24 and 36 h.

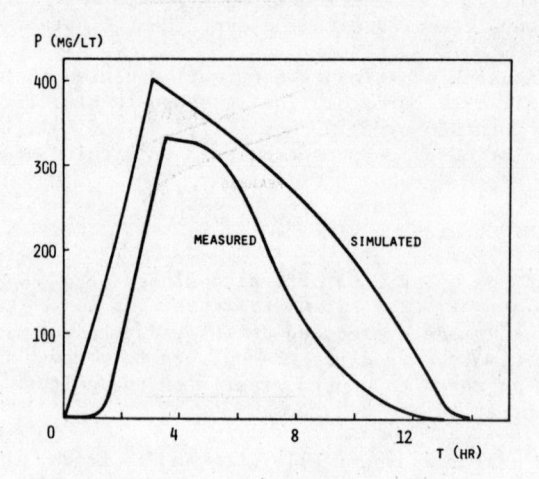

Fig. 16. Comparison of the Simulated Results with the Measured Data from Fig. 13 Between t = 24 and 36 h

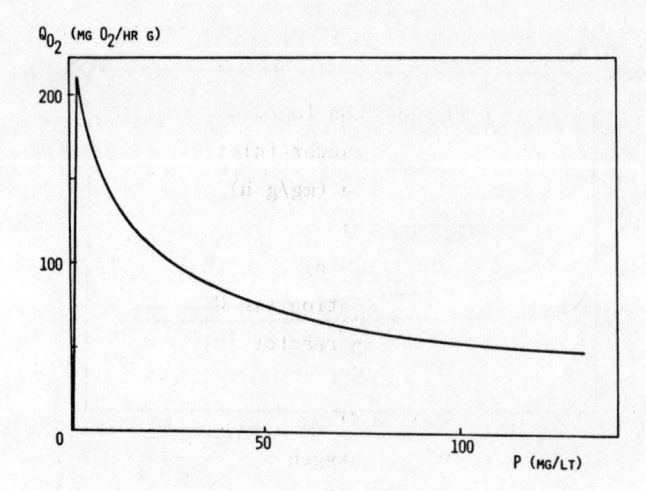

Fig. 17. Data Replotted from Fig. 7 Showing the Non-linear Relation Between Q_{O_2} and P

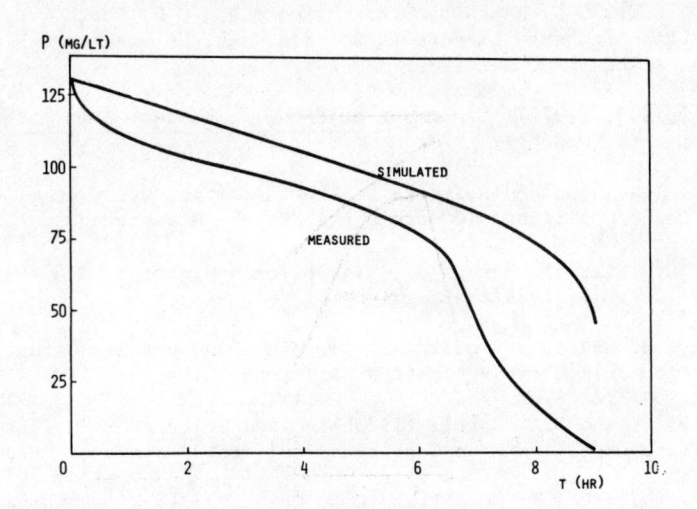

Fig. 18. Comparison of the Simulated Results with Experimental Data from Fig. 7

NOMENCLATURE

C = Constant in Eq. (3) (mg/g-h)

F_I = Reactor outlet flow (ml/h)

F_P = Phenol feed rate (ml/h)

k = Constant in Eq. (3) (ℓ/h-g)

P = Phenol reactor concentration (mg/ℓ)

P_I = Initial phenol concentration (mg/ℓ)

P_o = Phenol concentration in reactor inlet (mg/ℓ)

Q_{O_2} = Specific oxygen uptake rate (mg/g h)

r_p = Phenol uptake rate (mg/ℓ h)

r_S = Substrate uptake rate (mg/ℓ h)

S_I = Initial substrate concentration (mg C/ℓ)

S_o = Substrate concentration in reactor inlet (mg C/ℓ)

V = Reactor Volume (ℓ)

X = Biomass concentration (g/ℓ)

$Y_{P/O}$ = Yield coefficient phenol/oxygen

ACKNOWLEDGEMENT

The authors would like to thank Prof. J.R. Bourne for his support of this work.

REFERENCES

Holladay, D.W., C.W. Hancher, D.D. Chilcote and C.D. Scott (1978). Bio-Degradation of Phenolic Waste Liquors in Stirred Tank, Columnar and Fluidized Bed Bio-Reactors. *AIChE Symp. Ser. 172, 74,* 241.

Lanouette, K.H. (1977). Treatment of Phenolic Wastes. *Chemical Engineering; Deskbook Issue, 99* (Oct. 17, 1977).

Mona, R., I.J. Dunn and R.J. Bourne (forthcoming). Dynamics of an Activated Sludge Process Using Continuous TOC and Oxygen Uptake Measurements. *Biotechnol. Bioeng.*

Paulson, E.G. (1976). How to Get Rid of Toxic Organics. *Chemical Engineering; Deskbook* (Oct. 18, 1976).

Pawlowsky, U. and J.A. Howell (1973a). Mixed Culture Bio-Oxidation of Phenol; I. Determination of Kinetic Parameters. *Biotechnol. Bioeng., 15,* 889.

Pawlowsky, U. and J.A. Howell (1973b). Mixed Culture Bio-Oxidation of Phenol; II. Steady-State Experiments in Continuous Culture. *Biotechnol. Bioeng., 15,* 897.

Pawlowsky, U. and J.A. Howell and C.T. Chi (1973c). Mixed Culture Bio-Oxidation of Phenol; III. Existence of Multiple Steady-States in Continuous Culture with Wall Growth. *Biotechnol. Bioeng. 15,* 905.

Yang, R.D. and A.E. Humphrey (1975). Dynamic and Steady-State Studies of Phenol Bio-Degradation in Pure and Mixed Cultures. *Biotechnol. Bioeng., 17,* 1211.

Zdybiewska, M. (1968). Microbiological Degradation of Phenol Compounds. Transl. from *Postepy Mikrobiologii, 7,* 161. ORNL-tr-4037.

DISCUSSIONS

Samsunlu, A. (Turkey) : Could you explain how the apparatus part of the system works and how the specific oxygen uptake rate (Q_{O_2}) was measured?

Yongaçoğlu, S. (Switzerland) : During the batch studies, the outlet to reactor B was closed and a pulse of phenol was injected directly into reactor A. From there our samples were automatically collected. At the same time the dissolved oxygen concentration was measured with an electrode located in the reactor and recorded. At 50% oxygen saturation, aeration automatically stopped. The oxygen uptake of the microorganisms during this period without aeration was then recorded as a curved line. From the slope of this line the uptake rate per minute ($Q_{O_2}X$) could be observed. By dividing this value into the biomass in the reactor, we obtained the specific oxygen uptake rate (Q_{O_2}). As soon as the oxygen concentration dropped to 20%, aeration started again and when it stopped, the second curve was obtained. In a system that worked well, one curve could be obtained per minute.

Samsunlu, A. (Turkey) : Didn't you try working with higher phenol concentrations than those given in your paper?

Yongaçoğlu, S. (Switzerland) : I worked over a long period with 5000 mg/ℓ of phenol. I even used such high concentrations that the microorganisms died and the system was completely destroyed. - at what concentration exactly I could not determine. In the present series of experiments, I have increased the phenol concentration to 2000 mg/ℓ, but the results are not included in this paper.

Samsunlu, A. (Turkey) : Were your experiments conducted only with phenol?

Yongaçoğlu, S. : We tested only with phenol.
  (Switzerland)

Curi, K. (Turkey) : Approximately how long did it take for the system to reach steady state and what difficulties were encountered, if any?

Yongaçoğlu, S. : Under optimum conditions, steady state could be attained
  (Switzerland) in 2 weeks. We considered it to have been reached when the biomass concentration and the quantity of microorganisms remained equivalent, when a determined oxygen uptake rate was obtained and when the amount of organic carbon remained constant. We found that the slightest change in the speed of our pumps could ruin the culture, system in one night. Furthermore, with a mixed culture we could never be sure that the quantity of microorganisms would remain the same.

Curi, K. (Turkey) : When you worked with phenol, were there any additional problems or extra precautions which had to be taken, in order to attain steady state?

Yongaçoğlu, S. : Generally shock loads of phenol were given. We tried to
  (Switzerland) reach steady state with a continual feed of phenol and substrate, not with phenol alone. It perhaps took a little longer, but no new problem arose.

Samsunlu, A. (Turkey) : In your experiments did you observe the oxygen consumption relative to the phenol concentration over a long period? Did you obtain information about the adaptation of the cultures to phenol and about the purification efficiency of the system?

Yongaçoğlu, S. : We kept the total organic carbon level at 5 or 6 mg/ℓ.
  (Switzerland) BOD and COD measurements were not made.

Samsunlu, A. (Turkey) : For how long did one experiment continue with a feed of domestic wastewater and phenol?

Yongaçoğlu, S. : We worked with this continuously in one system for one
  (Switzerland) day. The total duration of the experiments was at least a month. On one occasion we also worked continuously with phenol alone. We were hampered by having only one system for phenol. In our new series of experiments we hope to increase the number.

The Investigation of the Equation Describing the Substrate Removal Mechanism on an Inclined-Plane Biological Filter

H. URÜN

Civil Engineering Department, Karadeniz Technical University, Trabzon, Turkey

ABSTRACT

In this study, the second-order partial differential equation describing the mechanisms of organics removal in the inclined plane biofilms is written in terms of dimensionless variables.

In order to obtain the optimum solution, the required stability condition in the case of the finite difference method is numerically and graphically given.

Finally, the partial differential equation is solved numerically using a computer for the appropriate and sufficient boundary conditions.

INTRODUCTION

In biological treatment units, in order to investigate the dipersion and diffusion phenomena and the concentration gradients of wastewater, which are very important from the environmental point of view, it is necessary to know the physical parameters of the surface on which the wastewater flows, such as slope and properties like velocity, flow rate and temperature.

Inclined-plane models are generally preferred, because of the fact that their construction is easy and the results obtained are reasonable.

To simplify the problem, the events occurring in the biofilm, which is very thin, are taken as boundary conditions.

MATHEMATICAL MODEL

It is known that for a plane surface reactor, laminar flow occurs when Reynolds number Re is less than 140 (Maier, Behn and Gates, 1967). This kind of flow model is shown in Fig. 1.

In this model a mass balance on an elemental volume is given by the following partial differential equation for conservative materials (Ürün, 1978):

$$\underbrace{\frac{\partial c}{\partial t}}_{(a)} + \underbrace{u \frac{\partial c}{\partial x} + v \frac{\partial c}{\partial y} + w \frac{\partial c}{\partial z}}_{(b)} = \underbrace{D \left(\frac{\partial^2 c}{\partial x^2} + \frac{\partial^2 c}{\partial y^2} + \frac{\partial^2 c}{\partial z^2} \right)}_{(c)} \qquad (1)$$

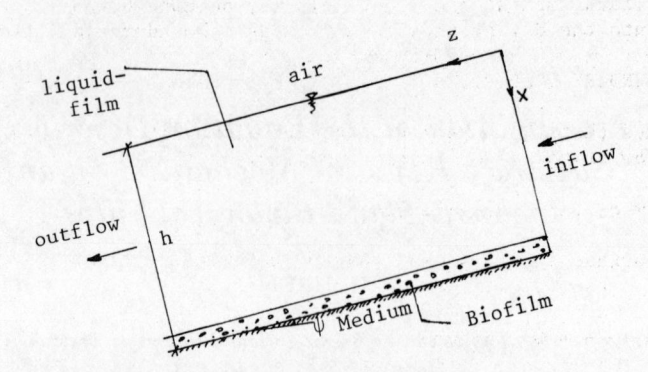

Fig. 1. The Inclined-Plane Flow

in which (a), (b), and (c) are, respectively, the concentration gradient with time, the convection and the diffusion term. D is the diffusion coefficient.

It is assumed that the convective transport occurs only in the Z direction and that there is no diffusion in the y direction.

In this case the equation can be simplified as,

$$\frac{\partial c}{\partial t} + w \frac{\partial c}{\partial z} = D \left(\frac{\partial^2 c}{\partial x^2} + \frac{\partial^2 c}{\partial z^2} \right) \qquad (2)$$

$\frac{\partial^2 c}{\partial z^2}$ is neglected, when compared with the convection term in the same direction. (Maier Behn and Gates, 1967). Thus, Eq. 1 can be written for the steady-state as follows

$$w \frac{\partial c}{\partial z} = D \frac{\partial^2 c}{\partial x^2} \qquad (3)$$

The velocity distribution in flow direction z is,

$$w = \frac{\gamma \sin\psi}{2\mu} h^2 (1 - x^2/h^2), \quad w_{ave} = \frac{\gamma \sin\psi}{3\mu} h^2, \quad w_{max} = \frac{3}{2} w_{ave} \qquad (4)$$

The combination of Eqs. 3 and 4 yields,

$$w_{max} (1 - x^2/h^2) \frac{\partial c}{\partial z} = D \frac{\partial^2 c}{\partial x^2} \qquad (5)$$

Eq. 5 is given for conservative materials. When biological treatment plants are taken into consideration, it is necessary to develop models which can be applied to non-conservative materials such as organics. This may be done by substituting a biological process term into the differential equation as a boundary condition.

BOUNDARY CONDITIONS

If it is assumed that the substrate concentration distribution is uniform and equal to c_0 at the inflow of the model considered, then

$$c = c_0, \quad \text{when} \quad z = 0 \tag{6}$$

At the water surface it is assumed that no material is trasferred. Thus,

$$D \frac{\partial c}{\partial x} = 0, \quad \text{when} \quad x = 0 \tag{7}$$

Also, the following boundary condition can be written for the bottom of the reactor:

$$-D \frac{\partial c}{\partial x} = k_s c, \quad \text{when} \quad x = h \tag{8}$$

Eq. 8 describes the metabolic activities of the microorganisms, and Eq. 3, along with boundary conditions, Eqs. 6 - 8, can be applied to non-conservative materials as in the biological reactors. This means that the substrate is removed at the bottom of the reactor by microorganisms in the biofilm.

The boundary condition of Eq. 8 can also be written as,

$$c = 0, \quad \text{when} \quad x = h \tag{9}$$

That is, as soon as the substrate is transferred to the bottom of the reactor it will be removed completely by the microorganisms (Maier, Behn and Gates, 1967).

The following substitutions have been made in order to express Eq. 3 and the boundary conditions in the dimensionless form:

$$X = x/h, \quad Z = D \cdot z/h^2 \cdot w_{max}, \quad C = c/c_0 \tag{10}$$

Hence, Eq. 3 will be:

$$(1-X^2) \frac{\partial C}{\partial Z} = \frac{\partial^2 C}{\partial X^2} \tag{11}$$

The boundary conditions in terms of dimensionless coordinates are:

$$C = 1, \text{ when } Z = 0; \quad \frac{\partial C}{\partial X} = 0; \text{ when } X = 0;$$

a) $\dfrac{\partial C}{\partial X} = \dfrac{-ksh}{D} C = -\eta C$ or

b) $c = 0, \text{ when } X = 1$ \hfill (12)

FINITE DIFFERENCE EQUATIONS

Eqs. 11 and 12 will not be expressed in terms of finite differences. The flow depth is divided into N equal intervals, each having a thickness of ΔX (Fig. 2).

The boundaries between increments are numbered from $I = 1$ on the surface ($X = 0$) to $I = N+1$ at the bottom ($X = 1$). The distance from the inflow of the reactor is indicated by J, starting with $J = 1$ at $Z = 0$; thus $Z = (J-1)\Delta Z$, assuming that the mesh size is ΔZ along the Z axis. At any vertical section, the value of the dimensionless point concentrations between the increments, I and I-1, is defined as $C(I,J)$. The average value of the dimensionless concentrations in the increment between J and $J + 1$ is denoted as C_{ave} and will be computed later by using the trapezoidal rule.

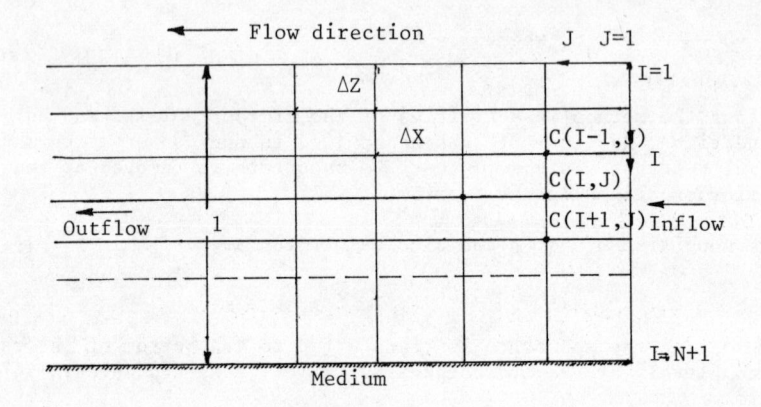

Fig. 2. The Grid for the Finite Difference Equation

Several difference schemes were evaluated, and the one with the best fit will be shown after inclusion of the following approximations:

$$\frac{\partial C}{\partial Z} = \frac{C(I, J+1) - C(I, J)}{\Delta Z} \quad , \quad \frac{\partial C}{\partial X} = \frac{C(I+1, J) - C(I, J)}{\Delta X} \quad (13)$$

$$\frac{\partial^2 C}{\partial X^2} = \frac{C(I+1, J) - 2C(I, J) + C(I-1, J)}{\Delta X^2} \quad , \quad T = \frac{1}{1 - X^2} \quad (14)$$

By incorporating Eqs. 11, 12, 13 and 14, the following difference equations are obtained:

$$I = 1 \; ; \; C(I, J+1) = C(I,J) + \frac{\Delta Z \cdot T}{\Delta X^2} \{C(I-1, J) - C(I, J)\} \quad (15)$$

$$1 < I < N \; ; \; C(I, J+1) = C(I,J) + \frac{\Delta Z T}{\Delta X^2} \{C(I+1, J) - 2C(I, J) + (I-1, J)\} \quad (16)$$

$$I = N \; ; \; C(I, J+1) = C(I,J) + \frac{\Delta Z \cdot T}{\Delta X^2} \{C(I-1, J) - C(I, J)(\Delta + X.\eta)\} \quad (17)$$

STABILITY ANALYSIS

Richtmeyer and Morton (1966) have concluded that if finite difference equations are stable, then when mesh sizes approach zero, the solutions of the difference equations approach the true solutions. The stability conditions must be known in order to obtain stable solutions. Therefore, this condition was derived by applying linear stability analysis.

Eq. 16 is expanded into the Taylor series around $(I-1)\Delta X$ and $(J-1)\Delta Z$, assuming Eq. 1 to be unconditionally consistent with Eq. 16.

In linear stability analysis, the following Fourier series is used (Sarıkaya, 1977).

$$C(I, J) = \sum_{k}^{k} S_k^J \exp(ikmI\Delta X) \quad (18)$$

in which $i = \sqrt{-1}$. In order to examine the behavior of a component of Eq. 18 between $J\Delta Z$ and $(J-1)\Delta Z$, the considered component is substituted into Eq. 16, and the following relationship is found between

S_k^{J+1} and S_k^J :

$$S_k^{J+1} = S_k^J \{1 + T \cdot \frac{\Delta Z}{\Delta X^2} (\exp(ikm\Delta X) + \exp(-ikm\Delta X) - 2)\} \quad (19)$$

The term in square brackets is called the amplification factor, A. This factor must be smaller than or equal to one in order to have stable difference equations (Ricthmeyer and Morton, 1966) Thus,

$$|A| = 1 + T \cdot \frac{\Delta Z}{\Delta X^2} \cdot \{\exp(ikm\Delta X) + \exp(-ikm\Delta X) - 2\} < 1 \quad (20)$$

Exponential functions in Eq. 20 can be replaced with trigonometric functions:

$$\exp(ikm\Delta X) = \cos km\Delta X + i \sin km\Delta X, \; \exp(-ikm\Delta X) = \cos km\Delta X - i \sin km\Delta X \quad (21)$$

when Eqs. 21 are substituted into Eq. 20,

$$|A| = 1 - 4 \frac{\Delta Z}{\Delta X^2} \cdot T \cdot \sin^2 \left(\frac{km\Delta X}{2} \right) \tag{22}$$

Since the amplification factor is obtained in the form of a complex number, its absolute value should satisfy Eq. 20. Thus,

$$-1 < A < 1 \tag{23}$$

The right-hand side of Eq.23 can be satisfied for every value of ΔX, ΔZ and T as follows

$$-4 T \frac{\Delta Z}{\Delta X^2} \cdot \sin^2 \left(\frac{km\Delta X}{2} \right) < 0 \tag{24}$$

The stability condition may be derived, when the left-hand side of Eq. 23 has been taken into account:

$$\frac{\Delta Z}{\Delta X^2} \leq \frac{1}{2 T} \tag{25}$$

as given by Richtmeyer and Morton (1966) for the explicit difference equation corresponding to the one-dimensional Fickian diffusion equation.

SOLUTIONS OF THE DIFFERENCE EQUATIONS

The finite difference equations have been solved by employing an IBM 360 computer for uniform velocity distribution, assuming that the initial concentration distribution is uniform at the inflow of the reactor.

In order to solve the finite difference equations, Eqs. 15, 16 and 17, under the boundar conditions of Eq. 14, it is necessary to know the values of ΔX, ΔZ and η.

In this case, the equations have been solved by accepting the values given in Table 1, for various values of η. It is seen both numerically and graphically that for the biological efficiency of % 95 at Z=1, the best fitting values are, $\Delta X = 0.05$ and $\Delta Z = 0.0001$ (Fig. 3).

When mesh sizes are taken smaller than those selected in Table 1, the solutions would show no partical change (Ürün, 1978).

Computer programs have been written to compute all point and average dimensionless concentrations for every Z section.

AVERAGE CONCENTRATIONS

For the unit flow of an inclined plane model, the following equation is given by Ürün (1978):

$$q = c_{ave} \cdot h \cdot w_{ave} = \int_0^h c(1.dx) \cdot w \tag{26}$$

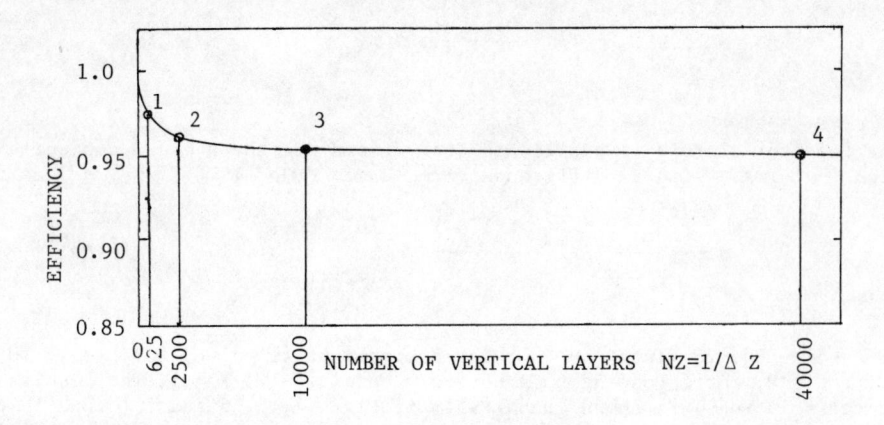

Fig. 3. Values of ΔX and ΔZ Given Graphically

Table 1 Selection of the Fitting X and Z Values

Efficiency %	Z	ΔX	ΔZ	NZ Z/ΔZ
0.974	1	0.2	0.0016	625
0.961	1	0.1	0.0004	2500
0.954	1	0.05	0.0001	10000
0.950	1	0.025	0.000025	40000

in which, q = unit flow, c_{ave} = average concentration, h = flow depth, w_{ave} = average velocity, c = point concentration, dx = finite element, w = velocity.

If the following are substituted in Eq. 26,

$$w = w_{ave}(1 - \frac{x^2}{h^2}), \quad c = C.c_0, \quad c_{ave} = C_{ave} c_0, \quad x = X.h \quad dx = dX.h \quad (27)$$

Then the dimensionless average concentration is found to be,

$$C_{ave} = \frac{3}{2} \int_0^1 C(1 - X^2) \, dX \tag{28}$$

If the trapezoidal rule is applied to Eqs. 27 and 28, the average concentration may be written in terms of finite difference forms as in Ürün (1978):

$$C_{ave} = \frac{3}{2} \Delta X \left\{ \frac{C(1, J)}{2} + \sum_{I=2}^{20} C(I, J)\{ 1 - (I-\Delta)^2 \Delta X^2 \}\right\} \tag{29}$$

After feeding Eq. 28 into the computer program, it is solved for $\eta = 1$, 10, 100 and also for $C = 0$, corresponding to a boundary condition of $X = 1$. The results of the numerical solutions are shown graphically in Figs. 4, 5, 6 and 7 (Ürün, 1978).

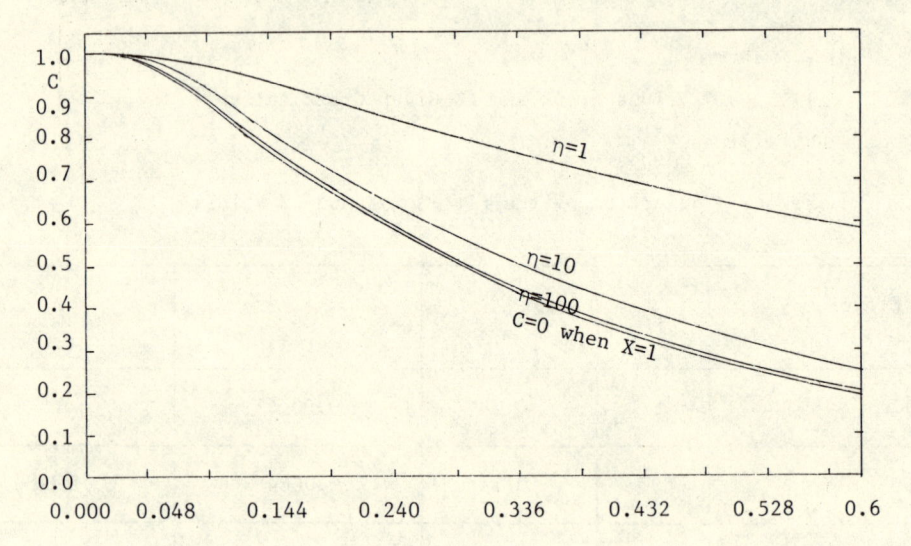

Fig. 4. Concentrations in Z Direction, when $X = 0$.

CONCLUSIONS

The differential equation representing the dispersion of the substrate in the flow of an inclined plane model has been solved by the finite difference method. The stability condition of the finite difference equations is derived by applying linear stability analysis.

For simplicity, dimensionless parameters have been used and the equation has been written in its dimensionless form. Under appropriate boundary conditions, several solutions have been obtained. It has been shown that the solution for $C = 0$, when $X = 1$, is the same as the one for $\eta = 100$.

The results have been compared with the analytical solutions obtained by Maier, Behn and Gates, 1967. A good agreement has been obtained between the solutions.

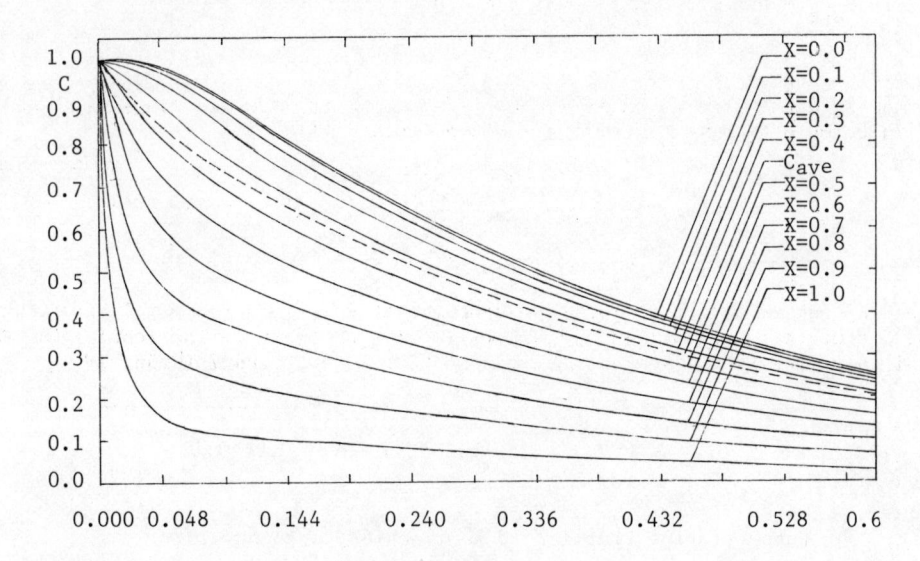

Fig. 5. Concentrations in Z Direction, when $\eta = 10$.

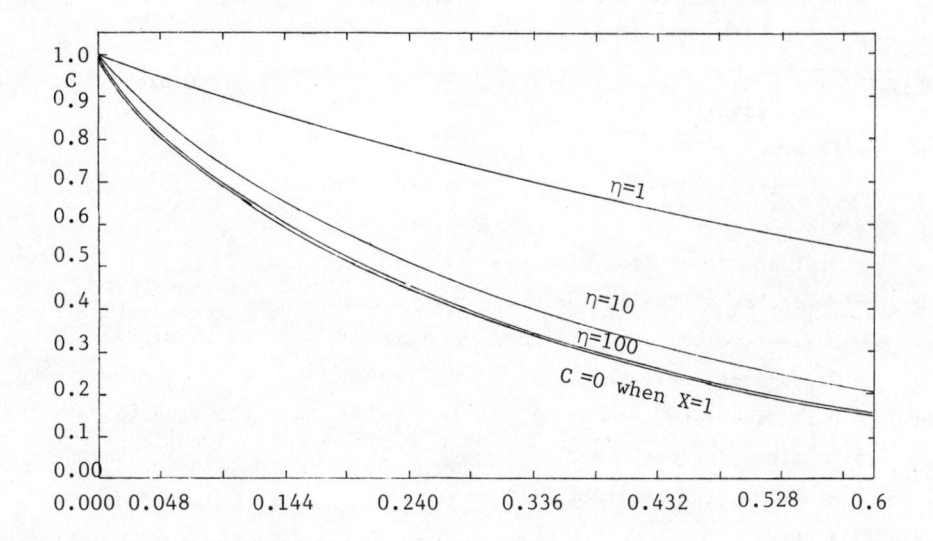

Fig. 6. Average Concentrations in Z Direction

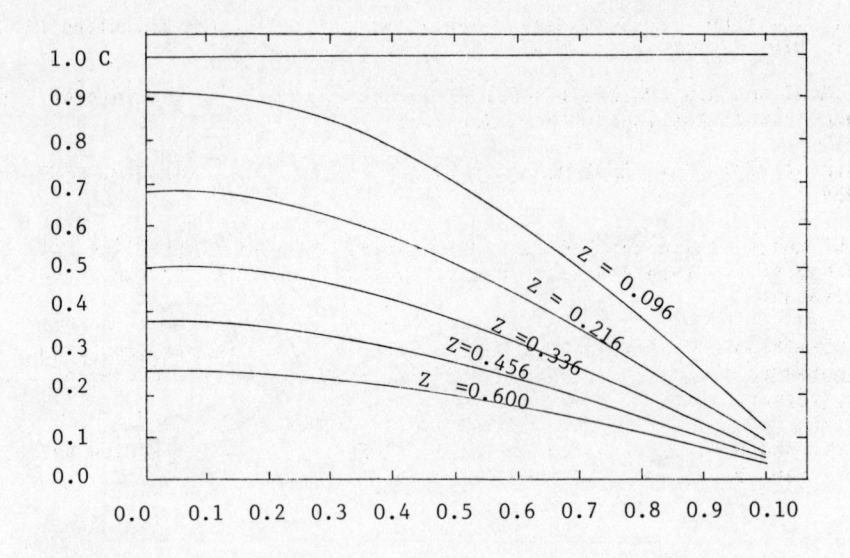

Fig. 7. Concentration Profiles in X Direction, when η = 10.

NOMENCLATURE

A	= Amplification factor
C	= c/c_0
c	= Substrate concentration, (mg/l)
c_0	= Substrate concentration, at inflow, (mg/l)
c_{ave}	= Cross-sectional average concentration, (mg/l)
C_{ave}	= c_{ave} / c_0
t	= Time, (sec)
h	= Flow depth, (cm)
u,v,w	= Velocities in x, y, z directions, (cm/sec)
X	= x/h
x	= Distance along flow direction from inflow, (cm)
D	= Molecular diffusion coefficient, (cm^2/sec)
w_{max}	= Maximum velocity in z direction, (cm/sec)
w_{ave}	= Cross-sectional average velocity, (cm/sec)
q	= Unit flow rate, (cm^3/sec)
γ	= Density, (g/(cm^2) (sec)
k_s	= Reaction rate constant, (cm^2/sec)
η	= $k_s h/D$, (cm)
Z	= $zD/h^2 w_{max}$

REFERENCES

Maier, W.J., V.C. Behn, and C.D. Gates (1967). Simulation of the Trickling Filter Process. *Env. Div. ASCE, 93,* SA 4, 91.

Richtmeyer, R.D. and K.W. Morton (1966). *Difference Methods for the Initial Value Problem.* Interscience Publishers Inc., New York.

Sarıkaya, H.Z. (1977). Numerical Model for Discrete Settling. *Hid. Div. ASCE, 103,* No. HY 8, 865.

Swilley, E.L. and B. Atkinson (1963). A Mathematical Model for Trickling Filters. In *Proceedings of the 18th Annual Industrial Waste Conference.* Purdue University, Lafayette, Indiana.

Ürün, H. (1978). *Biyolojik Film Reaktörlerinin Kinetiği Üzerine Bir Araştırma.* Thesis presented to the Technical University of Istanbul for the degree of Ph. D. Istanbul, Turkey.

Williamson, K. and P.L. McCarty (1976). A Model of Substrate Utilization by Bacterial Films. *Wat. Pol. Con. Fed., 48,* 9.

DISCUSSIONS

Uslu, M. (Turkey) : It is generally accepted that from the point of view of stability, implicit finite difference schemes are better than explicit schemes. Why didn't you use an implicit scheme to solve the equations numerically?

Ürün, H. (Turkey) : Solution of the equations was not possible with an implicit scheme.

Uslu, M. (Turkey) : I think that the numerical solution of any differential equation can be tried implicitly or explicitly. In your case, you were dealing with numerical stability problems which are difficult to solve by explicit schemes. Using an implicit scheme would have diminished these problems.

Atımtay, A. (Turkey) : You have mentioned that your results agree well with the analytical solutions obtained by Maier and others. What is the difference between the methods used by you and Maier and others?

Ürün, H. (Turkey) : Maier solved the problem analytically, whereas I solved it numerically. I applied the inclined plane model to spherical filter materials arranged cubically. By finding the variation in concentration of the inclined plane flow over each succeeding sphere, I arrived at a numerical solution.

Atımtay, A. (Turkey) : Did you conduct any experiments?

Ürün, H. (Turkey) : Experiments were carried out and the results were found to agree well with the analytical solutions of Maier and others. Fuller details can be found in reference no. 5.

Atımtay, A. (Turkey) : It would perhaps have been better to use graphs to show the correspondence between theoretical and experimental results.

The Response of Activated Sludge to High Salinity

R. Y. TOKUZ

R. L. Goodson, Jr., Inc., Dallas, Texas, USA

SUMMARY

The response of activated sludge to high salinity was examined. Bench-scale units and a glucose-peptone synthetic substrate which contained sodium chloride and/or sodium sulfate were used. The results indicated that for the sodium chloride system, effluent quality in terms of suspended solids was excellent up to a 35-40 g/l level of salinity. High sodium sulfate levels did not affect the effluent suspended solids and chemical oxygen demand concentrations at all. At a salinity level of 50 g/l comprising 25 g/l sodium chloride and 25 g/l sodium sulfate, the kinetic parameters for design were determined at the kinetics of the system was found to be similar to that of the freshwater systems.

INTRODUCTION

In some localities, the scarcity of freshwater has made the use of saline waters for domestic purposes necessary. High salt content wastewaters are often encountered in industrial effluents.

A review of the literature indicates that the response of the activated sludge process to high salinity is characterized by a loss in organic removal capability and an increase in the concentration of effluent suspended solids. But some researches disagree. They indicate that in some cases, after acclimation, the process effluent quality may improve, although the final product is always lower in quality.

The common method of investigation of the response of the system to high salinity, as used by most of the researchers, is to select several salinity levels and examine the system caharcteristics at such predetermined salt levels.

The objective of this study was to determine the response of activated sludge to gradually increasing salinity. A method of acclimation was also sought, which would prevent effluent quality from deteriorating abruptly.

LITERATURE REVIEW

The effects of sodium chloride on bacterial growth have been examined by several researchers. For example, Winslow (1931) and Ingram (1939, 1940) indicated that low concentrations of sodium chloride were stimulatory to the growth of *Escherichia coli* and *Bacillus cereus*, respectively. The ranges of concentrations used were 0.3 to 14.6 g/l for *Escherichia coli* and 0.3 to 11.7 g/l for *Bacillus cereus*. Winslow (1939) indicated that 58.5 g/l sodium chloride was toxic to the bacteria and none survived.

Doudoroff (1940) examined the response of several freshwater bacteria to salinity, including *Bacillus subtilis* and *Escherichia coli*. He reported that with a steady increase in the salinity of the growth medium the viable count remained constant until a certain sodium chloride concentration was reached; thereafter the count dropped with increasing salt concent. The ability to survive and grow in saline media was a function of the developmental stage of the culture; it was lowest in the early logerithmic period and greatest during the early stationary phase. Doudoroff also indicated that maximum adaptation was obtained when the salinity of the medium was gradually raised by allowing exposurer of sufficient duration to each intermediate salt concentration.

Krul (1977) examined the floc formation characteristics of *Zooglea ramigera,* when affected by different salinity levels. He reported that sodium chloride at 0.1, 0.6, and 1.2 g/l concentrations stimulated the aggregation of resuspended cells. The highest concentration was the most effective one.

In the wastewater treatment field, the early investigators studied the response of treatment systems to shock loadings of salinity. Lawton and Eggert (1957) studied the effects of salinity on trickling filters. They reported that the sudden application of high salinity caused adverse effects. In general, the larger salt concentrations had greater effect on the system and the time required for acclimation was longer. Even after acclimation, the biological oxygen demand (BOD) reduction for a high salinity (50 g/l) system was about 15 percent lower than that for the salt-free system.

Mills and Wheatland (1962) reported that constant application of 6.6 g/l sodium chloride did not upset the biological removal efficiency of trickling filters. But intermittent application, 9 a.m. to 5 p.m. daily, Monday through Friday, caused an increase in the effluent BOD.

Stewart, Ludwig, and Kearns (1962) studied the effects of salinity on the extended aeration process and reported that a combination of high organic loading, BOD shock loading and a rapid change in salinity affected the effluent quality significantly.

Ludzack and Noran (1965) reported that no detectable changes occured in the activated sludge performance when chloride concentrations were below 5 to 8 g/l (8.3 to 13.2 g/l of sodium chloride) even when the system was subjected to intermittent (5 days a week) feeding. When the chloride concentration was alternated weekly from 0.1 to 20 g/l or vice versa, effluent quality deteriorated.

Kincannon, Gaudy, and Gaudy (1966) and Kincannon and Gaudy (1966, 1968) studied the response of batch and continuous-flow activated sludge systems to salinity. They reported that high salinity shocks caused a drastic decrease in the substrate removal rate, but acclimation restored this capability. They examined the settling characteristics of the sludge. The sludges developed at 30 and 45 g/l salt concentrations were dispersed systems and did not flocculate. But they reported that in one of their earlier studies the solids flocculated well.

Burnett (1974) studied the response of the activated sludge process to alternating shocks of freshwater sewage and high-salinity sewage, and found that such shocks caused a reduction in BOD removal efficiency, an increase in effluent turbidity and solid losses, as well as changes in the mixed liquor microorganism populations.

Adams, Eckenfelder, and Novotny (1975) studied the treatability of a high salinity wastewater which contained considerable quantities of chlorides plus high concentrations of complex organics. Due to poor settling of activated sludge solids, they recommended a treatment system which did not include the activated sludge process.

Hockenbury, Burstein, and Jamro (1977) reported the results of treatability studies conducted on various industrial wastes. The dissolved salts in those wastes typically consisted of sodium chlorides, and sulfates. They observed that high salinity caused an increased in the effluent suspended solids and a decrease in the organic removal rate for activated sludge systems. For example, increasing salinity from 1.5 g/l to 5.0 g/l caused to concentration of effluent suspended solids to increase from 125 mg/l to 350 mg/l and the Total Organic Carbon (TOC) removal efficiency to drop from 90 to 75 percent.

To summarize, the researchers agree on the unfavorable response of activated sludge to high salinity. But there is considerable disagreement on the degree and nature of the adverse effects. While some investigators indicate that the organic removal efficiency is affected, not everyone agrees. Some claim persistently an increase in the effluent suspended solids and a decrease in the ability of the activated sludge to settle. Others indicate that the solids settle well. Nobody seems to be interested in determining the approximate minimum level of salinity above which there exists an abrubt change in effluent quality.

MATERIALS AND METHODS

Three plexiglass, continuous-flow activated sludge units were used. The total volume of each unit was 10 liters; the aeration chamber held 8 liters, the remaining 2 liters being the capacity of the settling basin. The aeration chamber and settling compartment were separated by an adjustable baffle which enabled one to control the amount of sludge being circulated from the clarifier into the aeration basin. One of the units was operated as the control and was fed the substrate alone (Table 1). The other two units, after acclimation to the synthetic substrate, received varying amounts of sodium chloride and sodium sulfate. The substrate had a chemical oxygen demand (COD) of about 1100 mg/l and a BOD of approximately 400 mg/l. The units were started with the activated sludge taken from a local municipal wastewater treatment plant and acclimated to the synthetic substrate for a period of more than a month. After completion of the acclimation, sodium chloride was added to the inflow at gradually increasing concentrations.

During the first three months of the study, the adverse effects of salinity on the activated sludge effluent quality in terms of high concentrations of effluent suspended solids and chemical oxygen demand were observed. Another problem encountered was the loss of activated sludge solids with the effluent due to so-called "deflocculation" of the sludge. After experiencing these problems, a method of acclimation to salinity was

Table 1 Synthetic Substrate Composition

Component	Concentration (mg/l)
Glucose	500
Peptone	500
Potassium phosphate (monobasic)	527
Potassium phosphate (dibasic)	1070
Ammonium sulfate	235
(Tap water to volume)	

devised, and the concentration of inorganic salt in the influent was increased gradually (1 to 5 g/l increments). The acclimation to the intermediate salt concentration was assumed satisfactory when the shock effects of the step increase in salinity disappeared, and the salt concentration was increased further. In some instances, the duration of acclimation to the intermediate salt concentration exceeded one month (Tables 2 and 3).

Upon reaching a 25 g/l concentration of sodium chloride in unit 2, the salt level was decreased gradually and a switch was made to sodium sulfate (Tables 2 and 3). In the meantime, unit 1 continued to receive substrate at increasing levels of salinity.

The food-to-microorganism ratio (F/M ratio) was approximately 0.3 mg COD/day/mg MLVSS* at the begining of the experiments. The hydraulic detention time was about two days. Since mixed liquor solids were permitted to build up, the F/M ratio decreased to 0.1 gradually. No attempt was made to maintain the F/M ratio at 0.3 by wasting sludge (wastage only occurred when samples were collected for analysis), because of the problems encountered with "deflocculating" sludge and the resulting solid losses at the initial phases of the experiments.

Table 2 Periods of Acclimation to Sodium Chloride

Date	Sodium Chloride Concentration (g/ℓ) Unit 1	Sodium Chloride Concentration (g/ℓ) Unit 2
1-18-77	8.0	5.0
2-10	15.0	10.0
3-21	20.0	15.0
4-18	25.0	20.0
5-22	30.0	25.0
6-2	30.0	20.0
6-5	30.0	15.0
6-12	37.5	10.0
6-19	40.0	
6-29	38.0	
7-4	40.0	
7-7	41.0	
7-14	42.0	
7-19	43.0	
7-21	44.0	
7-25	45.0	
7-28	46.0	
8-1	47.0	
8-4	48.0	
8-8	49.0	
8-11	50.0	

*MLVSS: Mixed liquor volatile suspended solids.

In the second phase of the study, the response of activated sludge to different organic loadings at a constant salinity level was studied and kinetic parameters such as a, b, a´, b´, and K were determined. Three different organic loading rates were used: an F/M ratio of 0.3, 0.6 and 0.9. The salinity level was maintained at 25 g/l sodium chloride and 25 g/l sodium sulfate.

The routine analyses on the samples, mamely, for BOD, COD, total and volatile suspended solids (TSS and VSS), etc. were made following the procedures described in Standard Methods (1971).

The dissolved oxygen (DO) measurements were made with a membrane electrode. The DO levels in the reactors were maintained at about 8 mg/l during the first phase of the experiments and the temperature of the units was kept at $20.5 \pm 2^{\circ}C$. In the second phase, the dissolved oxygen was kept above 2.0 mg/l and the temperature of the units was $21 \pm 0.5^{\circ}C$.

Table 3 Periods of Acclimation to Sodium Sulfate

Date	Sodium Sulfate Concentration (g/ℓ)
7-6-77	8.0
7-7	10.0
7-11	12.0
7-14	14.0
7-21	15.0
7-25	17.0
7-28	20.0
8-1	23.0
8-4	26.0
8-8	30.0
8-11	35.0

RESULTS AND DISCUSSION

It was found that when the system was subjected to gradually increasing salinity and if an adequate time of exposure was allowed to each intermediate salt concentration, the response of activated sludge to high salinity was much more favorable than was anticipated.

It was also discovered that the effect of salinity on the effluent suspended solids was almost negligible up to a sodium chloride level of 35-40 g/l (Fig.1). The effluent COD was affected adversely by increasing salinity (Fig. 2). But in comparison, sodium sulfate did not affect the effluent quality at all in terms of suspended solids and COD (Fig. 3 and 4).

The adverse effect of chlorides on the organic removal capacity was attributed to the ability of the chloride ion to penetrate cells much more readily than the sulfate ion. Comar and Bronner (1960) found that the sulfate ion was virtually non-penetrating compared to the chloride ion in the experiments made with frog-skin cells. In another

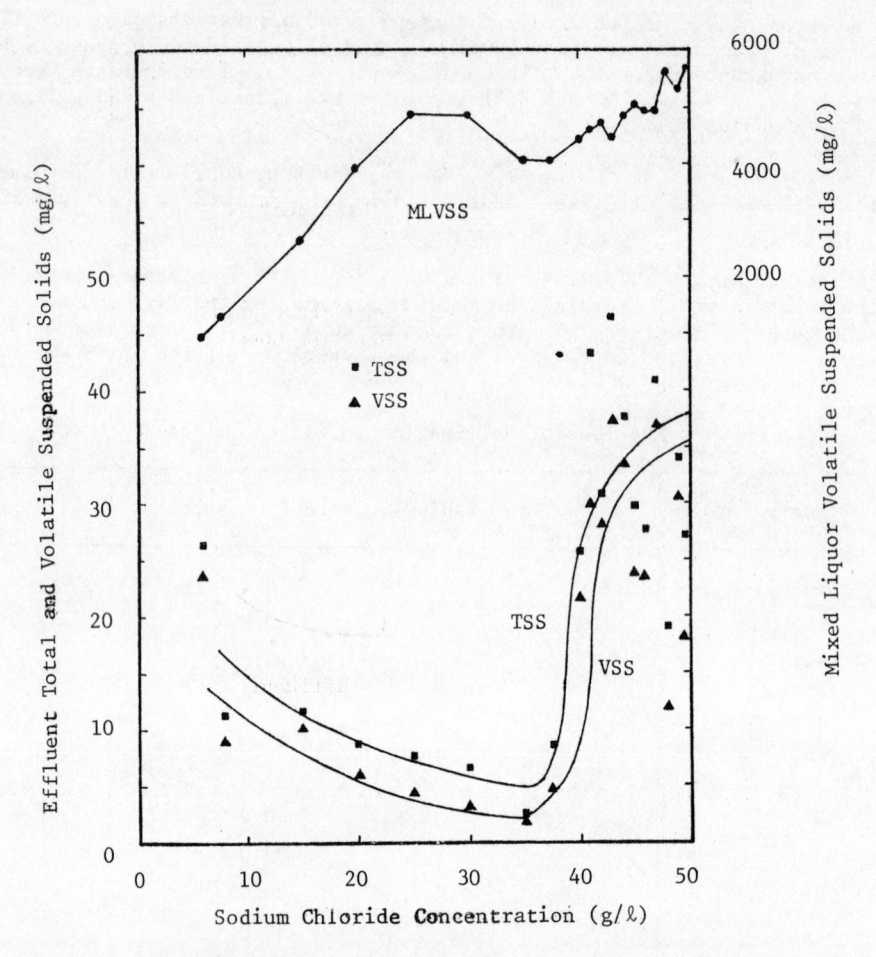

Fig. 1. The Effect of Sodium Chloride on the Mixed Liquor Volatile Suspended and Effluent Suspended Solids (Unit 1)

study, Comar and Bronner (1962) noted: "... the striking ability of the intestine to absorb Cl⁻ from an isotonic mixture of NaSO₄ and NaCl until the Cl⁻ concentration falls to a few milliequivalents per liter is also suggestive of a primary transport of Cl⁻ ..." Thus it was concluded that due to the ability of chloride ions to penetrate cells more easily and disrupt the normal functioning of the microorganisms, the sodium chloride system was affected more than the sodium sulfate system.

It was also suspected that the chloride ion present in such large concentrations may have interfered with the COD analysis. The COD test was modified to eliminate chlorides from the samples prior to analysis, but due to the presence of chloride ions in such large quantities it was highly probable that some interference had occurred, resulting in effluent COD values higher than normal.

During the initial three months of the study several difficulties were encountered. One of these problems was the inability to maintain a healthy population of activated sludge microorganisms. On several instances the sludge would suddenly lose its flocculation capability and large amounts of solids would be discharged with the effluent. In such cases the only remedy was to discard the sludge and start with a new batch (Table 4).

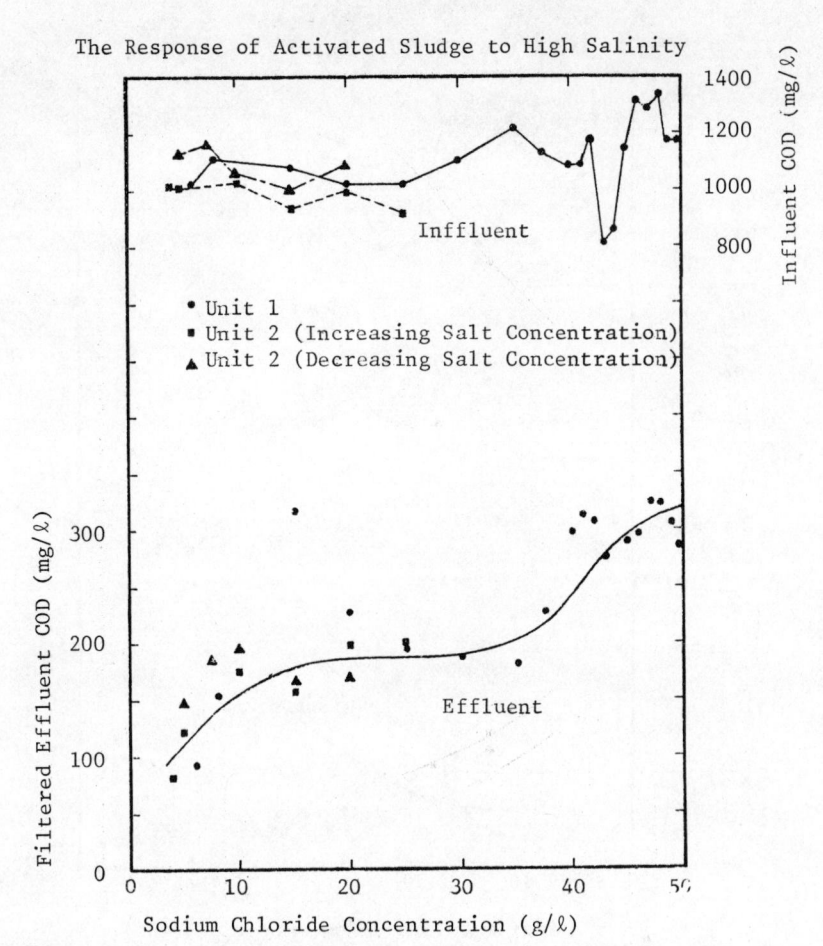

The Response of Activated Sludge to High Salinity

Fig. 2. The Effect of Sodium Chloride on Filtered Effluent COD (Units 1 and 2)

Table 4 The Data on "Deflocculation" of Activated Sludge

Date	Salinity	TSS*	VSS*	MLSS*	MLVSS
		Unit 1			
10-4-76	11.0	8.7	8.7	1809	1641
10-9	11.0	21.8	19.0	1277	1168
10-12	11.0	29.3	24.5	781	709
10-13	11.0	36.0	32.2	853	770
10-14	11.0	44.6	41.8	984	879
		Unit 2			
10-18-76	1.0	29.7	26.9	1442	1332
10-19	1.0	29.9	26.8	1234	1122
10-21	1.0	57.3	53.9	1112	1022
10-23	1.0	71.0	67.2	914	845
10-28	1.0	55.6	51.2	738	689

*TSS: Total suspended solids; VSS: Volatile suspended solids; MLSS: Mixed liquor suspended solids (all in mg/l).

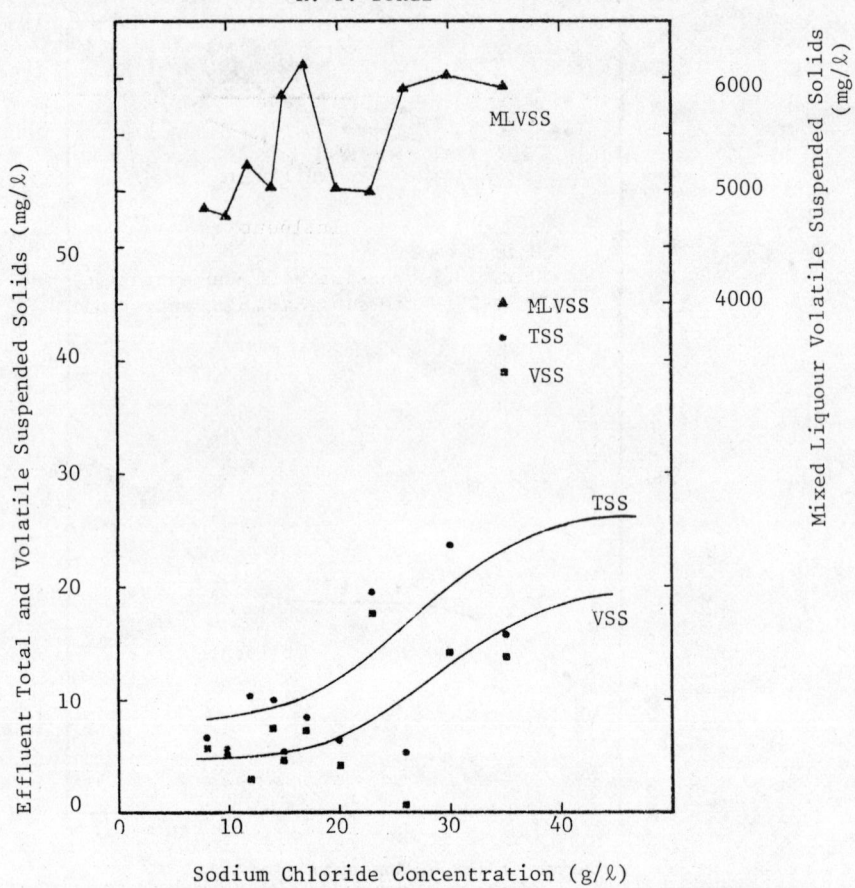

Fig. 3. The Effect of Sodium Sulfate on the Mixed Liquor Volatile Suspended and Effluent Suspended Solids

In the second phase, the effect of organic loading was investigated. Salinity was kept at 50 g/l, comprising 25 g/l sodium chloride and 25 g/l sodium sulfate. Three different food-to-microorganism ratios were used: 0.3, 0.6 and 0.9. The summary of the results are given in Table 5.

A food-to-microorganism ratio of 0.3 yielded a fair quality effluent (TSS = 56 mg/l, COD = 202 mg/l). Higher loading rates caused TSS and COD increases in spite of higher organic removal rates.

The data for phase two were used in determining the kinetic parameters; sludge yield parameters a and b, oxygen utilization coefficients a´ and b´, and the reaction rate coefficient K. The kinetic parameters are presented in Table 6. The kinetic models derived for freshwater activated sludge systems were used in the determination of kinetic models for the salt water systems. As can be from Figs. 5 - 7, a perfect fit between models and data was obtained.

CONCLUSION

It was concluded that the activated sludge process can be used for the treatment of

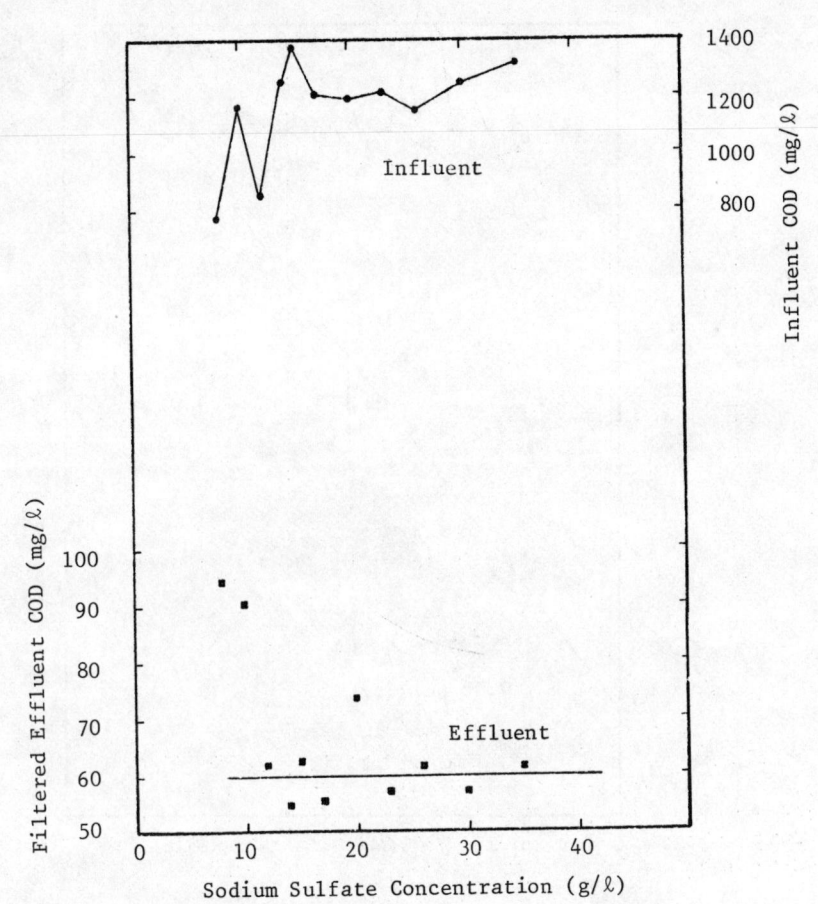

Fig. 4. The Effect of Sodium Sulfate on the Filtered Effluent COD

Table 5   The Summary of Results for Phase Two

	F/M	0.3	0.6	0.9
Total suspended solids, mg/ℓ		56	121	181
Volatile suspended solids, mg/ℓ		42	94	152
Influent COD, mg/ℓ		1191	2229	3414
Filtered effluent COD, mg/ℓ		202	229	288
Percent removal		83	90	92
Oxygen uptake, mg/hr/g MLVSS		6.1	8.9	13.9
Biodegrable fraction		0.69	0.73	0.74

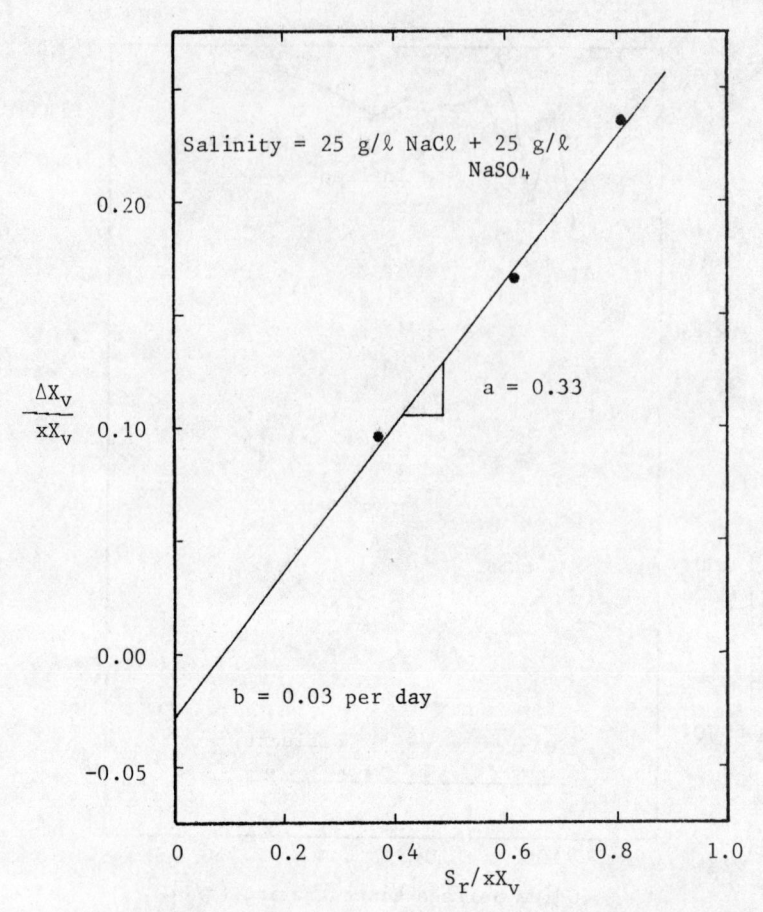

[Figure 5 plot: y-axis $\Delta X_v / x X_v$ from -0.05 to 0.20; x-axis $S_r / x X_v$ from 0 to 1.0. Salinity = 25 g/ℓ NaCℓ + 25 g/ℓ NaSO$_4$. a = 0.33, b = 0.03 per day.]

Fig. 5. Determination of Sludge Production Parameters, a and b

Table 6 The Kinetic Parameters

Parameter	Value
a	0.33
b	0.03 per day
a´	0.41
b´	0.05 per day
K	19.2 per day

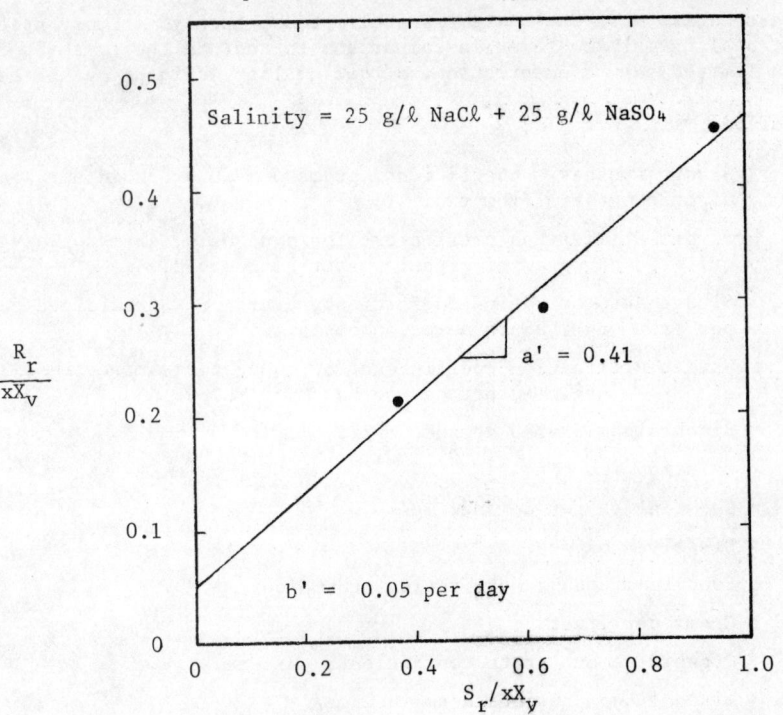

Fig. 6. Determination of Oxygen Utilization Coefficients, a' and b'

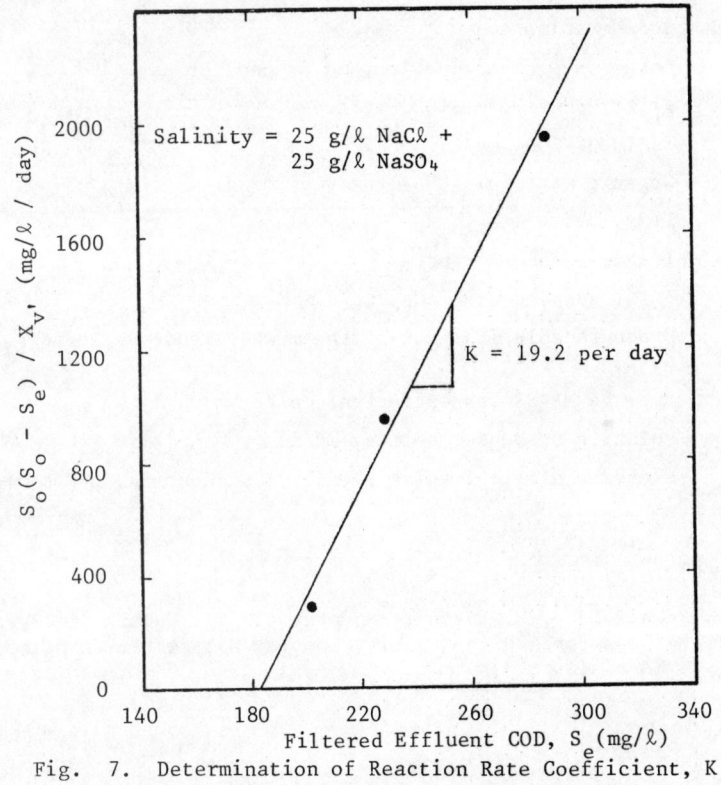

(Salinity = 25 g/ℓ NaCℓ + 25 g/ℓ NaSO₄; a' = 0.41; b' = 0.05 per day)

Fig. 7. Determination of Reaction Rate Coefficient, K

(Salinity = 25 g/ℓ NaCℓ + 25 g/ℓ NaSO₄; K = 19.2 per day)

high salinity wastewaters when the system is acclimated properly. In most cases the process effluent quality will be lower in comparison to that in the freshwater system. However, up to a limited salt concentration, a fair quality effluent can be expected.

NOTATION

a	Sludge synthesis coefficient, grams of VSS produced per gram of organic material removed.
a	Oxygen utilization coefficient for synthesis, grams of oxygen utilized per gram of organic material removed.
b	Sludge auto-oxidation coefficient, grams of VSS oxidized per day per gram of MLVSS in aeration basin.
b	Oxygen utilization coefficient for endogenous activities, grams of oxygen utilized per day per gram of MLVSS.
BOD	Biochemical oxygen demand, mg/ℓ.
Cl^-	Chloride ion
COD	Chemical oxygen demand, mg/ℓ.
DO	Dissolved oxygen, mg/ℓ.
F/M	Food-to-microorganism ratio, mg COD/day/mg MLVSS.
g/ℓ	Grams per liter.
K	Organic removal rate coefficient, per day.
MLSS	Mixed liquor suspended solids, mg/ℓ.
MLVSS	Mixed liquor volatile suspended solids, mg/ℓ.
NaCl	Sodium chloride.
NaSO$_4$	Sodium sulfate.
R_r	Total oxygen utilization, mg oxygen per day.
S_e	Filtered effluent COD, mg/ℓ.
S_o	Influent COD, mg/ℓ.
S_r	Organic material (COD) removed, mg/d.
t	Time, days.
TOC	Total organic carbon.
TSS	Total suspended solids, mg/ℓ.
x	Biodegradable fraction of the mixed liquor volatile suspended solids.
X_v	Average MLVSS concentration, mg/ℓ.
VSS	Volatile suspended solids, mg/ℓ.
ΔX_v	Excess biological volatile solids production, grams of VSS per day.

REFERENCES

Adams, C.E., Eckenfelder, W.W. and V. Novotny (1975). Equalization and Biological Treatment Techniques for a High-Salinity Complex Wastewater. *Progress in Water Technology, 7*, Nos. 3/4, 635.

APHA-AWWA-WPCF (1971). *Standard Methods for the Examination of Water and Wastewater.* Thirteenth Edition. American Public Health Association, Washington, D.C.

Burnett, W.E. (1974). The Effect of Salinity Variations on the Activated Sludge Process. *Water and Sewage Works, 121,* 37.

Comar, C.L., and F. Bronner, eds., (1960). *Mineral Metabolism.* Volume I, Part A; Chp. 6: Ion Transport. Academic Press, New York.

Comar, C.L., and F. Bronner, eds., (1962). *Mineral Metabolism.* Volume II, Part B; Chloride. Academic Press, New York.

Doudoroff, M. (1940). Experiments on the Adaptation of *Escherichia coli* to Sodium Chloride. *J. Gen. Physiol., 23,* 585.

Hockenbury, M.R., Burstein, D. and E.S. Jamro (1977). Total Dissolved Solids Effects on Biological Treatment. *Proceedings of the 32nd Industrial Waste Conference,* Purdue University, West Lafayette, Indiana.

Ingram, M. (1939). The Endogenous Respiration of *Bacillus cereus.* II. The Effects of Salts on the Rate of Absorption of Oxygen. *J. Bacteriology, 38,* 613.

Ingram, M. (1940). The Endogenous Respiration of *Bacillus cereus.* III. The Changes in the Rate of Respiration Caused by Sodium Chloride in Relation to Hydrogen Ion Concentration. *J. Bacteriology, 40,* 83.

Kincannon, D.F., and A.F. Gaudy (1966). Some Effects of High Salt Concentrations on Activated Sludge. *J. Water Poll. Cont. Fed., 38,* 1148.

Kincannon, D.F., Gaudy, A.F. and E.T. Gaudy (1966). Squential Substrate Removal by Activated Sludge After a Change in Salt Concentration. *Biotechnol. Bioeng., 8,* 371.

Kincannon, D.F. and A.F. Gaudy (1968). Response of Biological Waste Treatment Systems to Changes in Salt Concentrations. *Biotechnol. Bioeng., 10,* 483.

Krull, J.M. (1977). Some Factors Affecting Floc Formation by *Zooglea ramigera,* Strain I-16-M. *Water Res., 11,* 51.

Lawton, G.W. and C.V. Eggert (1957). Effect of High Sodium Chloride Concentration on Trickling Filter Slimes. *J. Water Poll. Cont. Fed., 29,* 1228.

Ludzack, F.J. and D.K. Noran (1965). Tolerance of High Salinities by Conventional Wastewater Treatment Processes. *J. Water Poll. Cont. Fed., 37,* 1404.

Mills, E.V. and A.B. Wheatland (1962). Effect of Saline Sewage on the Performance of Percolating Filters. *Water and Waste Treatment, 9,* 170.

Stewart, M.J., Ludwig, H.F. and W.H. Kearns (1962). Effects of Varying Salinity on Extended Aeration Process. *J. Water Poll. Cont. Fed., 34,* 1161.

Winslow, C-E.A. (1931) The Specific Potency of Certain Cations with Reference to Their Effect on Bacterial Viability. *J. Bacteriology, 22,* 49.

DISCUSSIONS

Sarıkaya, H.Z. (Turkey) : How do you explain the decrease in suspended solids concentration (Fig. 1) in the effluent, which was observed up to a 35 mg/ℓ concentration of sodium chloride?

Tokuz, R. Y. (U.S.) : We attributed this decrease to the decrease in the F/M ratio from 0.3 to 0.1.

Sarıkaya, H. Z. (Turkey): What was the cause of the deterioration in effluent quality with respect to suspended solids when the sodium chloride concentration exceeded 35 mg/ℓ?

Tokuz, R. Y. (U.S.) : At a concentration of 35 mg/ℓ the microorganisms are healthy and active. But beyond this concentration, between 35 and 40 mg/ℓ, the protozoa begin to die off. When their polishing effect disappears, the turbidity of the effluent increases and settling of the solids becomes impossible.

Atımtay, A. (Turkey) : You have mentioned that high sodium sulphate levels did not affect the COD and suspended solids concentration at all, whereas sodium chloride did. Could this effect be attributed to the chloride ion? Did you conduct any experiments using magnesium chloride and magnesium sulphate instead of sodium chloride and sodium sulphate?

Tokuz, R. Y. (U.S.) : No. Only sodium chloride and sodium sulphate were used. Although sodium also has an effect on the microorganisms, we attributed the greater effect of sodium chloride on the activated sludge system directly to the chloride ion. This was mentioned in the results and discussion section of my paper.

Velioğlu, S. G. (Turkey): How was the interference of the chloride concentration in the COD experiments prevented?

Tokuz, R. Y. (U.S.) : It is possible to prevent such interference by increasing Ag_2SO_4. Let me remind you that all experiments were conducted in accordance with the 1971 edition of *Standard Methods*. The new edition emphasizes this problem in particular.

Biodegradability of Aqueous Solutions of Ecotoxic Organic Chemical Compounds

F. SENGÜL and A. MÜEZZINOGLU

Environmental Engineering Department, Faculty of Civil Engineering,
Ege University, Izmir, Turkey

ABSTRACT

Biodegradability is a varying property for organic compounds that are practically non-biodegradable; some are partially or fully decomposable. The most abundantly found group of organic pollutants is detergents.

In this study, to determine the biodegradation of aqueous solutions several compounds are tested.

Two different test methods are applied during the program, in order to be able to compare the results on a different basis. The percentages of biological decomposition achieved each day are calculated and tabulated. The results thus obtained are discussed.

INTRODUCTION

The use of chemicals in everyday life has increased considerably and will continue to do so, which means their residues and their transformation will increase in diversity and concentration in the air, water, soil, foodstuff and living tissues. The persistence in the environment of recalcitrant compounds like DDT, polychlorinayed biphenyls (PCB) and some heavy metals has caused serious concern. Investigations concerning the environmental degradation reaction kinetics of the chemicals and the impact of their persistence in several media are required.

Degradation of chemicals in the environment through chemical, photochemical, microbiological means should be thoroughly evaluated, because some compounds may become more toxic after environmental transformation. These investigations and evaluations are based on data related to the chemical and toxicological properties of compounds, health statistics, medical experience, epidemiology and environmental monitoring, and ecological surveillance.

Simulation of the natural biodegradation processes by rapid laboratory bioassay systems is a very promising field of investigation. These systems vary from fish tests to bacterial degradability determinations. Such tests are being developed at present. With the use of these stimulating models, ecological effects may be studied by the use of laboratory ecosystems created during the tests, so that disturbances of the balance and output of materials can be detected.

Common ecotoxic materials might either be in the form of waste mineral materials or organic recalcitrant[1] molecules. Mineral materials like heavy metals, cya-

---

[1] Recalcitrant materials are materials which do not readily degrade by enymatic action.

nides, arsenic, selenium etc. are a matter of extensive concern as major chronic toxicants in the environment.

Most of the organic recalcitrant molecules are toxic to certain biological cycles. Among these compounds are two important groups extensively used in modern life styles: organic pesticides and detergents.

These common organics are more or less biodegradable within a given period of time provided that the appropriate microorganisms and environmental conditions are available. These can only be understood after the chemical is released into the natural environment.

These are more general rules, however, utilized to estimate the biodegradability of a large molecule:

1. Straight chain hydrocarbons are more easily degraded by enzymatic action.
2. Compounds containing one methyl branch are attached only when the molecule has a sufficiently long unbranched chain.
3. Branched chains having alkyl groups larger than methyl or a group of methyls are not degradable.
4. The addition of methyl groups to the chain often makes the molecule more resistant to biodegradation.
5. A quaternary carbon atom, especially if it is near the terminals, renders the molecule more recalcitrant. This is both true for methyl and phenly groups.

Synthetic anionic detergents are branched-chain molecules. The extent and location of branching vary for several detergent active substances. Each raw material has a different level of activity as a surfactant, the strength of which may be determined by the methylene blue active substance (MBAS) method. The more commonly used alkylbenzene sulfonate (ABS) is hard to degrade biologically, and this is quite understandable when its structural formula is studied in view of the 5 general rules stated above:

$$\underset{SO_3}{\bigcirc}-C-C-\underset{C}{\overset{C}{C}}-C-\underset{C}{\overset{C}{C}}-C-\overset{C}{C}-C$$

That is the reason this common detergent is classified as "hard".

Another detergent molecule more recently put into commercial production is the linear alkylbenzene sulfonate (LAS). Its structural formula is

$$\underset{SO_3}{\bigcirc}-C-C-C-C-C-C-C-C-C-C-C-C$$

As is theoretically expected in view of the biodegradability rules, LAS is more biodegradable than the hard ABS molecule, thus classified as "soft".

The surface activity of detergents causes foaming in water solutions, even at concentrations as low as 1 ppm. That is why commercial detergents contain less than 30% active material, the rest being fillers, water-softening and whitening agents, etc. Almost all these additional chemicals, but especially phosphates, nitrogen and boron, are the cause of another problem in the environment.

Biodegradability of Aqueous Solutions

In this paper some detergent raw materials and commercial cleaning materials available in Turkey, examined from the point of view of pure hydrocarbons utilized as detergent raw materials for comparison purposes only, are studied with respect to biodegradability rates.

METHODS FOR THE BIODEGRADABILITY DETERMINATION OF WATER-SOLUBLE ENVIRONMENTAL ORGANIC CHEMICALS

One can apply one of the following tests to determine the biodegradability of chemicals in water:

(a) MITI Test (Japan)
(b) Closed Bottle (BOD) Test (USA)
(c) Sturm CO_2 Production Test
(d) Zahn Wellens Test (German)
(e) AFNOR T90/302 Test (French)
(f) OECD Screening Test

In this study the tests shown in (b) and (f) are used. The test shown in (f) is applied as a modified test, following some of the requirements of the Zahn Wellens Test. That is, COD determinations are made use of in determining the degree of oxygen demand, although DOC determinations by means of the TOC apparatus are required by the original OECD screening test. Surface active material loss is traced by applying the MBAS test according to Standard Methods (1965), parallel with the closed-bottle test for the pure chemicals and with COD determination for the commercial cleaning compounds.

The following organic ecotoxic materials are studied by means of these tests:

(1) Tetrapropylene benzene sulfonate, TPBS, a "hard" detergent raw material.

(2) LAS, a "soft" LAS-type detergent raw material

(3) Aniline, a "soft" reference material

(4) N-Sodium 4-Acetylamino benzene sulfonate, a "hard" reference material.

(5) A native, home-made olive-oil soap

(6) A well-known commercial brand shampoo

(7) A well-known commercial brand detergent powder for laundry

(8) A well-known commercial brand liquid detergent for kitchen use

After the commercial test, materials are dried at 105^0C, and 1 g/ℓ of stock solutions of the chemicals are prepared. The test concentrations for the closed-bottle and MBAS tests, 5 mg/ℓ, are chosen for all materials. During preparation of the stock and diluted standard concentrations, carbon contents and dissolved organic carbon amounts are not taken into consideration. However, in OECD tests applied to the pure chemicals, a dissolved carbon concentration of 5 mg/ℓ is always maintained by calculated additions of inoculum and stock solution. In these calculations the DOC/COD ratio is taken as 3/10 (Taken from the Zahn-Wellens Test). The concentration applied to the commercial cleaning materials is 5 mg/ℓ, regardless of the DOC content. This is a necessity as the formula for the surface active material, as well as the percentages of other additives in detergents, is not known.

CLOSED BOTTLE TEST (OECD, 1978)

The heart of the method is the determination of biochemical oxygen demand (BOD) approximately as it is described in the method H5 of German Standard Methods for the Analysis of Water and Waste Water. The biodegradability of any organic material may thus be measured in a concentration range which is usually inaccessible by chemical analysis. MBAS determinations are according to Standard Methods (1965).

Analytical Methods

(a) Oxygen determination according to Winkler (1965).

(b) In the case of anionic surfactants: MBAS determination according to Standard Methods (1965).

Test Materials

1% stock solutions whatever the solubility with respect to anhydrous active substance. In the case of anionic surfactants, the reference is MBAS activity.

Dilution Water

Distilled water air saturated by strong aeration at room temperature for at least 20 minutes. It is ready for use after the addition of appropriate amounts of nutrients, minerals and trace elements, and after standing at 20^0C for approximately a day.

Inoculum

Biologically treated fresh effluent from a treatment plant processing only municipal wastes is used as inoculum. Its bioactivity is maintained by continuous oxygenation after the effluent is taken to the laboratory. Collection of the effluent is made as the test is prepared.

The type of the waste treatment plant is a Pasveer-type oxidation ditch.

To prepare the inoculum, the sample is filtered through a coarse filter and the first 200 ml discarded. The rest of the filtrate is kept aerobic until used.

Preparation of the Experiment

Direct comparisons with the same solutions and seeding materials are always necessary for biotests. Therefore, groups of parallel bottles are prepared for the determination of the BOD of the test material (with ecotoxic material and inoculum) and control material (with ecotoxic material but without inoculum) in a simultaneous experimental series. When chemical analyses are performed simultaneously in a sufficient number of bottles-including the control material for the inoculum (without test material) and the blank (without test material and inoculum)- have to be prepared. Enough parallel bottles are prepared for the test material for each of the 0, 5, 15, and 30-day tests after a sufficient volume in large bottles. These large bottles are first filled by hose to about 1/3 of their volume with distilled water. Then the individual salt stock solutions are pipetted into these bottles according to the final volume, and the respective test or control materials are added in sufficient amount for final concentrations of 5 mg/l to be attained. Subsequently, the solution for the test series is inoculated with 5 ml per liter of final volume and likewise the inoculum control. Finally, the solution is made up to volume with distilled water by a hose which reaches down to the bottom of the flask. The same procedure is repeated for the control series excluding the inoculum control. Each prepared solution is filled immediately into the respective group of bottles by hose from the lower quarter of the bottle, and the zero controls are analyzed for DO. The remaining parallel bottles are placed in an incubator at 20^0C and removed after 5, 15 and 30 days, respectively, to be analyzed for DO. The resulting BOD changes are compared with BOD_T (theoretical BOD calculated from the structure formula of the compound). Each series is accompanied by a complete parallel series for the determination of the BOD of the control, the oxygen depletion without inoculation, a blank and a zero test. The blank, with inoculation but without the chemical, is subtracted from the oxygen depletion of the test for the calculation of the true degradability. After division by the concentration (W/V) of the test material, the oxygen depletion is obtained in mg O_2/mg of active material. For a given day, this value is transformed to the % BOD_T (percentage BOD_T depletion) according to the following formula:

$$\% \; BOD_T = \frac{mg \; O_2/mg \; AS}{BOD_T} \times 100$$

If the structure or elemental composition of the test material is unknown, the chemical oxygen demand (COD) might serve as a reference instead of the BOD_T.

$$\% \; BOD/COD = \frac{mg \; O_2/mg \; AS}{mg \; COD/mg \; AS} \times 100$$

Substances which are only partially degraded or those which have a low BOD_T are advantageously tested in parallel experiments at initial concentrations of 5 or even 10 mg/ℓ.

OECD SCREENING TEST (MODIFIED TO UTILIZE THE COD METHOD AS IN THE ZAHN-WELLENS TEST)

The screening test is limited to the examination of water-soluble organic materials. The test concentration is 5 mg AS/ℓ.

Materials

1. Deionized water: deionized or distilled water free from toxic substances, for general use as a solvent. Water which has been deionized by distillation or ion exchange is suitable.

2. Nutrient Solution: to the distilled water, nutrients, minerals, trace elements, as well as a small amount of Difco yeast extract, are added in calculated amounts (OECD, 1978).

The effluent of a sewage treatment plant dealing predominantly with domestic sewage is used to inoculate the mineral nutrient solution with aerobic polyvalent microorganisms. The amount of inoculum which caused the chemicals to degrade in the proper ranges is employed in the actual tests. The selected concentration, 5 mg/ℓ of test solution, was found by experience.

Procedure

The test materials are evaluated simultaneously in duplicate in accordance with the hard and soft standards, and a control test with inoculation but without test material is conducted for the determination of COD blanks.

To 1500 mℓ of nutrient solution is added so much of the stock solution (containing 1 g/ℓ) of the test material that a carbon concentration of 15 mg COD/ℓ is attained. The test series is inoculated. A rounded filter paper is used to cover the mouth of each vessel in such a way that the exchange of air between the water and the surrounding atmosphere is not unduly impeded. In the course of the biodegradation test, the COD concentrations are determined in triplicate, at the beginning and at least on the 5th, 14th and the 19th days. The biodegradability test is finished after 19 days, provided the standards exhibit degradation rates within the specified ranges and times.

Calculation of Results

The degradation at the time, t, is calculated from the means of the determinations of the COD concentrations at the beginning (C_o) and the time, t, (C_t) according to:

$$A_t = (1 - \frac{C_t - C_{B1-t}}{C_o - C_{B1-0}}) \times 100$$

where:

A_t = percentage degradation at time, t

C_o = mean starting COD concentration of the culture medium (mg COD/ℓ)

C_t = mean COD concentration of the culture medium at time, t (mg COD/ℓ)

C_{B1-0} = mean starting COD Blank of the mineral nutrient solution with inoculation but without test material (mg COD/ℓ)

C_{B1-t} = mean COD blank of the mineral nutrient solution with inoculation but without test material at the time, t (mg COD/ℓ)

RESULTS

The findings of the closed-bottle and OECD tests indicate the degree of biodegradability with respect to time by the BOD, COD values, and for anionic surfactants, MBAS-activity loss. The results for the pure chemicals used as detergent raw materials and for reference are expressed as a percentage decrease with respect to the indicators tabulated (Table 1 to 4). The same figures are plotted as time-biodegradability graphs in Figs. 1 and 2.

The 4 commercially available cleaning materials, soap, powder and liquid detergents, shampoo, are only evaluated with the OECD Test with COD modification as recommended by the Zahn-Wellens Test. The biodegradability is followed by COD and MBAS surface activity determinations. The biodegradability percentages with time are tabulated in Tables 5 to 8, while graphical representation are shown in Figs. 3 to 5.

Table 1
Test Material   : TPBS
Applied Concentration: 5 mg/ℓ
BOD-determination

Time (days)	Oxygen Consumption (mg O_2/ℓ)			Biodegradation (%)
	Sterile	Blank	AS	
0	5.5	6.7	6.7	0.0
5	5.4	6.5	6.5	1.85
15	5.4	6.4	6.3	4.95
21	5.3	6.1	6.0	5.50
30	5.3	6.0	5.8	11.0

MBAS Determination

Time (days)	MBAS (mg/ℓ) Sterile	Blank	AS	Biodegradation(%)
0	0.20	4.5	4.5	0.0
5	0.20	4.5	4.5	0.0
15	0.20	4.5	4.5	0.0
21	0.20	4.5	4.5	0.0

COD Determination

Time (days)	COD (mg/ℓ) Sterile	Blank	AS	Biodegradation(%)
0	16.1	16.48	56.5	0.0
5	16.0	16.40	56.48	0.5
14	16.0	16.40	56.0	0.8
19	16.0	16.40	56.0	0.8

Table 2
Test Material   : LAS
Applied Concentration: 5 mg/ℓ
BOD-Determination

Time (days)	Oxygen Consumption (mg O_2/ℓ) Sterile	Blank	AS	Biodegradation(%)
0	5.5	5.5	5.6	0.0
5	5.4	5.3	3.6	17.0
15	5.4	4.7	0.5	43.0
21	5.3	3.0	0.1	48.0
30	5.3	3.0	0.0	50.0

MBAS Determination

Time (days)	MBAS (mg/ℓ) Sterile	Blank	AS	Biodegradation(%)
0	0.20	2.85	2.25	0.0
5	0.20	1.80	1.35	40.0
15	0.20	0.75	0.0	86.65
21	0.20	0.75	0.0	100.0

COD Determination

Time (days)	COD (mg/ℓ) Sterile	Blank	AS	Biodegradation(%)
0	16.1	16.48	40.4	0.0
5	16.0	16.36	34.9	24.2
14	16.0	16.30	20.0	83.5
19	16.0	16.30	6.1	85.0

Table 3
Test Material   : Sodium 4-acetylamino-benzene sulfonate
Applied Concentration: 5 mg/ℓ
BOD-Determination

Time (days)	Oxygen Consumption (mg O$_2$/ℓ) Sterile	Blank	AS	Biodegradation (%)
0	5.5	5.3	5.5	0
5	5.4	5.1	5.3	9
15	5.4	4.1	4.0	20.4
21	5.3	4.0	4.0	20.4
30	5.3	4.0	3.0	38

MBAS-Determination

Time (days)	MBAS (mg/ℓ) Sterile	Blank	AS	Biodegradation (%)
0	0.2	0.80	0.60	0
5	0.2	0.80	0.50	16.6
15	0.2	0.78	0.47	21.6
21	0.2	0.35	0.47	21.6

Table 4
Test Material   : Aniline
Applied Concentration: 5 mg/ℓ
BOD-Determination

Time (days)	Oxygen Consumption (mg O$_2$/ℓ) Sterile	Blank	AS	Biodegradation (%)
0	5.5	6.0	5.7	0
5	5.4	2.5	2.0	30.7
15	5.4	1.0	0.3	47
21	5.3	1.0	0.1	47
30	5.3	0.1	0.0	48

MBAS-Determination

Time (days)	MBAS (mg/ℓ) Sterile	Blank	AS	Biodegradation (%)
0	0.20	0.45	0.40	0
5	0.20	0.20	0.0	100
15	0.20	0.0	0.0	100
21	0.20	0.0	0.0	100

COD-Determination

Time (days)	COD (mg/ℓ) Sterile	Blank	AS	Biodegradation (%)
0	16.1	16.5	40.4	0.0
5	16.0	4.12	10.1	75.0
14	16.0	2.1	5.2	87.2
19	16.0	0.0	0.0	100.0

Table 5
Test Material   : Powder detergent
Applied Concentration: 5.0 mg/ℓ
MBAS-Determination

Time (days)	MBAS (mg/ℓ) Sterile	Blank	AS	Biodegradation (%)
0	0.70	1.5	2.0	0.0
5	0.60	0.0	0.55	72.5
15	0.60	0.65	0.48	76.0

COD-Determination

Time (days)	COD (mg/ℓ) Sterile	Blank	AS	Biodegradation (%)
0	8.20	126	128	0.0
5	8.20	120	60	53.1
15	8.0	100	50	61.0

Table 6
Test Material   : Liquid detergent
Applied Concentration: 5.0 mg/ℓ
MBAS-Determination

Time (days)	MBAS (mg/ℓ) Sterile	Blank	AS	Biodegradation (%)
0	0.70	5.60	6.0	0.0
5	0.60	0.58	4.0	33.3
15	0.60	0.52	0.8	86.6

COD-Determination

Time (days)	COD (mg/ℓ) Sterile	Blank	AS	Biodegradation (%)
0	8.20	48.8	49.4	0.0
5	8.20	48.0	40.0	19.0
15	8.0	44.0	20.0	59.5

Table 7
Test Material   : Shampoo
Applied Concentration: 5.0 mg/
MBAS-Determination

Time (days)	MBAS (mg/ℓ) Sterile	Blank	AS	Biodegradation (%)
0	0.70	7.95	8.50	0.0
5	0.60	6.30	6.15	28.0
15	0.60	1.30	0.50	94.1

COD-Determination

Time (days)	COD (mg/ℓ) Sterile	Blank	AS	Biodegradation (%)
0	8.20	42.0	49.5	0.0
5	8.20	24.0	20.0	59.5
15	8.0	20.6	8.0	83.8

Table 8
Test Material   : Soap
Applied Concentration: 5.0 mg/ℓ
MBAS-Determination

Time (days)	MBAS (mg/ℓ) Sterile	Blank	AS	Biodegradation (%)
0	0.70	0.95	0.95	0
5	0.60	0.90	0.90	5.2
15	0.60	0.90	0.85	10.5

COD-Determination

Time (days)	COD (mg/ℓ) Sterile	Blank	AS	Biodegradation (%)
0	8.20	50.0	50.0	0.0
5	8.20	50.0	49.8	0.4
15	8.0	49.8	49.4	1.2

DISCUSSION of RESULTS and CONCLUSIONS

This study is aimed at establishing the possibility of biodegradation, under the conditions mostly encountered in nature, of some complex organic chemicals used for household-cleaning purposes. The percentage biodegradation of these ecotoxic compounds is investigated by determining the organic material remaining after 5, 15, 20 and 30 days with respect to the (0) day values.

The results of this study show that the following pure chemicals are in the order of susceptibility to biodegradation:

1. Aniline
2. A LAS-type soft detergent raw material
3. Na-N-Acetyl Sulfonilate
4. TPBS, Tetrapropylene benzenesulfonate

This classification is in decreasing order of "softness" or increasing order of "hardness". The hardest detergent in the list, TPBS, is very resistant to natural decomposition conditions, its decomposition rate being in the order of 1-5% in 30 days. Aniline, on the other hand, is the easiest to degrade: though not a detergent, it is included in the study as a reference material for soft detergents. The LAS-type detergent, however, is a good surfactant and its decomposition is just next to aniline, 84% degradability in 20 days' time. There is a large gap between this and the next material on the list, Na-N-acetyl sulfonilate, which degrades very little within the 30-day period. Figs. 1, 2, 5 and Tables 1-4 show the biodegradation percentage with respect to days elapsed in terms of BOD, COD, and MBAS activity loss.

In the second part of this study, some commercial detergents and other cleaning materials are tested for biodegradation by the same methods. In decreasing order of "softness", these materials are

1. Shampoo
2. Liquid detergent
3. Laundry (powder) detergent
4. Soap (Native olive-oil soap)

The first three are famous brands in the market today, and the first two are found to be quite acceptable for use both with respect to MBAS activity and biodegradability.

The laundry (powder) detergent, however, in spite of its well advertised name, is rather insoluble in water leaving a cloudy mass floating on the surface. That might be the reason for the low MBAS activity and the discrepancy between (0) day control and test detergent activities. That might also be the reason for the high level of recalcitrancy of this famous cleaning material. Whatever the reason, this powder detergent is a very objectionable material to use. In this paper the trade names of the commercial materials used in the study are not disclosed for obvious reasons.

In discussing the results of this study, the rate of degradation of the active part of the detergents and the inactive chain are evaluated by calculating and plotting the percentage COD values divided by the mg active substance remaining in aqueous solutions on the corresponding days (Fig. 5). From this illustration it is obvious that with easily biodegradable materials there is a loss of surface activity within shorter periods. (Examples are LAS and Shampoo, which appear to have similar chemical forms) with "hard" materials the loss of surface activity does not easily take place (Examples are TPBS, Na-N-Acetyl Sulfonilate and pure olive oil soap, which all contain at least one or more unsaturated carbon bonds). Liquid detergent is more biodegradable than the "hard" detergents discussed above, but is harder than the LAS and shampoo.

The hardness found in olive oil soap was a surprise for us, as it is a compound of pure natural origin and should have been one of the easiest to decompose by enzymatic action. The real reason for its hardness could not be determined but it might be related with alkalinity levels in its aqueous solutions.

More organic chemicals used in everyday life in this country will be continually investigated using the same methods as in this study to get a more complete picture of the effects of modernized life styles on our environment. As for the detergents, their use has exponentially increased in Turkey, reaching 134 700 tons in 1976 (TCBDPT, 1977).

If one considers that only 1 ppm of detergent active material in wastewater may cause foaming, this total production-consumption figure is obviously the cause of a great environmental hazard. Also in Turkey, hard detergents are generally produced, which means that at least a couple of months are required even for the biological decomposition of 1 ppm concentrations under natural conditions to be effective, if the material decomposes at all.

We think that detergent consumption is a big problem and that the necessary measures should be taken by the authorities to alleviate this problem in Turkey.

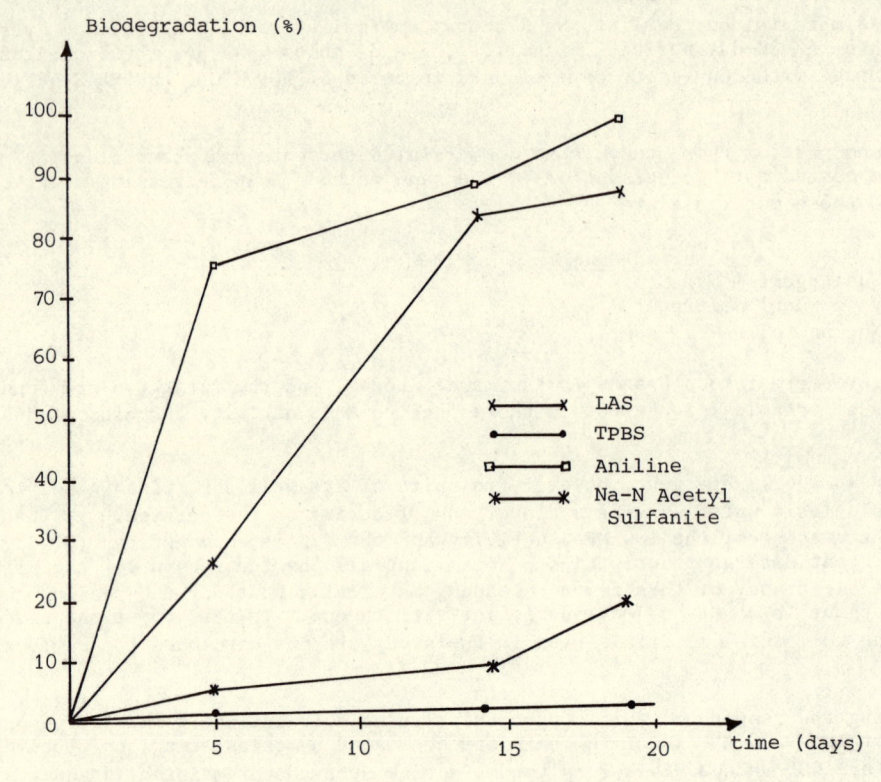

Fig. 1: Time-biodegradability graph.

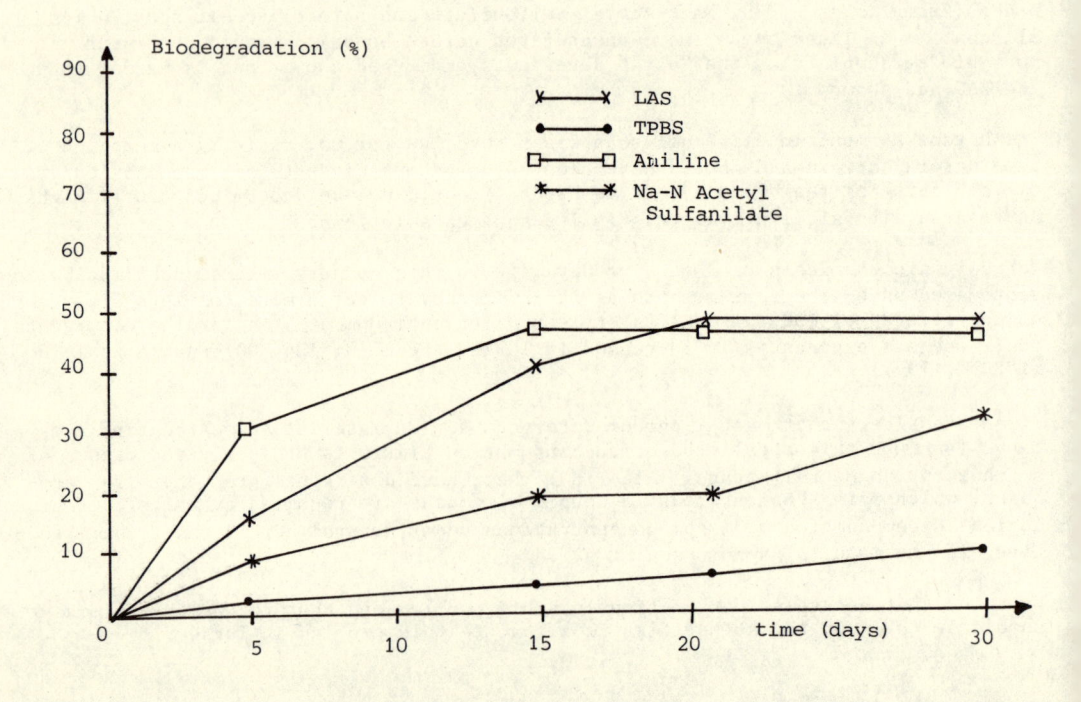

Fig. 2: Time-biodegradability graph.

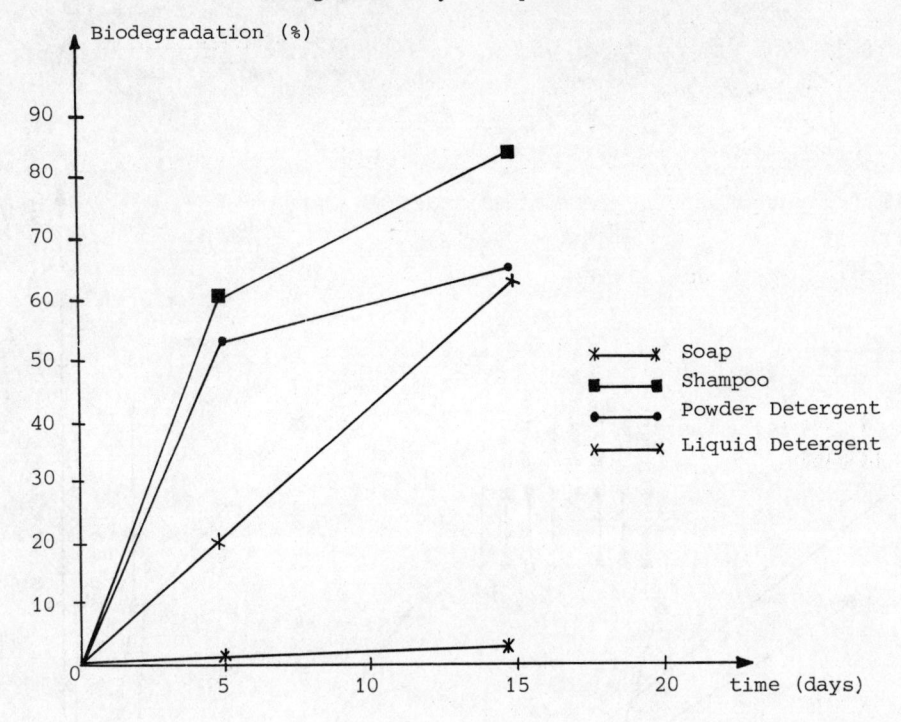

Fig. 3: The Biodegradability percentages with time.

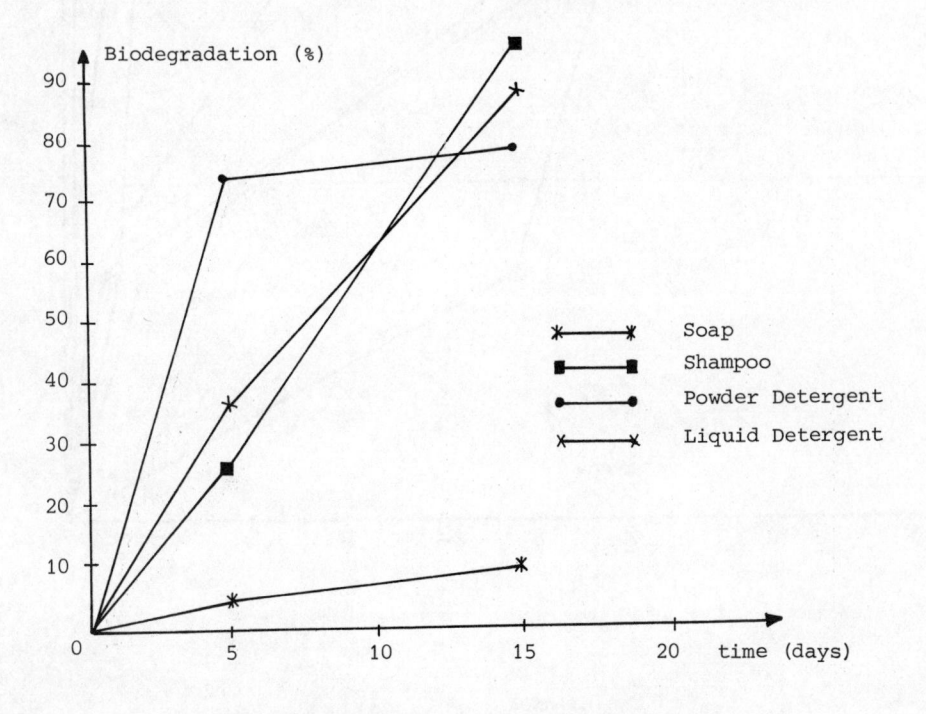

Fig. 4: The biodegradability percentages with time.

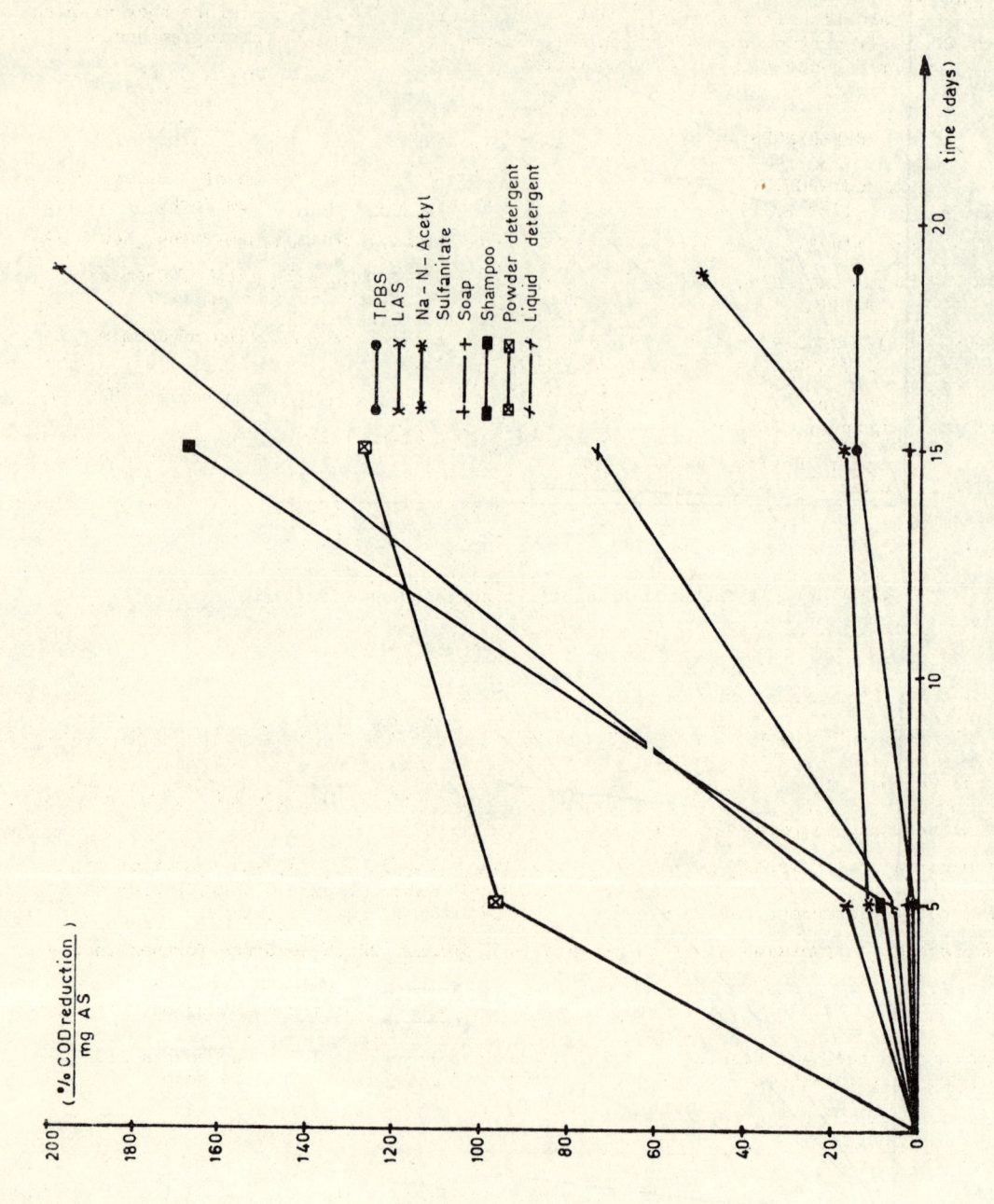

Fig. 5: The Biodegradability percentages with time.

ACKNOWLEDGEMENT

In the first part of this study, the authors have made use of the data they produced for an international ring test program they participated in. Therefore they wish to express their gratitude to OECD, which sponsored and directed the program and supplied the reference materials and methods.

REFERENCES

APHA, AWWA and WPCF. (1965). *Standard Methods for the Examination of Water and Wastewater*. 12th Edition. American Public Health Association, New York.

Dugan, R.P. (1972). *Biochemical Ecology of Water Pollution*. Plenum Press, New York.

OECD (1978). *OECD-Ring Test Programme on Detecting Biodegradability of Chemicals in Water*. Berlin.

OECD (1978). *OECD-Ring Test Programme on Detecting Biodegradability of Chemicals in Water. Closed-Bottle Test*. Berlin.

Sawyer, N. C. and L. P. McCarty (1967). *Chemistry for Sanitary Engineers*. McGraw-Hill Book Company, New York.

T.C. Başbakanlık Devlet Planlama Teşkilatı (1977). *Yayın No. DPT: 1603-ÖIK 278*. Ankara, Turkey.

DISCUSSIONS

Rüffer, H.R. (Germany) : It is surprising that soap was found to be non-biodegradable. I cannot believe that this is due to alkalinity because you used a very low concentration of 5 mg/ℓ, i.e. a pH of not higher than 9. Couldn't it be that the screening test is limited to soluble substances and that the soap is partially insoluble?

Müezzinoğlu, A. (Turkey): No. The soap apparently dissolved as a result of shaking. As I mentioned earlier, we had this problem with the powder detergent, which was insoluble and floated on the surface.

Rüffer, H. (Germany) : I think this is a problem which requires further investigation.

Müezzinoğlu, A. (Turkey): It is a point we intend to continue working on. Throughout our investigation we made double tests, both of which gave the same results. Perhaps the problem is related to the adaptability of the microorganisms or the unsaturated styrene groups present in the soap.

Anaerobic Treatment of High-Strength Industrial Wastewaters

G. K. ANDERSON*, T. DONELLY** and D. J. LETTEN***

*University of Newcastle upon Tyne, UK
**Northumbrian Water Authority, Newcastle upon Tyne, UK
***BioMechanics Ltd., Ashford, Kent, UK

ABSTRACT

The basic principles of anaerobic contact digestion processes are described with emphasis being placed upon the importance of hydraulic retention time and solids retention time, with the latter being suggested as the basis for the design and operation of such systems. Digestion is essentially a two-stage biological process with the overall rate of reaction being controlled by the second stage, namely, the conversion of volatile acids into methane and carbon dioxide gases. It is also essential to control those environmental factors which affect these biological reactions including temperature, volatile acid concentration, pH, and the levels of both essential nutrients and harmful toxins.

It is always important to initiate a research project with controlled bench-scale studies and in this case both semi-continuous and continuous systems were investigated under laboratory conditions. These were used as the basis for evaluating the kinetic data which were in turn used to initiate bench-scale feasibility studies using a wide range of industrial wastewaters, mainly in the food and drinks indusries where the greatest potential appears to be for the process.

To illustrate the full-scale potential of contact digestion, two full-scale plants are described, as well as two pilot-scale systems. The basic conclusion from these studies is that the process does indeed have a great potential for the future, but it is also pointed out that considerable efforts must be made to evaluate each industrial waste with exhaustive bench-scale and pilot studies in order to determine the optimum loading and environmental conditions. A cost comparison indicates the suitability of anaerobic processes for the treatment of high-strength wastes.

INTRODUCTION

Almost all naturally occuring organic matter and many synthetic organic compounds can be fermented (digested) anaerobically and if the process is carried to completion, the end products are usually CH_4 and CO_2 from carbonaceous matter and NH_3 from organically combined nitrogen.

The anaerobic process is, in many ways, ideal for waste treatment, having several significant advantages over other available methods. It has been used for many years for the stabilization of municipal wastewater treatment plant sludges and has considerable potential for the treatment of many industrial wastewaters. Unlike sewage sludge, which always contains a large population of microorganisms, most industrial wastes contain organic matter largely in solution, and it has been found that, in order to induce satisfactory digestion, it is essential to add and then maintain a biological phase. The control and operation of digesters treating industrial wastes must, therefore, be concerned with providing an optimum environment for the maintenance of an active digesting biomass.

CHOICE OF PROCESS

The first question which arises from the proposed anaerobic treatment of industrial wastes could well be - "Why employ anaerobic methods when the technology of aerobic methods is so well developed?"

In aerobic treatment, as represented both by the activated sludge and fixed film reactor processes, the waste is mixed with large quantities of micro-organisms together with oxygen or air. Micro-organisms use the organic waste for food and use the dissolved oxygen to burn a portion of this food to form carbon dioxide and water for the production of energy from the oxidation, thus enabling rapid growth to take place; consequently a large portion of the organic waste is converted into new cells, i.e. synthesized. However, this converted organic matter remains unstabilized and the problem is merely translated from one of soluble organic disposal to solid (sludge) disposal.

Furthermore, for these processes, there is a limit to the organic load which can be applied to the system due to the relatively low rate of oxygen transfer from the gas to liquid phase. It can be demonstrated that in the activated sludge process, for a conventional retention time of 6-8 hours, the highest concentration of BOD_5 which may be applied is of the order of 2000 mg/l and at this high level the degree of treatment is considerably reduced. Past research into the variation of unit cost with organic strength has concluded that anaerobic digestion becomes an economic proposition at an effluent concentration of 4000 mg/l COD, i.e. a BOD_5 of approximately 2000 mg/l and that the economic advantage of anaerobic systems at very high concentrations is clearly apparent where, for example, concentrations in excess of 10000 mg/l BOD_5 have a unit cost of approximately one quarter of that for equivalent aerobic treatment.

At these concentrations, the advantages of anaerobic treatment are: no oxygen transfer limitation, an extremely low sludge production, low nutrient requirements and the production of an energy source (methane gas).

PROCESS FUNDAMENTALS

Despite the widespread use of anaerobic digestion in sludge treatment, the basic microbiology and biochemistry of the process are still poorly understood. A comprehensive review indicating the current state of knowledge in this field has been presented by Donnelly (1979). The process relies on two major groups of bacteria, namely those which convert soluble organic matter into acids and those which convert these acids ultimately into methane and carbon dioxide. In addition, protozoa have frequently been observed in small numbers along with large populations of fungi.

For successful growth of these groups of micro-organisms, a number of process environmental requirements must be followed which may be summarized thus:

    i) Anaerobic conditions.

    ii) Constant temperature, although three ranges of temperatures are possible, namely, psychrophilic, mesophilic and thermophilic generally an operating temperature of 35^0C has been found most suitable.

    iii) pH in the range 6.4 to 8.0 with an optimum mean 7.0.

    iv) Absence of materials above their toxic level, in particular the salts of sodium, potassium, ammonium, calcium and magnesium and also heavy metal ions.

v) Presence of all nutrients in sufficient quantity, in particular nitrogen phosphorus, together with traces of K, Na, Ca, Mg, Co and Fe.

THE ANAEROBIC CONTACT PROCESS

The concept of solids retention time (or sludge age) has been firmly established in the activated sludge (aerobic) process and in order to provide satisfactory criteria for anaerobic digestion of the more soluble industrial wastewaters, it has been necessary to apply this same concept. This naturally led to the development of an anaerobic activated sludge process, followed by the anaerobic filter, both of which are known as contact digesters.

In a conventional digestion process the solids retention time (SRT) and the hydraulic retention time (HRT) are equal and the limiting detention time is reached when the bacteria are being removed from the system faster than they can reproduce themselves. In the case of sludge digestion, the input of bacteria in the feed can offset a certain degree of washout but in the case of industrial wastewaters in which the organic matter is largely in solution, long retention times would be necessary in the digester to ensure that washout did not take place, thus requiring very large tanks. As a result of this, the solids retention time needs to be controlled, independent of the hydraulic retention time, by means of solids seperation after digestion and the subsequent recycle of solids. By definition,

$$\text{Solids Retention Time} = \frac{\text{Suspended Solids in System}}{\text{Suspended Solids Removed per Day}}$$

Hence, SRT becomes the average retention time of solids in the system under steady-state conditions, and it has been found that the use of suspended solids to represent the concentration of micro-organisms is normally adequate. The usefulness of SRT as a design and control parameter for contact digesters has been discussed by Anderson and Donelly (1977), but here it is only necessary to indicate two important concepts derived from the SRT. The minimum solids retention time (SRT_{min}) is the point at which those micro-organisms responsible for treatment are removed from the system faster than they can reproduce. Under these conditions, process failure occurs and the effluent substrate concentration from the reactor equals the influent substrate concentration. Dague, McKinney and Pfeffer (1966) found that SRT_{min} for wastewater sludges was 3 days at $35°C$ but indicated that a higher value of SRT was also of importance. This was defined as the time below which the concentration of organic intermediates in the digester contents begins to increase due to loss of one or more key organisms. The effluent volatile acids concentration and gas production levels pointed to this value being close to 10 days; it is generally referred to as critical solids retention time (SRT_c).

PRELIMINARY LABORATORY STUDIES

The plant developed for these studies consists of a 25-litre completely mixed reactor with a level control system and solids recycle, a splash degasifier operated under a partial vacuum and a settling column from which the clarified effluent overflows to disposal and from which the thickened solids are recycled. In this way high concentrations of biomass have been maintained in the reactor, thus allowing the system to cope with high-strength organic inputs.

To date, the system has proved simple to operate with mixed liquor suspended solids

Table 1 Performance Data for Laboratory Anaerobic Contact Digesters (Donnelly, 1979)

Feed COD (mg/l)	Flow Rate (litre/d)	Loading Rate kg COD/kg MLVSS.d	HRT (d)	SRT (d)	Effluent COD (mg/l)	% COD Removal
2050	0.79	0.12	5.06	110	70	97
2050	1.03	0.31	2.43	43	100	95
2050	1.12	0.27	2.23	68	68	97
2050	1.24	0.21	1.26	82	78	96
2050	1.40	0.21	2.86	52	95	95
2050	1.74	0.54	1.44	24	115	94
2050	1.86	0.26	1.34	71	81	96
2050	2.52	0.62	0.99	15	480	77
14750	2.16	0.33	7.04	28	600	96
14750	2.23	0.16	6.81	172	270	98
14750	2.66	0.18	5.71	146	260	98
14750	2.25	0.24	6.77	61	440	97
14750	0.95	0.17	16.00	80	350	98
14750	2.10	0.47	7.23	15	910	94
14750	0.69	0.11	22.58	120	340	98
14750	1.48	0.50	10.27	23	700	95

Table 2 Wastewater Characteristics

Wastewater	pH	SS (mg/l)	COD (mg/l)	BOD$_5$ (mg/l)	TOC (mg/l)	NH$_3$-N (mg/l)	TKN (mg/l)	PO$_4$-P (mg/	SO$_4$ (mg/l)
Edible Oil Yeast-	6.2	1000	2500	1520	690	NIL	4.5	10300	17440
Molasses	5.5	600	59000	33600	24400	350	1680	-	5500
Potato (Blanch)	4.7	580	11000	4800	3200	-	93	15	-
Dairy (Cheese)	5.1	960	4400	2470	-	187	218	50	-
Cider	5.8	760	1500	900	-	6	14	2	-
Wine Distillery Waste	4.4	2200	50000	26400	25000	196	1400	-	5400

concentrations of up to 20000 mg/l and underflow concentrations (sludge recycle) of up to 30000 mg/l. Due to the ease with which this high MLSS concentration is maintained and the relatively low growth rate of the anaerobic biomass, wastage from the system needs to be comparatively small, thus giving rise to the possibility of very long solids retention times. This has the further effect of forcing the operational growth rate to be much less than the maximum growth rate, thus indicating that the system should be able to withstand shock organic loads.

The system has been operated at SRT's in the range 15-320 days and at a HRT in the range 1-23 days. Some of this data (Donelly, 1979) is presented here in Table 1 and from this it can be seen that the system is well able to respond efficiently to the full range of operation. Work is still progressing on the lower operating levels but it is felt that HRT's in terms of hours rather than days could be adopted while retaining sufficiently high removal efficiencies in many circumstances, e.g. partial treatment prior to discharge to Municipal Sewage systems or prior to a further treatment unit.

Due to the kinetic behaviour of the anaerobic biomass, reduction of the SRT to very low levels is both unnecessary and undesirable, but work is being carried out at present to determine an optimum SRT at which the system responds most favourably.

FEASIBILITY STUDIES

In order to evaluate further the potential of the anaerobic contact process for treating essentially soluble industrial wastewaters, a series of semi-continuous digesters have been operated at the University Laboratories (but in conjuction with Biomechanics Ltd.), using a wide range of industrial waste sources. The semi-continuous system has a 5-litre capacity digester maintained at 35^0C. Ten such digesters were used, each having the capacity to feed waste on a daily (batch) basis with a similar arrangement for mixed liquor withdrawal. Gas production was measured and a range of control parameters used to evaluate the system after steady-state conditions were established. The currently available data is shown in Tables 2, 3 and 4.

The results shown in Tables 2-4, together with on-going tests, have convincingly demonstrated that anaerobic contact digestion has a considerable role to play in the treatment of industrial wastewaters. For each individual waste, however, a pilot plant operated on a continuous basis is always recommended in order to evaluate design criteria for full-scale operation.

FULL-SCALE ANAEROBIC CONTACT PROCESS

A number of full-scale treatment plants are now in operation, which embody the principles of anaerobic digestion outlined earlier. Many authors make reference to a problem inherent in most digestion processes, namely, that of effective and economic solid/liquid separation. This separation is essential for two reasons, firstly to ensure an adequate recycling of the biomass to the reactor and secondly to remove as much solid matter as possible from the effluent prior to its final discharge. It is this solid/liquid separation technology, more than any other, which differentiates one particular process - the Bioenergy Process - from all other contact processes.

Basically, this process has been developed from the classical anaerobic digesters which have been in use since the beginning of this century and is characterized by:

   i) The conversion of organic matter to gas (via an acidification stage).

   ii) The use of a sealed tank, making the process odourless.

   iii) The production of very small quantities of sludge (which, it should be noted, may account for up to 50% of the total cost in aerobic biological treatment).

   iv) A low nutrient requirement.

   v) An operating temperature of about 35^0C.

The wastewater is fed continuously into the biological reaction unit where it is

Table 3 Digester Operating Data (Steady State)

Wastewater	pH	Temperature °C	MLSS (mg/l)	HRT (d)	SRT (d)	Feed Rate ml/d	Notes
Edible Oil Yeast	7.7	35	5000	30	150	170	H_2S Stripping
Molasses	7.2	35	11500	30	200	170	H_2S Stripping
Potato (Blanch)	7.3	35	5000	30	150	170	
Dairy (Cheese)	7.1	35	2500	30	300	170	
Cider	7.0	35	4000	30	280	170	
Wine Distillery	7.4	35	12000	30	30	170	H_2S Stripping

Table 4 Steady-State Results for Digester

Wastewater	Organic Loading kg BOD_5/ kg MLSS d	VA (mg/l)	COD (mg/l)	BOD_5 (mg/l)	% BOD Red^n	Gas (mg/l)	CH_4 : CO_2	Settled Volume (%)
Edible Oil Yeast	0.1	30	700	135	91	130	70:30	30
Molasses	0.1	50	9000	2000	93	4400	65:35	35
Potato (Blanch)	0.1	-	1000	285	94	650	90:10	-
Dairy (Cheese)	0.033	12	600	50	97	320	75:25	20
Cider	0.01	10	350	50	95	130	65:35	40
Wine Distillery	1-0.06	150	16500	1800	93	4500	65:35	38

completely mixed with the biomass, which itself is fed to the digester from the seperator unit.

The function of the biomass is to provide the various micro-organisms necessary for the various biological reactions.

Gas recirculation is used to accomplish complete mixing within the reactor. After the necessary retention time (based upon the hydraulic and solids retention time requirements), the mixture of treated wastewater and anaerobic micro-organisms is fed to the "Separator" system consisting of a heat exchange cooling system and settling unit. Finally, the effluent is discharged either to a watercourse or to a further treatment unit for additional treatment, if deemed necessary by discharge consent conditions.

In order to control the process under the optimum operating conditions, it may be expedient to periodically waste any excess biomass which has been produced and to remove gas (methane and carbon dioxide) on a continuous basis for burning, thus producing heat which in turn is used both for factory heat and to control the digestion temperature.

At this stage it may be useful to compare the process with that of the more traditional applications of anaerobic digestion, namely, sewage digestion. Sewage sludges are produced in very large quantities at sewage treatment works and are the result of the sedimentation of solids from the raw wastewater and solids produced during the aerobic biological treatment stage. The nature of sewage sludge is such that a large portion (up to 80%) of the solids may be in suspension, thus requiring solubilization before anaerobic digestion proper. This is turn leads to a very large hydraulic retention time, thereby a high capital cost and its resultant unpopularity with many engineers, scientists and plant operators. The Bionergy Process, however, has greatly enlarged the field of application to cover mainly wastewaters which are largely soluble (to the extent where wastes with a high insoluble content are not really suitable for contact systems). In particular, and as examples, high-strength soluble effluents such as those from the food and drinks industry are very amenable to the anaerobic biological contact process. The reactor tank will be generally very much smaller (and consequently less expensive) than the equivalent traditional digester and in the majority of applications there will be a large net production of gas which can be, and is, used in the factory to reduce the consumption of expensive fossil fuels. Compared to this, sludge digesters will produce a relatively small net energy which is normally wasted by burning, except at very large installations where it may be used either to generate electricity or may be sold as gas. The major breakthrough with respect to the Bioenergy System has been the ability to design the reactor based upon solids retention time, which consequently gives a greater stability and treatment efficiency than for conventional digestions since these criteria are independent of reactor size, whereas they are totally linked to reactor size for the traditional process. This advantage has been achieved by concentrating the biomass in the digestion tank, by means of the seperator and by preventing an unwanted loss of anaerobic bacteria from the system. Previously, this had been a major technological problem, but the development of a simple and inexpensive seperation method, based upon applying a temperature shock which in turn encourages flocculation and thereby good settling of the sludge mass, has overcome this.

It is well established that anaerobic treatment processes operate most efficiently at a steady temperature in the range of 33^0 to 38^0C. In the traditional sludge digestion process, the heat required for maintaining this temperature is obtained by burning the gas produced by the biological reaction, but the nature of the waste is such that little excess energy remains for further use. In the majority of cases where the Bioenergy System would be applicable, the factory itself has sufficient low-grade waste heat at 40^0 to 50^0C for maintaining the digestion temperature, thus leaving the biologically produced gas available as a high-grade heating fuel for use in the factory.

Finally, there is the question of sludge production, which, as has been referred to earlier, may account for 40-50% of the total capital and operating costs in waste treatment systems employing aerobic methods. The Bioenergy Process, by its very nature, is suitable mainly for soluble wastes and will have a relatively low biomass (sludge) yield, which in turn is both reasonably stable and odour free and would not be a significiant burden on the overall plant operation.

INSTALLATIONS

At present (1979), there are two full-scale installations using the system which may be cited as examples:

Starch Gluten Effluent (UK)

The factory processes wheat flour into starch gluten and produces effluent with a pollution load equivalent to that produced by 40.000 people. It was calculated (at 1977 prices) that approximately ₤500,000 would be required to extend the already overloaded municipal sewage treatment plant. After considerable investigation of a number of alternative processes including protein recovery, reverse osmosis, activated sludge and biofiltration, the Bioenergy System was selected as the most attractive 'on-site' treatment method.

The operating criteria and costs of the process are:-

i) A conversion of 2.7 ton BOD per day to fuel gas with a saving of 1.35 ton fuel oil per day, i.e. an annual saving of ₤ 34,000 at 1977 prices.

ii) A low operating cost (20kw electricity).

iii) A low construction cost of ₤ 180,000 at 1977 prices.

iv) No odour or noise problem - an important factor considering its location adjacent to a housing development.

v) No sludge handling or disposal since in this particular case discharge to the municipal sewer is acceptable.

No further treatment is necessary since discharge of the effluent is direct to the municipal sewer. This type of operation would have many similar applications in the U.K., where industry may have to choose between, on the one hand, discharge of effluents with high BOD and payment of a high charge to the Regional Water Authority (RWA), and on the other hand, installment of 'on-site' treatment, removing at low cost, say, 90% of the BOD with final discharge of a relatively low BOD and payment of a low charge to the RWA.

Food Products (France)

This plant is designed to receive 5000 kg BOD/day, representing a gas production of 4080 m^3 /day with a financial saving of ₤ 195 per day at oil prices of 30 pence per gallon. The energy input required to achieve this gas production and 95% BOD removal is 38.5 kw per hour, representing a 90% lower requirement than that of activated sludge treating the same effluent, assuming an oxygen transfer efficiency of 0.89 kg O_2 per 750 watts per hour, and an oxygen requirement of 2 kg O_2 per kg BOD applied. The plant in France differs somewhat from that in the U.K., in that the final effluent needs to be of a higher standart. Consequently, two Bioenergy units are in parallel use, followed by an activated sludge polishing system, after which discharge direct to the River Garonne is possible. In the French system the feed BOD is 13980 mg/l and the digester effluent 500 mg/l, figures which would not be possible with conventional aerobic treatment. The effluent of 500 mg/l BOD approximates the level in domestic sewage, making aerobic treatment at this stage, prior to final discharge, a viable proposition.

The capital cost of the Bioenergy System is similar to that of a comparable aerobic system but, whereas the aerobic process will entail high running costs, the Bioenergy Process operates at a profit, which with the correct application could lead to a capital cost payback within 3 years.

Pilot Plant, Seville, Spain

A large pilot plant was assembled in the U.K. and transported to Seville, Spain in order to investigate the amenability of alpechin (olive oil waste) to the Bioenergy Process. The raw waste has a BOD of the order of 90,000 mg/l but after storage in a lagoon this falls to 20,000 to 25,000 mg/l. The pilot plant was fed with super-

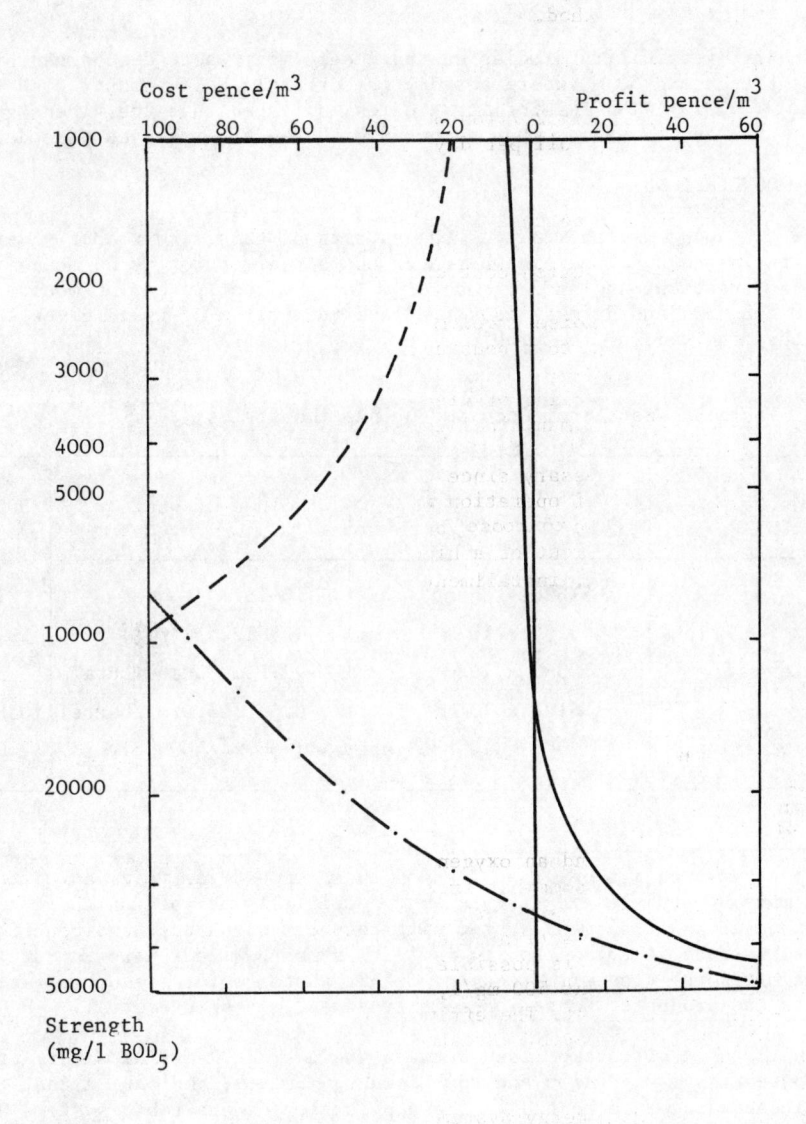

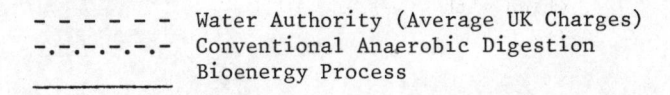

Anaerobic Treatment of High-Strength Industrial Wastewaters 139

- - - - - - Water Authority (Average UK Charges)
-.-.-.-.-.- Conventional Anaerobic Digestion
─────── Bioenergy Process

Fig. 1 Net Cost of Treating Effluents of Various Strengths.

natant from the lagoons and although found to be lacking in nutrients, the waste was readily treated by contact digestion after nutrient addition. The olive oil industry is seasonal; consequently the plant has been shut down until production begins again later this year.

Cheese-Whey Waste, Appleby, Cumbria

After laboratory feasibility studies on the cheese/whey waste discharged by a dairy, a pilot plant treating 60 m^3 waste per day is currently being constructed (due for commissioning in 1979). The feasibility studies indicated that the Bioenergy Process was suitable for such a waste, when compared with the alternatives considered.

CONCLUSIONS

There is little doubt in the minds of the authors of this report that a very large market exists in the U.K. and elsewhere for contact digestion as an economic means of wastewater treatment and energy recovery. The greatest potential would lie in the wastes from the food and drinks industries and the following table gives some indication of the magnitude of the problem.

Table 5 Waste from Food and Drinks Industries

Industry	U.K. Production per annum	Volume of Waste	BOD MG/1
Sugar refining	518280 tonnes	10-18 m^3/tonne	210-1700
Dairies (milk)	6269 x 10^{3}m^3	125-1225 m^3/m^3	1000-2000
Distilleries	343 x 10^{3}m^3	2270 m^3/bushel grain	5000
Breweries	6183 x 10^{3}m^3	0.6 to 3.9 m^3/barrel	500-1300
Poultry processing	300 x 10^6 broilers	1-55 m^3/1000 birds	100-2400

Considerable pressures are being applied on these industries to clean up their wastewaters and many companies are carrying out a critical analysis of their current or proposed disposal and treatment system with respect to the cost and benefits of the various alternatives. Local authority facilities may be inadequate in rural areas and this is often the case with food and drinks wastes, which are often generated close to the raw products. Hence, on-site treatment may be essential.

Figure 1 indicates a cost comparison between the average Regional Water Authority Charges, Conventional Digestion and the Bioenergy System, indicating that there are considerable advantages to be had in the application of anaerobic systems and that, as the strength of the wastewater increases, further advantage is to be gained when using anaerobic digestion.

REFERENCES

Anderson, G.K. & T.Donelly (1977). Anaerobic Digestion of High Strength Industrial Wastewaters. *J.Inst.Public Health Engrs.*, May 1977.

Dague, R.R., R.E. McKinney & J.T. Pfeffer (1966). Solids Retention in the Anaerobic Contact Process. *J. Wat. Pollut. Contr. Fed.* 38, 220.

Donnelly, T. (1979). *The Kinetics of Mathematical Modelling of a Contact Digester*. Ph.D. Thesis, University of Newcastle upon Tyne, Newcastle.

DISCUSSIONS

Samsunlu, A. (Turkey) : Up to what BOD limit can anaerobic treatment be applied?

Anderson, G.K. (U.K.) : We have found that there is no real upper limit. Our pilot plant in Seville, where olive-oil wastewater is being treated, has an anaerobic lagoon in which the BOD can be reduced from 90,000 mg/ℓ to 20,000 mg/ℓ. Further reduction to 10,000 mg/ℓ is technically feasible. If the municipal sewage system accepts a BOD of 1000 mg/ℓ, then anaerobic treatment is satisfactory. In our lab it is quite simple to reduce the unfiltered BOD of the effluent from an anaerobic system to a BOD of about 100 mg/ℓ. So, if it is acceptable a one-stage anaerobic process is sufficient. With aerobic processes there is no real difficulty in getting down to a BOD of 15 or 10 mg/ℓ, or even less. When a two stage process is used, however, there is an important danger, namely, that of trying to use conventional design criteria for the effluent into the aerobic process after it has been anaerobically treated. A BOD, for example, of 500 mg/ℓ from the anaerobic process is not the same as a BOD of 500 mg/ℓ in domestic sewage, because the anaerobic process will have removed the more readily biodegradable material first. The effluent from the anaerobic process is, therefore, less biodegradable that the BOD indicates. We are currently working on this problem.

Samsunlu, A. (Turkey) : Have you any experience of using a vertical column with the anaerobic and aerobic processes being carried out at the bottom and top, respectively.

Anderson, G.K. (U.K.) : We have never used this system.

Taygun, N. (Turkey) : Could you explain further how thermal shock causes the anaerobic microorganisms in the reactor to flocculate?

Anderson, G.K. (U.K.) : As a civil engineer, I cannot explain the microbiological aspects of this process. It is, however, a fairly well understood phenomenon that microorganisms will flocculate at lower temperatures. We pass the mixed liquor from the reactor through a heat exchanger, reducing the temperature quickly to about 10°C, at which the best flocculation occurs. Each industry would have to determine its own operating criteria. I would say a detention time of 5-15 minutes at the cold temperature is sufficient.

Taygun, N. (Turkey) : What is the mixed liquor suspended solids concentration in the reactor? Can you also give figures for the ratio of nitrogen and phosphorus to BOD in the anaerobic treatment process?

Anderson, G.K. (U.K.) : Depending on the strength of the wastewater going into the reactor, the biomass concentration varies from 5000 mg/ℓ normally to 15,000 mg/ℓ in full scale systems. To express it another way; this particular process works best at .1 mg BOD per kg of biomass in the reactor. With regard to the BOD:N:P ratio in anaerobic systems, we found that with all our wastes 150:5:1 was adequate. In aerobic systems we normally talk about 100:5:1. So in anaerobic systems 50% or 30% less nitrogen can be used.

Knapp, N. (Australia) : Do you know if anyone has investigated the possibility of using flotation rather than settling as a means of separating the biomass after anaerobic treatment?

Anderson, G.K. (U.K.) : We have investigated the flotation of sludges at Newcastle. Aerobically it is quite easy to carry out. Anaerobically, a way of inducing the process has to be found. If it is to be a gaseous system, you have to add a gas that is not oxygen or compressed air; otherwise, aeration will take place and the anaerobic concept will be destroyed.

Knapp, N. (Australia) : Can't the dissolved methane in the system be used as the flotation agent?

Anderson, G.K. (U.K.) : Methane would be a possibility. However, let me say that from my own experience flotation systems tend to be not as simple and certainly not as efficient as conventional settling.

Interactions Between Ferrous Iron Oxidation and Phosphate

H. Z. SARIKAYA

Environmental Engineering Division, Istanbul Technical University, Istanbul, Turkey

ABSTRACT

The rate of ferrous iron oxidation by atmospheric oxygen in the bicarbonate solutions of a heterogeneous nature due to high amounts of initial ferrous iron concentrations, as applied in phosphorus removal practice, is investigated. The autocatalytic reaction rate is observed when a certain amount of ferric hydroxide flocs are established in the solution. While phosphate accelerates the oxidation rate at the beginning of oxidation, the retarding effect of phosphate is observed towards the end of the oxidation. This differs from previously reported experimental findings with small initial Fe(II) concentrations. A model is suggested for the retarding effect of phosphate, which is explained by the surface charge of mixed hyroxo-phosphate precipitate.

INTRODUCTION

The oxidation of ferrous iron by atmospheric oxygen finds its application in iron removal, in limnology and in the chemical treatment of wastewater, particularly in phosphorous removal. In the past, several investigators have studied the oxidation of ferrous iron by aeration in order to achieve better iron removal from natural water. Thus, the reaction medium and the range of initial ferrous iron concentrations have been chosen to simulate the natural water.

Since several wastewater treatment plants started utilizing ferrous iron (e.g., in the form of waste pickle liquor from the steel industry) in order to precipitate phosphorus as ferric phosphate, $FePO_4$, following oxidation of Fe(II) to Fe(III), a better understanding of ferrous iron oxidation in wastewater treatment process has become important.

The kinetics of ferrous iron oxidation in homogenous bicarbonate buffer solutions has been extensively studied and the results are well formulated. This study is focused on the ferrous iron oxidation by atmospheric oxygen with high initial ferrous iron concentrations, as applied in chemical treatment of wastewater and in phosphorus removal. Consequently, an appreciable amount of $Fe(OH)_3$ formation creates a heterogeneous reaction medium.

Although the catalytic effect of $Fe(OH)_3$ flocs has been reported previously (Holluta and others, 1964; Lerk, 1965), the results are not sufficient to derive a definite relationship for the kinetics of the ferrous iron oxidation in the heterogeneous systems. In addition, ferrous precipitation and the pH dependence of the reaction rate constant require additional investigation.

ANALYTICAL AND EXPERIMENTAL APPROACH

Ferrous iron determination in the presence of ferric iron is the main concern of this study. In the literature, different methods can be found, but the most suitable one is the so-called "bathophenanthroline method", as defined by Lee and Stumm (1960) with pertinent details. Since this study will be extended to cover the effect of organic matter on the ferrous iron oxidation, a modification of the bathophenanthroline method, as suggested by Theis and Singer (1973), was applied, where samples were not acidified to extremely low pH values and were not boiled because it is known that many organic substances reduce ferric iron if the pH is low enough. Therefore, the pH of the samples was adjusted to pH 4 with sodium acetate-acetic acid buffer solution.

Experiments were carried out in a 1.5 ℓ volume reaction vessel under controlled pH, temperature and oxygen partial pressure. The temperature of the reaction solution was kept constant at (25±0.5)°C in all the experiments by means of a constant temperature bath. The reaction solution was intensely mixed, and the desired mixture of (Oxygen + Nitrogen + CO_2) gas was introduced into the solution by means of a porous tube. The pH of the solution was controlled by varying the CO_2 flow or sometimes by adding acid or base solution (especially at high pH values). The partial pressure of oxygen was controlled by an oxygen electrode connected to an O_2 meter and also by gas flow meters as shown in Fig. 1. The pH of the solution was measured with a combined glass electrode and pH meter (Electrofact 36200) with an accuracy of ±0.01 pH unit. The calibration of gas flow meters was achieved by a gas meter, which measures the total volume of the gas passed during a certain time period.

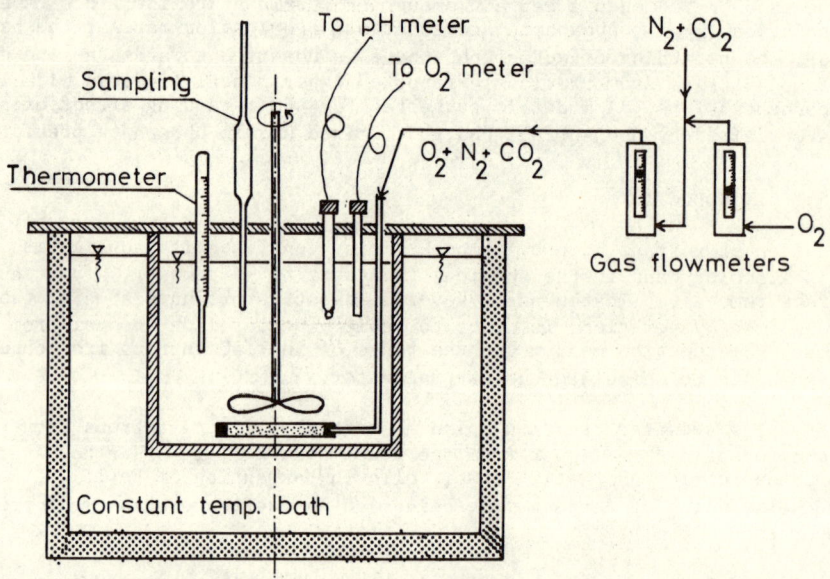

Fig. 1: Experimental Set-Up

By affecting the CO_2 flow, the pH was controlled within ±0.05 pH unit, even in many cases ±0.02 pH unit. Although there was a high drop in pH at the start of the experiments due to the acidity of the Fe(II) solution, it was possible to increase the pH of the solution to a desired level within 30 seconds by affecting the CO_2 flow. However, the recovery of the desired pH level in 30 seconds may affect the results at high pH values (e.g. higher than 7.0), due to the fact that 30 seconds is not negligible as compared to oxidation time. Thus, at high pH values a dosing of base or acid solution was also employed in order to reduce the recovery period.

Fe(II) solution was prepared by dissolving ferrous ammonium sulphate in demineralized water which contains 1.2 ml of concentrated H_2SO_4 per liter. After the addition of Fe(II) solution to the reaction vessel, a stopwatch was started and samples were withdrawn at desired times. These were put into pH buffer solution to quench the oxidation and were analysed for Fe(II) contents.

PRECIPITATION of FERROUS IRON in WATER CONTAINING BICARBONATE

Ferrous iron may precipitate as $FeCO_3$ or $Fe(OH)_2$ in water containing bicarbonate. Detailed information can be found elsewhere regarding the solubility of iron (Ghosh and others, 1967; Singer and Stumm, 1970; Stumm and Lee, 1960; Stumm and Morgan, 1970). By employing the most recent solubility product of $FeCO_3$ given by Singer and Stumm (1970) and considering the activity corrections, Fig. 2 was prepared in order to determine the maximum soluble ferrous iron, depending on pH and alkalinity. Other stability constants needed for solubility calculations were assumed the same as those given by Singer and Stumm (1970) and by Stumm and Lee (1960).

If the ferrous iron concentration range, as applied in phosphorus removal practice (10 mg/l to 90 mg/l), is compared with the maximum soluble ferrous iron concentrations determined from Fig. 2., one can see that ferrous iron should precipitate. However, many investigators have reported the oversaturation of natural waters with regard to Fe(II). For instance, Illinois ground waters were reported to be supersaturated 20 to 30 times when activity and temperature corrections were ignored (Ghosh and others, 1966). But, even if these corrections had been accurately applied, they would still have been supersaturated by factors of approximately 6 to 9 (Winklehaus and others, 1966). Later, Singer and Stumm (1970) attributed this supersaturation to the inaccuracy of the solubility product of $FeCO_3$ reported by Latimer (1952), based upon the experimental work of Smith (1918). But, even if the solubility product given by Singer and Stumm (1970) were used, Illinois ground waters would still be supersaturated by factors of 2-3. Although supersaturation was noted, precipitation of $FeCO_3$ was reported too (Cleasby, 1975; Ghosh and others, 1966; Morgan and Birkner, 1966); Olson and Twardowski, 1975). Therefore, the magnitude of ferrous iron precipitation during the oxidation process needed to be investigated. This was accomplished, qualitatively, by turbidity measurements and quantitatively by the determination of precipitated ferrous iron. The increase in turbidity due to ferrous precipitation in the bicarbonate buffer solutions under nitrogen atmosphere was measured. The results thereof proved that ferrous precipitation was not significant within the oxidation time for pH<8 and initial ferrous iron concentration Fe(II) 100<mg/l . In order to get quantitative results, the amount of ferrous iron precipitated under nitrogen atmosphere was measured as follows:

After the desired experimental conditions (temperature, pH, pO_{2_-} -0) were attained, a certain amount of ferrous iron was added into the reaction solution which contained a known amount of alkalinity. By allowing the ferrous iron to precipitate for a time period roughly equal to the oxidation time under prevailing conditions, 25 ml of samples were withdrawn and immediately filtered through 0.2 µ membrane filter. The filter paper was then rinsed with a constant volume of distilled water after each filtration. Following rinsing, the filter paper was immediately immersed in the pH 4 buffer solution to dissolve Fe(II) precipitate, and the Fe(II) content was determined by the bathophenathroline method. Because the size of $FeCO_3$ crystals was 5µ - 100µ in diameter (Olson and Twardowski, 1975), any formed $FeCO_3$ particle had to be retained on 0.2µ membrane filter. Experimental results proved that the percentage of ferrous iron precipitated as $FeCO_3$ within the oxidation time was not significant. This statement is not in agreement with Fig. 2. However, one should consider that Fig. 2 is given for equilibrium conditions but we are interested in what happens within the oxidation time under experimental conditions. In addition, the following ferrous iron oxidation experiments clearly indicated that Fe(II) precipitation was not important,

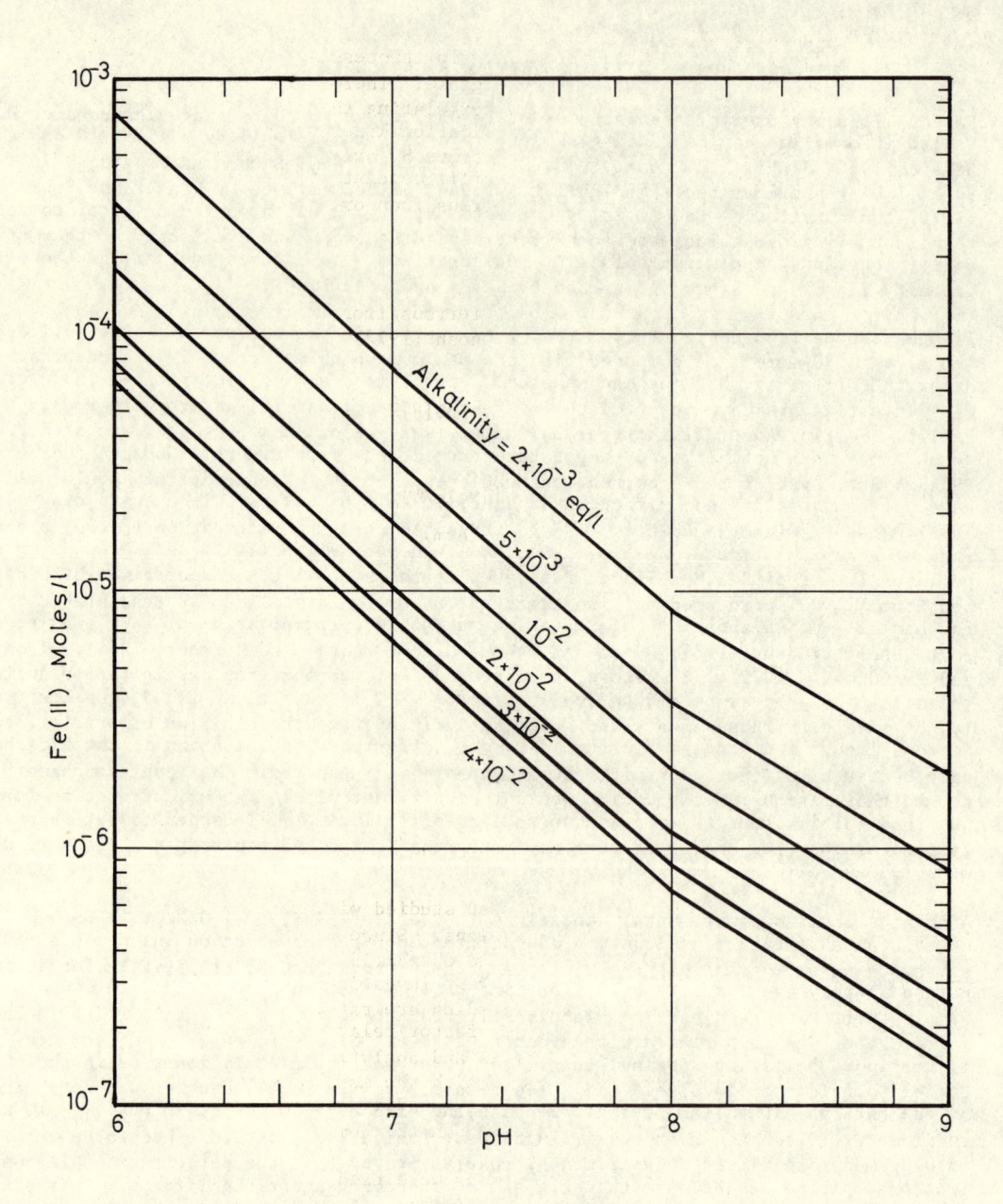

Fig. 2: Solubility of ferrous iron in bicarbonate-containing water

because the initial rate of oxidation was not smaller than the oxidation rates in a homogeneous medium. If there were considerable $FeCO_3$ precipitates, there should also be considerable reduction in the oxidation rate, since it is known that oxidation of the ferrous carbonate is extremely slow or may not even occur at all. (Ghosh and others, 1967; Kolthoff and Elving, 1963).

It is not the scope of this study to investigate the kinetics of $FeCO_3$ precipitation, but it can be concluded that it is a rather slow process and the amount of precipitated Fe(II) within oxidation time is negligible. Therefore, the $FeCO_3$ precipitation phenomenon will not be considered in formulating the Fe(II) oxidation kinetics. Then, if there is any deviation from the oxidation kinetics in a homogenous medium, it will be due to the $Fe(OH)_3$ solid phase formed by oxidation, and its amount and effect will be correlated with the initial Fe(II) concentration. Thus, attention was focused on the effect of $Fe(OH)_3$ on the ferrous iron oxidation rate.

KINETICS of FERROUS IRON OXIDATION

The most recent formulation of the rate of ferrous iron oxidation in homogeneous bicarbonate solutions is given by Jobin and Ghosh (1972) as follows:

$$-\frac{d\{Fe(II)\}}{dt} = k \{Fe(II)\}\{OH^-\}^2 \; pO\{\beta\}^n \tag{1}$$

where

   $\{Fe(II)\}$: Ferrous iron concentration $\{Mole/\ell\}$

   pO_2 : Partial pressure of oxygen $\{Atm.\}$

   k : First order reaction rate constant $\{T^{-1}\}$

   β : Buffer intensity $\{eg/pH\}$

When

   $\beta < 4.10^{-3}$ eq/pH $n=0$
   $\beta > 4.10^{-3}$ eq/pH $n=0.5$

Equation (1) can be rewritten for $\beta > 4.10^{-3}$ eq/pH as

$$-\frac{d\{Fe(II)\}}{dt} = k\{Fe(II)\}\{OH^-\}^2 \; pO_2 \cdot \{\beta\}^{0.5} \tag{2}$$

The rate of ferrous iron oxidation was first studied with 3 mg/ℓ initial ferrous iron concentration to obtain homogeneous conditions. Since the effect of alkalinity was not definitely formulated prior to Jobin and Ghosh (1972), it was not possible to compare the oxidation rates given by different investigators at different alkalinity values. Thus, the results of different studies are rather scattered as can be seen in Fig. 3. But alkalinity is not the only factor related to different oxidation rates, the effect of different experimental and analytical procedures can not be excluded. However, this study compares well with the most recent one by Jobin and Ghosh (1972) for the same alkalinity value. In addition, rather reproducible results were obtained for the ferrous iron oxidation in the homogeneous medium.

When high initial ferrous iron concentrations were used, the oxidation rate was not first order, as was the case in homogenous systems. A typical oxidation rate curve is illustrated in Fig. 4, which indicates that the oxidation curve is a straight line at the initial stage of oxidation (Stage I). The slope of this straight line is approximately equal to that given by Stumm and Lee (1961). Unless a certain amount of ferrous iron is converted to the ferric form, then oxidation rate is first order with

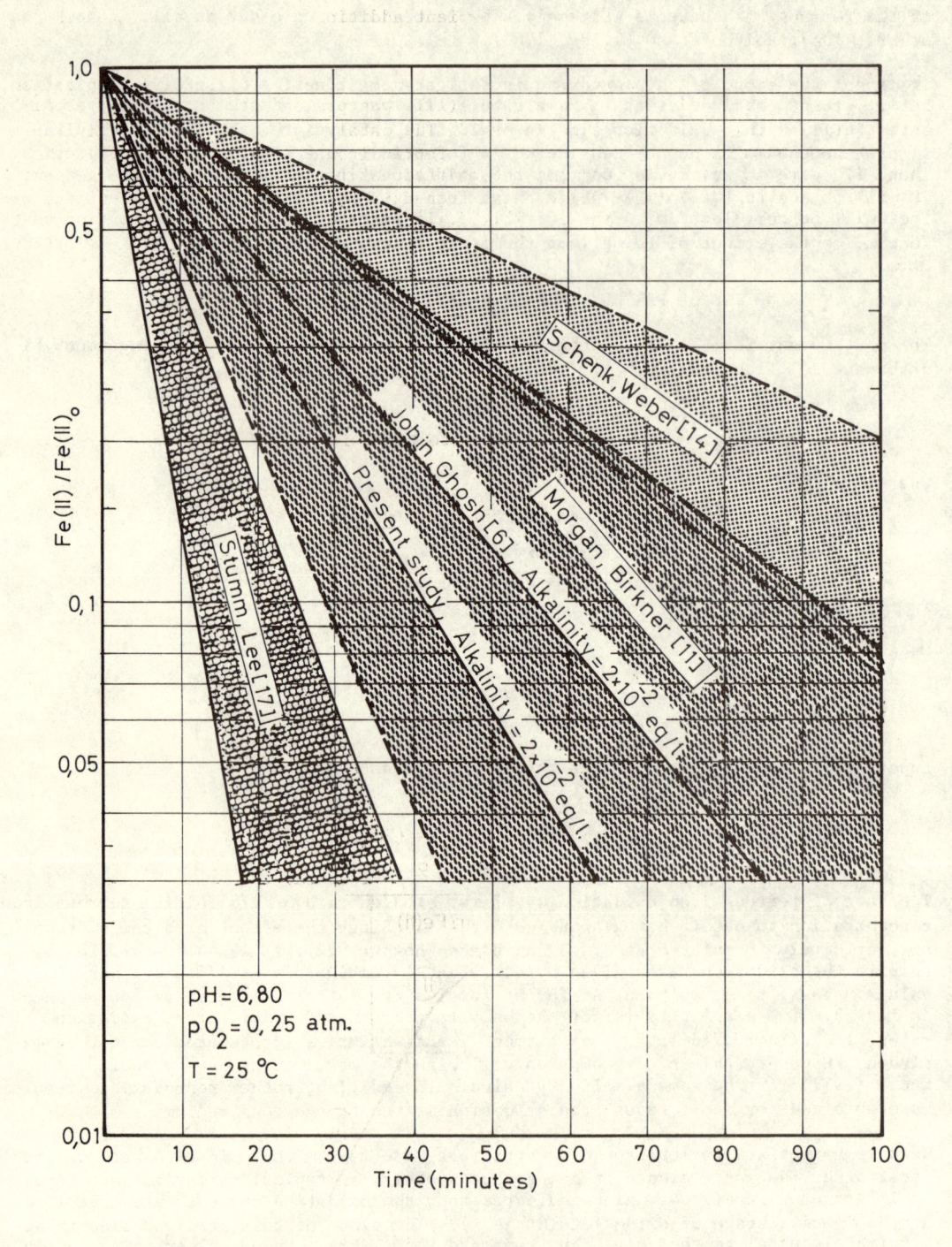

Fig. 3: Comparison of the ferrous iron oxidation rates

Interactions Between Ferrous Iron Oxidation and Phosphate

regard to ferrous iron concentration. The minimum amount of ferric iron requried in order to start a catalytic effect on the ferrous iron oxidation is called "min.Fe(III)" in this study. The existence of such a minimum concentration is in accord with the study of Stumm and Lee (1961), where incipient addition of 10^{-4} M ferric iron did not alter the oxidation rate.

When the ferric iron concentration was more than min. Fe(III), the oxidation process was catalysed by $Fe(OH)_3$ flocs, because Fe(III) was practically in the form of $Fe(OH)_3$ within the pH range of this investigation. The catalytic effect of ferric hydroxide was supported by the incipient addition of ferric iron, as illustrated in Fig. 5. Similarly, Fig. 6 indicates the catalytic effect of ferric hydroxide when alkalinity is 3×10^{-2} eq/ℓ. The kinetics of ferrous iron oxidation in the catalytic stage (Stage II. in Fig. 4) can be formulated in Fig. 7, is examined, where the plot of $\log(\{Fe(II)\}/\{Fe(III)\})$ versus time yields fairly straight lines, particularly in stage II. Thus, one can mathematically write that

$$\ln \frac{\{Fe(II)\}}{\{Fe(III)\}} = -K_o^1 t - K_o^{11}$$

where K_o^1 and K_o^{11} are constants.

If Eq. (3) is differentiated with respect to time, it becomes

$$\frac{1}{\{Fe(II)\}} \cdot \frac{d\{Fe(II)\}}{dt} - \frac{1}{\{Fe(III)\}} \cdot \frac{d\{Fe(III)\}}{dt} = -K_o^1$$

By assuming that the experiment is started with an initial addition of $\{Fe(III)\}_o$ and $\{Fe(III)\}_o$, then the concentration of ferric iron will be,

$$\{Fe(III)\} = \{Fe(III)\}_o - (\{Fe(II)\}_o - \{Fe(II)\}) \tag{5}$$

Substituting Eq. (5) into Eq. (4) and by rearranging it,

$$\frac{d\{Fe(II)\}}{dt} = \frac{K_o^1}{\{Fe(II)\}_o - \{Fe(III)\}_o} \{Fe(II)\}\{Fe(III)\} \tag{6}$$

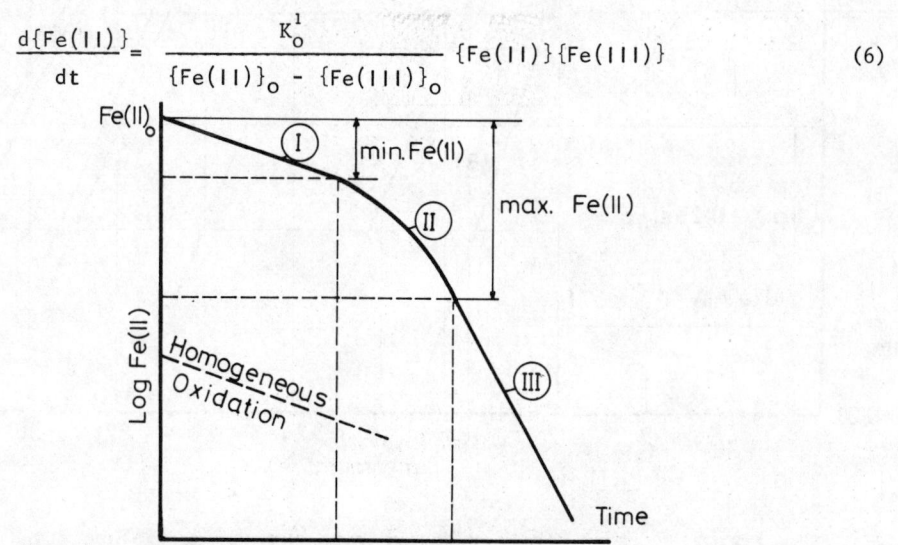

Fig. 4: A typical ferrous iron oxidation rate curve in heteregenous medium.

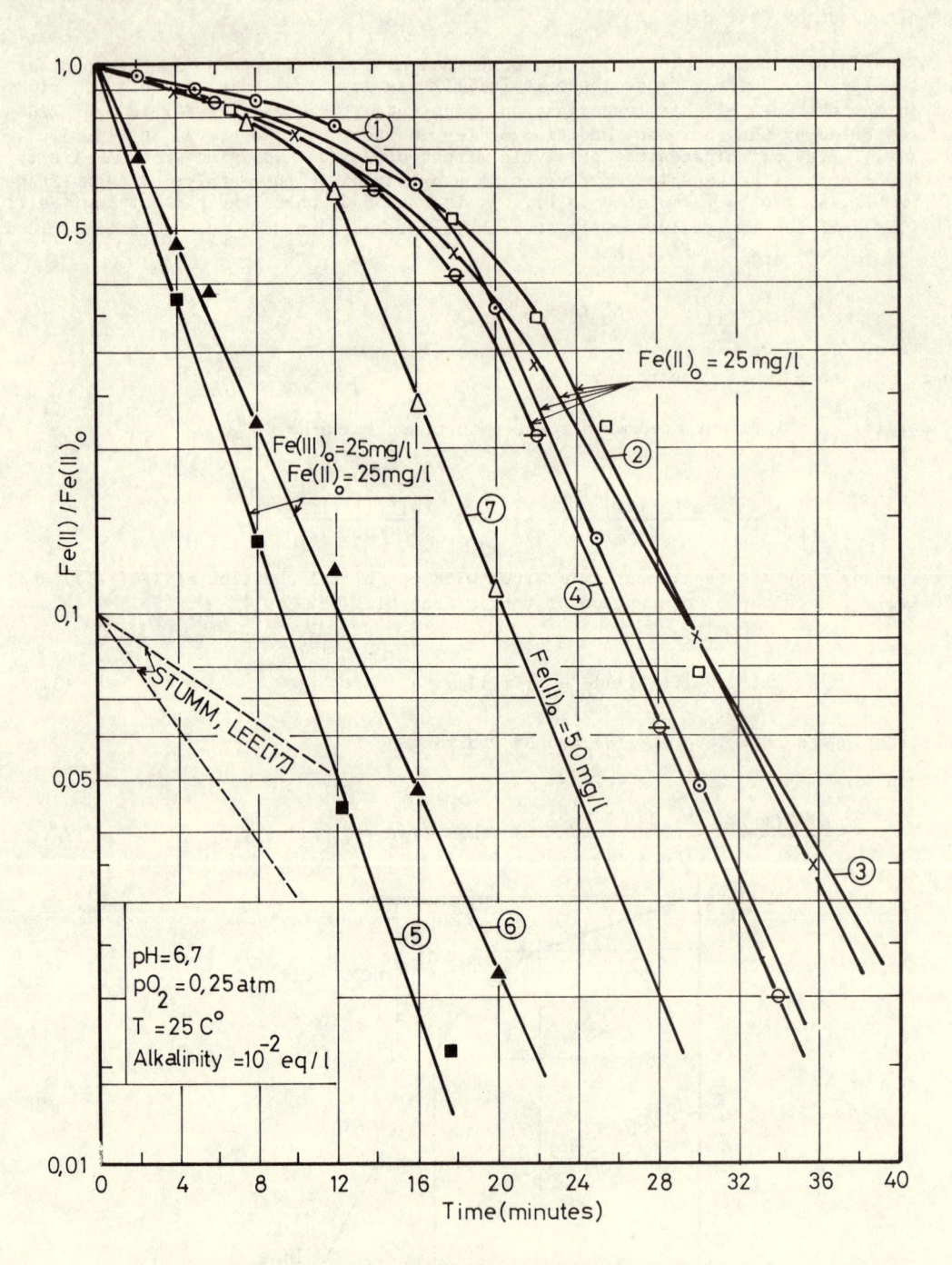

Fig. 5: Ferrous iron oxidation rate curves for the specified conditions

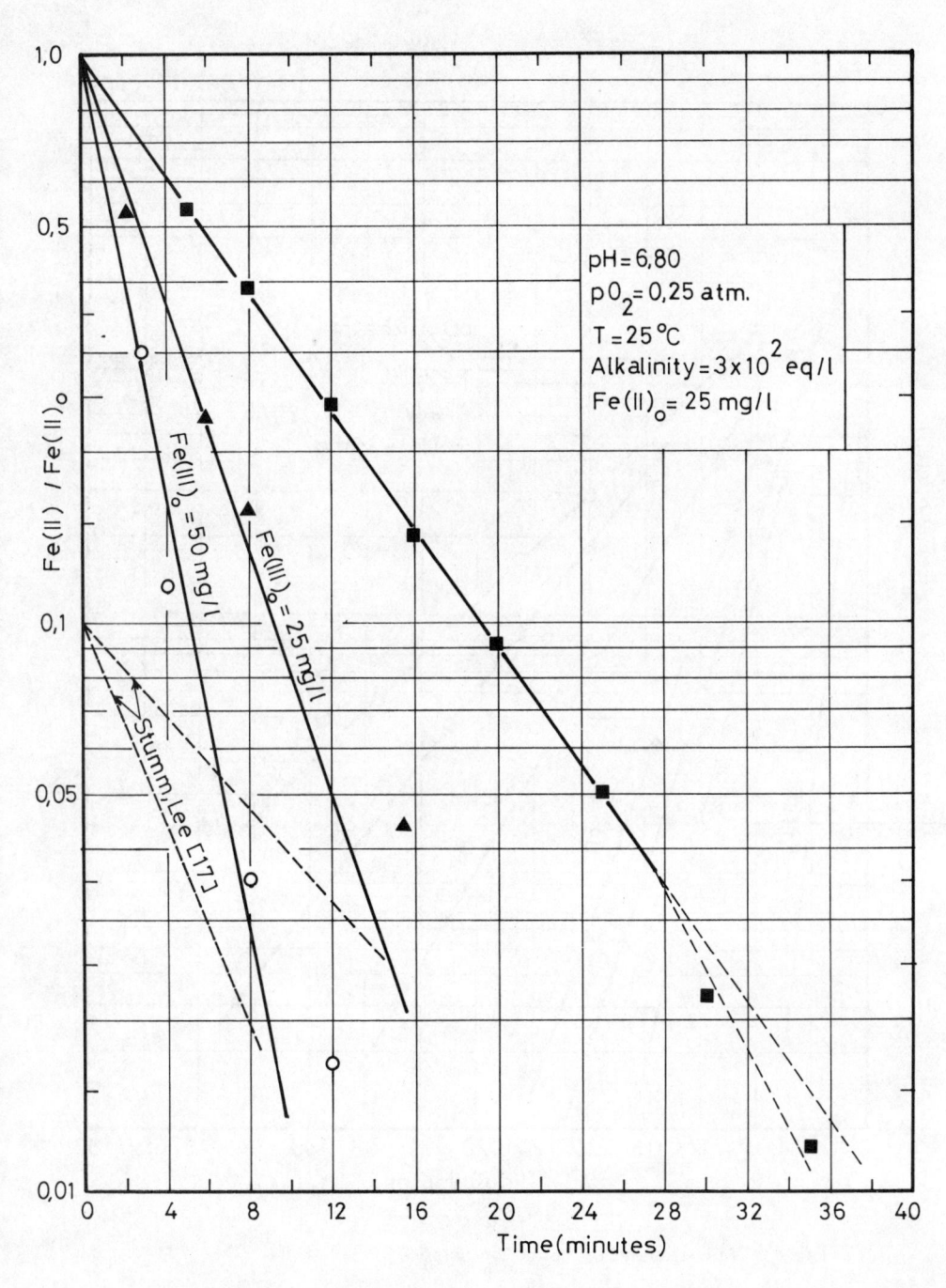

Fig. 6: Ferrous iron oxidation rate curves for the specified conditions

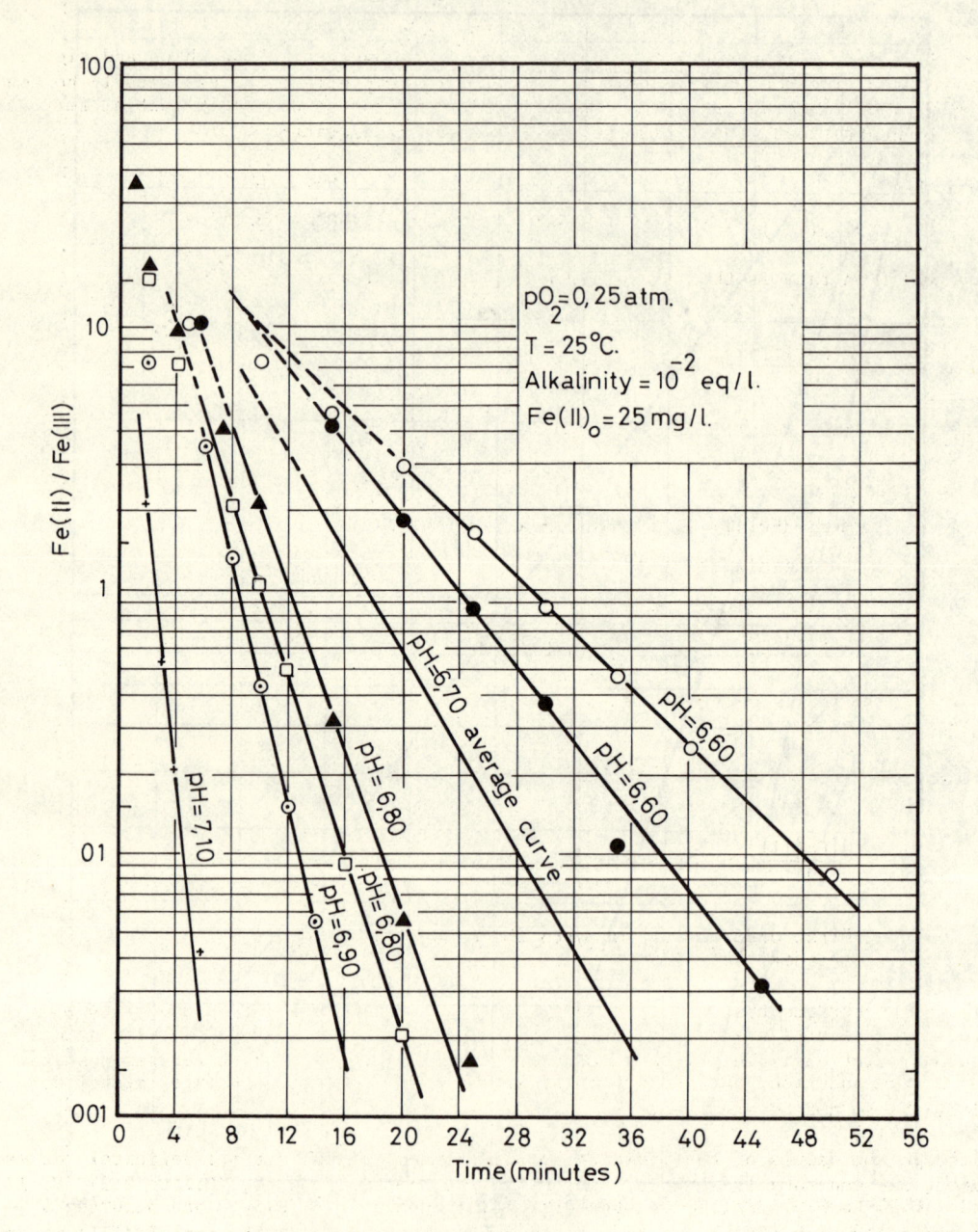

Fig. 7: Autocatalytic reaction rates

If the term in curly brackets in Eq. (6) is called overall autocatalytic reaction rate constant, K_o, Eq.(6) becomes

$$\frac{d\{Fe(II)\}}{dt} = -K_o \{Fe(II)\}\{Fe(III)\} \quad (7)$$

A maximum ferric iron concentration has also been observed, above which additional ferric hydroxide does not contribute to the catalytic effect, as can clearly be seen from Fig. 5. In order to summarize the oxidation process with high initial ferrous iron concentrations, the following three stages are distinguished:

Stage (I):

$$\{Fe(III)\} < Min. \{Fe(III)\}$$

The oxidation rate is the same as in homogeneous systems

$$\frac{d\{Fe(II)\}}{dt} = -k_o \{Fe(II)\} \quad (8)$$

where k_o is the overall first order oxidation rate constant.

Stage (II):

$$Min. \{Fe(III)\} < \{Fe(III)\} < Max.\{Fe(III)\}$$

Autocatalytic region

$$\frac{d\{Fe(II)\}}{dt} = -K_o\{Fe(II)\}\{Fe(III)\}$$

Stage (III):

$$\{Fe(III)\} > Max.\{Fe(III)\}$$

$$\frac{d\{Fe(II)\}}{dt} = -K_o \cdot Max. \{Fe(III)\}\{Fe(II)\} \quad (9)$$

First order oxidation with the rate constant equal to $(K_o \cdot Max. Fe(III))$.

Fig. 5 illustrates that as different oxidation curves are obtained under the same conditions, the results are not strictly reproducible, as they were in the case of homogeneous solutions. This irreproducibility is attributed to the heterogeneous character of the solution, but the effect of slight differences in the experimental conditions should not be excluded.

Although the values of the limiting Fe(III) concentrations are not definitely determined, one can roughly say that Min. Fe(III) concentrations are (0.2 to 0.3) $\{Fe(II)\}_o$, and (0.95 to 1.0) $\{Fe(II)\}_o$, when alkalinity equals 10^{-2}, $2 \cdot 10^{-2}$ and $3 \cdot 10^{-2}$ eq/ℓ, respectively, for $\{Fe(II)\}_o = 25$ mg/ℓ and pH about 6.7-7.1. The max. Fe(III) is approximately 17.5 to 20 mg/ℓ if Fig. 5 is considered. The catalytic effect of Fe(OH)₃ particles can be explained by the pH increase due to the specific adsorption of OH⁻ ions in the diffuse layer, since ferric hydroxide particles will have a positive charge under applied experimental conditions. The pH of the reaction solution is less than the zero point of charge of ferric hydroxide solution, which is about 8.5 (Stumm and Morgan, 1970). This possibility has also been reported by Holluta and Kölle (1964).

Alkalinity has an accelerating effect on the ferrous iron oxidation after a certain amount of alkalinity has been exceeded (Eg. (1)). But, while alkalinity increases oxidation, the minimum Fe(III) required to start the autocatalytic process increases too. This delaying effect of alkalinity is explained by the reduction of the diffuse layer thickness with increasing alkalinity.

In order to determine the pH dependence of the autocatalytic reaction rate, the constant, $K_o^!$, is plotted versus pH in Fig. 8, and a fairly straight plot is obtained. The slope of this line establishes the relation between $K_o^!$ and hydroxide ion concentration as second order

$$K_o^! \; \alpha \; \{OH^-\}^2 \tag{10}$$

However, one must keep in mind that Eq. (10) is valid only within a pH range of 6.6-7.0, and extrapolation to values beyond this range needs further study.

EFFECT of PHOSPHATE on FERROUS IRON OXIDATION

The accelerating effect of phosphate on ferrous iron oxidation in homogeneous bicarbonate solutions was reported in previous studies (Stumm and Lee, 1961). Ferrous iron oxidation experiments were carried out in the presence of ortho-phosphates to check whether the accelerating effect of phosphate was also observed in heterogeneous bicarbonate solutions created by high initial Fe(II) concentrations. Several experiments were done with phosphate; however, only three of them are illustrated in Fig. 9. Although phosphate accelerated the oxidation rate at the initial state of oxidation, the later retarding effect of phosphate was observed (Curve 2 and 3 in Fig. 9). This was not the case with relatively small initial Fe(II) concentrations (Curve I in Fig.9).

The accelerating effect of phosphate has already been explained by the strong affinity of Fe(II) to PO_4^{3-} anion. Since experiments proved that there was no significant ferrous phosphate precipitation within the oxidation time, as was also the case with $FeCO_3$, attention was focused on the surface charge of the mixed ferric hydroxo-phosphate precipitates. The surface charge of the precipitate was measured by a zeta meter (Rank-Brother Eng.) and the negative surface charge was found for the pH range of experiments. The retarding effect of phosphate may be explained by this negative surface charge of the precipitate. Due to specific adsorption of the H^+ and Fe(II) species in the diffuse layer, ferrous iron will be concentrated in the layers where pH is lower than the pH of the solution. Thus, the oxidation rate of ferrous iron will be lowered. Since the retarding effect of phosphate will be proportional to the amount of mixed ferric hydroxo-phosphate precipitate, whose amount depends on the Fe(III) oxidized from Fe(II). The oxidation kinetics of Fe(II) in the presence of phosphate can be expressed as follows:

$$\frac{d\{Fe(II)\}}{dt} = -K_p \frac{\{Fe(II)\}}{\{Fe(III)\}} \tag{11}$$

where K_p is the overall oxidation rate constant in the presence of phosphate. If oxidized Fe(II) is the only source of Fe(III), one can write

$$Fe(III) = Fe(II)_o - Fe(II) \tag{12}$$

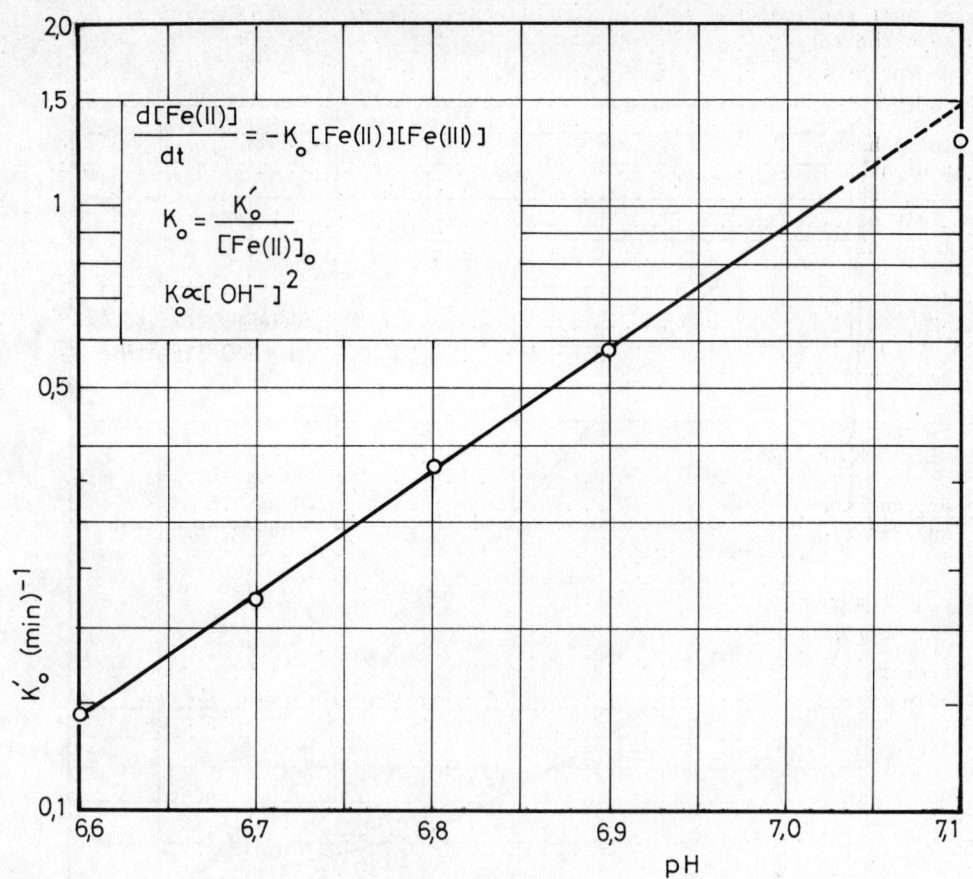

Fig. 8: pH dependence of autocatalytic reaction rate constant

Substituting Eq. (12) into Eq. (11)

$$\frac{d\{Fe(II)\}}{dt} = -K_p \frac{\{Fe(II)\}}{\{Fe(II)_o - Fe(II)\}} \tag{13}$$

or by integration, the following relationship is obtained:

$$\frac{Fe(II)}{Fe(II)_o} - \ln \frac{Fe(II)}{Fe(II)_o} - 1 = \frac{K_p}{\{Fe(II)_o\}} t \tag{14}$$

If Eq.(11) is valid, one should obtain a linear plot between the left hand side of Eq.(14) and time. Indeed, when experimental results are plotted in Fig. 10, approximately linear relationsihps are obtained for the data corresponding to the retarding phase of ferrous iron oxidation.

If we consider the practical impact of Eq. (11), we should allow a longer aeration time to utilize practically all the Fe(II) added to remove phosphate from the wastewater.

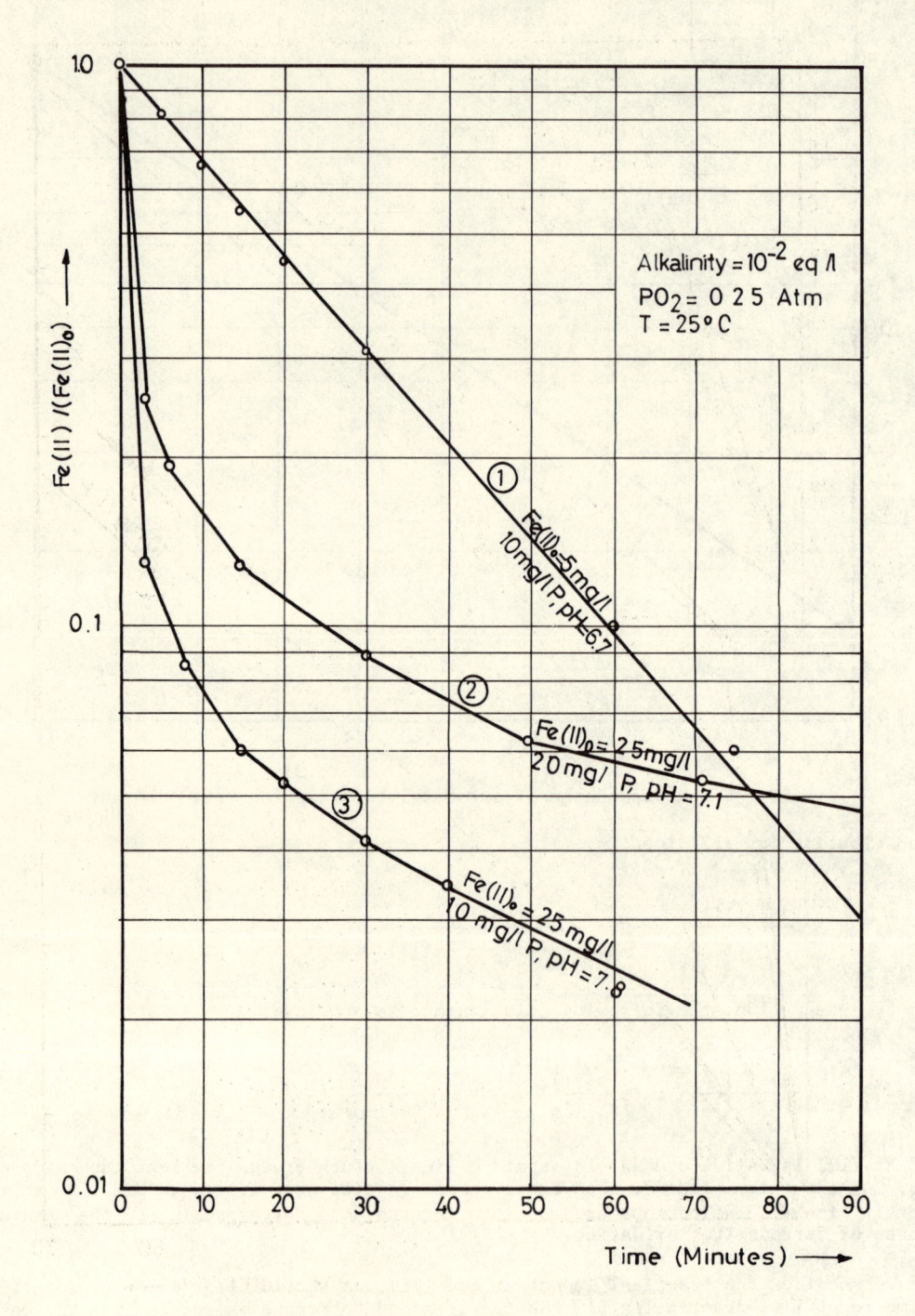

Fig. 9: Effect of phosphate on ferrous iron oxidation

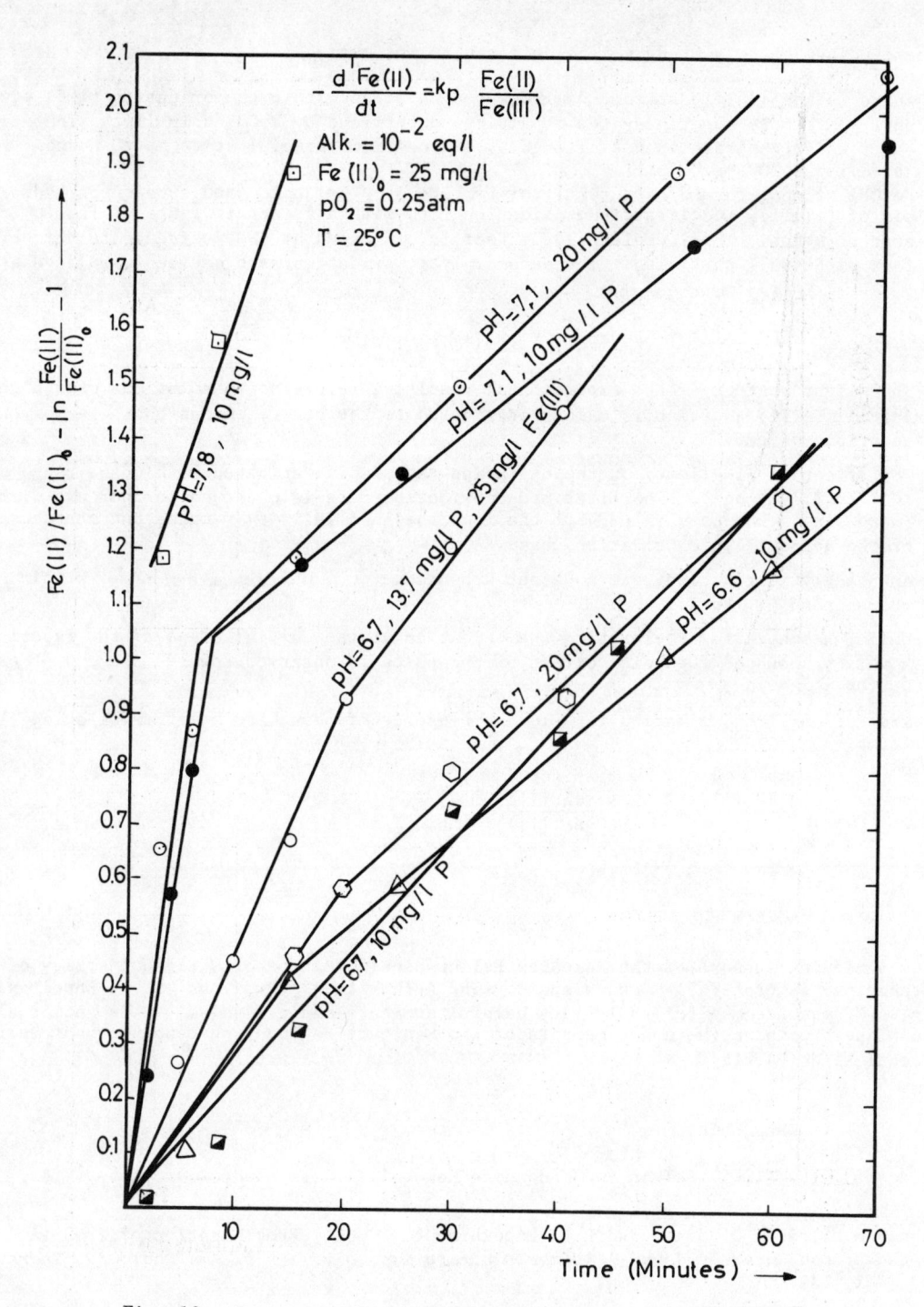

Fig. 10: Ferrous iron oxidation kinetics in the presence of phosphate

CONCLUSIONS

The findings of this study can be outlined as follows:

- There is no significant ferrous iron precipitation within the oxidation time, even under extremely supersaturated conditions determined by $FeCO_3$ solubility. The first order oxidation rate is not altered by ferrous iron precipitation at pH<8 and $\{Fe(II)\}_o < 100$ mg/ℓ.
- $Fe(OH)_3$ flocs, formed as a result of ferrous iron oxidation and subsequent hydrolysis of Fe(III), accelerate the oxidation rate within a certain range of ferric iron concentration. This accelerating effect is explained as the surface catalytic effect of $Fe(OH)_3$ particles. The autocatalytic oxidation rate can be formulated as

$$\frac{d\{Fe(II)\}}{dt} = -K_o \{Fe(II)\}\{Fe(III)\}$$

- The slight scatter of the experimental results in bicarbonate solutions with high initial Fe(II) addition is mainly attributed to the heteregeneous character of the reaction medium.
- The effect of alkalinity on the oxidation rate is investigated within the range of 10^{-2} to 3.10^{-2} eq/ℓ. The first order oxidation rate is observed to increase with increasing alkalinity value, but the increase in alkalinity retards the starting of the autocatalytic oxidation phase.
- Autocatalytic reaction rate constant dependence on $\{OH^-\}$ is of second within a pH range of 6.7 to 7.0.
- While phosphate is accelerating the oxidation at the initial stage of the reaction, a very pronounced retarding effect of phosphate is observed towards the end of oxidation.
- The ferrous iron oxidation rate in the presence of phosphate has the following form:

$$\frac{d\{Fe(II)\}}{dt} = -K_p \frac{\{Fe(II)\}}{\{Fe(III)\}}$$

ACKNOWLEDGEMENTS

The financial support of the Research Fellowships Committee of Delft University of Technology is gratefully acknowledged. The author thanks Prof. Dr. H. J. Pöpel and Ir. A.N. van Breemen for their very helpful comments and criticisms. The author also wishes to thank the other members of the Sanitary Engineering Laboratory of Delft Technical University.

REFERENCES

Cleasby J. L. (1975). Iron and Manganese Removal -A Case Study *Journal A.W.W.A.*, *67*, 147.

Ghosh M.M., J.T. O'Connor and R.S. Engelbrecht (1966). Precipitation of Iron in Aerated Ground Waters. *Journal of the Sanitary Eng. Div., Proc. Am. Soc. Civ. Engrs.*, *92*, SA1, 199.

Ghosh M.M., J.T. O'Connor and R.S. Engelbrecht (1967). Bathophenanthroline Method for the Determination of Ferrous Iron. *Journal A.W.W.A.*, *59*, 878.

Holluta V.J. and W. Külle, (1964). Über die Oxydation von zweiwertigem Eisen durch Luftsauerstoff. *GWF, 105,* Heft 18, 471.

Just G. (1908). Kinetic Investigation of the Autoxidation of Ferrous Bicarbonate in Aqueous Solution. *Zeitschrift für Physikalishe Chemie, 63,* 385.

Jobin R. and M.M. Ghosh (1972). Effect of Buffer Intensity and Organic Matter on the Oxygenation of Ferrous Iron. *Journal A.W.W.A., 64,* 590-595.

Kolthoff I.M. and P.J. Elving (1963). *Treatise on Analytical Chemistry.* Part 1, Vol. 1. Wiley-Interscience, New York.

Latimer W.M. (1952). *Oxidation Potentials.* Prentice-Hall, New York.

Lee G.F. and W. Stumm (1960). Determination of Ferrous Iron in the Presence of Ferric Iron with Bathophenanthroline. *Journal A.W.W.A., 52,* 1567.

Lerk C.F. (1965). Enkele Aspectan Van de Ontijzering Van Grondwater. Ph.D. Thesis (chapter 2). Uitgeverij Waltman, Delft.

Morgan J.J. and F.B. Birkner (1966). Discussion of "Precipitation of Iron in Aerated Ground Waters" by Ghosh M.M., O'Connor J.J. and Engelbrecht R.S. *Jour. San. Eng.Div., Proc. Am. Soc. Civ. Engrs. 92,* SA6. 137.

Morgan J.J. and W. Stumm (1964). The Role of Multivalent Metal Oxides in Limnological Transformations as Exemplified by Iron and Manganese. *Advances in Water Pollution Research,* Proc. of Second Int. Conf., Tokyo. p. 103.

Olson L.L. and C.J. Twardowski (1975). $FeCO_3$ and $Fe(OH)_3$ Precipitation in Water-Treatment Plants, *Journal A.W.W.A. 67,* 150.

Schenk J.E. and W.J. Weber (1968). Chemical Interactions of Dissolved Silica with Iron (II) and (III). *Journal A.W.W.A., 60,* 199.

Singer, P.C. and W. Stumm (1970). The Solubility of Ferrous Iron in Carbonate-Bearing Waters. *Journal A.W.W.A., 62,* 198.

Smith J.J. (1918). On the Equilibrium in the System: Ferrous Carbonate, Carbon Dioxide and Water. *Journal of the Am. Chem. Soc., 70,* 879.

Stumm W. and G.F. Lee (1961). Oxygenation of Ferrous Iron. *Industrial and Engineering Chemistry, 53,* 143.

Stumm W. and G.F. Lee (1960). The Chemistry of Aqueous Iron. *Schweizerische Zeitschrift für Hydrologie,* 295.

Stumm W. and J.J. Morgan (1970). *Aquatic Chemistry.* Wiley - Interscience, New York.

Theis T.L. and P.C. (1973). The Stabilization of Ferrous Iron by Organic Compounds in Natural Waters, In P.C. Singer (Ed.) *Trace Metals and Metal-Organic Interactions in Natural Waters.* Ann Arbor Science Publishers, Ann Arbor, Mich.

Weiss J. (1953). The Autoxidation of Ferrous Ions in Aqueous Solution, *Experianta, 9,* 61.

Winklehaus C., F.A. Digiano and W.J. Weber, Jr. (1966). Discussion of Precipitation of Iron in Aerated Ground Waters, by Ghosh M.M., O'Connor J.T. and Engelbrecht R.S. *Jour. San. Eng. Div., Proc. Am. Soc. Civ. Engrs. 92,* SA6, 129.

DISCUSSIONS

Atımtay, A. (Turkey) : It seems that you added to the reaction vessel a fixed amount of phosphate solution. Did you work with different phosphate concentrations?

Sarıkaya, H. (Turkey) : The various P concentrations used can be found in Fig.10. These ranged from 10 to 20 mg/ℓ.

Atımtay, A. (Turkey) : Was the pH of the system controlled manually or automatically?

Sarıkaya, H. (Turkey) : The pH of the system was controlled by two methods: for a pH below 6.8 or 6.9, by changing the CO_2 flow, which did not affect the partial pressure of oxygen; or parallel to this, by adding an acid or base solution when the pH was 7 or 7.1. In the latter case, a delay of 30 sec. in regulating the pH could lead to considerable error, since the oxidation time was about 5 min. Such a delay was not so significant when the pH was 6.7 or 6.8, because then the oxidation time was about one hour.

Effect of Recycle Ratio on the Activated Sludge Treatment of Metal-Containing Wastes

D. ORHON and O. TÜNAY

Maçka Civil Engineering Faculty, Technical University of Istanbul, Turkey

ABSTRACT

In this study, the dynamic modelling of the activated sludge process treating metal-containing wastes is developed. The adverse effects of metallic ions on the system are investigated using this model. Control of the adverse effects is achieved by adjusting the recycle ratio and wastage ratio of the system.

INTRODUCTION

The activated sludge process is effectively used to treat metallic industrial wastes as well as mixtures of such wastes with domestic sewage. Since industrial waste discharges generally follow irregular patterns, the evaluation of the response of the process to metallic waste transients is of great importance as far as predicting the performance of the process is concerned. It is far more important, however, to have a full understanding of the complex removal mechanism so that the adverse effects of these transients on the process may be controlled.

Metallic wastes, especially heavy metal salts, are very common in most industrial wastes, i.e. metal-finishing industries, tanneries, etc. The concentration of metal ions in these wastes is usually far above the safe limit, beyond which severe adverse effects are experienced in biological treatment (*Koziorowski and Kucharski*, 1972; *Barth and others*, 1965). The normal steady-state operation of the activated sludge process is very difficult to maintain when treating industrial wastes, since they exhibit substantial time fluctuations in both flow rate and waste-water strength. The nature of the fluctuations is often closely related to the industrial process concerned.

This paper attempts to illustrate how and to what extent benefits can be gained by manipulation of the main operational parameters, namely, recycle rate and wastage rate, when the system is subjected to inhibitory transients. This approach inevitably calls for a full grasp of the behavior of activated sludge under metallic waste transient loadings. A dynamic modelling reflecting the inhibitory effects of such transients on the activated sludge process is also presented as the reference point of evaluation.

DYNAMIC MODELLING OF INHIBITION
FOR THE ACTIVATED SLUDGE PROCESS

Physical Interpretation of the
Metal Inhibition Mechanism

The adverse effects of metallic ions on the activated sludge process are mainly attributed to their growth-hindering action on the suspended biomass. Inhibition is triggered off by the blocking effect of inhibitory substances on the active enzyme sites, essential for their catalytic activity in the conversion of the substrate to metabolic products. This phenomenon is reflected in an overall decrease in the growth rate of the system. Enzyme-inhibitor attachment may occur competitively or non-competitively, depending on the type of inhibition. The extent of inhibition depends upon the substrate and inhibitor concentrations as well as their affinity to the enzyme in the case of competitive inhibition, whereas the inhibitor concentration alone governs non-competitive inhibition, regardless of the substrate concentration present in the system. The majority of metallic ions are non-competitive inhibitors, except for a few which form organo-metallic complexes.

Mathematical Interpretation of
the Metal Inhibition Mechanism

The multitude of mathematical models developed to reflect the substrate removal mechanism in the activated sludge process solely depend on the substrate as the unique growth-controlling factor. While the effluent quality and substrate reduction efficiency of different systems may be predicted within acceptable approximations when no hindering effect is present, substantial discrepancies immediately appear in the prediction potential when inhibitors are introduced. This is to be expected since these models simply lack the basic parameters to reflect changes occurring in the removal mechanism due to the effects of inhibition.

Any mathematical attempt to define inhibition should start with an adequate description of the relevant reaction rate. It is shown that the basic enzyme kinetics

$$E + s \rightleftarrows |E.s| \rightleftarrows E + P$$
$$E + s \rightleftarrows |E\,I|$$
(1)

can be taken as a reference point for formulating the reaction rate equation for non-competitive inhibition (*Tunay*, 1978):

$$\mu = \left[\frac{k_i}{k_i + I}\right] \hat{\mu} \frac{s}{K_s + s}$$
(2)

where

$$k_i = \frac{|E|\,|I|}{|E.I|}$$

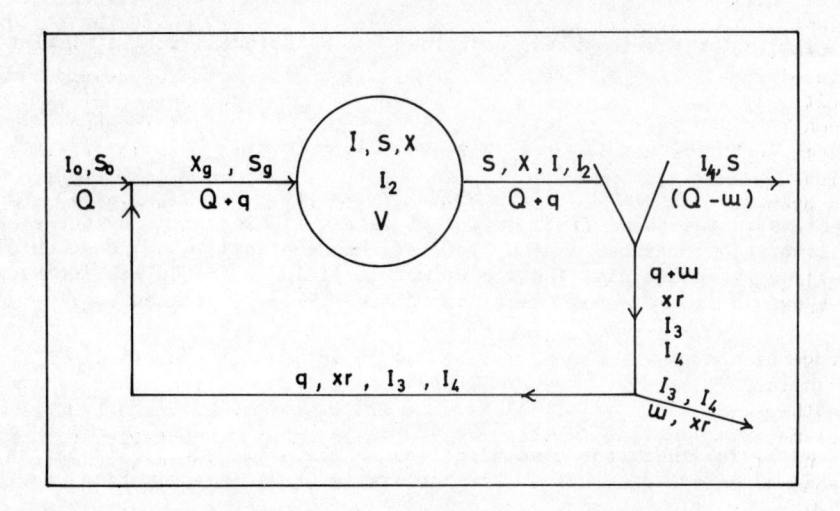

Fig.1. Activated Sludge Process Treatment Scheme

When the above rate equation is incorporated into the mass balance with respect to the process schematized in Fig.1., the relevant continuity equation for the substrate can be formulated as follows:

$$V \frac{ds}{dt} = s_0 Q - (Q+q) s - V \frac{\hat{\mu} k_i}{Y k_i + I} \frac{sx}{K_s + s} \qquad (3)$$

given that $r = q/Q$ and $\bar{t} = V/Q$, Eq. (3) becomes

$$\frac{ds}{dt} = \frac{s_0}{\bar{t}} - \frac{s(1+r)}{\bar{t}} - \frac{\hat{\mu} k_i}{Y k_i + I} \frac{sx}{K_s + s} \qquad (4)$$

A similar approach for biomass continuity with respect to the process yields

$$\frac{dx}{dt} = \frac{r}{\bar{t}} x_r - \frac{(1+r)}{\bar{t}} x + \hat{\mu} \frac{k_i}{k_i + I} \frac{sx}{K_s + s} - k_d x \qquad (5)$$

given the following mass balance for the settling basin,

$$x_r = \frac{1+r}{r+b} \qquad (6)$$

where

$$b = w/Q$$

Eq. (5) is converted into

$$\frac{dx}{dt} = \frac{r(1+r)}{\bar{t}(r+b)}x - \frac{(1+r)}{\bar{t}}x + \hat{\mu}\frac{k_i}{k_i+I}\frac{sx}{K_s+s} - k_d x \qquad (7)$$

The mass balance for the inhibitor concerned can be formulated in a similar way as follows:

$$V\frac{dI}{dt} = I_0 Q + (I_3 + I_4)q - (q+Q)I - (q+Q)I_2 \qquad (8)$$

Investigations on the phase distribution of metals in biological cultures have shown that metals are concentrated in the biomass. Mathematical interpretation of experimental work on the phase distribution of metals yields a Freundlich isotherm type equation (*Neufeld and Hermann*; *Cheng and others*, 1975), expressed as

$$\frac{I_b}{I_1} = cx^k \qquad (9)$$

where I_b and I_1 are inhibitor concentrations in the solid and liquid phases, respectively, and c and k are constants. For the system under investigation, Eq.(9) can be written as

$$\frac{I_2}{I} = cx^k \qquad (10)$$

and

$$\frac{I_3}{I_4} = cx_r^k \qquad (11)$$

On the other hand, I_3 and I_4 can be expressed in terms of I and/or I_2 establishing another mass balance for the settling basin as follows:

$$(q+Q)I + (q+Q)I_2 = (Q-w)I_4 + (q+w)(I_3+I_4) \qquad (12)$$

Using Eqs. (10), (11) and (12), Eq. (8) can be rewritten thus:

$$\frac{dI}{dt} = \frac{I_0}{\bar{t}} - \frac{r(1+cx_r^k)(1+cx_r^k)I}{t\left[1+\frac{r+b}{1+r}cx_r^k\right]} - \frac{(1+r)(1+cx_r^k)I}{\bar{t}} \qquad (13)$$

Eq.(4), (7) and (13) define the time-dependent, in other words, dynamic behavior of the activated sludge process under inhibitory transient loadings.

OPERATIONAL CONTROL OF INHIBITION EFFECTS

Control Parameters

The efficiency of the activated sludge process depends on sludge age, θ_c which is a widely accepted design and control parameter. Severeal investigators have shown that θ_c has a substantial effect on the degree of inhibition in the system. This effect is more pronounced in the case of non-competitive inhibition (*Tunay*, 1978; *Neufeld and Hermann*, 1975).

In a conventional activated sludge system, θ_c can easily be controlled by changing either the recycle ratio, r, or the sludge wastage ratio, b. Sometimes a combination of changes in both r and b may be appropriate.

Theoretical Evaluation of
Inhibition Effects

In this study, inhibition effects on the activated sludge process are evaluated employing the dynamic modelling of the system presented in this paper. For this purpose a system working with assumed values of the kinetic and operational parameters is envisaged. Nickel ion is chosen as the inhibitor present in the influent sewage because of its strong inhibitory effect and the availability of its phase distribution data in the literature. In Table 1, the assumed parameter values used in the computations are shown.

Table 1: Kinetic and Operational Parameters of the Activated Sludge System as used in the Computations

μ d^{-1}	K_s mg/l	θ_c d	Y	$\bar{t}$ h	k_d d^{-1}	r	k_i mg/l	s_0 mg/l	c	k
0.5	100	10	0.5	8	0.05	0.3	2.3	400	0.72	1.3

To evaluate both the response of the system to inhibitor loading and the operational parameter changes required to compensate or at least to decrease the adverse effect due to this loading, the following procedure is used.

First, a 1.5 mg/ℓ concentration of nickel is introduced as a step loading to the system. After a new steady state has been reached, two different approaches are taken into consideration to adjust the system so as to diminish the efficiency decrease taking place in the system. Firstly, it is proposed to return the system to the effluent substrate concentration value obtained under the initial steady-state conditions. Secondly, an intermediate effluent substrate concentration value is chosen in such a way that the necessary adjustment of the parameters to reach this value can be more readily carried out within the system and this new concentration will be less or equal to the standards of the region concerned.

Discussion of the Results and Conclusions

An activated sludge system operated with the assumed values of the parameters has an effluent substrate concentration of 42.8 mg/ℓ under steady-state conditions. Following the inhibitor step loading of 1.5 mg/ℓ, an efficiency decrease of from 89.3% to 80.9% is observed when the effluent substrate concentration increases to 76.4 mg/ℓ under new steady-state conditions. Thereafter, the values of b and r are changed from 0.008 to 0.004 and from 0.3 to 0.774, respectively, to obtain the effluent substrate concentration of 42.8 mg/ℓ (Fig.2). As a second approach, the value of b and r are changed to 0.005 and 0.44, respectively, to obtain a 50 mg/ℓ effluent substrate concentration (Fig.3).

The presence of inhibitory wastes in industrial and domestic wastewaters causes various important problems in biological treatment. The decrease in substrate removal efficiency is of primary importance. As can be seen from the above examples, the decrease in substrate removal efficieny can to a certain extent be maintained at resonable levels by making use of the appropriate control mechanisms within the system.

However, another problem is that the system response to any change in both influent characteristics and operational parameters is rather slow. This fact can easily be seen from the time lapse in reaching the two steady states of the system. For instance, the first steady-state was attained following inhibitor transient loading after about one month, while the second steady-state was reached after r or b changes in 20 days. These periods may be too long for temporary solutions to inhibitory transients. However, they seem to be feasible in the case of long-term inhibitor loading for which adjustments may require to be made over a longer period. Furthermore, when short-term inhibitor loading is expected, these adjustments can also be carried out within a very short time if much greater changes than necessary are made in the r and b values.

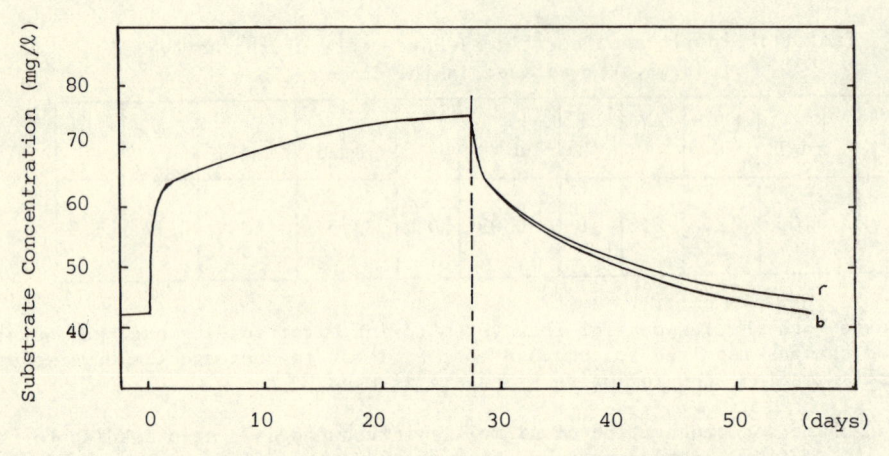

Fig. 2: Response of system to inhibitor step loading and changes in r and b.

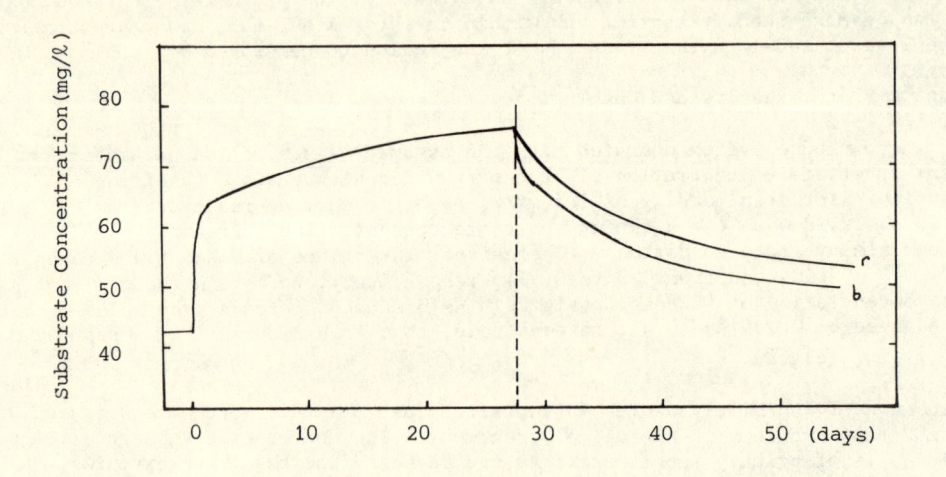

Fig. 3: Response of system to inhibitor step loading and changes in r and b.

NOTATION

b	:	Wastage ratio
c	:	Inhibitor distribution constant
E	:	Enzyme
E.I.	:	Enzyme-inhibitor complex
E.s.	:	Enzyme-substrate complex
I_b	:	Solid phase inhibitor concentration, M/L^3
I_1	:	Liquid phase inhibitor concentration, M/L^3
I_0	:	Influent inhibitor concentration, M/L^3
I	:	Reactor inhibitor concentration (liquid phase), M/L^3
I_2	:	Reactor inhibitor concentration (solid phase), M/L^3
I_3	:	Settling basin inhibitor concentration (liquid phase), M/L^3
I_4	:	Settling basin inhibitor concentration (solid phase), M/L^3
k	:	Inhibitor distribution constant
k_d	:	Activated sludge decay coefficient, $1/T$
k_i	:	Enzyme-inhibitor dissociation constant, M/L^3
K_s	:	Half velocity constant, M/L^3
P	:	Enzyme product
Q	:	Influent flow rate, L^3/T
q	:	Recycle flow rate, L^3/T
s	:	Reactor substrate concentration, M/L^3
s_0	:	Influent substrate concentration
s_g	:	Reactor feeding flow substrate concentration, M/L^3
$\bar{t}$	:	Reactor hydraulic retention time
x	:	Reactor activated sludge concentration, M/L^3
x_e	:	Effluent activated sludge concentration, M/L^3
x_g	:	Reactor feeding flow activated sludge concentration, M/L^3
x_r	:	Recycle flow activated sludge concentration, M/L^3
V	:	Reactor volume, L^3
w	:	Wastage flow rate, L^3/T
Y	:	Yield coefficient
θ_c	:	Sludge age, T
μ	:	Activated sludge net growth rate, $1/T$
$\hat{\mu}$	:	Activated sludge maximum growth rate, $1/T$

REFERENCES

Barth, E.F. and others (1965). Summary Report on the Effects of Heavy Metals on the Biological Treatment Process. *J. Water Poll. Control Fed.*, 37, 1.

Cheng, M. and others (1975). Heavy Metals Uptake by Activated Sludge. *J. Water Poll.*

Control Fed., 47, 2.

Koziorowski, B. and J. Kucharski (1972). *Industrial Waste Disposal*. Pergamon Press, London.

Neufeld, R.D. and E.R. Hermann (1975). Heavy Metal Removal by Acclimated Activated Sludge. *J. Water Poll. Control Fed.*, 42, 2.

Tünay, O. (1978). *Effect of Inhibition on the Activated Sludge Process*. Ph.D. Thesis. Technical University of Istanbul.

Laboratory Scale Studies as a Design Aid for Industrial Wastewater Treatment Plants

I. SEKOULOV

University of Stuttgart, Stuttgart, Germany

ABSTRACT

In spite of the development of a great number of theoretical and mathematical design methods in the past years, the layout of a wastewater treatment plant is still based primarily on experimental data.

An engineer has to rely on design parameters collected from tests carried out on the wastewater.

Trials on different scales (laboratory, pilot-and full scale plant), closely connected with the questions which have to be solved, might be necessary.

The construction of models according to the theory of hydraulic similarity is not feasible here. But with laboratory scale tests run under optimum conditions, a proposed wastewater purification method can be closely followed.

This paper deals with a laboratory study used to test a treatment plant layout, designed for purifying wastewater from a factory producing fertilizers. The results confirmed the chosen first stage- biological treatment for 8 hours in a completely mixed tank with a pure oxygen supply. The second stage -biological treatment for 24-hours in a completely mixed tank with normal aeration - failed. Nutrification was only 20 to 40% of the expected efficiency. Thus, for the second stage a fixed bed reactor was proposed.

The use of laboratory studies before the construction of a treatment plant for industrial waste can avoid a lot of disappointments and unnecessary expenses.

INTRODUCTION

In the past years a great number of mathematical models and theoretical design methods for sewage and industrial wastewater treatment plants have been developed. They have helped to give considerable insight into the biochemical purification processes, but the final layout of a plant is still based primarily on experimental data.

This is because the elaborated design methods can be successfully applied only to the purification of sewage or industrial wastewater with components having similar and well-known biodegradation characteristics. Additional parameters, such as climate and type of sewage collection system, have also to be taken into account. In fact, it is unlikely that all these conditions will be satisfied. This makes direct use of foreign design experience and data rather limited. The existing theories and models are not predictive without experimental data. Consequently, the engineer cannot design a treatment plant without applying coefficients, which have to be determined empirically on the wastewater concerned. He has to rely in all new cases on design parameters collected preferably from pilot studies in situ. If large-scale trials are unfeasible from a technical or economical point of view, laboratory scale tests can be also very helpful. They are carried out mainly as a preliminary study to avoid sole hazard, but an improvement in plant efficiency and a reduction in total costs are also possible at this stage.

This paper will deal with the laboratory device used to check the layout of a biological treatment plant, designed to purify wastewater from a factory producing fertilizers. The results obtained will be discussed briefly.

BACKGROUND

Trials on different scales (laboratory, pilot plant and full scale), closely related to the engineering problems involved, are necessary. Because in this field methods for constructing scaling-up devices have not yet been developed, the desire to achieve reproducible results leads to the construction of pilot plants on as large a scale as possible (in extreme situations of real dimensions).

Due to difficulties such as unavailability of wastewater, transport and preservation of wastewater, as well as technical-economical considerations, there are not many opportunities for large-scale trials in practice. Pilot plants with a reaction tank of 1 to 2 m^3 are mostly used.

In general, experimental data from laboratory scale (1 to 200 ℓ) tests are thought to be not sufficiently predictive for the situation in practice. But laboratory scale tests are very often the only possible studies on the wastewater before building the treatment plant. This shows how important they can be. Therefore, it is necessary to learn how to construct and operate small devices which will prove reliable on an enlarged scale.

The simple operations for calculating the likelihood ratios of the hydraulic characteristics of the laboratory model to those of the technical device (Horvath, 1978) are valid only for one-phase fluid systems (e.g. water). The fluid conditions in biological reactors are two-phase (water and activated sludge) or three-phase systems (water, activated sludge and air). Thus, the construction of a biological laboratory reactor using the theory of hydraulic similarity is not feasible. This explains to some extent the great variability in the forms and volumes of laboratory equipment used.

The questions which arise here are, which parameters measured during a wastewater purification process on a laboratory scale are representative also for a technical unit and whether there are rules for construction and operation which might help to improve the similarity between trial and practice.

Based on the experience of scaling-up models developed by bio-engineers, the parameters which are relatively independent of reactor volume and geometry can be summarized as follows:

- substrate removal rate
- growth rate of microorganisms
- respiration rate
- effluent quality
- pH changes due to biochemical reactions
- Self-developed microorganism species induced by the substrate (wastewater composition) and applied technology.

This means that primarily process rates or rates of changes in a biochemical reactor are reproducible. However, the reactor type (total mixed, plug flow, fixed bed, etc.), as well as environmental and operational conditions (temperature, oxygen concentration, oxygen supply, mixing, loading, hydraulic detention time, sludge retention time and sludge age must be simulated in laboratory studies as close as possible to the later situation in practice.

Prior to the construction and operating of laboratory devices, the system parameters must be divided into predominantly static (conservative) ones:

- reactor type
- temperature
- oxygen supply and concentration
- detention time
- loading (hydraulic and organic)
- sludge return ratio
- sludge age

and dynamic ones:

- mixing
- flow pattern
- diffusion
- share forces.

Although in some cases rather complicated automatic apparatus is necessary, adaptation of the static parameters to the desired real conditions is not difficult. The problems of similarity are primarily related to the dynamic parameters. These cannot be predicted or calculated. They have to be determined experimentally, not each parameter separately (this is not possible) but as an overall dynamic characteristic of the reactor. It is accepted that the residence time distribution curve (RTD) describes satisfactorily the desired dynamic conditions. This requires knowledge of the time distribution curve, not only of the laboratory unit but also of the later treatment plant device, which will be the basis of a similarity factor.

In contrast to bioengineering practice, where reactor types have well-known time distribution characteristics, in wastewater treatment practice, biological units are constructed rather arbitrarily and subjected to flow changes, generally without known dynamic parameters. Because of the lack of this information, laboratory units cannot be fitted to technical equipment and results have low predictivity. In such cases, models are run under optimal conditions to give some idea of the maximum possible biochemical and related reactions, but not for scaling-up.

In the future, much more attention will have to be paid to the dynamic conditions in biological treatment units, if success is desired with laboratory scale tests. Now, our models can be constructed with respect to all the static reactor parameters, using, for the dynamic characteristic, theoretical or literature data.

The distribution of the residence time in a reactor can be described quantitatively by two distribution functions which are closely related, the F - and the E- function (Fig. 1).

Although residence time distribution is of great importance to the design of process equipment, our discussion will be restricted to the principle. A more comprehensive treatment is given by Levenspiel (1972).

The residence time parameters are usually measured by trace stimulus response techniques. Uranin (samples are controlled photometrically at λ = 490 nm) or sodium chloride (samples are controlled with conductivity measurement) are applied as tracer. For easier determination of the retention time parameters, pulse response input of sodium chloride as tracer is preferred.

The mean residence time is:

$$\bar{t} = \frac{t}{\Theta} \tag{1}$$

where Θ is total residence time.

According to the E-function, the mean residence time is given by

$$\bar{t} = \frac{\int_0^\infty t \cdot c \cdot dt}{\int_0^\infty c \cdot dt} \tag{2}$$

If the distribution curve is only known at n discrete time values, t_i, then

$$\bar{t} = \frac{\Sigma t_i \cdot C_i \cdot \Delta t_i}{\Sigma C_i \cdot \Delta t_i} \quad \text{for } i = 1, 2, \ldots n \tag{3}$$

The next most important descriptive quantity is the spread of distribution. This is commonly measured by the variance, σ^2, defined as

$$\sigma^2 = \frac{\int_0^\infty (t-\bar{t})^2 \cdot C \cdot dt}{\int_0^\infty c \cdot dt} = \frac{\int_0^\infty t^2 \cdot C \cdot dt}{\int_0^\infty c \cdot dt} - \bar{t}^2 \tag{4}$$

again in discrete form,

$$\sigma^2 \simeq \frac{\Sigma t_i^2 \cdot C_i \cdot \Delta t_i}{\Sigma C_i \cdot \Delta t_i} - \bar{t}^2 \tag{5}$$

From the variance (spread) number (σ^2) of the pulse response, the dispersion number (Peclet, Bodenstein) can be calculated as follows:

$$\sigma^2 = 2 \left[D/u \cdot L \right] \tag{6}$$

where
- D = eddy diffusivity (called also longitudinal or axial dispersion coefficient)
- u = superficial flow velocity
- L = axial distance along the reactor

The dimensionless group $\left(\frac{D}{u \cdot L}\right)$ describes quite satisfactorily the flow pattern:

$\frac{D}{u \cdot L} \to 0$ negligible dispersion, hence plug flow

$\frac{D}{u \cdot L} \to \infty$ large dispersion, hence mixed flow

The use of scaling-up models for wastewater treatment plant design will save money and time. It is a pity that the construction of reliable laboratory units still remains underdeveloped. For this purpose, better knowledge of the residence time distribution in the laboratory and in the later technical device is necessary.

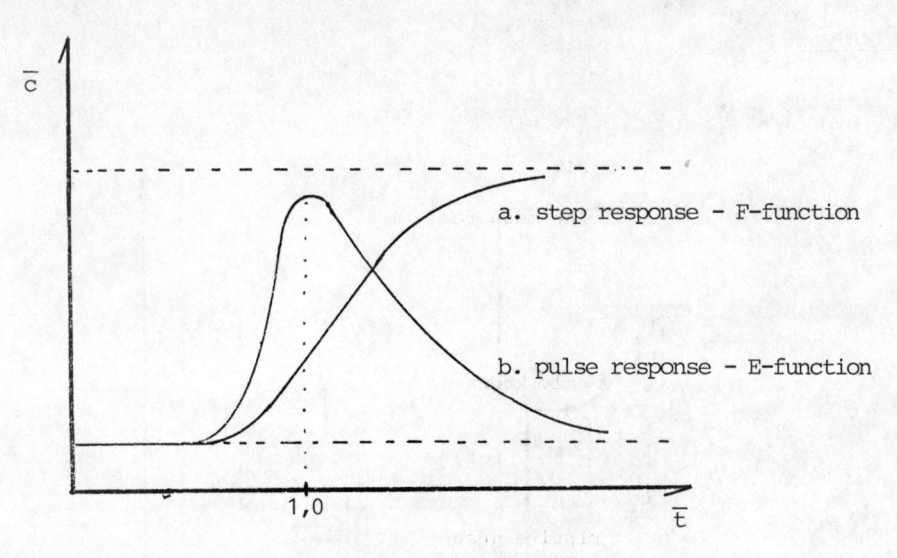

Fig. 1: Typical residence time distribution curves

a. step response - F-function
b. pulse response - E-function

WASTEWATER CHARACTERISTICS

In the designing of a factory producing fertilizers, a wastewater treatment plant also had to be provided. The expected process water and wastewater to be purified in the treatment plant were calculated by the technologist as follows:

Sewer A (*domestic sewage, storm water, industrial wastes*)

Q = 500 m³/h
F_L = 18 T/d BOD₅
L_o = 1500 mg/ℓ BOD₅
F_C = 45,6 T/d COD

L_C = 3800 mg/ℓ COD
F_N = 1920 T/d N-NH₄⁺
L_N = 160 mg/ℓ N-NH₄⁺
pH = 9 to 10

Sewer B (*process wastewater*)

Q = 108 m³/h
L_C = 40 mg/ℓ COD
F_N = 1,2 T/d N-NH₄⁺
L_N = 463 mg/ℓ N-NH₄⁺
t = 18°C

Requirements of the effluent quality are:

L_e = 75 mg/ℓ   BOD
L_C = 760 mg/ℓ COD
L_N = 16 mg/ℓ   N-NH₄⁺
L_s = 50 mg/ℓ   Suspended solids
pH = 6,5 to 8,5

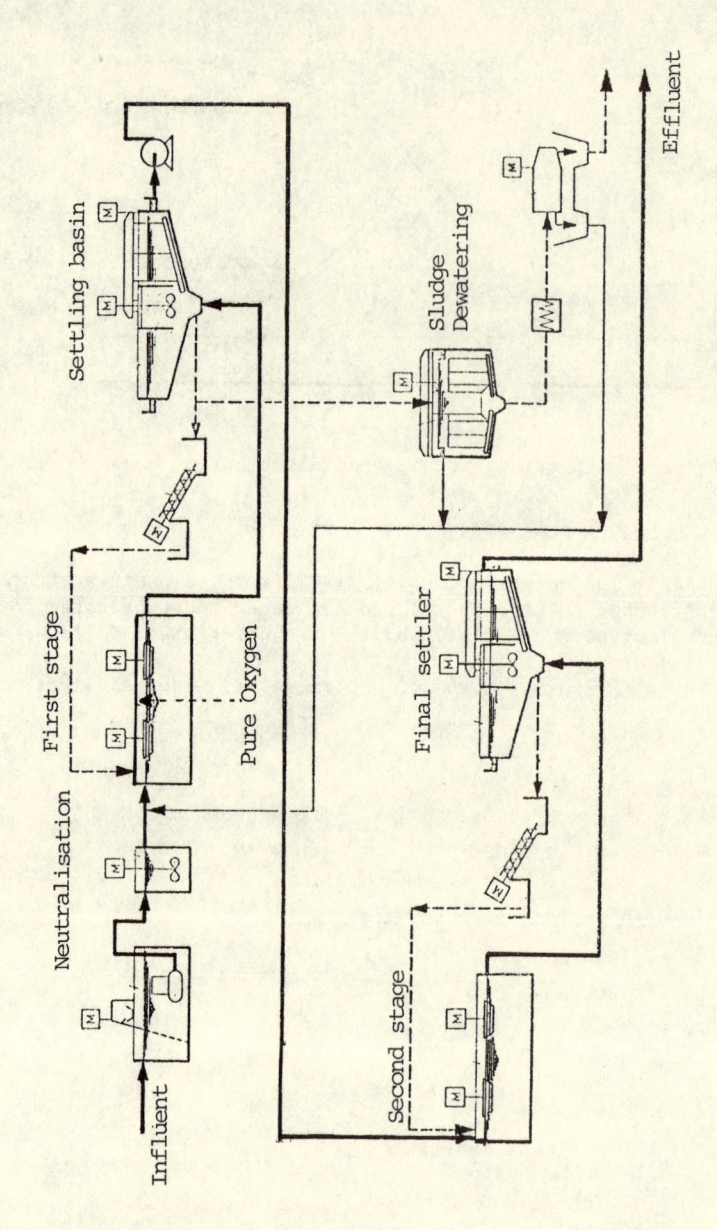

Fig. 2: Flow scheme of the biological wastewater treatment plant

After a technical-economical study of different purification methods (physical-chemical treatment, ion exchange, $N-NH_4^+$ -stripping), a combined treatment of the wastewater discharged from sewers A and B in a biological treatment plant was decided upon.

PROCESS DESCRIPTION

The combined biological treatment plant was designed as a two-stage completely mixed system with two separate activated sludge cycles. The first stage, a pure oxygen (UNOX) basin, is followed by a conventional aeration basin (Fig. 2).

Design parameters of the first pure oxygen basin (BOD and COD removal) are:

B = 0.6 kg BOD/kg SS·d
t = 8.0 h detention time
SS = 4.1 g/ℓ activated sludge (dry matter)

and of the second aeration basin (nitrification) are:

B = 0.1 kg BOD/kg SS·d
t = 24 h detention time
SS = 2.1 g/ℓ activated sludge (dry matter).

Each biological stage has a separate settling basin and its own sludge recirculation system.

LABORATORY DEVICE

Because of uncertainties related especially to the nitrification efficiency of the second stage, a trial was made before determining the final layout of the treatment plant. In this case only laboratory studies with synthetic wastes, carried out under optimal conditions, were possible (compounds and concentrations obtained from the technologists).

The laboratory device used is shown in Fig. 3.

The trials were run for three months at a constant temperature ($20^0 C$) and constant flow rate, allowing the prescribed detention times of 8 hours in the first stage and 24 hours in the second one to be held. Oxygen was supplied by blowing pure oxygen and normal aeration, respectively.

The mixing of the reactors, wasting of excess biological solids, return sludge ratios and dissolved oxygen concentration in the reactors were adjusted to optimal operation conditions. The flow pattern and time distribution curves of the full-scale devices were unknown, which prevented a simulation of the dynamic parameters from being made.

RESULTS and DISCUSSION

At the beginning of the laboratory studies, both biological reactors were filled with activated sludge from a domestic sewage treatment plant (LFKW-Busnau, B=0.15 kg BOD/kg SS·d; activated sludge SS= 3.5 g/ℓ).

First stage - In the first biological reactor supplied with pure oxygen, a new activated sludge was developed after three weeks. It remained without further changes till the end of the experiments (3 months). Characteristics of this sludge were:

respiration rate A = 68 to 89 mg/g·h O_2
activated sludge concentration SS = 4,1 to 8,5 g/ℓ

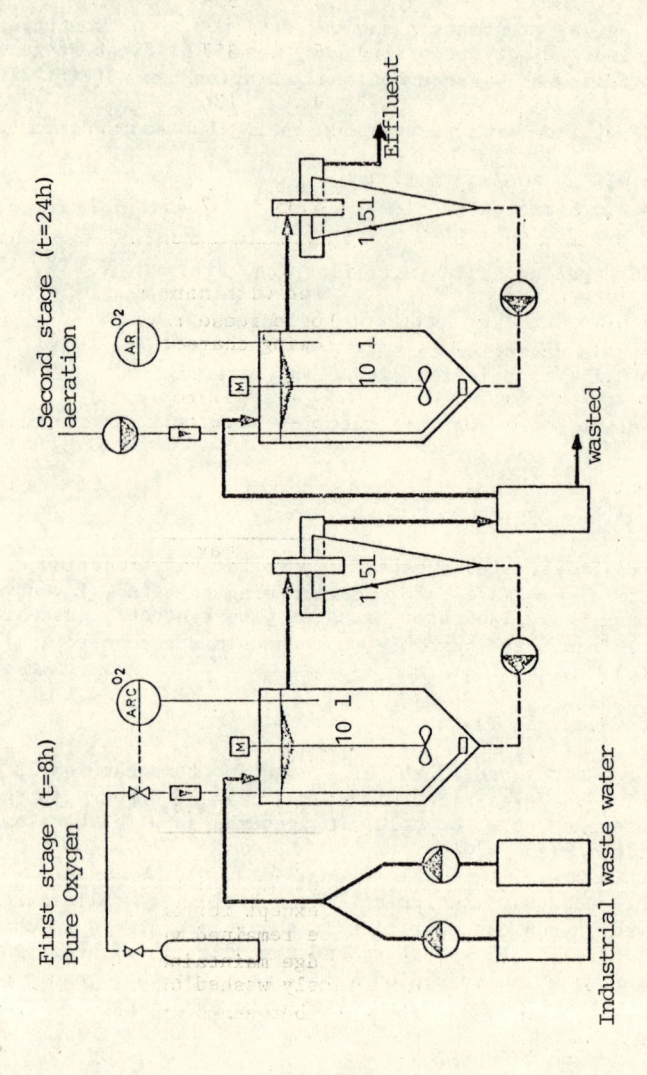

Fig. 3: Laboratory device used for proving efficiency of a two-stage biological treatment plant

The quality of the effluent, which was at the same time influent to the second stage, was:

	min.	max.	average	
L_{e1}	33	359	159	mg/l BOD
L_{c1}	306	357	328	mg/l COD
L_{N1}	49	134	84	mg/l $N-NH_4^+$
pH			7.5	
Alkalinity	4.35	7.45	3.0	mVal/l

Second stage - The second stage was aerated with normal air. The initial activated sludge concentration of 3.5 g/l started to decrease rapidly during the first two weeks. Afterwards it was stabilized and the following characteristics were measured:

respiration rate A_o = 4 to 8 mg/g·h O_2
activated sludge concentration SS = 1.6 to 2.1 g/l

The final effluent quality was:

	min.	max.	average	
L_{e2}	8	65	30	mg/l BOD
L_{c2}	215	215	215	mg/l COD
L_{N2}	33	93	59	mg/l $N-NH_4^+$
pH	5.65	7.70	6.70	
Alkalinity	0.15	4.50	1.63	mVal/l

The results obtained confirmed the effluent qualities expected with the proposed two-stage treatment plant for all parameters except for ammonia oxidation. Even under optimal conditions, the nitrification rate remained very low. This was mainly due to the low concentration of activated sludge maintained in the system. Suspended solids up to 20 or 30 mg/l were continuously washed out from the final settling tank provided for establishing a sludge age of between 5 and 7 days, necessary for the nitrifying microorganisms.

CONCLUSIONS

The developed mathematical and theoretical design methods for sewage and industrial wastewater treatment plants have helped to give considerable insight into biochemical purification processes, but the final layout of a plant is still based primarily on experimental data.

An engineer needs for the design of a treatment plant coefficients and data which have to be determined empirically on the wastewater concerned.

Trials on different scales (laboratory, pilot plant and full scale), closely related to the specific engineering problems, might be necessary.

Tests on a laboratory scale are very often the only possible studies before building the treatment plant.

There are no feasible methods for constructing laboratory devices based on theories of hydraulic similarity.

The retention time distribution can be used as a basis for determining the likelihood ratios between model and technical device. Measurements of retention time distribution in treatment plants in operation are rare. But without this information, laboratory studies cannot be used for scaling-up purposes and the predictibility of the results obtained will remain rather low.

Laboratory devices constructed only on the basis of conservative parameters (reactor type, loading, retention time etc.) and operated under optimal conditions (not achievable on full scale) can be very useful for testing the layout of a wastewater treatment plant and observing the purification process.

Even at this stage, important changes in the planned purification reactors and the operation can be made. By carrying out laboratory studies before completion of the treatment plant design, money and time can be saved.

For the purification of wastewater from a factory producing fertilizers, a two-stage completely mixed activated sludge plant was planned. The first stage was provided for BOD and COD removal and the second stage for nitrification.

The treatment plant was simulated on a laboratory scale with 10-1 reactors and 1.5-1 settling basins. The composition of the wastewater was prescribed by the technologist of the factory. Experiments were run as closely as possible to the needs of the future technical plant.

The results showed clearly that with the proposed treatment plant the desired effluent quality, except for nitrification, can be achieved.

It was further evident that the low nitrification rate of the second stage was due to the wash-out of suspended solids from the system.

A fixed bed reactor which is less sensitive to the wash-out of nitrifying bacteria was proposed as an alternative to the second-stage completely mixed reactor. In further studies this suggestion was confirmed and in the final layout of the treatment plant two stages were again proposed - in the first one, pure oxygen was supplied with an 8-h detention time, but for the second stage a fixed bed reactor was now designed.

REFERENCES

Horvath, I. (1978). *Some Questions Related to Scale-Up Aeration Systems*. IAWPR-Specialised Conference on Aeration, RAI Congress Center, Amsterdam, p. 77.

Levenspiel, O. (1972). *Chemical Reaction Engineering*. John Wiley & Sons, Inc.

DISCUSSIONS

Şentürk, H. (Turkey) : The retention time distribution or dispersion number is a good basis for comparing the results obtained with different laboratory-scale units. Can some degree of similarity be obtained between the dispersion numbers in the model and in the full-scale plant?

Sekoulov, I. (Germany) : Since in the future we will be confronted more and more with industrial process water, we must try to utilize units already existing in the industry; for example, in the pharmaceutical industry old fermentation reactors can be used. These have dispersion numbers and time distribution curves already accurately measured. It will, however, be extremely difficult to relate these distribution curves to the normal treatment plant with large aeration basins.

Samsunlu, A. (Turkey) : Is it possible to use laboratory-scale tests to obtain exact design parameters?

Sekoulov, I. (Germany) : In one of our projects due to be completed in 1981, we have been working on this question. We have found that, because of the daily variations in flow in the large treatment plant and the difficulty of measuring the parameters, our laboratory-scale tests, are not sufficiently predictive for the results to be transferred as design data for the large plant.

Taygun, N. (Turkey) : Why, were such high concentrations of oxygen used?

Sekoulov, I. (Germany) : This treatment plant was designed by UNOX, a company designing pure oxygen basins. We in the laboratory were trying to model the treatment plant according to the parameters given by the designers. Whether it was optimum to work with 16 mg/ℓ of oxygen was not our problem. We did not choose the concentration. We were only testing to see if the bacteria would withstand this high concentration in this basin. This reactor will in the end probably work with 4-8 mg/ℓ of oxygen.

Assistance to Small Industries in the Treatment of their Wastewater

U. SESTINI

Monza & Brianza Ind. Assn., Italy

ABSTRACT

Factories of small and medium size located in the highly industrialized area north of Milan are faced with serious problems related to the monitoring of their wastewater and the selection of technically and economically sound pollution abatement processes. The Monza and Brianza Industrial Association has set a precedent in Italy in establishing a chemical laboratory as well as advisory services to assist these industries in complying with the recently issued State pollution control law.

HISTORICAL DEVELOPMENT of THE MONZA & BRIANZA INDUSTRIAL
ASSOCIATION PROGRAM FOR POLLUTION MONITORING AND CONTROL

The topics which are covered and debated in this important symposium extend beyond the borders of our countries, as they are common to all nations striving to improve their conditions and standards of living. It is by mutual cooperation and help that our pollution problems can be solved most effectively.

One of the most important of these problems is related to the monitoring of the actual situation in specific areas where rampant industrialization and urbanization have taken place. The rapid expansion and diversification of industry, together with the parallel rise in population and the lack of efficient pollution control measures in the past, have resulted in a serious deterioration in the quality of inland and coastal water.

The Monza & Brianza Industrial Association is faced with such a situation, being located in the hinterland north of Milan. This area covers approximately 400 km^2 and has a population of around 1 million people living in 61 towns and villages where more than 600 industries are located. The Lambro river flows through the area and practically all wastewater, both domestic and industrial, are eventually discharged into this river either directly or through the municipal sewers. The factories in the area, which are of mainly small or medium size, and manufacture a wide variety of goods: wooden furniture, textiles, as well as mechanical, plastic and chemical products.

Beginning in 1971, the association provided the industries with only administrative and bureaucratic assistance as part of a rather serious public campaign, against pollution, but soon it became evident that the opportunity should be taken to parallel such assistance with an efficient technical structure, by organizing and operating a chemical laboratory to which the associated industries could refer in order to have their particular pollution problems identified and to obtain advice about possible ways to solve and control them. In this respect, it is worthwhile mentioning that the local law enforcement agencies were somewhat lacking in organization and understaffed. Moreover, they did not advise the industries on the best methods of abating pollution.

Another consideration pressed the association into the decision to organize a technical branch in their pollution control, that is, the difficulties met by the small-sized industries in obtaining offers from the large and most experienced wastewater treatment suppliers, who are mostly interested in large-scale projects. Consequently, these industries (comprising around 75% of the total) are compelled to give their tenders to waste treatment plant suppliers who in most cases lack a satisfactory level of technical knowledge, experience, craftmanship, as well as manpower. Another problem arises because these industries, after assessing the alternative offers, usually base their final decision not on sound technical judgement, but just on the lowest bid.

The chemical laboratory started operation in 1974, provided with the essential equipment and instrumentation, and year by year it has grown to its present size and capacity, with further expansion under way. It should be mentioned that the technical service has already expanded into the fields of air pollution and noise problems, both being of great interest to the associated industries. The taking of such an initiative has attracted the attention of other industrial associations, with which contacts are being made to encourage collaboration in providing a similar service in other parts of the country. Such a technical service sponsored by private industry was at first looked upon with some degree of suspicion by the local government agencies. However, this initially negative attitude is now gradually changing and eventually it is hoped that an amicable and mutually beneficial relationship will develop.

It should be pointed out that the service available from the association is fully independent of any third party. Furthermore, it is not normally concerned with the details of wastewater plant construction. Obviously, contacts have to be maintained with qualified wastewater treatment suppliers in order to keep up-to-date with the rapid technological developments in the field.

Another important service available to the associated industries is the periodical checking and control of the effluent quality, once operation of the plant has commenced.

POLLUTION CONTROL LAW

Natural water resources are under the direct control of the Regional Public Works Agency, which authorizes their use in terms of quantity. Pollution control, however, was covered up to 1976 only in a rather general and confused way by Common Law, with respect to problems such as the poisonous water sources, the causing of damage to others and the disturbing of fish. The absence of specific laws and a national program for pollution control resulted in a rather chaotic situation, the proliferation of administrative bodies and the imposition of confused and unrealistic effluent standards in the belief that pollution could be prevented and abated almost overnight.

Following long debates in Parliament, a Bill drafted by the legislative and technical committee, was at last made law in May, 1976. This law established guidelines on the action and responsibilities of the involved parties, both public and private, as well as on the reorganization of administrative and law-enforcing agencies. In addition, it set national standards for the discharge of effluents into waterbodies and municipal sewer systems.

At the same time, the National Research Center (CNR) issued through its Institute for Water Research (IRSA) an official standard methodology setting out the procedures to be followed in the sampling and analysis of wastewater effluents and to be applied by the technical control agencies. This methodology is very similar to the Standard Methods issued in the U.S.A. by the A.S.T.M. and the A.W.W.A. Obviously, the ideal system of pollution abatement, the optimal law and administrative procedures have still to be found. It cannot be expected that these will be efficiently established and satisfactorily tested within a short period of time.

Some significant drawbacks in the present law are related to the actual availability of funds and financial resources, to difficulties in the reorganization and the coordination of administrative and technical bodies, as well as to the most important problem of final disposal of waste sludge. Nevertheless, clear point of reference has now at last been made available to the involved parties, and it is foreseen that the present drawbacks and defects in the law will be solved in the near future.

Discharge Limits for Industrial Wastewater

The following table gives the permissible limits according to law for a composite sample of effluent discharged into waterbodies and municipal sewers.

The first column (1) gives the limits applicable to effluents from factories already existing and operating by May, 1976, while the second column (2) applies to newly-established industries.

Composite sampling should be carried out over a minimum of three hours are in accordance with the official sampling procedure.

In certain specific cases, it is extremely difficult to obtain some of the listed values, for example, for Cu, Zn, F, SO_4, Cl, COD, BOD_5, total P and N. Therefore, a revision of certain parameters is envisaged in the very near future, taking into account not only such practical difficulties in complying with the present limits but also the results of practical experience.

Generally speaking, the ultimate aim of the law is to reduce all effluents to the limits of the second column, by means of either separate treatment plants or combined private-public wastewater works. It should be added that the official methodology for the analysis of certain parameters is not yet available, for example, gas chromatography for organic solvents. It is hoped, for example, it will be issued as soon as approved.

The limits, especially with regard to suspended solids, BOD_5 and COD, will be somewhat relaxed for industrial effluents discharged into municipal sewer systems which are linked to or expected to be linked in the future to centralized waste treatment plants operated by special public agencies. In such cases, an agreement will be drafted between the public agency and the private users, and an appropriate fee will be paid by industry to refund the costs of the joint treatment of such wastes. One of these agencies (Consorzio Provinciale Alto Lambro) already exists in the outskirts of Monza. The area of its present physical treatment plant (sedimentation, sludge digestion and drying beds) will be expanded and a secondary biological process (activated sludge type) possibly with sludge incineration and power recovery, will be added.

It is unfortunate that one of the most serious problems related to pollution abatement, namely, the ultimate disposal of sludge originating from the wastewater treatment plants, is still pending a solution, both legal and technical. Such a situation is an obvious drawback in the efficient attainment of the objectives of the pollution. Several Draft Bills are, however, being examined at present by the authorities, and it is expected that a law concerning sludge will be issued in the near future.

Table 1: Effluent Quality Standards for Industrial Wastewater

Parameter	(1)	(2)
pH	5.5-9.5	
Sediment (2 h), ml/l	2	0.5
Suspended matter (filtrable)	200	80
BOD_5 (mg/l)	250	40
COD ($K_2Cr_2O_7$ - 2 h), mg/l	500	160
Oil (ether extractable), mg/l	10	5
Cadmium (mg/l)	0.02	0.02
Chromium (III) (mg/l)	4	2
Chromium (VI) (mg/l)	0.2	0.2
Iron mg/l	4	2
Nickel mg/l	4	2
Lead mg/l	0.3	0.2
Copper mg/l	0.4	0.1
Zinc mg/l	1	0.5
Total CN mg/l	1	0.5
Residual Chlorine mg/l	0.3	0.2
Sulphides mg/l as H_2S	2	1
Sulphates mg/l	1000	1000*
Chlorides mg/l	1200	1200*
Fluorides mg/l	12	6
Total Phosphorus mg/l	10	10**
Total Ammonia mg/l	30	15**
Nitrites mg/l	0.6	0.6**
Nitrates mg/l	30	20**
Phenols mg/l	1	0.5
Aldehydes mg/l	2	1
Surfactants mg/l	4	2
Aromatic solvents mg/l	0.4	0.2
Nitrosolvents mg/l	0.2	0.1
Chlorinated solvents mg/l	2	1

*Not applied to effluents discharged directly into the sea.
**Discharges located within 10 km from internal lakes should generally have total phosphorus and total nitrogen values of 0.5 mg/ and 10 mg/, respectively.

SCOPE AND ACTIVITY OF THE LABORATORY

Wastewater analysis is a fundamental part of the effort to define the nature and extent of pollution and is of vital importance in the development of an effective pollution control program. The type of waste treatment a factory effluent will require is determined by comprehensive sampling and analysis, followed by accurate measurement of the impurities present in the waste.

While the larger industries have already developed their own organizations and laboratories for the study of the problems related to their own factories, the small and medium-sized industries generally lack, both technically and financially, the means of establishing and running such facilities. In order to assist the affiliated industries in solving their problems, the Monza and Brianza Industrial Association

- at present the only one in the country- has sponsored the development of a centralized laboratory with the relevant services.

The wastewater section of this laboratory, which covers an area of approx. 300 m^2 is involved in the following activities:

1. Sampling of wastewater at the proper points and times within the factory and at the final point of discharge to the various waterbodies, this sampling only being performed after a discussion with the factory staff. Whenever opportune, the wastewater samples are duly preserved to avoid possible deterioration or changes in quality during the time lapse between sampling and actual analysis in the laboratory location.

2. Chemical analysis of the samples for all potential pollutants bearing in mind the industrial process from which the waste has originated. A full report is then handed to the factory officials for examination and further discussion.

3. Laboratory test runs on pilot plants in order to determine the potential efficiency of the most common treatment processes. The most widely used pilot plant studies are performed using the jar test (chemical precipitation), sedimentation tubes, activated carbon and ion-exchange columns or a biological reactor. At the same time valuable information can be gained related, for example, to the optimum chemical dosages, the best types of coagulants, the optimal type of adsorption media, aeration requirements and the amount and type of sludge. Such information, together with the results of the pilot plant studies, is also discussed with the factory staff.

4. Giving advice and assistance, if requested, with regard to the best way of issuing tenders and of assessing offers received from potential treatment plant suppliers.

5. Certifying the plant, on its completion, as fully operational following a trial run, and subsequently checking periodically the effluent quality and thereby the efficiency of treatment.

Another important service available is the provision of an up-date technical library containing books of mostly American, British or French origin. The laboratory staff is at present following a training course in Germany in order to have access also to the most valuable German publications in the field.

The future program of the service will probably be extended also to include technical training courses for the treatment plant operators.

LABORATORY INSTRUMENTATION

Instrumental analysis of wastewater samples is now routine and it has almost completely replaced the classical gravimetric and volumetric manual methods, both for precision as well as for speed.

Following the issue of the official methodology by the authorities, the range of instruments available in the laboratory has been extended to meet the growing requirements of the associated industries. A molecular absorption spectrometer, colorimeter, conductometer, as well as equipment for determining pH, were among the first instruments to be used in the laboratory during the initial period of operation.

The subsequent introduction of an atomic absorption spectrometer, extended the scope and range of the above instruments, especially with regard to their sensitivity to toxic metal traces. The consequent awareness of the problems caused by organic micropollutants and solvents later led to the installation of a gas chromatograph, together with a flame ionizer and electron capture detectors.

The future program includes another gad chromatograph of more advanced design and a computerized integrator, as well as I.R. instrumentation for the characterization of sludges. In the meantime, in case of necessity, the laboratory will have access to a gas chromatograph and mass spectrometer as a result of an agreement with another private organization in Milan.

The expansion of the service will very soon necessitate relocation of the laboratory in new and larger premises in Monza. The project is already under way, with completion expected by the end of the present year.

IMPORTANCE of QUALITY CONTROL

The basic role of the chemical laboratory is to provide useful and reliable data to facilitate decision-making by the industry staff. Such data include information on the waste characteristics, their variability range in time, the most appropriate treatment process, as well as its predicted efficiency.

It is, therefore, of the utmost importance to check the actual reliability of the analytical work by establishing an efficient quality control program. In other words, it is essential to apply some type of routine standard procedure to control the measurement processes and results. Also, the degree of analytical precision and accuracy must be determined.

A program of quality control is at present being considered by the laboratory staff and it is expected to be established at the earliest opportunity both for the standard procedures as well as for gas chromatography.

DISCUSSIONS

Göçer, İ. (Turkey) : In your paper mercury concentration was not mentioned, in spite of the dangers to human health due to its accumulation in fish.

Sestini, U. (Italy) : The list given is actually an abstract from the original one issued by the law, and so it does not include all parameters.

Curi, K. (Turkey) : Does your organization have the right to impose penalties?

Sestini, U.(Italy) : It is the duty of the government law-enforcing agencies to impose penalties. Industries discharging raw wastewater have to pay a rather high fine.

Curi, K. (Turkey) : What are your standards, for the discharge of industrial wastewater based on? Are they receiving water standards or effluent quality standards?

Sestini, U. (Italy) : The standards are based only on the effluent quality. The discharge of diluted wastewater is, for example, permitted only if it originates from the cooling process. Except for discharges to lakes and obviously to drinking-water tables underground, no limits' are set for the quality of the receiving waterbody. Standards based on this would have taken too long to establish.

Karlsson, L. (Sweeden) : Are there any governmental control agencies in Italy which check the quality of industrial effluents and the waterbodies into which they are discharged?

Sestini, U. (Italy) : There are such agencies. They are organized locally and, after making their investigations, they report to the relevant administrative bodies.

Treatment of Olive Oil Production Wastes

K. CURI[*], S. G. VELIOGLU[**] and V. DIYAMANDOGLU[***]

*Department of Civil Engineering, Bogazici University, Istanbul, Turkey
**Grad. Student, Department of Civil Engineering, University of California,
Irvine, Ca., USA

ABSTRACT

The treatability of wastewater from olive oil extraction factories in the Gemlik area has been investigated. The principal objectives of the work were:

i. To regain oil from the wastewater prior to disposal, using gravity separation and dispersed air flotation.

ii. Treatment (coagulation and acidification) of the waste for disposal, without considering recovery of oil.

iii. Color removal by activated carbon adsorption.

The results of the experiments carried out have shown that the wastewater in question can be treated satisfactorily using conventional treatment methods. Based on the observations, suggestions are made for practical applications in individual factories.

INTRODUCTION

Visible oil floating on the surface of rivers, lakes and sea has always presented aesthetic problems, whether the quantity of oil is sufficient to interfere with the beneficial uses of the water or not. Heavy surface films of oil also interfere with the natural processes of aeration and photosynthesis and directly contribute to organic pollution.

Turkey leads the world in olive production (DPT, 1977) and is fourth in olive oil production, which amounts to 103,000 tons per year. The Gemlik area provides approximately 7% of the total yearly production (DPT, 1977).

The wastewater from the olive oil production industries in Turkey is directly discharged either to the municipal sewerage system or into the sea. Oil spills are unavoidable during the very primitive processes used to extract oil from the olives, which suggests that this oil pollution may best be controlled at the source.

The purpose of the study is to determine the most effective and practical method for the in-plant removal of oil from the wastewaters of the oil production factories in the Gemlik area prior to final discharge into the sea. The investigations performed for this purpose were carried out in two directions:

1. Treatment of the wastewater coupled with oil recovery.
2. Treatment of the wastewater without oil recovery.

In the first case, gravity separation and dispersed air flotation were the methods applied. In the latter case, certain chemical processes, the addition of coagulants, and several combinations of these were examined. Finally, to remove the color of the wastewater, a combination of centrifugation, vacuum filtration and activated carbon adsorption was applied.

CHARACTERISTICS OF THE WASTEWATER

The surveys performed in the area under investigation have shown that the methods used in olive oil production are primitive and too manpower-dependent. The production season for olive oil is between October and January, peak production being in the month of December. It has been observed that the olive oil production methods for all the factories in the Gemlik area are virtually the same. The process can be summarized as follows:

i. *Washing*: Olives are washed thoroughly with hot water.

ii. *Extraction*: Washed olives are placed in cloth bags, of about 25 to 30 kg capacity, and then are pressed to extract the oil. During this process, in order to facilitate extraction, the bags are washed with hot water and as a result, some water as well as some solid material from the olives is mixed with the extracted oil.

iii. *Separation*: To separate the oil from the impurities, the effluent from the above process is taken to a series of pools, where it is rested for 3 to 4 hours. During this resting period, the oil is collected on the surface (due to density difference) and is scraped off intermittently.

The water which remains in the pools is referred to as "black water" due to its dark brown color. According to the information obtained from the olive oil producers, the amount of "black water" generated varies on the average from 3.3 to 4.2 liters per liter of olive oil produced, although values as high as 10 l/l are reported. This "black water", being the main effluent of the oil extraction facilities in the Gemlik area, is examined in the present study.

The wastewater used in this study was obtained from a typical olive oil production factory in Gemlik, currently discharging its waste directly into the sea without any treatment. The initial observations were that the wastewater was highly turbid, rich in suspended solids and dark brown in color, and that it had the characteristic odor of olives.

The results of a series of experiments carried out to determine the characteristics of the wastewater in question are summarized in Table 1. The table also contains, for purposes of comparison, results obtained from experiements performed on olive oil extraction wastes in Portugal (*Raimundo and Oliveira*, 1976).

The average BOD: COD ratio observed for the wastewater in question is about 0.37, proving biodegradability to a certain extent; thus, biological treatment can be regarded as feasible.

However, the values observed for the residues and turbidity are quite high, necessitating the removal of suspended matter prior to biological treatment or any other appropriate disposal method, for example, sea disposal.

Table 1. Characteristics of the Olive Oil Production Wastes

Parameter	Current Study Range	Average	Raimundo and Oliveira (1976)
pH	6.4-7.0	6.7	5.2
Turbidity	36000-45000	42000	65000
Conductivity (mhos/cm)	11000-13600	12400	22500
Total Residue (g/l)	66-72	68	56
Total Fixed Residue (g/l)	40-47	45	N.A.*
Total Volatile Residue (g/l)	25-26	25.5	N.A.
Total Nonfiltrable Residue (g/l)	0.39-0.44	0.42	N.A.
Oil (g/l)	5.0-14.5	12.2	N.A.
Total Nitrogen (mg/l)	105-142	125	N.A.
Orthophosphate (mg/l)	7.1-9.2	8.5	N.A.
BOD_5 (mg/l)	12000-41000	22000	9600
COD (mg/l)	50100-67800	59000	97740
Color (APHA Co-Pt Units)	52500-60500	56270	180000

* Not available

METHODS OF OIL REMOVAL WITH RECOVERY

The Gravity Separation Experiments

These experiments were carried out in a sedimentation column (Fig. 1) equipped with sampling orifices located at depths of 5, 35, 65 cms measured from the surface.

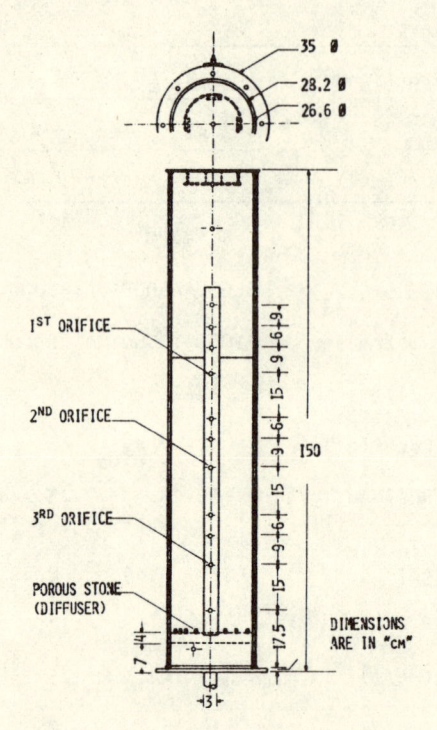

Fig.1: The Column Used In Gravity
Separation and Dispersed
Air Flotation

The column was made of plexiglass. The wastewater was introduced into the column after thorough mixing, and samples were taken at different time intervals. The initial oil concentration ranged from 8.5 to 9.0 g/l. Seven to eight samples were taken from each orifice and were tested for oil concentration. The duration of the experiments was fixed at 48 hours in order to have thorough understanding of the behavior of the waste during gravity separation. The variation in oil concentration, at different depths, with time is plotted in Fig.2.

As can be observed from this figure, the concentration increase in the top layer is compensated by a decrease in the bottom layer. After about 24 hours of gravity separation, the rate of change in oil concentration with time is rather small for the three layers. Within the first 30 minutes, the concentration at all sampling depths increases with time. This is attributed to the oily particles rising, which results in a temporary concentration increase, as the particles pass by the sampling points.

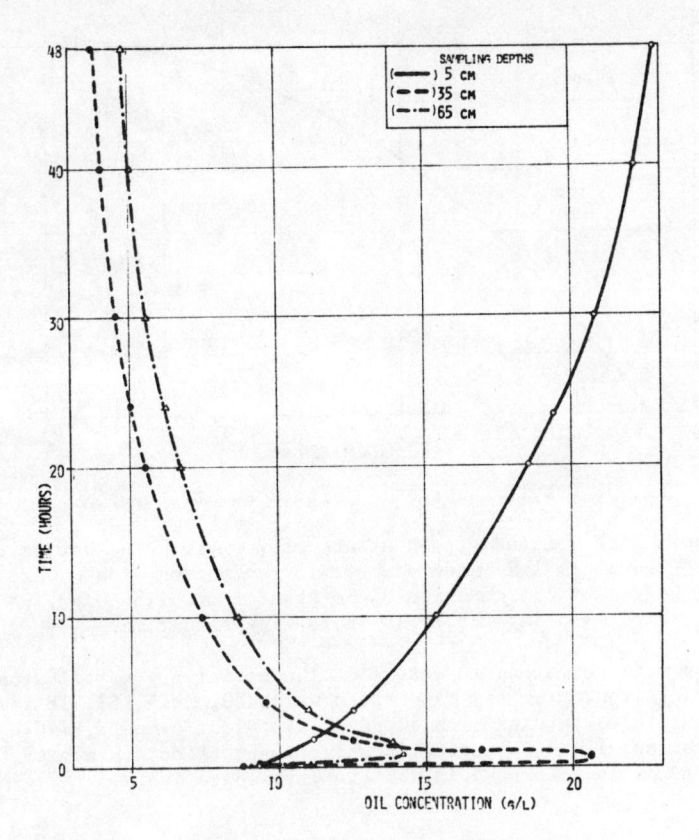

Fig.2: Gravity Separation

Another important observation, as represented in Fig.2, is that the oil concentration in the bottom layer is higher than that in the intermediate layer. This is thought to be due to the adsorption of oil on the settling solids.

The percentage purification efficiency, as defined by

$$\text{purification efficiency (\%)} = \frac{C_0 - C}{C_0} \times 100$$

where :

C_0 = initial oil concentration

C = oil concentration at time "t",

is given in Fig.3 for the intermediate layer, where the percentage purification is plotted against time. As seen, the percentage purification levels off after about 40 hours, which implies that a further increase in the experiment duration beyond 48 hours is not justified.

Dispersed Air Flotation Experiments

These experiments were run in the sedimentation column used for the gravity separation experiments (Fig.1). The air dispersion was accomplished by using porous stones

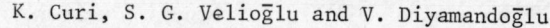

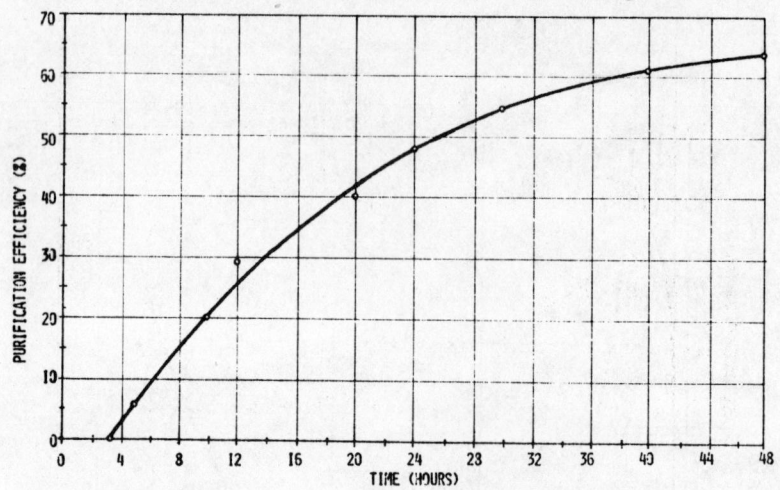

Fig.3: Gravity Separation: Intermediate Layer

placed on a wooden disk located at a distance of 7 cm from the bottom of a column having a diameter equal to its inner diameter. The porous stones were cylindrical in shape with a height of 2.5 cm and a diameter of 1 cm. The experiments were run for seven hours, with sampling intervals increasing with time.

Four experiments were performed to determine the efficiency of oil separation by dispersed air flotation. The air flow rates were 120, 76.5, 51, 18 $1/min\ m^2$, respectively. The results obtained, as shown in Fig.4, 5, 6 and 7, indicate that a better clarification, i.e., reduction in oil concentration, is achieved in the intermediate layer, as was the case in the gravity separation.

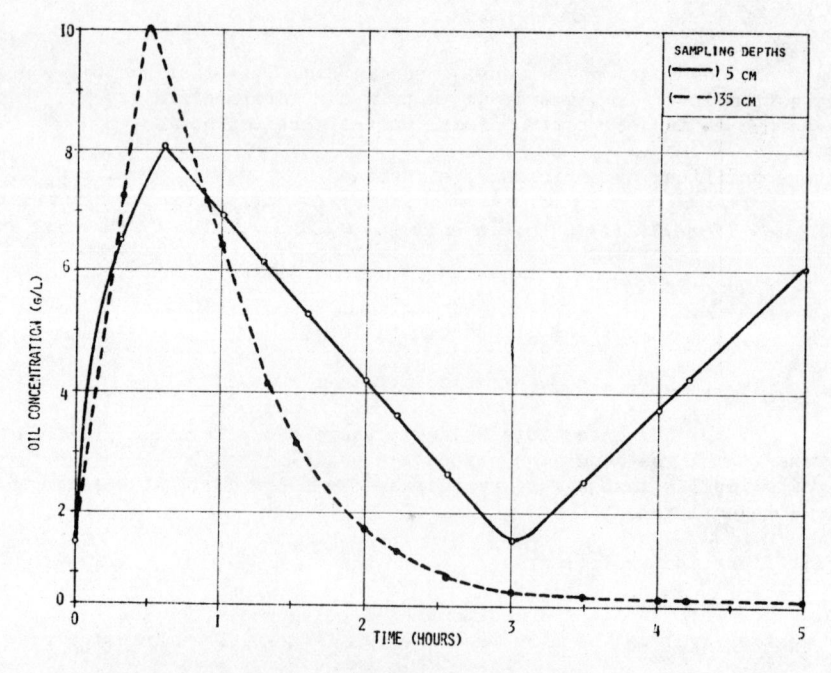

Fig.4: Dispersed Air Flotation: Q_{AIR} = 120 $1/min/m^2$

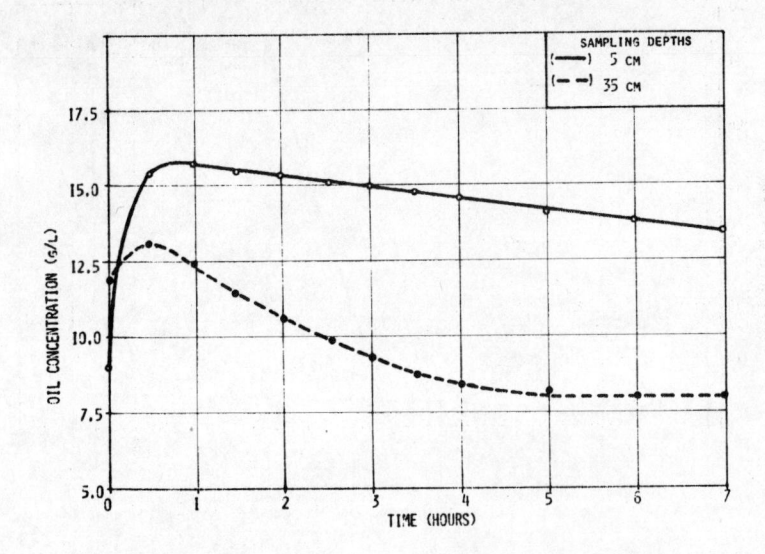

Fig.5: Dispersed Air Flotation: Q_{AIR}=76.5 1/min/m^2

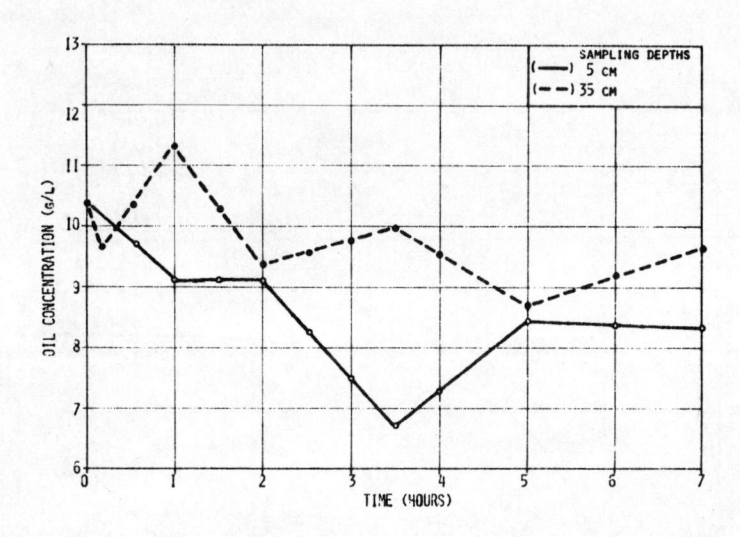

Fig.6: Dispersed Air Flotation: Q_{AIR}= 51 1/min/m^2

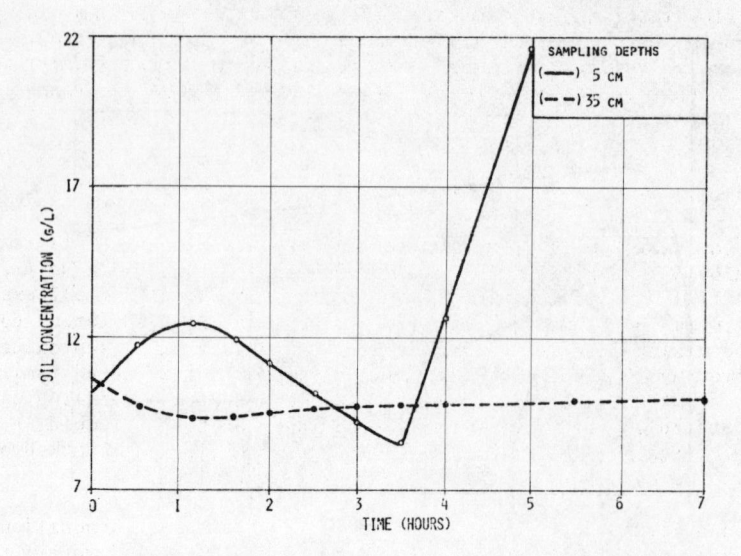

Fig.7: Dispersed Air Flotation: $Q_{AIR} = 18$ $1/min/m^2$

The variation in oil removal efficiency with different air flow rates for the middle layer of the experiments is shown in Fig.8.

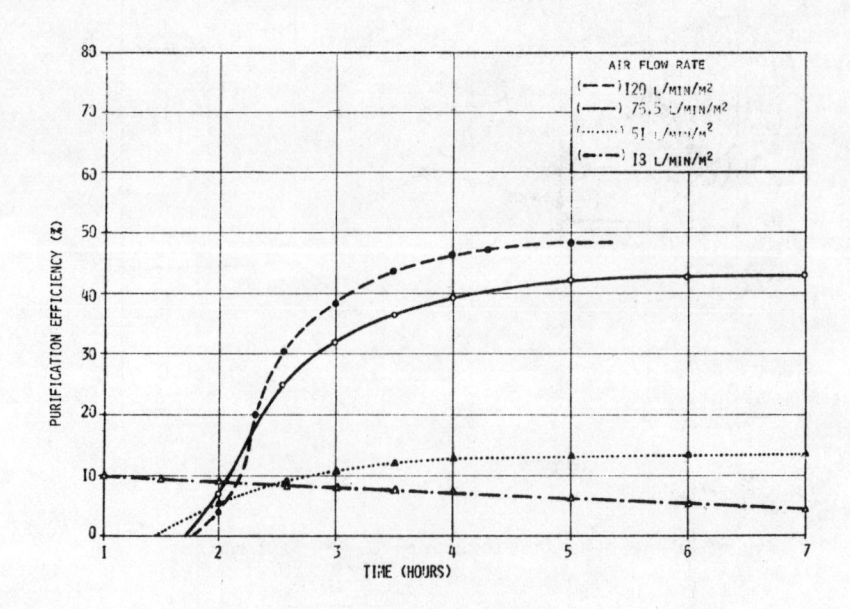

Fig.8: Dispersed Air Flotation

As can be observed from this figure, the removal efficiency increases with an increase in air flow rate; however, an air flow rate below 51 1/min/m^2 is not justified. Furthermore, after approximately the fourth hour, no significant variation in the removal efficiency can be noted. Hence, detention times greater than 4 hours in the flotation unit do not seem to pay off for the waste in question.

METHODS OF OIL REMOVAL WITHOUT RECOVERY

Coagulation

The standard "jar tests" were made to determine the efficiency of oil removal by chemical coagulation using the most commonly used coagulants, such as $Al_2(SO_4)_3$ and $FeCl_3$. The initial oil concentration was between 5 to 6 g/l for all experiments, which were run using a volume of one liter of waste for each different dosage of coagulant. The mixing duration was one minute of rapid mixing (approximately at 100 rpm) followed by 40 minutes of slow mixing (approximately at 40 rpm). This was followed by an eight-hour settling period. At the end of the settling period, three distinctly separated layers were formed: a top layer of turbid liquid, a bottom layer consisting of the settleable solids and a rather solid-free, dark brown intermediate layer. The oil content of each layer was determined.

Since Al(III) and Fe(III) salts tie up hydroxyl ions, (OH^-), the addition of $Al_2(SO_4)_3$ and $FeCl_3$ to water is similar to an acidimetric titration of the water; thus, the pH of the solution after addition of these coagulants depends upon the coagulant dosage and the alkalinity of the wastewater. Although a quantitative determination of the alkalinity was not made for the wastewater in question, it was observed that the initial pH of 6.4 to 7.0 (see Table 1) dropped 1 to 2 units at the applied coagulant dosages, and it was concluded that the alkalinity of the wastewater was low. In the coagulation experiments, since the wastewater had a high colloid concentration and a low alkalinity, only one chemical parameter - the optimum coagulant dosage - was tested, as suggested by Weber (1972).

The purification curves for the top layer are given in Fig.9.

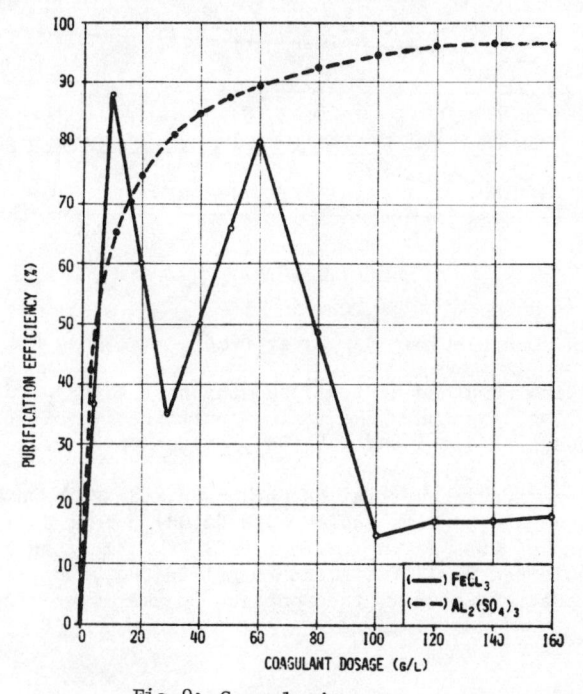

Fig.9: Coagulation: Top Layer

As can be observed from this figure, the maximum purification achieved by $Al_2(SO_4)_3$ is about 96% at a dosage of 150 g/l and by $FeCl_3$ about 88% at a dosage of 10 g/l. The required optimum coagulant dosages are relatively high, as expected, due to high colloid concentrations. Incidentally, the decreasing purification efficiency with increasing $FeCl_3$ dosage can be attributed to particle restabilization inhibiting aggregation of particles as a consequence of overdosing, as pointed out by several researchers (*Hahn and Stumm*, 1968; *Stumm and O'Melia*, 1968).

In the light of the fact that colloid destabilization using Al(III) and Fe(III) salts takes place through adsorption of polymers produced from these chemicals leading to charge neutralization (*Weber*, 1972), it was decided to examine the effect of aiding the destabilization by yet another factor, namely, enmeshment in a precipitate. When a rapid precipitation of a metal hydroxide (e.g., $Al(OH)_3$, $Fe(OH)_3$) or a metal carbonate (e.g., $CaCO_3$) is achieved, colloidal particles can be enmeshed in these precipitates as they are formed. Thus, colloid stabilization can be achieved.

The effect of aiding adsorption and charge neutralization by enmeshment in a precipitate was studied by using $FeCl_3$ and $Ca(OH)_2$ simultaneously. $Ca(OH)_2$ was chosen for economic reasons in order to provide hydroxyl ions and thus facilitate a rapid precipitation of ferric hydroxide, although it was not expected to result in a rapid precipitation of calcium carbonate (due to low alkalinity). Experiments were first carried out to find the appropriate dosage combinations which would yield meaningful results. This was followed by another set of experiments, results of which are summarized in Fig.10, where purification efficiency is plotted against different combinations of $FeCl_3$ and $Ca(OH)_2$ dosages. The pH of the solution for all different dosage

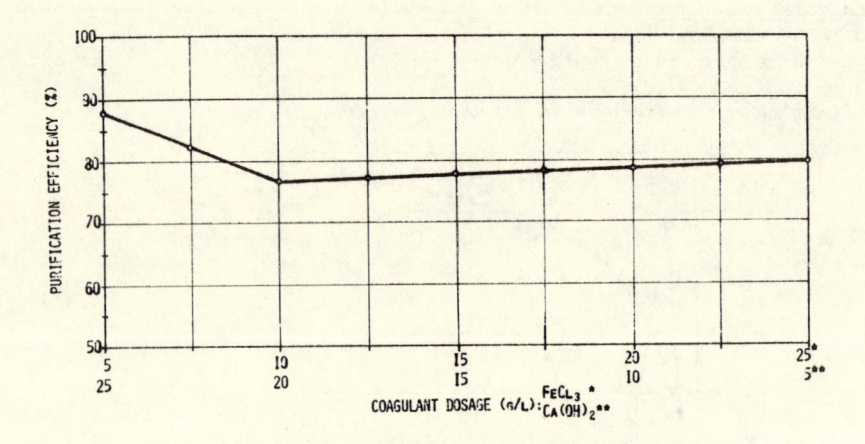

Fig.10: Coagulation: Top Layer ($FeCl_3 + Ca(OH)_2$)

combinations more or less remained neutral, showing that $Ca(OH)_2$ addition provided additional alkalinity, as expected. The maximum purification efficiency obtained was 88% at a $FeCl_3$ dosage of 5 g/l and a $Ca(OH)_2$ dosage of 25 g/l.

Given that purification efficiency remained unchanged, there is an economic trade-off between using $FeCl_3$ alone and in combination with $Ca(OH)_2$. That is to say, to get a purification efficiency of 88%, if the cost of $FeCl_3$ is more than 5 times that of $Ca(OH)_2$, then one should use 5 g/l $FeCl_3$ and 25 g/l $Ca(OH)_2$; and if the cost of $FeCl_3$ is less than 5 times that of $Ca(OH)_2$, then one should use 10 g/l $FeCl_3$.

Acidification

Acidification usually breaks the bonds between oil and particles to which oil has adhered and thus facilitates flotation. HCl was used for this purpose in different amounts; rapid mixing for five minutes was followed by an 8-hour settling period in the settling column (Fig.1). The initial oil concentration of the samples was 5 g/l.

The experiment resulted in the formation of three layers at the end of the settling period. However, the bottom layer, being short in height, did not lend itself to sampling. The oil concentrations obtained from the top and intermediate layers for an acid addition of 20 ml/l (a value found to yield best results,(see Fig.12) are given in Fig.11. As apparent from the figure, the oil concentration in the intermediate layer was less than that in the top layer, as expected. However, the variation in these two layers is not compatible; that is, an increase in the concentration

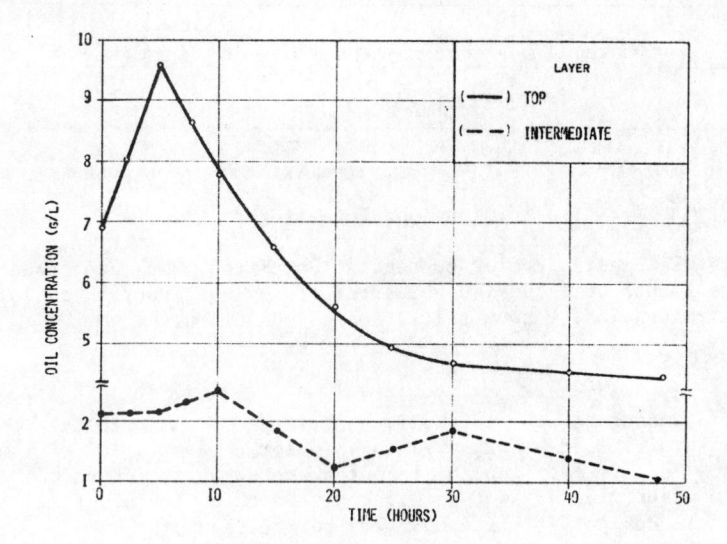

Fig.11: Effect of Acid Addition on Oil Concentration

of the top layer is not followed by a decrease in the concentration of the intermediate layer and vice-versa. It was not possible to give a conclusive explanation of this observation without knowing the concentration of the bottom layer.

The oil removal efficiency achieved in 48 hours in the intermediate layer for different HCl concentrations is given in Fig.12. The decrease in the purification efficiency observed in the intermediate layer at higher dosages of HCl is thought to be due to the release of oil from the settled bottom blanket of the column. When compared with Fig.3, Fig.12 reveals that acidification definitely enhances gravity separation up to acid addition levels of 30 ml/l.

REMOVAL OF COLOR BY ACTIVATED CARBON ADSORPTION

To express the distribution of the color-causing solute between the liquid and solid phases at a fixed temperature (i.e., the adsorption isotherm), the two most widely-known models proposed for isothermal adsorption relations, the Langmuir and Freunlich models, were used. Two sets of experiments were run to investigate the color removal efficiency of powdered activated carbon (activkohle rein, Art. 2183, by Merck).

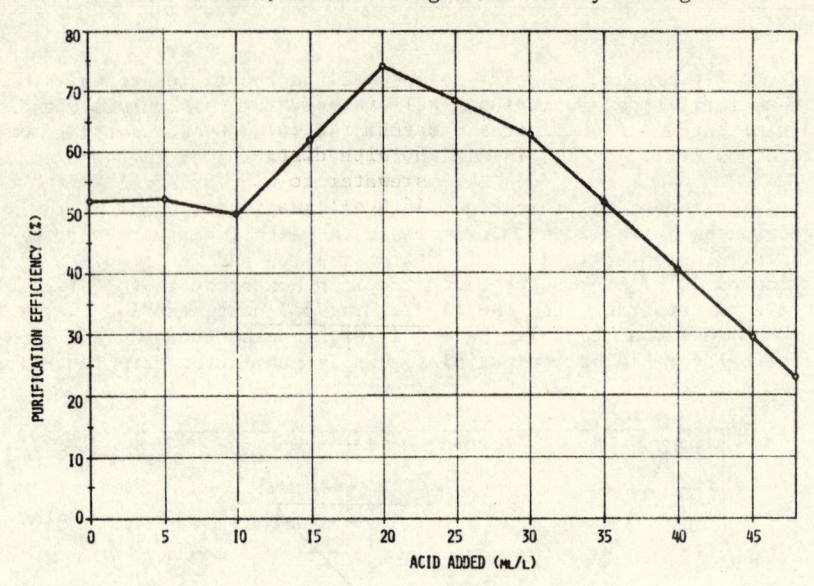

Fig.12: Acidification: Intermediate Layer

Due to the very high initial color concentration of the wastewater (see Table 1), adsorption experiments were carried out after achieving reduction in color concentration by centrifugation, vacuum filtration and dilution.

Experiment Set No.1

In this set, the wastewater was initially centrifuged. The decantate obtained was diluted 1:50 with distilled water, yielding a color concentration of 940 APHA Cobalt-Platinum units. To this solution, different powdered activated carbon dosages were applied. The contact time was set at 1.5 hours based on the observations that by running a constant dosage for different time periods equilibrium was reached. Increasing the contact time beyond 1.5 hours did not yield any further color removal. Next, the samples were filtered through Whatman (student model, d= 12.5 cm) filter papers for the removal of carbon from suspension. The resulting adsorption isotherm is given in Fig.13.

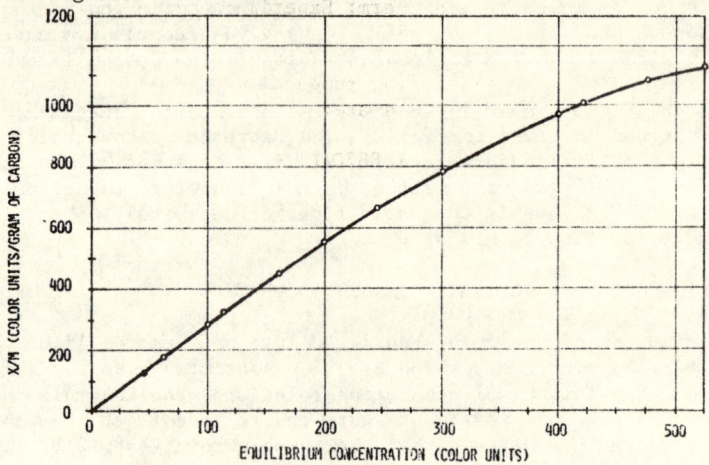

Fig.13: Adsorption Isotherm: Experiment No.1

Experiment Set No.2

In this set, the wastewater was initially filtered through Whatman (student model, d = 12.5 cm) filter papers, by applying vacuum filtration through a Buchner funnel. The filtrate obtained was then diluted 1:50 with distilled water. This treatment reduced the color concentration of the wastewater to 680 units. A wide carbon dosage range, varying between 0.004 g/l and 16.0 g/l, was examined at different dosages. The contact time was again kept at 1.5 hours. The resulting adsorption isotherm is shown in Fig.14.

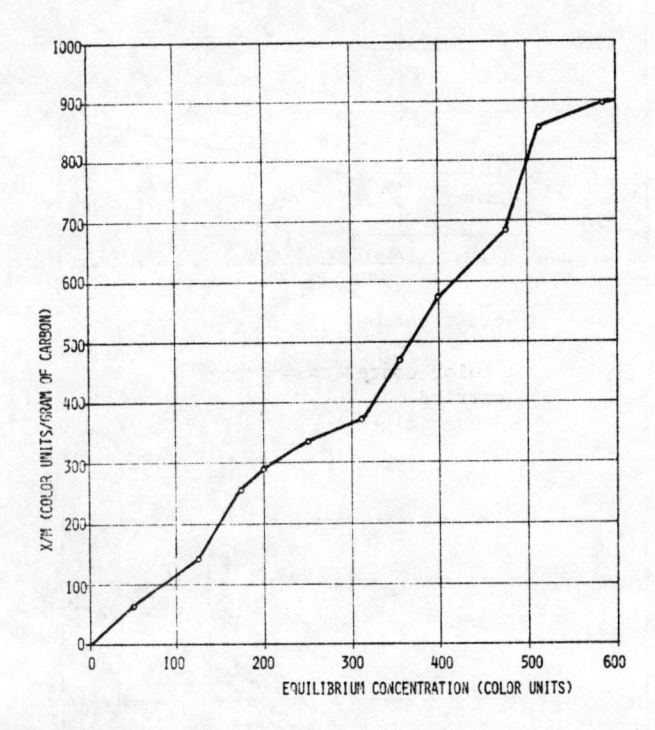

Fig.14: Adsorption Isotherm: Experiment No.2

The linearized forms of the Langmuir and Freundlich isotherms for the two sets of experiments are given in Fig.15 and Fig.16, respectively. The analytical expressions obtained for the second set of experiments are:

$$\text{Langmuir} : X/M = 1.887C/(1 + 00123 \, C)$$

$$\text{Freundlich} : X/M = 10.9 \, C^{1+1\sim 47}$$

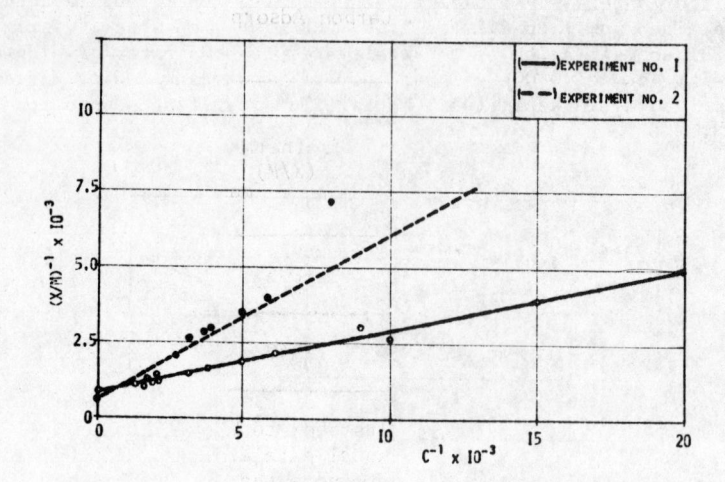

Fig.15: Langmuir Isotherms

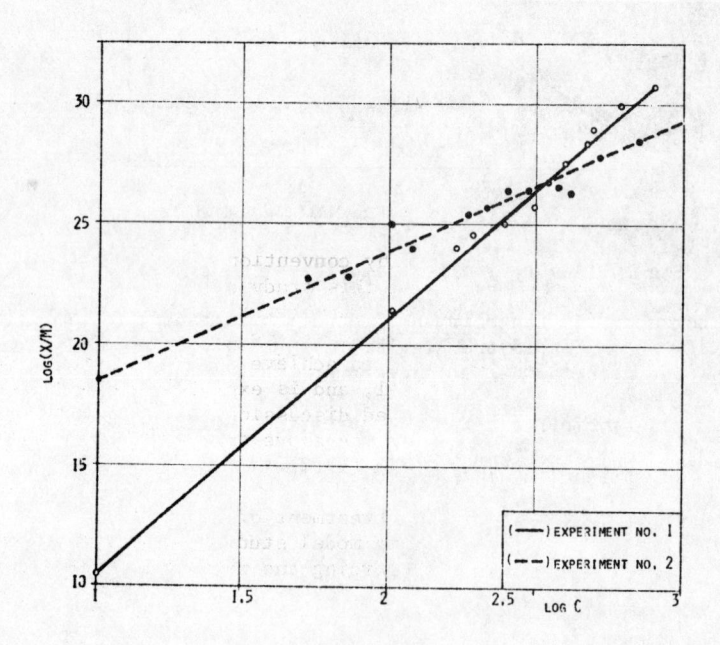

Fig.16: Freundlich Isotherms

The limiting capacity of adsorption (X/M) as "maximum possible color removal per gram of activated carbon" is summarized in Table 2.

Table 2: Results of Activated Carbon Adsorption Experiments

Experiment No.	Adsorption Isotherm $(X/M)_{max}$	Langmuir Isotherm $(X/M)_{max}$	Freundlich Isotherm $(X/M)_{C_0}$
1	1225	1205	1106
2	900	1540	1115

As can be seen from Table 2, all the results seem to be in pretty good agreement. However, it can be noted that the experimental data are best described by the Freundlich model. It is also interesting to note that a 38% change in the initial color concentration did not significantly affect the limiting capacity, which is observed to be in the neighborhood of 1000 color units per gram of carbon. Given that the original color concentration is about 56,000 units, huge quantities of carbon will be required to achieve satifactory color removal levels. A simple economical analysis, besed on the 1978 prices for imported activated carbon has shown that at a powdered activated carbon dosage of 15 g/l, a color removal of about 80% will be achieved and an approximate expenditure of 3.000.- TL per ton of waste required.

A question which still needs to be answered is the effect of higher initial concentrations on the removal efficiency of color, using powdered activated carbon.

CONCLUSIONS

The results obtained from the different methods examined in this study show that the treatment of olive oil production wastes by conventional methods is posible. In view of the results of the experiments of this study and the average volume of "black water" produced in the Gemik area, a scheme involving gravity separation tanks in series with a detention time of 48 hours followed by activated carbon treatment is recommended. Such a system is estimated to achieve about 85% purification of oil and a considerable amount of color removal, and is expected to satisfy effluent requirements prior to sea disposal. Detailed discussion of the results and experimental data leading to the recommendation of the above treatment scheme within an economic perspective is given by Diyamandoglu (1978).

Further investigations on the biological treatment of the wastewater in question are being carried at Boğaziçi University using model studies, at the end of which more conclusive results should be obtained regarding the most appropriate final disposal scheme.

REFERENCES

Diyamandoglu, V. (1978) *Source Control and Treatment of Wastewater from Olive Oil Production* (Unpublished) M.S. Thesis, Dept. of Civil Engineering, Boğaziçi University, Istanbul.

DPT, Devlet Planlama Teşkilâtı (1977). *IV Beş Yıllık Kalkınma Planı*. Meyvecilik Özel İhtisas Komisyonu, Turunçgiller ve Subtropik Meyveler Grubu, Zeytincilik ve İnceleme Grubu Raporları, Yayın No. DPT 1585 - Ö.İ.K., 268, Ankara.

Hahn, H.H. and W. Stumm (1968). Kinetics of Coagulation with Hydrolyzed Al (III). *J. Coll. and Interface Sci.*, 28, 134-144.

Raimundo, M.C. and J.S. Oliveira (1976). Pollution from Industrial Extration of Olive Oil in Portugal. *In Theory and Practice of Biological Treatment*. Nato Advanced Study Institute Preprints, Boğaziçi University, Istanbul.

Stumm, W. and C.R. O'Melia (1968). Stoichiometry of Coagulation. *J. Amer. Water Works Assn.*, 60, 514.

Weber, W.J. (1972). *Physico-chemical Processes for Water Quality Control*. Wiley - Interscience, New York.

DISCUSSIONS

Yılmaz, C. (Turkey) : Did you observe any emulsion in your wastewater?

Velioğlu, S. G. (Turkey): No. The water was well mixed.

Atımtay, A. (Turkey) : What will be the effect, with respect to olive oil recovery, of lengthening the detention time in the gravity settling tank?

Velioğlu, S. G. (Turkey): If we begin with 6-12 g of olive oil per liter of black water, we can get 70% recovery by extending the detention time from 4 to 48 hours. On average, 5 g of olive oil per liter of black water can be recovered.

Atımtay, A. (Turkey) : From the curve in Fig. 10, it would seem that a higher purification efficiency could be obtained by decreasing the coagulation dosage. It is obvious that the less coagulant added, the more economical the process is. Wouldn't it have been better to start with smaller concentrations of coagulant materials?

Velioğlu, S. G. (Turkey): Yes. This disturbed me too. We should have tried lower coagulant doses.

Sestini, U. (Italy) : Do the oil values shown in the table refer to total oil, free oil or emulsified oil?

Velioğlu, S. G. (Turkey): They refer to total oil.

Sestini, U. (Italy) : When you added the acid, did you heat it a little? We use sulphuric acid at temperatures as high as 60°C, but then materials like thin steel are unsuitable.

Velioğlu, S. G. (Turkey): The acid was added at room temperature. In the literature it is reported that thermal motion, rather than fluid motion, is important in the flocculation of olive oil. We conducted experiments heating the acid, but our results showed little difference.

Sestini, U. (Italy) : I would also like to mention that we recover most of the free oil from waste by using a very simple device: a large strip of stainless steel which is partially immersed in the waste-collecting tank and rotated slowly. The oil adheres to the surface of the strip, where upon it is removed by a scraping blade. This is also very useful in reducing the load of the subsequent unit.

Taygun, N. (Turkey) : Why did you add calcium hydroxide? What was the pH?

Velioğlu, S. G. (Turkey): The pH was neutral, around 7. Calcium hydroxide was added to the aluminium and ferric salts to achieve particle destabilization, the main cause of coagulation. The negative charge of the destabilization process was neutralized, and a great removal level was obtained. This destabilization can also be achieved by adsorption, bridging and enmeshment in a precipitate. Calcium hydroxide was especially chosen to speed up the precipitation of the salts and the removal of the colloidal particles. Calcium carbonate would not have produced such good results on account of its low alkalinity.

Taygun, N. (Turkey) : If with calcium hydroxide the pH is increased to around 11 and magnesium chloride is added to it until the pH is 10.5, I believe that an even greater increase in the settling rate, as well as a change in colour, will be observed.

In Situ Preparation of Self-Cleaning Ultrafiltration Membranes

Ö. VELICANGIL and J. A. HOWELL

Department of Chemical Engineering, University College of Swansea, Wales, UK

ABSTRACT

To circumvent the severe flux losses encountered during ultrafiltration of macromolecular solutions, enzymes were immobilized on the membranes to hydrolyze the deposited solute molecules. This resulted in 30 to 50% flux enhancements, averaged over a 24 h period, during a daily run. A mathematical model was developed to explain gel formation and the action of the enzyme precoat on the membrane surface.

INTRODUCTION

The separation and concentration of macromolecular solutions and colloidal suspensions by ultrafiltration suffers from a major drawback, significant flux losses due to the formation of a gel layer. The effect of cleaning the membrane surface with conventional detergents or with dilute bases wears off after a relatively short period, and around two thirds of the total subsequent flux drop occurs within the first 5 h of a routine 20 h when ultrafiltering cheese whey (Velicangil and Howell, 1977).

To provide further insight into the formation of the gel layer, a comprehensive mathematical model was developed. A parallel investigation was carried out on the preparation of self-cleaning membranes by attaching proteolytic enzymes on to the surface of UF membranes so that the clogging proteins were hydrolysed as they were deposited and the hydraulic permeability of the gel layer was improved as a result.

MATERIALS AND METHODS

Ultrafiltration membranes were obtained from Amicon Ltd., High Wycombe (types PM-10, PM-30, XM-50). Papain (E.C.No.3.4.4.10), twice crystallized from papaya latex, and Bovine Albumin (Cohn Fraction V) were purchased from Sigma Chemical Co. Ltd., Poole, Dorset. Corolase S100 (an industrial papain product from carica papaya, purified and activated by a special process) was kindly donated by Röhm GmbH, Darmstadt, West Germany. Cheddar cheese whey was regularly obtained from Unigate Ltd., Johnstown, Carmarthen.

The flat ultrafiltration cell (150 mm in diameter) was of the thin-channel spiral type with a 1.5 mm channel depth. It was connected to a centrifugal pump which delivered the feed solution at 5 l/min up to a pressure of 4 ats. The permeate and retentate were returned to the system and thoroughly mixed in the feed reservoir. Temperature and pH were controlled by probes which were installed in the feed tank. To simulate operational conditions in the plant, experiments with Cheddar cheese whey were carried out at 50°C and in the turbulent region (R_e = 15200) with PM-10 membranes. Those experiments with 0.5% Bovine Albumin were conducted at 25 or 30°C and at R_e = 6000 employing PM-30 and XM-50 membranes.

Two separate attachment procedures were used. The general procedure adopted to attach refined papain onto the membrane was as follows. Prior to immobilization, the membrane surfaces were etched in 3N HCl at 45°C, the solution having been stirred for 3 h. Then they were perfused with the protease in buffer containing 0.15 M KCl for about 1 h at 4°C. This was followed by circulation of 0.1% glutaraldehyde solution (pH 8.5) in the module at room temperature for 50 minutes. The membranes were then treated further with ice-cold 0.05 M $NaBH_4$ for 20 minutes to reduce the remaining aldehydic groups and thus to prevent their covalent cross-linking to whey proteins in subsequent experiments. Finally, the membranes were washed free of non-crosslinked enzymes with 1N NaCl.

For those experiments with crude proteases, a very simple immobilization procedure was employed. The enzyme was dissolved in buffer to a concentration identical to the protein concentration in cheese whey (0.6%). This solution was ultrafiltered through the membrane at 5-10°C until the permeate flux dropped to the values previously observed with cross-linked enzyme.

The amount of bound protein was determined by measuring the difference using the method of Lowry and others (1951). The proteolytic activities of the free and immobilized proteases were assayed by the standart Anson method with haemoglobin as substrate (Anson, 1938).

MATHEMATICAL MODEL

Ordinary Membranes

Although the recently developed UF equipment can to a certain extent avoid the shortcomings of stagnant filtration cells by employing high recirculation rates and minimizing the channel dimensions, thus increasing the shear at the membrane surface, it cannot completely overcome the effects of concentration polarization.

Existing models of ultrafiltration (Blatt and others, 1970; Porter, 1972) attribute membrane fouling to the buildup of a uniform gel layer on the membrane surface by a concentration polarization mechanism. Where the surface concentration is below that of the gel, it is postulated that there is a resistance to the solvent flow relative to the solute, which affects the flux as though it were a solid layer. When cheese whey is filtered, the gel layer is found to be complex, being made up of components of widely differing molecular weights ranging from lactalbumin at around 10 Kdaltons to residual caseinate complexes at around 300 Kdaltons.

Analysis of the concentration polarization mechanism is carried out by means of the conservation equation for solute across the boundary layer, ignoring flow and gradients in the plane of the membrane,

$$\frac{\partial C}{\partial t} = \mathcal{D}\frac{\partial^2 C}{\partial x^2} - u\frac{\partial C}{\partial x} \tag{1}$$

with the initial and boundary conditions   t = 0   C = C_b

$\qquad\qquad\qquad\qquad\qquad\qquad$ x = 0   C = C_b

$\qquad\qquad\qquad\qquad\qquad\qquad$ x = L $\frac{\partial C}{\partial x}$ = uC

C_w is the solution concentration at the wall which is given at $x = L$ by

$$\frac{C_w}{C_b} = e^{\frac{uL}{D}} + \Sigma\, a_n\, e^{-\lambda n^2 t}\, x_n \left(\frac{\lambda_n^2 D}{u^2}, \frac{uL}{D} \right) \tag{2}$$

where λ_n and x_n are the eigenvalues and eigenfunctions respectively of the Sturm-Liouville problem associated with Equation (1).

The flux across the membrane is given by

$$u = \frac{\Delta P}{\mu(R_m + \frac{\ell}{P_g})} \tag{3}$$

The permeability of the gel layer P_g is calculated from the Carman-Kozeny equation

$$P_g = \frac{d^2}{180} \frac{\varepsilon^3}{(1-\varepsilon)^2} \tag{4}$$

Reasonable values for the physical parameters were assumed, listed as in the Nomenclature section at the end of this paper. These quantities relate specifically to our experiments on the ultrafiltration of cheese whey. As the flux decreased rapidly over the first minute and as the solution quoted assumed a constant velocity the time constants of the solution were first established. The first and smallest value of λ^2 for a transverse velocity of 3.72×10^{-5} ms^{-1} (which was reached at the end of the first minute) was 0.22s^{-1}.

The eigenvalue was larger at higher fluxes, and thus the maximum time constant for the full effect of concentration polarization to occur would be 5 seconds. The experimental technique employed did not permit any flux measurement for periods shorter than six seconds, and thus the flux decline due to the concentration polarization effect could not be observed. In fact, it is likely that there would be an even faster effect in view of the limited area occupied by the pores on the upper membrane surface. This space was shown by H. J. Preusser (1972) to be less than 0.05% of the total membrane area; thus, velocities through the pores would be 2000 times or more higher than those towards the membrane. Concentration polarization could be expected to occur in the pores within a small fraction of a second. If one then calculated the thickness of the gel layer required to cause the flux through the pores to drop from that observed with distilled water to that observed after a ten-second ultrafiltration of protein, assuming that the Carman-Kozency law held, the resulting thickness would be of the order of a few Angstroms, in other words, of molecular dimensions (thus, the Carman-Kozeny law was unlikely to be valid). The result suggested, however, that initial flux drops could be caused by the blocking of a percentage of the pores by a single protein molecule convected there during the first moments of ultrafiltration. If the distribution of pore sizes in an XM-50 membrane as measured by H. J. Preusser (1972), were taken and if it were assumed that the pores of sizes close to the diameter of the solute molecule would be blocked but that the remainder would still pass solvent, then the resultant flux drop would match closely that observed over the first stages of flux decline.

To verify this hypothesis, three different protein solutions were ultrafiltered through an XM-50 membrane (molecular cut-off: 50,000 daltons), Papain (M.W. 21,000),

Ovalbumin (M.W. 45,000), and Serum Albumin (65,000). The observed 1st minute flux drop with each of these runs compared well with the predicted values derived from the percentage area of the pores likely to be blocked when each particular solution was passed through the membrane. For this purpose, the pore size distribution function of the membrane (Preusser, 1972) was integrated over the appropriate molecular dimensions.

To determine whether the gel concentration was actually reached at the membrane, the steady-state transverse velocity, u_s, was evaluated from the steady-state solution of Eq. (1) with a wall concentration equal to the gel concentration:

$$u_s = \frac{D}{L} \ln \frac{c_w}{c_b} \qquad (5)$$

where $c_w = c_g$.

The value for $k_m = D/L$ was calculated from the Dittus-Boelter correlation in thin channels. The predicted value of u_s was very close to the water flux, but significantly higher than the earliest measured flux with cheese whey. This gel polarization based on the total membrane surface could not account for flux drops below this.

Further calculation then showed that the flux declined below that which would cause a gel concentration to be present at the membrane surface by concentration polarization; yet the calculated gel layer was much smaller than that observed at later stages. It was therefore postulated that further buildup of the gel could occur by adsorption of protein molecules on to the membrane surface to form a monolayer. Chemisorption then took place, continuing for several hours and building up an increasing thickness of gel layer. The kinetics of such chemisorption should be observable; it was assumed in this work that it occured as an nth order reaction and it was attempted to fit the data with the best value of the index, n.

The analysis of the various flux data of the UF process yielded a first order relationship between the rate of gel layer growth and the wall concentration, c_w:

$$\frac{dl}{dt} = k_r c_w \qquad (6)$$

Self Cleaning Membranes

To explain the enchanced fluxes when using membranes pretreated with enzymes, an enzyme activity term was incorporated into Eq. (6):

$$\frac{dl}{dt} = k_r c_w - k_1 \exp(-k_2 t) \qquad (7)$$

By substituting Eq. (5) into Eq. (7) the following expression was obtained for the rate of gel layer growth:

$$\frac{dl}{dt} = k_r c_b \exp(u_s / k_m) - k_1 \exp(-k_2 t) \qquad (8)$$

By means of a computer program (non-linear optimization by least squares), the parameters k_1 and k_2 of Eq. (8) were evaluated. They compared well with those for free solution activity of the enzyme. Although the overall activity of papain (k_1) was decreased by 90% upon immobilization, the activity decay constant, k_2, was also decreased by 50% (from 0.1 to 0.05), thus doubling the enzyme half-life.

ULTRAFILTRATION WITH IMMOBILIZED ENZYME MEMBRANES

Separation of 0.5% Bovine Albumin Solution

Food-grade crude papain was introduced onto the PM-10 and PM-30 membranes by dissolving the enzyme to a 0.5% concentration (actual protein content: 0.2%) in acetate buffer and recirculating this solution in the system for 5 to 12 minutes prior to the UF of albumin. For some experiments a glutaraldehyde solution was subsequently circulated to cross-link and fix the enzyme, whereupon the excess glutaraldehyde was reduced by perfusion with $NaBH_4$. At the end of a 20 to 22-h UF run, the exhausted enzyme was removed from the membrane by detergent cleaning.

In these experiments either the same membrane was used alternatingly as prototype and then as control in consecutive runs, or two different membranes with identical history were employed as control, respectively. The flux enhancements for both cases were the same. In one such set of experiments, consisting of 21-h runs, 26% flux improvement was obtained with an enzyme membrane processing native albumin and 23% improvement, when processing heat denatured albumin. In another attempt where papain was fixed onto the membrane with glutaraldehyde and no cleaning was performed between the two 24-h runs, by the end of the second day the cumulative permeate was 214% higher than that of the control.

The net protein loss through the membrane due to the cleavage of filtered albumin by the active enzyme was found to be between 5 to 7% of the processed protein.

The results of similar experiments employing the same membrane as prototype and as control are presented in Fig. 1. A papain adsorbed membrane (activated by adding to the albumin solution, 0.005 M cysteine and 0.002 M EDTA) exhibited 44.4% improvement in the cumulative permeate over the control. The flux loss for the control over a 22-h period was 55%, but the prototype showed only 17.6% flux decline over the same period. In another run without activators, a 31.2% increase in permeate yield was obtained, although after the first 6 h the flux decline accelerated presumably due to the rapid inactivation of papain, and the final flux value matched the 22-h value of the control. The fourth experiment was designed to distinguish whether the flux enhancements with various prototypes were due to the physical effect of the adsorbed papain layer as a prefilter coat or due to its biochemical action. To this end, 0.006% H_2O_2, instead of the activators, was introduced into the feed solution to inactivate the papain completely. After 6 hours, the feed solution was replaced by a fresh batch containing activators and no H_2O_2. Although up to this point the rate of flux decline was identical with the control, a sudden recovery was observed and the cumulative permeate was 13.4% higher within the next 16 hours.

Cheese Whey Processing

UF of whey with PM-10 membranes was carried out in the presence of the activators for papain (crystalline grade). The control was operated for 20 hours every day. The remaining 4 hours were spent cleaning with 0.1N NaOH, H_2O and then with 2% Dymex solution at 50°C. In contrast, immobilized enzyme membranes were used continuously during the five days of a run. The PM-10 control membrane suffered a sharp loss over the first 5 hours of a 20-h run. It had then to be actively cleaned to restore a reasonable flux. In comparison with the control, enzyme-modified membranes exhibited 60% and 278% higher permeate yields. Initially, flux loss due to immobilization of papain amounted to 40% for the enzyme loading of 0.0048 mg/cm^2. The flux of this membrane matched the continuosly dropping flux of the control in the 7th hour.

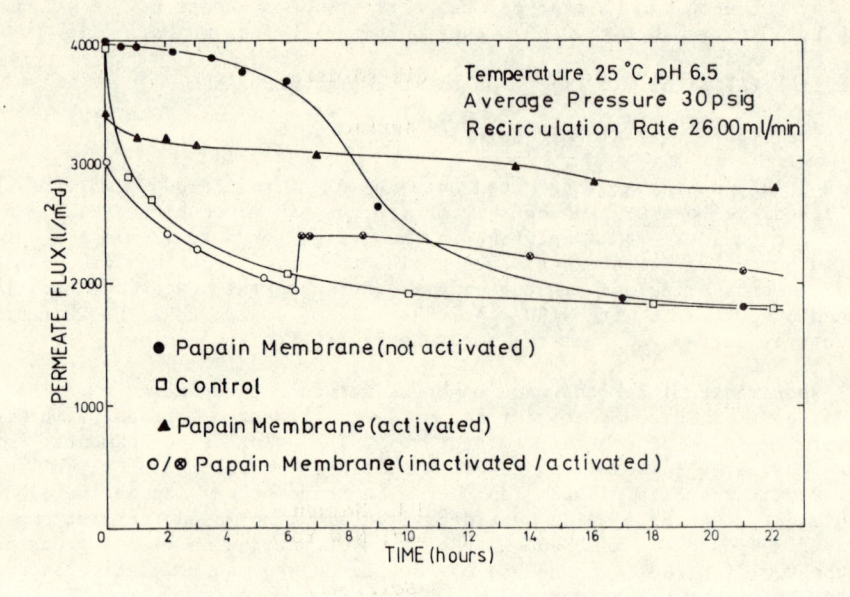

Fig. 1: Ultrafiltration rates on ordinary and pretreated PM-30 membranes for 0.5% Bovine Albumin Feed Solution

CONCLUSIONS

Initial flux drop in ultrafiltration of cheese whey or other proteins was probably due to local convective deposition of molecules of protein close to or in the pores. This process, complete in less than a second, caused the flux to drop below the level which would cause gel polarization at the surface by convection. Flux drop in the succeeding minute occurred too slowly to be due to gel polarization but was much faster than the subsequent flux decay over a period of hours. The former decay was attributed to adsorption of a monolayer at the membrane surface, and the latter decay was attributed to the reversible polymerization of protein to gel, or chemisorption. The process was first order with respect to the wall concentration, c_w.

A reproducible and inexpensive method was developed to introduce proteases onto the UF membranes, and 23% to 44% flux improvement was obtained during a standart 22-h run when processing 0.5% albumin solution. This technique was completely compatible with the requirements of the dairy industry and did not interfere with the routine in a whey-processing plant. On the other hand, the simple adsorption technique of crude enzyme did not prove to be effective when processing cheese whey, probably due to the high polymers of lactoglobulin, residual caseinate complexes etc., building a gel layer of a different nature from that produced by single protein solutions (Lee and Merson, 1976). Cheese whey ultrafiltration was effectively improved when improved recrystallized enzyme was covalently cross-linked onto the membrane.

NOMENCLATURE

D	$50^\circ C$	Diffusivity of proteins in whey	$16.1 \times 10^{-11} m^2 s^{-1}$
d		Average diameter	3×10^9
c_b		Solute concentration in the bulk solution	$0.5\% - 0.6\%$
c_g		Gel concentration	60%
c_w		Concentration at the membrane surface	
k_m		Mass transfer coefficient	
k_r, k_1, k_2		Rate constants (Eq. 6, 7, 8)	
l		Gel layer thickness	
ΔP		Transmembrane pressure	$1.58 \times 10^5 N\ m^{-2}$
P_g		Gel permeability	$2 \times 10^{-20}\ m^2$
R_m		Membrane resistivity	$13 \times 10^{-11}\ m^{-1}$
ε		Voidage of gel layer	0.5

REFERENCES

Anson, M.L. (1938). *J. Gen Physiology*, 22, 79

Blatt, W.F., A. Dravaid, A.S. Michaels and L. Nelsen (1970). In *Membrane Science and Technology*. Ed. J.E. Flinn. Plenum Press, New York, 1970, pp. 47-97.

Lee, D.N. and R.L. Merson (1976). *J. Food Sci.*, 41, (2), 403.

Lowry, O.H., J.N. Rosebrough, A.L. Farr and R.J. Randall (1951). *J. Biol. Chem.*, 193, 265.

Porter, M.C. (1972). *Ind. Eng. Chem. Prod. Res. Develop.*, 11 (3), 234.

Preusser, H.J. (1972). *Kolloid-Z.u.Z. Polymere*, 250, 133.

Velicangil, O. and J.A. Howell (1977). *Biotech. and Bioeng.*, 19, 1891.

A Case Study of Food Production Wastewater Difficulties

M. SUERTH

University of Notre Dame, Notre Dame, IN, USA

ABSTRACT

A case study is here presented of a specialty cheese plant, located in the dairy belt of the United States, which experienced wastewater difficulties. The plant's attempts at solving the problem are delineated along with the results of its efforts. Included is an adaptation of the Benedict's solution test, which in this particular situation is able to give an immediate indication of the BOD_5 level.

INTRODUCTION

The wastewater difficulties encountered by a specialty cheese plant located within the dairy belt of the United States are the focus of this study. The dairy belt of the United States is formed by a band approximately two states north to south, beginning in the East in the states of New York and Pennsylvania and travelling westward through Ohio; Michigan and Indiana; Wisconsin and Illinois; then into Minnesota, Iowa and Missouri. The cheese industry is located in the dairy belt because of the perishability of the milk, as well as the presence of surplus milk, loss of bulk which occurs in the cheesemaking process, and labor considerations, which include the possibility of lower money wages and the greater probability of finding the all-important highly skilled cheesemaker in the locale.

The firm on which this case study is made produces American blue cheese and is located on a high portion of land overlooking a horseshoe bend in the mighty Missippi River in the small town (population about 1200) of Nauvoo, Illinois. The cheese firm has been there since the 1930's, when the American blue cheese process was first perfected. The site was chosen because it was within the dairy belt having all the positive locational benefits delineated above and also because of the presence of limestone caves, which at that time were thought necessary for the aging of blue cheese. These caves were in fact used for the aging of the cheese at that site until 1978.

THE RISE OF WASTEWATER POLLUTION

Originally, cans of milk were brought by the farmers to the "milk door"ofthis specialty cheese plant which is located within the dairy belt of the United States. At the back of the plant the cans were picked and then filled with whey which was taken back to the farms and fed to the animals. All the cheesemaking was done by hand.

As business grew and as Federal Government sanitation regulations became more restrictive, it became imperative to go out to the farms with tank trucks to pick up the milk and bring it to the plant. A critical environmental juncture was here reached as it was no longer feasible to give the whey to the farmers. It was then that the whey was disposed of by sending it down the drain in the wastewater which went to the sewage plant and on into the Mississippi after some cleansing of the water. Thus, a contribution was made to the pollution of the river's water.

The general trend, then, of the expanding technological society can be seen in this individual firm: greater industrial growth, increased government regulations, changes in production methods, and increased pollution of the natural environment. All these factors were interdependent. At that point in time, however, almost complete ignorance of the problem of pollution prevailed. The factors people were more cognizant of were that they were wasting material which had some food value, and that it cost the manufactures more to dispose of the whey by using the town sewage facilities than it cost to give it to the farmers. However, no other alternative could be envisioned in the situation.

Meanwhile, as a slight but steady growth marked the passage of years, some small amount of automation was introduced into the cheesemaking process in Nauvoo. But the whey continued to be sent down the drains in the wastewater. As Federal sanitation regulations increased, more case than ever was taken to keep the floor, other surface and all equipment clean. But when the waste materials were collected, they were disposed of through wastewater drainage. Consequently, the pollutant level of the river was constantly rising. This story was repeated throughout almost the entire technological world.

FIRST ATTEMPTS TO ALLEVIATE WATER POLLUTION

Slowly it became apparent that the threshold of the water's inability to cleanse itself further was being reached. By this time it was generally recognized that whey was a major pollutant of the natural waterways. Consequently, in Nauvoo's cheese plant whey collectors, much new piping and a huge drier were installed to permit the drying of the whey. The dried whey was then sold as cattle feed. The United States Government tried to develop dried whey into a product which could be given or sold cheaply to developing countries as a food supplement. Though it was used to some extent, this was never a great success. However, dried whey is used in many manufactured food products in the United States. Most of this dried whey comes from the very large cheese companies such as Kraft and Borden. All of the dried whey produced in Nauvoo is still used as animal feed. Selling the dried whey eventually offset the cost of the process of drying it, so that this was no longer a sustained economical drain on the firm. Most importantly, it kept the major portion of the whey out of the wastewater and out of the river.

The waterways, though, were not recovering and were in fact deteriorating. In part, this was because all the factors involved were not recognized and also because many companies were hampered by inadequate tecnical knowledge and/or ability to cope with the problems. In some cases, industries were not trying to lessen pollution.

COMMUNITY CONFLICT

Because we were becoming more aware of the problem of pollution and because the apparent pollution seemed to be increasing, the United States Environmental Protection Agency regulations restricting the contents of wastewater became tighter. These regulations first applied to water entering the natural waterways and consequently, were most immediately felt by the water treatment plants.

Thus, in the early to mid-70's a conflict developed within the Nauvoo community between the water treatment plant officials on the one hand and the cheese firm manager on the other. Though extra precautions were taken by the cheese plant not to spill milk, whey or curd, and to keep them as much as possible out of the wastewater, the BOD_5 level, particularly, was claimed by the water treatment officials to be grossly above that acceptable, some times exceeding 4000 ppm.

Parenthetically, it should be noted that the BOD_5 test measures the biochemically caused demand for oxygen made on the water by a substance during the first five days of its presence in the water. It is generally agreed that by the end of five days approximately 65-90% of the demand will have been made. Since oxygen is vital for most forms of life in the waterways, a substance which demands a high amount of the oxygen in the water can be lethal to the life forms therein. Pure whey has a BOD_5 of about 37,000 to 42,000 mg/ℓ ppm. The United States Environmental Protection Agency regulations at this time would allow only 350 ppm to enter the waterways legally.

Nor it was agreed just what the BOD_5 levels of the plant effluent actually were since even independent testing firms, to which samples of the water were sent in an effort to resolve the discrepancy, did not achieve consistent results. This, of course, was due in part at least to the time lapse before testing and also to the lack of true composite samples.

TESTING AND ANALYSIS

After much discussion it was agreed that the cheese plant would hire someone to test the BOD_5 of a 24-hour composite sample every day. Equipment to provide the 24-hour composite sample and lab equipment to do the testing was purchased and installed in the cheese plant. Since the town was small, without professional chemists or lab technicians, and since there was little attraction in a position which, as described, would require about one hour per day all seven days a week, finding someone to perform the tests proved to be one of the most difficult aspects of this phase. Finally, the author, then teaching science in an academy in the community, was asked to perform the necessary tests. Because of the potential of helping to alleviate an environmental as well as a community problem, the author agreed to do the testing.

During the first week or two, only the BOD_5 test was run. Then a rather exhaustive study was made on the town water and on well water from the plant property, both of which were used in the production and processing of the cheese. The results of these tests were compared with the results of the same tests run on the wastewater. It was decided that ten tests would be run daily on the 24-hour composite sample of wastewater. The tests were chosen because of the known significant effect of the quality or substance on the wastewater or because of the test's potential to indicate possible courses to action to alleviate the wastewater problem. The tests included were: BOD_5, suspended solids, settleable solids, total solids residue on evaporation, total dissolved solids, apparent color, chlorides, pH, turbidity and nitrogen, ammonia. Daily records had been and continued to be kept on water usage and water flow.

By daily testing for several weeks, the existing levels of these substances in the plant effluent were established. The BOD_5, suspended solids and settleable solids were then considered to be the most important at this plant. The BOD_5 at this time usually varied between 2200 and 2700 mg/ℓ, averaging about 2400 mg/ℓ. Occasionally, however, it would exceed 3200 mg/ℓ. While a few months previously the water treatment plant officials claimed the cheese plant effluent BOD_5 levels were approximately 4000 mg/ℓ, as noted above, these tests were not performed on true composite samples.

Comparison of the existing levels with those desired gave some indication of the immense task ahead. An awareness campaign was carried out on the plant workers by dissemination of information and by daily posting of the results of the wastewater testing. This comparison of fluctuations in the daily levels with changes in the existing conditions in the plant brought to light areas and activities which tended to cause deterioration in the wastewater quality. Innumerable surveys of every portion of the plant were made in an attempt to discover changes which, if made, would help alleviate the problem.

As each change was made, test results showed an improvement in the wastewater. The BOD_5 level started to decline. Most other undesirable substances and characteristics were well within Environmental Protection Agency regulations or reacted favorably to the first or simpler changes made. It gave immediate feedback to the workers and an incentive to keep the levels down. Labor cooperation was very high and seemed to correlate with posted results.

Occurrences like milk spills were immediately apparent in the settleable solids level. Chlorides, pH, and nitrogen, ammonia were due primarily to cleaning agents and were readily kept within acceptable levels with minor, if any, changes. Since settleable solids are removed in the primary wastewater treatment, they were not in themselves the factor of major concern. Settleable solids varied most with milk spills and when one took place, there was also a moderate increase in the BOD_5. A change in whey content also produced a slight increase in settleable solids. Some correlation was found between the degree of turbidity and the concentration of suspended solids; whereas the latter correlated highly with fluctuations in BOD_5 (Table 1). The relation observed between BOD_5 and suspended solids was expressed by $BOD_5 = 2.60\,SS$. The correlation coefficient of this relation was 0.78.

Table 1: Correlation of BOD_5 and Suspended Solids (SS) Monthly Averages

Month	BOD_5 (mg/ℓ)	SS (mg/ℓ)	BOD_5/SS
Sept. 1973	3220	1296	2.48
October	2465	1043	2.36
November	1300	576	2.25
December	1154	506	2.28
Jan. 1974	806	376	2.14
February	987	438	2.25
March	737	340	2.16
April	723	308	2.34
May	749	328	2.28
June	996	333	2.99
July	962	324	2.96
August	1046	344	3.04
September	857	333	2.57
October	874	328	2.66
November	958	373	2.56
December	734	344	2.13
Jan. 1975	690	323	2.13
February	697	279	2.49

As mentioned earlier, the BOD_5 level declined as each change dictated by the study was put into effect. The BOD, nevertheless, was still at a level far above that which would be acceptable. A barrier to further improvement was found at this especially crucial BOD_5 point since five days were required to complete the test. It proved difficult for the workers to remember what conditions had prevailed five days previously. Continuous testing showed that the situation in this plant traced a lif or log curve in the five-day BOD test period. With experience one could predict the final results rather accurately by the third day, but the first and second day were unreliable predictors. Third day predictions were still not sufficient to allow people to remember former conditions.

ADAPTATION OF BENEDICT's SOLUTION TEST

It was discovered that a modification of the Benedict's solution test would in this particular situation give an immediate indication of the BOD_5 level which would be found five days later. Since this was a specialty blue cheese plant producing only that one product, the only simple sugar in the wastewater could be assumed to be lactose from the whey. Therefore, Benedict's solution, which tests for simple sugars, could be used to test for lactose and, hence, for whey in the wastewater. Though the test is usually a qualitative rather than a quantitative one, a solution can be made which gives a positive result for 1 part in 1600, that being equal to 625 ppm. If the wastewater gave a positive result on the addition of Benedict's solution, the wastewater was carefully diluted, then tested again. This process was repeated until the diluted wastewater just gave a positive result and further dilution gave a negative result. The ppm of assumed lactose from the whey could then be easily calculated by multiplying the dilution factor by 625 ppm; e.g., if the wastewater were diluted to twice the original volume, the assumed lactose level would be 2x625 ppm or 1250 ppm. Similarly, if the original wastewater gave a negative result, an amount in a graduated container was carefully concentrated by evaporation until it just gave the same positive result. In this case if the wastewater were concentrated to 4/5 the original volume, the assumed lactose level would be 4/5 x 625 ppm or 500 ppm.

This proved to be a major breakthrough in two respects. First, the results of this testing proved to correlate well with the BOD_5 test results (Table 2). The indication, therefore, was that by far the major factor in high BOD_5 levels was, in fact, whey content rather than milk, curd, aged cheese lost in packaging or some unknown factor. Secondly, though an exact numerical value for the BOD_5 could not be given, as factors other than lactose were involved in varying degrees, though to a lesser extent, one was able to determine immediately whether the BOD was high. With experience and in conjunction with other tests such as suspended solids and settleable solids, a rather close approximation to the numerical value of the BOD_5 could be given. This allowed for an immediate and much more direct tracing of result back to cause, which in turn indicated necessary changes. For example, a combination of lactose and apparent color tests with water flow charts indicated that one, often major, source of elevated BOD_5 counts was highly concentrated burned whey from the drier. Changes were made in the temperature and depth of the whey in the drier to help alleviate this problem.

IMPROVEMENTS

The plant manager, officials and the author, in close cooperation with the consulting engineering firm, were able to bring about several changes in addition to those mentioned above, such as in the drier controls, completely enclosed batches, new processes for filling the hoops with curd, addition of differently designed drain catchers, in-line pipe cleaning, differently designed packaging processes, and a modification in the cheese-making process, all of which resulted in the improvement of the quality of the wastewater, and, therefore, less pollution was caused to the environment.

An additional positive outcome was the improvement in relations between the town water-treatment officials and the cheese plant, which developed to the point at which wastewater samples were exchanged and test results compared amicably. This improvement in relations seemed to be dependent upon two factors; first, the availability of true 24-hour composite samples for which test results could be relied upon, and secondly, the evident decline in BOD_5 values as successive changes were made in the cheese plant.

Table 2: Correlation of BOD_5 and Lactose (LACT) Monthly Averages

Month	BOD_5 (mg/ℓ)	LACT (mg/ℓ)	BOD_5/LACT
Oct. 1973	2465	1003	2.45
November	1300	548	2.37
December	1154	465	2.48
Jan. 1974	806	278	2.89
February	987	322	3.06
March	737	257	2.87
April	723	249	2.89
May	749	260	2.88
June	996	331	3.01
July	962	339	2.83
August	1046	362	2.89
September	857	278	3.08
October	874	273	3.20
November	958	312	3.07
December	734	244	3.01
Jan. 1975	690	230	3.00
February	697	234	2.98

CONCLUSION

After approximately a year and a half, the BOD_5 level was one-third to one-fourth of its original level as a result of the changes made (See Table 2). All other factors were well within United States Environmental Protection Agency guidelines. Additionally, the quality control of the cheese, which was developed concomitantly with the improvement in the wastewater, allowed for a great expansion in the production and marketing of the cheese. Efforts to improve further the wastewater quality continue

DISCUSSIONS

Samsunlu, A. (Turkey) : You spoke about the BOD tests at first giving negative results and then later, after heating and evaporation, giving positive results. Could you explain this further?

Suerth, S. M. (U.S.) : This was in the modified form of the Benedict's solution test used to discover the cause of the high BOD levels. By determining the lactose and, therefore, the whey content in the wastewater and by comparing that with the BOD test, I found that the whey was the major factor contributing to high BOD levels. Changes were made accordingly in the plant.

Treatment of Yeast Factory Waste

H. M. RÜFFER

University of Hannover, FRG

ABSTRACT

The wastewater from yeast factories is highly concentrated with organics. Its discharge into a river or other receiving waterbody results in the wellknown disadvantages, except for a very high dilution. Treatment therefore, is in most cases indispensable.

A summary of the yeast production process will first be given. In Germant molasses from beet-sugar production is the main raw material. This summary will be followed by a synopsis of the wastewater quantity and quality. During modernization of the process the quantity has permanently been diminished and the concentration increased.

Finally, using samples from German factories, several possibilities of treatment with respect to disposal of such wastewater will be illustrated for instance, irrigation, anaerobic/aerobic biological treatment, and evaporation. Information will be given on efficiency, cost and specific problems.

INTRODUCTION

The waste from yeast factories is highly concentrated with organic substances. It is comparable with that from other fermentation processes like penicillin production, wine-producing and ethanol distillation, as well as the brewing process (Sierp, 1967; Meinck and others, 1968).

As in other industries, there has been a tendency during the last 10 years towards having larger factories, so we have now only 10 yeast-producing factories in Western Germany (Ruffer, 1977). The production ranges from 4 t/d yeast and 300 ℓ/d ethanol to 150 t/d and 25 000 ℓ/d, respectively. The greatest part of the yeast is used by baken; only a small part is used by pharmaceutical firms (e.g. ephedrine-production) and for yeast extract.

PRODUCTION

For about 100 years the aeration process in tanks has been used in Europe. The material of the tanks is now mostly stainless steel. The raw material is molasses; the residue of the sugar production process is a black syrup with about 50% saccharose. It is diluted with water, sterilized by heat and centrifuged to get rid of insoluble material, namely, "molasses sludge", which though of small volume, has a high content of oxidisable substances caused by sugar etc. Some factories use it for fermentation and for producing additional ethanol by distillation afterwards, i.e. without preseparation of the yeast. The yeast (saccharomyces cerevisiae) is grown in a substrate of molasses in water, completed by NH_4^+, PO_4^{---}, and H_2SO_4 with a pH of 4.5 to 5.5 in several stages (Fig. 1). In the first stage, and mostly also in the second, where a high degree of sterility is necessary, the air supply is reduced to force the cells to a partly anaerobic fermentation ($C_6H_{12}O_6 \to 2\ C_2H_5OH + 2\ CO_2 + 234$ k joule) besides the main process of sugar oxidation ($C_6H_{12}O_6 + 6\ O_2 \to 6\ CO_2 + 6\ H_2O + 2872$ k joule). In the last stages suspension is provided with surplus air. Cooling is necessary (Reiff, 1962), with respect to the exotherm reactions, to guarantee the optimal temperature of 30° to 32°C during the whole process of about 12 h. The yeast is separated then after each stage from the liquor by centrifuging. The 0.5 to 0.8% volume of ethanol in the liquor from the first stages is obtained by distillation discharging a hot, brown liquid residue still containing a high concentration of organics.

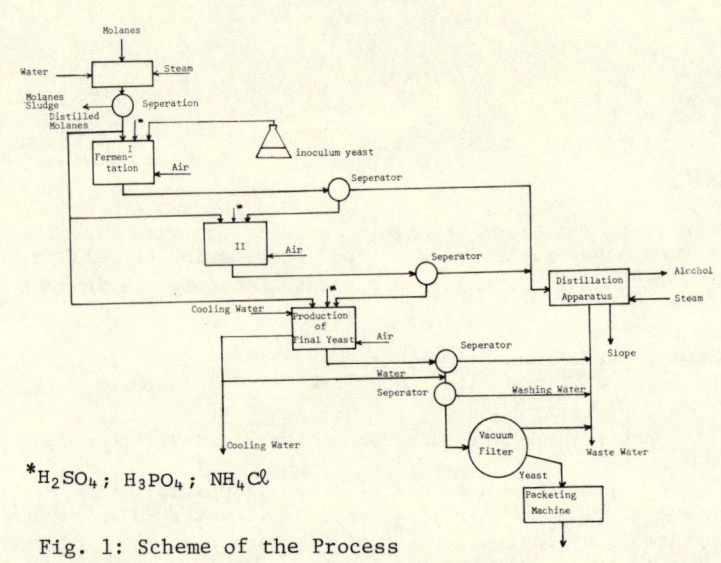

Fig. 1: Scheme of the Process

The liquid residue from the last stages forms the main part of the wastewater; it is also highly concentrated. Of a lower concentration is the washwater. (The separated yeast is diluted with water and again centrifuged, mostly 2 times. In modern factories the second wash water is used for the first washing).

The yeast is then suspended in water with about 15% dry matter, and is dewatered on vacuum filters to about 24% dry matter. At this concentration it is sent to the packeting machines.

AMOUNT and CONCENTRATION of WASTEWATER

Molasses show some differences in composition from year to year and also regionally. The main part of the oxidisable matter in the waste is the unfermented substances from the molasses, and additionally the byproducts of the metabolism of the yeast (Bronn, 1975). The concentration differs widely, because the concentration of the substrate varies from factory to factory. (Sierp, 1967; Meinck and others, 1968; Ruffer, 1977; Reiff, 1962; Bronn, 1975; Thommèl, 1969).

There is a tendency to use liquor of a higher concentration particularly because of the smaller amount of wastewater to be disposed of. The differences in effluent concentration during the day, a result of the discontinued release of the final liquors should also be taken into account. On the other hand, there should be a good conformity in the load (COD, BOD) with respect to the molasses originally used.

Table 1: Concentration of the Waste from Yeast Factories

COD (mg O /ℓ)	5 000 to 25 000
BOD$_5$ (mg O$_2$/ℓ)	4 000 to 20 000

Table 2: Load of Oxidisable Substances in Yeast-Factory Waste with respect to the Molasses Used

COD (kg O /t molasses)	140 to 250
BOD (kg O$_2$/t molasses)	120 to 220

(Bronn, 1975; Thommel, 1969)

Clean cooling water should be separately discharged. In the case of the economical use of water in the production process it is possible to obtain the following amount of wastewater.

Table 3: Amount of Wastewater in the Case of Using Highly Concentrated Substrate and Minimal Waste

liquid residue	4.3 m^3/t molasses
washing-water	4.0 m^3/t molasses
filtrate	0.6 m^3/t molasses
molasses-sludge } cleansing-water	0.6 m^3/t molasses
total wastewater	9.5 m^3/t molasses

(Bronn, 1975)

Of great importance is the content of sulfate. On the one hand it is damaging to concrete, when more than 400 mg/ℓ SO$_4^-$; on the other hand, the sulfate causes the production of H$_2$S under anaerobic conditions. The use of HCℓ instead of H$_2$SO$_4$ during the process is possible, but corrosion problems in the factory have prevented it in most cases. The composition of the wastewater is shown in Table 4.

Table 4: Analysis of Yeast-Factory Waste

Color	dark brown
Smell	typical of yeast prod.
Amount (m^3/t molasses)	10 to 40
Settl. solids (mℓ/ℓ)	0 to 5
pH	4.8 to 6.5
COD (mg/ℓ)	5000 to 25 000
COD (kg/ t molasses)	140 to 250
BOD$_5$ (mg/ℓ)	4000 to 20 000
BOD$_5$ (kg/ t molasses)	120 to 220
SO$_4^{--}$ (mg/ℓ)	600 to 1 200
N$_{total}$ (mg/ℓ)	500 to 1 200
P$_{total}$ (mg/ℓ)	10 to 50
K (mg/ℓ)	100 to 2 000

POSSIBILITIES FOR THE TREATMENT of THE WASTE

The wastewater is free of toxic material. The BOD:COD ratio of from 0.75:1 to 0.9:1 indicates a good degradability. This is found especially under anaerobic conditions (Ruffer, 1977;Kohler, 1973).

The wastewater from most German yeast factories is combined with the municipal sewers. In the case of a high dilution with sewage, there are relatively few problems. One has only to deal with the problems of high concentrations of SO_4, the heat of the waste, which sometimes causes odour problems, shock-loads of the treatment plant by the discontinuously discharged waste and a residue of brown colour in the effluent, combined with a somewhat higher COD.

In the FRG there are 4 factories with their own waste treatment facilities: the Friesische Hefe-und Spirituswerke Leer, Hesel; the Spiritus-und Presshefefabrik Ruttergut Falkenhardt, Diepholz; the Presshefefabrik J. Pleser Sohne, Darmstadt-Eberstadt; and the Uniferm Hefe-und Spiritusfabrik Monheim.

After the comparison of different possibilities, it was found that the only economic method for the 2 small factories in northern Germany was the irrigation of the waste on pasture land and meadows, which were available in these regions. This method has been used in Diepholz for over more than 12 years in Hesel for about 16 years. On the basis of a load of 20 kg BOD/ha.d, 25 ha in the first case (75 m^3/weekday) and 50 ha in the second case (145 m^3/weekday) were placed at their disposal. The pumping costs are very low, in the first case 1.5 DM/d. It is possible to irrigate warm wastewater also in wintertime. The waste does not contain matter of a doubtful character. If correctly used (only a few mm irrigation at once, no puddles), no smell develops. The fertilizing value is high (N, K); occasionally a small amount of P-salts has to be added. The investment costs are low (in the first case, 48 000 DM). Pollution in the river is reduced by 100%. The wastewater makes only into the first few centimeters of the soil and is used by microorganisms and plants. Wheat-corn-grass are grown in rotation e.g. in Germany. The only problem is the high content of potassium when the meadow is used by cattle. Sugar beets have the greatest uptake of K from the soil (Vetter, 1975).

The anaerobic treatment of this waste was first used by Merkel in Germany (Merkel, 1958; Pleser, 1965) after having been reported by Kiby (1959) from Kopenhagen in 1934. The firm of Pleser Sohne built such a treatment plant in Darmstadt 1955. The hot wastewater is digested in a two-stage digester (without heating). Detention time is 3 to 4 days. The BOD is diminished by 14 to 30%; the digester gas is of high heating value, produced at 0.5 m^3/kg BOD-removal. Difficulties arise from the high amount of H_2S in the digester gas and the digested waste. The latter is now pre-aerated in a closed tank before final treatment in trickling filters after dilution with (mechanically and biologically pretreated) sewage at a ratio of 1:14. The air in the tank is burnt with the digester gas. Efforts are still continuing to improve the cleansing effect of the plant. The investment costs of the digesters are estimated nowadays at about 1 dlio DM.

The most modern solution is that of the Uniferm Monhelm. This factory is situated near the river Rhein. The Lurgi Comp. constructed in 1976 a 6-stage evaporation plant. The last stage is doubled, one being charged, the other cleaned.

The very highly concentrated liquor (50 m^3/h) -without the washing water- is evaporated after preheating in a counter-current-heat exchanger, giving from 4% to 70-75% dry matter ($45^0 C$ = 318 K to $115^0 C$ = 388 K). The concentrated product, named vinasse, is sold to farmers, who can use it as additional feed, together with other fodder like sugar-beet pellets. The composition of the vinasse is shown in Table 5.

Table 5: Mean Composition of Vinasse from the Manheim Factory (Consulenschappen, 1974).

Dry matter	74.5 %
Moisture	25.5 %
Organic matter	50.2 %
Ash	24.3 %
Protein	26.3 %
Grease	0.1 %
Fiber	< 0.1 %
P_2O_5	0.2 %
K	6.11%
Na	1.9 %
Ca	0.5 %
Mg	0.8 %
Cℓ	1.7 %
pH inversion	5.2 -
Sugar (half) without inversion	2.8 %
with inversion	6.0 %
Starch units (CVB)	440. -

The costs of this treatment are high. 0.25 kg steam must be provided to 1 kg water distilled. The condensate and other diluted wastewater constitutes about 10% of the total BOD load. The biological treatment of this residue is being considered. The plant itself has been working for more than 2 years without any trouble.

The European sugar industry is still trying to increase the yield of sugar in the process. Methods are already being used, resulting in a smaller sugar content in the molasses, e.g. the luention-process with 42% sugar in the molasses. That means a greater amount of molasses used for the same yeast produced in the yeast factories; it means also a higher waste load (COD, BOD). As a result, the use of sugar in the future instead of molasses which would change the waste situation considerably, is already being considered in the yeast industry.

REFERENCES

Bronn, W.K. (1975). *Die Abwasserbelastung der Hefefabriken und Melassebrennereien:* die Branntweinwirtschaft.

Consulenschappen voor de Rundveehouderij en de Akkerbouw in Friesland (1974). Alvicoll, neprocoll en melasse als toevoeging aan gedrossgde pulp (brok)-Proevon bij Melkve op Vijf weidebedrijven. Gutachten.

Kiby (1959). *Chem. Ztg.*, *58*, 600. Zit. in Dietrich, K.R. *Ablaufverwertung und Abwasserreinigung in der Biochemischen Industrie.* Heidelberg, Germany.

Kohler, R. (1973). Anaerober Abbau von Hefefabrlkabwasser. *Wasser Luft Betrieb*, *17*, 342.

Meinck, F., Stoof, H., Kohlschutter, H. (1968). *Industrie-Abwasser.* 4. Auflage, Stuttgart.

Merkel (1958). Betriebserfahrungen bei der Abwasserreinigung einer Hefefabrik. *Wasser, Luft,* Betrieb *6*, 12.

Pleser, J. (1965). Technische Erfahrungen beim Ausfaulverfahren fur Abwasser *Vortrag Techniker-Tagung* 25./26.5.1965, Karlsruhe.

Reiff, F. u.M. (1962). *Die Hafen*, 2 Bde. Nurnberg.

Ruffer, H. (1976). Schwierigkeiten bei der Eindickung und Entwasserung von Abwasserschlammen aus Klaranlagen mit erhohtem gewerblichen Anteil unter besonderer Berucksichtigung von Brauereiabwasser. *Gewasserschutz Wasser Abwasser, 21,* 1.

Ruffer, H. (1977). Hochkonzentriertes Abwasser Beispiele wirtschaftlicher Reinigungsmethoden. *ISU,* (Aachen), *2,* 13.

Ruffer, H. (1977). Reinigung organisch hochbelasteter Abwasser aus der Nahrungsmittelindustrie *Münchn. Beitrage, 28,* 9.

Sierp, F. (1967). *Die gewerblichen und industriellen Abwasser.* 3. Aufl., Berlin.

Thommel, J. (1969). Abwasserprobleme in der Hefeindustrle. *Münchn. Beitrage, 16,* 233.

Vetter, H. (1975). Personl. Mitt. vom 18,6.1975 Landw. Unters. u. Forschungsanstalt der Landwirtschaftskammer Weser-Ems, Oldenburg.

DISCUSSIONS

Leentvar, J. (Holland) : I would like to know how effective anaerobic decomposition of the sludge in the second plant was in solving the wastewater problems. Were the anaerobic digestion units working as upflow reactors?

Rüffer, H. (Germany) : No, they were totally mixed reactors. Since this plant was established in 1955, efforts have been made to increase its efficiency. Recently, pilot plant studies have been made on improving the settleability of the sludge and on intensifying the decomposition of the sludge by returning it to the reactor.

A Study of Brewery Wastewater Characteristics and Treatment

S. MUTTAMARA and N. C. THANH

Environmental Engineering Division, Asian Institute of Technology, Bangkok, Thailand

ABSTRACT

The brewery under investigation produced approximately 100,000 m^3 of beer per annum. Because of flow and quality variations in the brewery wastewater discharges, equalization was recommended, by which part of the storage capacity could be used for sedimentation of suspended solids. Two alternative biological treatment systems for the combined brewery wastes were studied in the laboratory. Both the activated-sludge process and the high-rate biological filtration process using plastic media were found to be effective methods of treating the waste. The final choice would depend on an economic comparison of the two alternatives. Nitrogen and phosphorus supplementation were found to be necessary for both systems.

INTRODUCTION

The brewery under investigation is the largest brewery in Thailand, producing approximately 100,000 m^3 of beer per annum. It is located on the east bank of the Chao Phya River in Thailand and comprises two brewery plants, one larger and newer than the other.

Wastewater from the factory is, at the time of this study, discharged without treatment to the Chao Phya River as indicated in Fig. 1. As the final composite waste from any brewery is a combination of wastes from various batch-type operations, its strength and composition vary considerably. Before a wastewater treatment plant can be designed and constructed, it is essential to evaluate the flow and quality characteristics of the wastes and to consider alternative methods of treatment.

To purpose of this study was to determine the magnitude of variations in the strength and value of the total brewery wastes and to evaluate the amenability of appropriate techniques for the treatment of waste streams.

The method of investigation was to obtain representative samples of each source of effluent discharge so that from the physical, chemical and biological analysis, a composite picture of the brewery wastewater generated could be reproduced. This would provide basic criteria for the functional design of wastewater treatment processes. Recommendations on appropriate forms of treatment would result from the studies on waste treatability.

WASTEWATER SAMPLING STATIONS

Brewery liquid wastes can be classified as partially or highly contaminated process streams, clean and contaminated storm water, and sanitary waste streams. The eleven sampling points used during the study are indicated in Fig. 1. Looking at the sources of each point of discharge, these eleven points can be classified as follows:

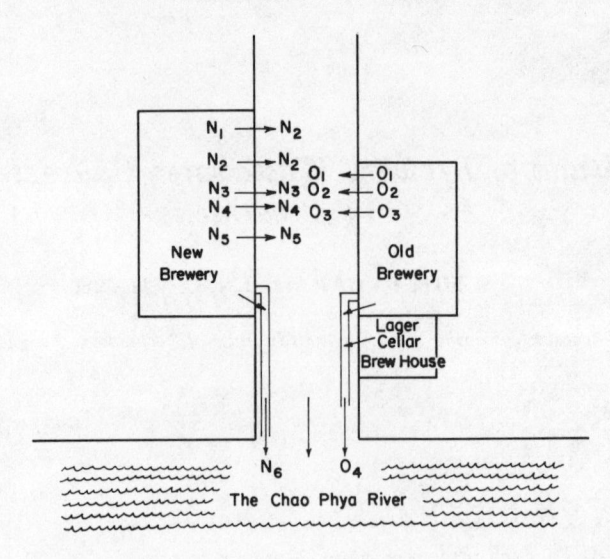

Fig. 1. Schematic Layout of the Wastewater Sources at the Brewery

O_1, O_2, O_3, N_1, N_2 — the wastewaters from unit operations which include bottle-washing, beer-filling, pasteurization and floor-washing.

N_3, N_4, N_5 — the wastewaters from the cooling unit and boiler spillage.

N_6, O_4 — the wastewaters from fermentation tanks, the lager cellar, the yeast tube, brewing, hop-washing and water discharged from the new and old breweries.

WORKING-DAY WASTEWATER FLOW

The period of major wastewater flows at the brewery extends beyond the production shif (7 a.m. - 4 p.m.) and is at least from 8 a.m. to 6 p.m. on 5 working days per week, Monday to Friday.

Wastewater from the two brewery plants at different points show fluctuations in flow on both an hourly and daily basis. The high variability of the flow of the waste strea is indicated in Table 1, where it can be seen that the mean flows vary significantly from day to day. Averaging these total mean wastewater flows over the four working day for which complete data are available gives average working-day hourly flows of 76.4 m /h from the new brewery and 32.0 m^3/h from the old brewery. These rates would give total quantities of wastewater of approximately 760 m^3 and 320 m^3 discharged from the new brewery and old brewery, respectively, during a 10-hour working day.

Because of the high variability of flow in all wastewater streams, it is desirable to prepare cumulative relative frequency diagrams for the working-day hourly flows at all sampling points. Using the data collected over the period from 5 August to 4 September

Table 1 Maximum, Minimum and Mean Hourly Flow Rates During Working Days, 8 a.m. - 6 p.m. (m^3/h)

Sample Station	Monday 5th Aug. 1974 Max	Min	Mean	Monday 19th Aug. 1974 Max	Min	Mean	Friday 23th Aug. 1974 Max	Min	Mean	Wednesday 28th Aug. 1974 Max	Min	Mean	Wednesday 4th Sept. 1974 Max	Min	Mean
N₁	21.6	7.2	14.6	32.1	28.8	29.9	20.1	10.8	15.3	32.4	25.2	29.3	27.0	13.5	20.4
N₂	27.0	9.0	18.7	39.2	22.9	27.8	26.3	18.0	22.0	25.2	22.3	23.9	36.0	15.4	21.0
N₃	4.8	1.8	3.2	6.2	2.1	5.8	6.0	2.4	5.8	7.8	3.7	6.8	5.8	5.2	5.5
	(21st Oct. 1974)			(23th Oct. 1974)			(25th Oct. 1974)			23rd Oct. 1974			(30th Oct. 1974)		
N₄	4.9	1.3	3.9	6.5	5.4	5.9	6.1	5.7	5.9	7.2	6.5	6.8	5.4	3.9	4.4
N₅	5.4	1.2	3.1	26.3	5.8	13.0	6.5	2.2	4.5	3.2	2.2	2.9	4.5	2.6	4.1
N₆	–	–	–	9.2	2.9	4.2	63.8	4.5	18.6	22.5	3.6	10.3	27.6	2.4	11.5
Σ Mean N	–	–	–	–	–	86.6	–	–	72.1	–	–	80.0	–	–	66.9
O₁	3.9	0.3	2.8	7.2	0.4	4.3	9.7	0.4	4.0	7.9	6.1	6.8	19.7	13.5	15.8
O₂	16.6	15.5	16.1	16.9	0.1	10.5	18.0	0.4	7.7	17.3	12.	15.5	9.0	3.8	6.7
O₃	5.6	4.7	5.3	3.2	0.1	1.4	7.9	0.4	5.8	6.8	4.3	5.9	1.3	0.1	0.6
O₄	–	–	–	24.0	3.9	12.0	8.4	3.9	6.5	24.0	3.9	11.0	24.0	4.9	13.4
Σ Mean O	–	–	–	–	–	28.2	–	–	24.0	–	–	39.2	–	–	36.5

1974, flow probability lines have been prepared, a summary of the working-day period from 8 a.m. to 6 p.m. The average working-day flow rates of the new and old breweries fall within the 60-70 percentile range and therefore could be expected to be exceeded about 35% of the time. The wastewater flow at the 90 percentile may be selected as the design parameter for the following reasons:

1. Higher flow increases the safety factor from overloading

2. The treatment plant will be large enough to receive the peak loading during operation.

Table 2 Probabilities of Flow at Different Wastewater Sampling Points During Working Day, 8 a.m. - 6 p.m.

Sampling Station	Wastewater Flow, m^3/h				
	50 Percentile	60 Percentile	70 Percentile	80 Percentile	90 Percentile
N_1	21.4	24.5	27.8	29.0	34.0
N_2	20.0	22.4	25.5	29.0	34.7
N_3	5.2	5.6	5.9	6.4	7.0
N_4	5.1	5.4	5.6	6.0	6.6
N_5	3.0	4.2	4.9	5.8	7.3
N_6	6.0	10.0	13.0	17.5	23.5
ΣN	60.7	72.1	82.7	93.7	113.1
O_1	5.6	7.0	7.9	10.8	13.5
O_2	8.5	11.0	15.0	21.0	33.0
O_3	1.4	1.7	2.0	2.5	3.3
O_4	8.5	10.0	11.5	13.0	15.5
ΣO	24.0	29.7	36.4	47.3	65.3

NIGHT FLOW AND WEEKEND FLOWS

From the end of one working day (6 p.m.) to the begining of the next (8 a.m.) during the week, wastewater flows do not completely cease, although the breweries are not in operation. Night flow from the breweries is mainly from washing machines and from partially open water taps but, in addition, spent hops are frequently discharged into wastewater streams, N_6 and O_4, for about half an hour.

Normally, the brewery works from Monday to Friday, but occasionally during periods of high demand, Saturday will also become a working day. However, even during normal operation there is always wastewater stream flow during the weekend. Approximately 75% of the male labour force are normally on duty on Saturdays for the weekly washing and maintenance of machines.

Measurements indicated that the average flow during the nights of working days accounted for 35% of the mean flow, while the average weekend flow represented 47% of the mean working-day flow.

During the weekend, the wastewater came from the same sources as the night flows. The average night and weekend flows were approximately 41% and 18%, respectively, of the average working-day flow. Both night and weekend flows will not play an important role in the design practice of the actual treatment plant.

WASTEWATER CHARACTERISTICS

In addition to the wastewater flow at each point being highly variable, the quality of the brewery-wastewater streams also varies significantly from hour to hour. Table 3 gives a review of the 50 and 80 percentile quality level parameters for the different brewery-wastewater streams. All analyses were conducted according to Standard Methods (1971).

From an examination of these results, it is apparent that the new brewery-wastewater streams, N_3, N_4 and N_5, met effluent standards imposed by the local Ministry of Industry, except for suspended solids. Without any treatment, these wastewater streams could be discharged, directly after mixing with effluent from a biological treatment plant which is usually low in suspended solids. Other brewery-wastewater streams have high organic contents, with N_6 and O_4 being the most highly concentrated wastes from the point of view of BOD_5 and COD. A composite sample of brewery wastes made up from hourly samples of wastewater streams, N_1, N_2. N_6, O_1, O_2, O_3, and O_4, contained 106 mg/l of total nitrogen and had a BOD/N ratio of 44/1, and BOD/P ratio of 120/1. For an aerobic biological treatment process to operate without problems, these combined brewery wastes would require nutrient supplementation of the influent to maintain BOD/N and BOD/P ratios of about 150:1, respectively, on a continuous basis.

Based on the information obtained, it is recommended that wastewater streams, N_1, N_2, N_6, O_1, O_2, O_3, and O_4, be combined for biological treatment. Table 4 gives a summary of the principal characteristics of present discharges, for an average working-day, from the brewery plant and relates them to the production of beer.

WASTEWATER TREATABILITY STUDIES

Aerobic biological processes are generally used to treat brewery wastewater to obtain the desired degree of removal of organic matter. Laboratory and pilot-scale studies of activated sludge and trickling filtration treatment processes were carried out on composite samples of waste streams, N_1, N_2, N_6, O_1, O_2, O_3 and O_4 combined. For the biological treatment studies, nutrients were added in the forms of urea and phosphate (K_2HPO_4 + KH_2PO_4) to maintain a COD/N ratio of 20/1 and a COD/P ratio of 150/1. Activated-sludge studies were separated into batch-process and continuous-process studies, while pilot-scale trickling filters were used to study biological filtration.

BATCH ACTIVATED-SLUDGE STUDIES

1.5-litre units were used as aeration tanks, being fed with 1 litre of the combined waste every 24 hours until the mixed liquor suspended solids (MLSS) had built up to a level of about 4,000 mg/l and were acclimatized to the waste. Then after settling the solids and drawing off 1 litre of supernatant, 1 litre of combined waste was fed

Table 3 Fifty and Eighty Percentile Quality Levels of Brewery Wastewater Streams

Wastewater Characteristics	Unit	New Brewery Sampling Stations														Old Brewery Sampling Stations							
		N₁		N₂		N₃		N₄		N₅		N₆		O₁		O₂		O₃		O₄	
		50%	80%	50%	80%	50%	80%	50%	80%	50%	80%	50%	80%	50%	80%	50%	80%	50%	80%	50%	80%
BOD₅	mg/l	100	310	600	920	2	4	0	3	2	10	1250	3200	185	400	52	95	86	190	700	2490
COD	mg/l	150	400	680	1100	15	50	7	24	28	74	2140	6700	400	1100	100	180	180	370	900	3400
Susp. solids	mg/l	66	130	120	230	55	105	27	62	49	98	600	1700	90	145	69	110	68	125	280	660
Total solids	mg/l	470	800	1350	1950	480	700	320	460	610	1300	1570	2800	840	1400	760	1100	580	790	1300	3000
pH	-	9.5	10.4	11.5	11.8	8.0	9.0	6.9	7.7	8.2	9.6	7.1	8.0	9.5	10.3	9.8	10.7	8.1	8.2	7.0	7.8
	mg/l as CaCO₃	400	650	1050	1240	270	275	270	285	300	400	150	220	450	615	460	530	240	290	185	280
BOD₅/COD (50 Percentile values)		0.66		0.88		-		-		-		0.58		0.46		0.52		0.47		0.77	

Table 4 Characteristics of Brewery Wastewater Discharges During the Working Day, 8 a.m. - 6 p.m.

Parameter	Value
Volume of beer bottled per working day, m^3/day	385
Mean wastewater discharged, m^3/working day	870
Mean wastewater discharged, m^3/m^3 of product	2.3
Alkanity (Median value), mg/l as $CaCO_3$	520
Suspended Solids (Median value), mg/l	255
BOD (Median value), mg/l	450
COD (Median value), mg/l	590
COD discharged (Median value at mean flow), kg/day	513.3
Population Equivalent (based on 50 gBOD/cap-d)	7250
kg COD/m^3 product	1.3
kg BOD/m^3 product	1.0

and aeration started. Over the next 24 hours, MLSS and COD levels were determined.

The results of the treatability study of brewery wastewater are shown in Fig. 2. It can be seen that, at this level of MLSS, the combined brewery wastewater was biologically degraded mainly within the first 2 hours, and 90% COD removal was achieved within 4 hours. The substrate removal rate over the first two hours was in the range of 1.4 to 1.9 g COD/g MLSS per day. Very little improvement in COD removal occurred after 4 hours of aeration.

CONTINUOUS-FLOW ACTIVATED-SLUDGE STUDIES

For the continuous-flow study, 8.5-litre units were operated as completely mixed aeration tanks at retention times of 4. 6. 8 and 24 hours. 4-litre sedimentation tanks received overflow from the 4, 6 and 8-hour aeration tanks, and sludge return pumps operated 30 seconds every 10 minutes, providing a recycle ratio of about 0.4. Every day, between 200 and 800 ml of excess sludge (at MLSS 4,000 - 7,000 mg/l) was wasted from these sedimentation tanks. The 24 hour-aeration tank was combined with a sedimentation section, which provided 100% sludge recycle, to operate as an extended aeration plant. The experiments were conducted at a room temperature of about 30^0C.

The results of the continuous-flow activated-sludge process operating under steady-state conditions are shown in Table 5. From this table, it can be seen that over 90 percent COD-removal efficiency was achieved at organic loadings in the range 0.2-4 kg COD/kg MLSS per day and that the effluent COD was usually less than 100 mg/l, acceptable to local authority effluent standards.

The choice of the design organic-loading rate will depend not only on the COD-removal rate but also on sludge settleability and production. At an organic loading approaching

Table 5 Biodegradability of Combined Brewery Wastewater by the Continuous-Flow Activated-Sludge Process

Influent COD, mg/ℓ Total	Influent COD, mg/ℓ Filtered	Influent Filtered BOD$_5$, mg/ℓ	Influent S.S., mg/ℓ	Effluent COD, mg/ℓ Super-natant	Effluent COD, mg/ℓ Filtered	Effluent Filtered BOD$_5$ mg/ℓ	Effluent S.S., mg/ℓ	Mixed Liquor Suspended Solids, MLSS, mg/ℓ	Return Sludge X_r mg/ℓ	Aeration Time, h	Organic Loading Rate Kg COD/Kg MLSS per day	COD Removal Efficiency,% Based on Super-natant	COD Removal Efficiency,% Filtered Basis	Sludge Volume Index SVI
1512	800	690	–	98	68	15	–	2200	5700	4	4.140	94	92	–
1320	910	–	200	120	82	–	56	2250	4000	4	3.520	91	91	373
1540	1256	–	287	136	86	–	64	2230	3500	4	4.150	91	93	358
1486	1212	–	96	76	–	–	–	2430	4000	4	3.660	94	95	358
1248	920	–	305	86	65	–	54	3000	6000	6	1.660	93	93	143
1544	1240	–	662	110	106	12	20	3400	7100	6	1.810	93	91	109
1884	1156	–	820	116	70	11	42	3200	6900	6	2.350	94	94	102
1560	1284	780	450	108	78	10	40	2500	7160	6	2.500	93	95	137
1128	1048	–	–	76	72	–	–	3790	–	8	0.892	93	93	–
1490	1228	–	200	84	62	–	53	400	6500	8	1.110	94	95	273
1060	800	–	335	100	74	–	64	3510	6700	8	0.906	91	91	250
1228	890	–	–	74	62	–	–	3580	6160	8	1.030	94	93	251
1060	780	–	–	50	35	–	–	4660	100%	24	0.227	95	96	195
1152	856	–	–	98	44	–	–	4310		24	0.167	92	95	185

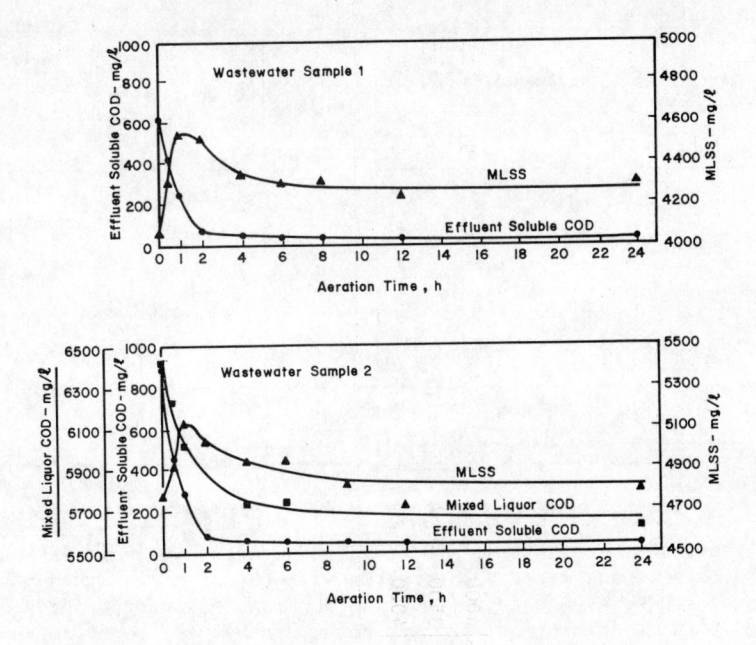

Fig. 2. Batch Activated-Sludge Studies on Combined Brewery Wastewater

4 kg COD/kg MLSS -d (4 h aeration time), the SVI was approximately 360 and the sludge was extremely bulky. With an organic loading of approximately 1.0 kg COD/kg MLSS -d (8 h aeration time), the SVI was about 260 and the sludge was slightly bulky. Although sludge sedimentation was not seriously impaired, sludge bulking might well be a problem from time to time, if this organic-loading level were accepted. However, the best sludge settleability was obtained at an organic loading approaching 2.0 kg COD/kg MLSS -d (6 h aeration time), when the SVI was around 125. This would be the most suitable design loading for a conventional activated sludge process. For the extended aeration system, the sludge settleability was acceptable at an SVI of 190.

Using the data given in Table 6, the sludge-yield coefficient, Y, has been calculated to be

$$Y = \frac{\overline{X}(1 + \alpha - \alpha C)}{S_i - \overline{S}} = \frac{3025 \times (1\ 0.4 - 0.4 \times 2.27)}{1559 - 105}$$

$$= 1.02 \text{ kg MLSS/kg COD removed}$$

where S_i, $\overline{S}$, $\overline{X}$, C and α have the meaning indicated in Table 6 (Gaudy and Ramanathan, 1969). This sludge-yield coefficient would result in 472 kg MLSS to be disposed of each day on the basis of the average flow and COD for the combined brewery wastewater. Assuming a sludge concentration of 1% solids in the secondary sedimentation tank, this would give 39 m^3 of sludge to be dewatered each day. Alternatively, the sludge-handling problem could be minimized by adopting the extended aeration process, but this would require four times the aeration tank capacity and increase the capital cost.

Table 6 Sludge-Yield Coefficient (Loadings of 1.6-2.5 kg COD/kg MLSS-d)

Influent COD, S_i (mg/ℓ)	Effluent COD, $\bar{S}$ (mg/ℓ)	MLSS, $\bar{X}$ (mg/ℓ)	Return Sludge Concentration, X_r (mg/ℓ)	$C = \dfrac{X_r}{X}$	Recycling Ratio α
1248	86	3000	6000	2.00	
1544	110	3400	7100	2.09	0.4
1884	116	3200	6900	2.16	
1560	108	2500	7160	2.86	
Average = 1559	105	3025	6790	2.27	

One important characteristic of the combined brewery waste is its changes in flow and quality at nights and weekends, compared with the average working-day flow. This will occur at weekends, even when an equalization tank is incorporated into the system, and will result in the biological process being starved for a period (say one day) and then being returned to its normal feeding level. Although the magnitude of the change is difficult to predict with the data available, it was decided to make an arbitrary test of the effects on the 6-hour aeration time activated-sludge process of a two day reduction of loading to 20% of the normal operating level. Fig. 3 shows the pattern of flow adopted for this test and the influent COD data. From the limited resuts obtained, it would appear that no adverse response to the severe step-wise change of organic and hydraulic loading occurs. The effluent COD level remained almost constant during the period of low loading, and shock return to normal loading had no noticeable effect on this parameter.

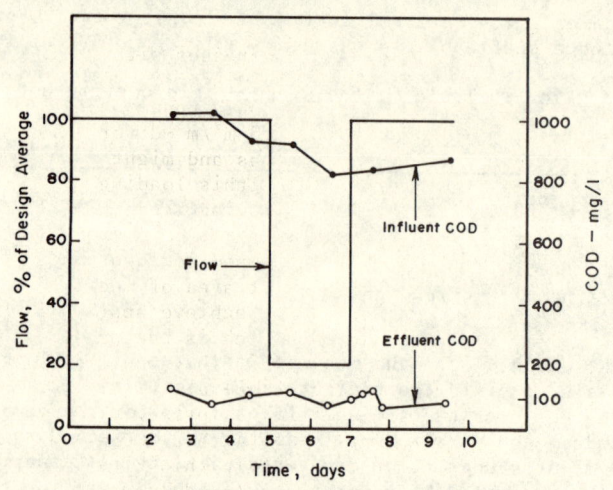

Fig. 3. Effect of Reduced Loading on Activated Sludge Effluent COD for a 6-Hour Aeration Time.

PILOT-SCALE TRICKLING FILTER STUDY

Four different types of filter media were used in this study: 5-7.5 cm nominal size rock, plastic balls 3.7 cm in diameter, corrugated PVC sheeting with wave length 77 mm, wave height 16 mm, thickness 1.58 mm and spacing 10-12 mm, and finally the same corrugated PVC sheeting but with the surface roughened. These media had specific surface values of 92, 86, 195 and 195 m^3, respectively. The trickling filters used were 13.5 cm^2 in section and 2.05 m deep. Brewery wastewater with COD concentration 400-1,000 mg/l was applied to these filters at hydraulic loading rates of 3, 6, 12 and 18 m^3/m^3-d. The results of this pilot-scale study are presented in Fig. 4.

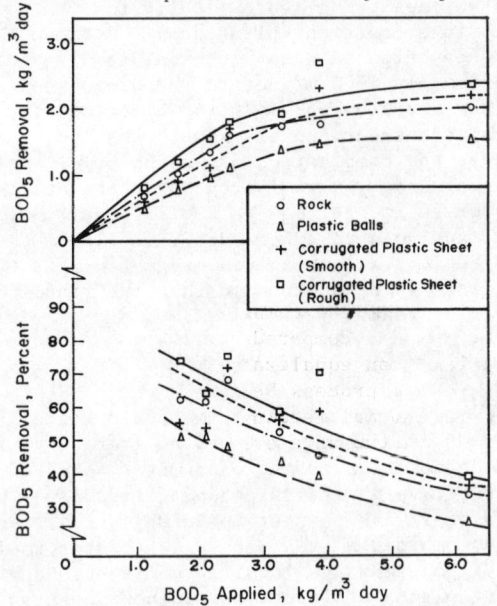

Fig. 4. Performance of Pilot-Scale Trickling Filter Treating Brewery Wastewater.

Although the efficiency of BOD$_5$ removal decreases with an increasing BOD$_5$ application rate, the upper curve in Fig. 4 suggests a discontinutiy at a loading of approximately 3.0 kg BOD/m^3-d for all media tested. It would appear that this level of BOD$_5$ loading, which gives a hydraulic loading rate of 5.0 m^3/m^3-d for a wastewater BOD$_5$ strength of 600 mg/l, is optimum for the test conditions and might well be the basis for the design of a full-scale trickling filter plant. At this loading rate, the efficiency of BOD$_5$ removal will depend on the type of medium being used in practice.

From Fig. 4 it can also be seen that roughened corrugated PVC sheeting gave best results at a lower organic loading per unit area of medium. With the commercial type of plastic medium it should be possible to achieve approximately 60% COD removal at a loading make of 2.1 x 10^{-2} kg COD/m^2-d, which is the same as 3.0 kg BOD$_5$/m^3-d for a medium with a specific surface of 195 m^2/m^3. This would result in a BOD$_5$ removal rate of approximately 2.0 kg/m^3-d or 65%. If high-rate filtration is adopted in a full-scale plant, the effluent will still require further treatment in an activated sludge unit or conventional-rate trickling filter with a BOD$_5$ loading of 1 kg BOD$_5$/m^3-d. With this two-stage filter unit, 80% of the COD in this brewery wastewater could be simply removed without any addition of nutrients. An advantage of this two-stage filter process would be its ability to operate stably under the variable organic loading which results from the brewery's intermittent production schedule.

RECOMMENDATIONS ON WASTEWATER FLOWS, CHARACTERISTICS AND TREATMENT

Wastewater Flow

Considering the combination of brewery-wastewater streams which require biological treatment, it is clear that the flow rate would fluctuate through the working day and drop at nights and the weekend. For treatment units with a relatively short retention time it is advisable to design for a high probability of flow, perhaps the 90 percentile level. In this case, designing hydraulically for the 90 percentile working-day flow rate would result in a treatment plant nearly 2 times the size of what it would be if the mean working-day flow rate were chosen. An alternative is to provide an equalization of approximately 1000 m^3 before the biological treatment units. This will satisfy the mean working-day flow (87 m^3/h for a period of 12 hours per day. Subsequen units in the biological treatment process should then be designed for a wastewater flo rate of 87 m^3/h. Because the combination of wastes will contain a significant amount of suspended solids, sedimentation will occur over the average 12 hours of retention and the equalization (or sedimentation) tank will require sludge-scraping and draw-off mechanisms. Therefore, due to high BOD$_5$ and COD levels in the combined waste, biologic activity is likely to deplete all dissolved oxygen in this tank and cause anaerobic odours. To prevent this, it is recommended that the final portion of the tank be designed as a pre-aeration tank.

Wastewater Characteristics

It is recommended that median values of the measured wastewater characteristics be adopted for the design of treatment plant units. Therefore, it is expected that the average BOD and COD concentration in the wastewater directed to biological treatment would be 450 mg/l, respectively. Its suspended solids level would average 225 mg/l and its alkalinity 520 mg/l with respect to CaCO$_3$, suggesting a pH of about 9.4. The average quality of the combined N$_3$, N$_4$ and N$_5$ wastewater would meet the requirement of effluent standards imposed by the local authorities and, particularly after dilution with the treatment plants effluents, would pose no problem on discharge to the watercourse.

Wastewater Treatment

On the basis of the Laboratory test it is recommended that wastewater streams, N$_1$, N$_2$, N$_6$, O$_1$, O$_2$, O$_3$ and O$_4$ be combined for biological treatment. All of these wastes must be discharged into a sump to allow pumping to the treatment plant site. Brewery-wastewater streams, N$_3$, N$_4$, and N$_5$, should be collected together and discharged probably after pumping, along with effluent from the biological treatment plant. In this way, the suspended solids concentration could be reduced through dilution and the total effluent would thus conform to the effluent standards of the local authorities.

For biological treatment the combined waste requires pumping to the treatment site as well as primary treatment in the sedimentation and pre-aeration unit processes, as shown in Fig. 5. This primary treatment unit also acts as an equalization unit to even out the variations in wastewater flow and quality. After primary treatment the wastewater will require one of the two alternative aerobic biological treatment processes shown in Fig. 5. The final decision on which alternative should be adopted will depend on the amount of capital, and a comparison of operating cost, imported equipment requirements, ease and reliability of operation, etc. For both alternatives, nitrogen and phosphorus nutrients will have to be added and in both cases the biological sludge produced would be handled in a dewatering process.

Alternative A is based on biological filtration and comprises a first-stage high-rate trickling filter followed by sedimentation and a second-stage low-rate trickling filter

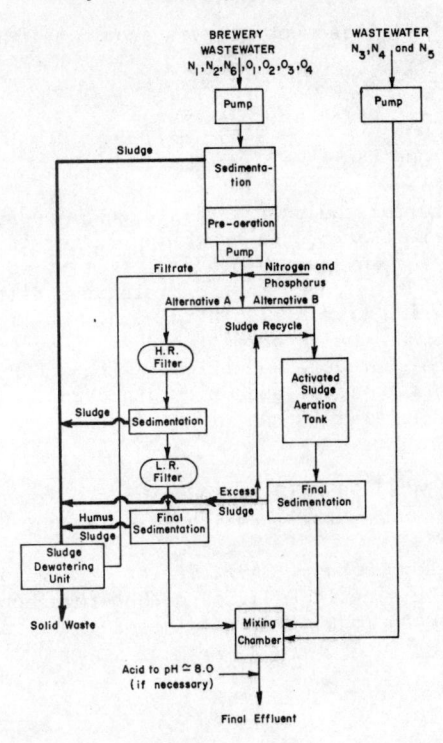

Fig. 5. Schematic Layout of Wastewater Treatment Plants for the Brewery Under Investigation

again followed by sedimentation. The principal advantage of this alternative is the ability of the filter system to perform efficiently under the variable loading conditions to which the biological treatment process would be subjected in this case. An additional advantage would be the lower sludge production from this system in comparison which the activated-sludge process, and this is worth considering when the difficulty of sludge disposal is taken into account. The first-stage high-rate filter could be constructed very deep, up to 6 or 7 m, and this would reduce the land area requirements for the treatment plant, but would increase pumping costs.

Alternative B is the activated-sludge process and includes an aeration tank and final sedimentation tank with sludge recycle. This process, alhough efficient, is known to be more difficult to control under variable loading conditions and to produce more sludge than a biological filtration process. Although the requirements for pumping will be fewer using the activated sludge system, the power demand for aeration will exceed that for additional pumping in the filter system. The activated-sludge process is likely totake up less land than the biological filter process suggested.

The design of the initial pumping of the combined wastewater to the treatment plant site should be based on not less than the 90 percentile working-day flow rate. The sedimentation tank section of the primary treatment unit should be designed to provide at least 2 h retention at the 90 percentile working-day flow rate.

In summary, the biological treatment process should be designed on the basis of the following:

>   Wastewater flow rate - 87 m^3/h
>   Wastewater BOD$_5$ - 450 mg/l
>   Wastewater COD   - 590 mg/l

The organic loading rate for the activated-sludge process is 2.0 kg COD/kg MLSS-d for 6 h retention time, while that of the trickling filter is 3.0 kg BOD$_5$/m^3-d. An acceptable loading rate for the low-rate biological filters would be 1 kg BOD$_5$/m^3-d. Intermediate and final sedimentation tanks in the biological filtration system would normally be designed for a retention time of 1.5 h in each. The final sedimentation tank in the activated-sludge system would also normally be designed for a retention time of 1.5 h. Both systems A and B would produce an effluent with a BOD$_5$ concentration of about 20 mg/l, which would conform to local authority standards.

REFERENCES

AWWA-APHA-WPCF (1971). *Standard Methods for the Examination of Water and Wastewater.* 13th Ed. American Public Health Association, New York.

Gaudy A.F. and M. Ramanathan (1969). Effect of High Substrate Concentration and Cell Feedback on the Kinetic Behavior of Heterogeneous Population in a Completely Mixed System. *Biotechnology and Bioengineering, II,* 207.

Environment-Protecting Installation for the Cellulose Industry

A. C. SAATCI

Technical Assistant and Special Engineer, Winterthur, Switzerland

THE PRODUCTION of CELLULOSE

The main raw material needed for the production of 1 ton of cellulose is about 6 m^3 wood. For this process an aqueous solution of calcium bisulphite is utilized. This is obtained from calcium and sulphur. Calcium is obtained from dust-fine ground limestone, and sulphur dioxide is made of scrap and liquid sulphur as well as pyrite.

The chemicals needed for bleaching the cellulose fibres are chlorine and soda lye; these are obtained from common salt.

Water is of very special importance, since for producing 1 ton of cellulose more than 300 m^3 of it is needed. The required water is taken from groundwater streams or from rivers. The river water is cleaned in big quartz sand filtration plants and upgraded to process water.

The lye, resulting from the cooking process and which contains about 50% of the set-in wooden substance, is used to obtain the following by-products:

- refined spirit for the chemical, pharmaceutical and cosmetic industries
- pure alcohol for pharmaceutical and cosmetic applications
- torula yeast as a component of food stuffs
- moistening agents for pigments and insecticides.

WASTEWATER LOADING

It has been well known for a long time that cellulose manufacturing plants are among the greatest water polluters. First, in-plant measures can be taken as, for example, the removal of the fall-off lye and its processing into yeast and/or alcohol, the condensing of the lye and its burning or its application in various other industries, and the closure of the water circuits.

All these measures can be considered only to a small extent as environment-protecting measures since they are generally coupled with returns on investment. But such measures may, if properly implemented, result in an enormous reduction in the pollutant level of the remaining wastewater. Without these "internal measures", efficient treatment of the remaining wastewater would be almost impossible.

Waste substances, which can not be removed by the in-plant measures remain, however, in the unusable process water.

242   A. C. Saatci

The three installations described below serve the purpose of treating the remaining
wastewater from the manufacturing process, and of removing the sludge which remains
after clarification of the sewage water.

MECHANICAL-CHEMICAL SEWAGE TREATMENT PLANT

The operation of the installation is shown in Fig. 1.

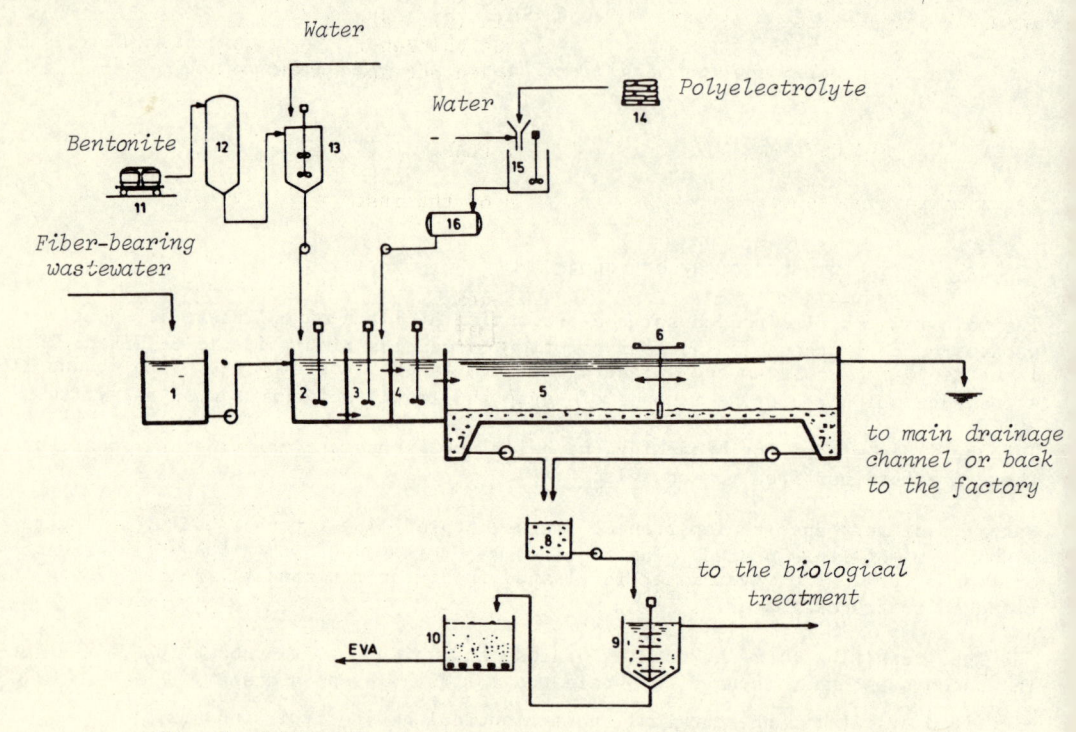

Fig. 1: Schematic representation of the mechanical-chemical
wastewater treatment plant.

The fiber-bearing wastewater, mostly coming from the manufacture of cellulose and from
the dewatering machines, is conducted to a reservoir (1), and from there, after deter-
minating its quantity, to a reaction basin (2). There a bentonite sludge dose is
added to the sewage water.

Bentonite is a clay mineral with a strong swelling capacity. In its sodium state(active bentonite) and in water, it can be divided into colloidal particles. Bentonite
can absorb many organic substances. On this point, the main data regarding the mechanical-chemical wastewater treatment plant are partly of chemical and partly of physical character. The bentonite is delivered in silo trucks (11) and blown into the
bentonite silo of the plant. The silo has a capacity of 100 m^3. In mixing container
(13) with a capacity of 60 m^3, a watery sludge of 10 to 20 g/ℓ bentonite is prepared
and conducted into the reaction basin (2).

The mixture of wastewater and bentonite flows subsequently into the reaction basin(3)
In this basin an anionic polyelectrolyte (polyacryl amide) is added and has the function of a flocculator. The polyelectrolyte is delivered in bags or in small cardboard boxes (14). In a dissolving container (15) the flocculator is dissolved in wa-

ter at a concentration of 0.5 g/ℓ. The flocculator solution is kept in the tank(16), which has a storage capacity of 10 m^3. In the flocculating basin (4), which is provided with a slow rotating stirring device, the flocculation of the sludge takes place. The mixture of wastewater and flocs flows through the sedimentation pre-chamber to the sedimentation pit (5). Here the sludge flocs are deposited on the bottom of the pit owing to natural gravity. The shield-type sweeping carriage transports the sedimented sludge alternately to the front and the rear sludge bog (7). The purified sewage water runs off through overflow conduits and can be fed to the main drainage waterway (Aere), or re-used for certain purposes at the plant.

The fiber sludge is drawn from the sludge bog (7) at a concentration of between 0.4 and 0.7% and conducted into the pumping basin (8), after which it is pumped into a sludge thickener (9). After passing into the thickener the sludge reaches a concentration of between 2.2 and 3.5%. Then it is taken out and stored in a ventilated storage container until further processing.

BIOLOGICAL WASTEWATER TREATMENT PLANT

A schematic representation of the operation of the installation is given in Fig. 2.

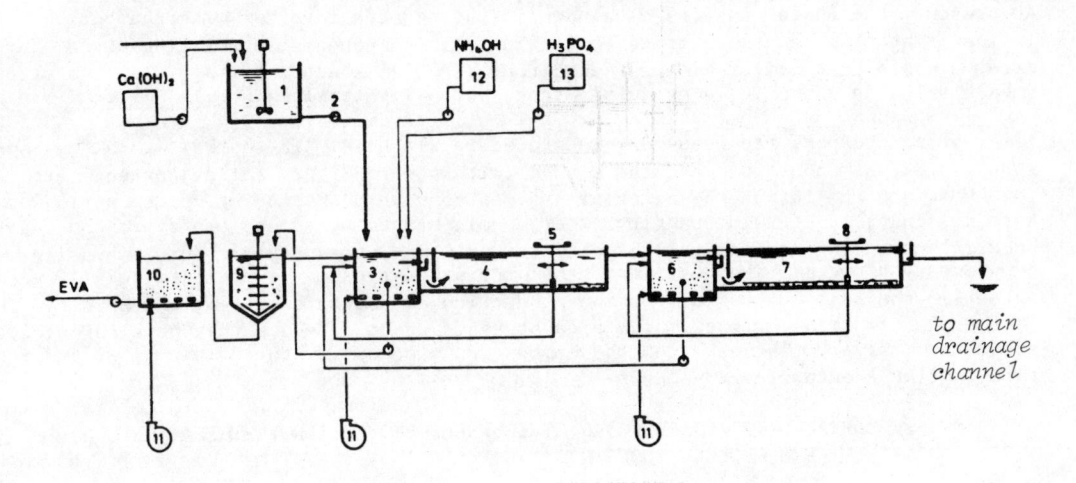

Fig. 2: Schematic representation of the biological wastewater treatment plant.

The collected, chemically and biochemically polluted wastewater, mainly from the bleachery and from evaporated condensates, is conducted to the neutralizing installation. There it is neutralized to a pH value of between 7.3 and 7.8 by means of ash from the combustion of sulphite lye and calcium sludge (1). The neutralized wastewater is pulped to the activated sludge basin of the first biological stage (3). There a decomposition process through micro-organisms takes place. The activated sludge thus formed is precipitated in the adjoining post-clarification basin (4). The sucking sweeper (5) returns the settled sludge to the activating basin (3) while the pretreated sewage water flows to the activating basin of the second biological stage(6). In the activating basin of the second stage, process similar to that in the first stage takes place, but under different conditions. In the second stage there is also a constant returning of sludge from the post-clarification basin (7) to the activating basin (6), also by means of a sucking sweeper (8). The treated water from the second stage is fed into the main drainage waterway.

The sludge from the biological process is carried over from the second to the first stage. From there it is pumped, together with the excess sludge from the first stage, to the sludge thickener and thickened to a dry solids content of 2 to 4%. The excess water from the thickener is conducted to the biological stage, and the thickened sludge is stored in a ventilated storage container until further processing. The air supply for the biological stage and for the storage container is provided by ventilators(11). Small quantities of nutritive substances, such as ammonia water (12) and phosphoric acid, are added to the biological process.

DEWATERING and INCINERATION PLANTS

Dewatering Plant

The dewatering installations consist of the following main parts: one thickener for fiber sludge, one storage container (ventilated) for fiber sludge, one storage container (ventilated) for mixed sludge, three dewatering presses with double sieve and one flocculator processing installation.

Incineration Plant

The incinerating plants are dimensioned to incinerate the waste bark and the sludge from the biological water treatment plant, as well as one third of the fiber sludge. At present, the whole quantity of sludge is incinerated together with the bark. The incinerating plant consists of the following main components: a whirling stream furnace, a waste heat boiler with forced circulation, an electro-filter (Elex), a stack, a bark silo, an ash silo, a crude oil tank, and a light oil tank.

After being chopped, the wood bark is stored in the silo. The bark (with a dry solids content of about 50%) and the sludge (with about 22% dry solids content) are fed together into the furnace by means of a screw conveyor. Depending on the ratio of bark to sludge, a certain quantity of crude oil has to be added. To start the furnace (after idling periods), light oil is used. The necessary air for combustion is supplied by blowers. The heat content of the exhaust gases is utilized to a large extent in the adjacent boiler for the production of saturated steam, which is fed into the plant's steam network. The exhaust gases are cleaned as they pass through the subsequent filter. The ash from the combustion process is stored in a silo, and from time to time transported to a dump.

GENERAL DATA REGARDING AN ENVIRONMENT-PROTECTION INSTALLATION for THE CELLULOSE INDUSTRY

The yearly consumption of a paper factory amounts to approx. 600,000 m^3 of wood for the production of about 100,000 tons of cellulose.

Dimensioning of the Installation
Mechanical-Chemical Wastewater Treatment Plant

Hydraulic capacity (two independent lines with a capacity of 1,250 m^3/h each)	2,500	m^3/h
Reaction basin for bentonite	2 x 138	m^3
Reaction basin for polyelectolyte	2 x 69	m^3
Flocculation basin	2 x 114	m^3
Sedimentation pre-chamber	2 x 52	m^3
Sedimentation basin: usable volume	2 x 2,067	m^3
Surface load	2.05	m^3/m^3h
Detention time	1.7	h

(continued..)

(continues...)

Operators	1 man
Purification grade (suspended matter)	93 to 98%
Electrical energy	0.1 kWh/m^3 wastewater
Bentonite	12 g/m^3 wastewater
Polyelectolyte	0.5 g/m^3 wastewater

Biological Wastewater Treatment Plant

Hydraulic Capacity

Wastewater flow rate	52,800 m^3/d
Hydraulic inhabitant equivalent (500 ℓ/I_{Eqh})	105,600
Mean Wastewater flow rate	2,200 m^3/h
Peak Wastewater flow rate (max. 2h)	2,500 m^3/h
Short-time peak flow rate (max. 30 min each 12h)	3,000 m^3/h

Biological capacity

BOD$_5$ daily inflow	17,500 kg O/d
BOD$_5$ concentration	332 mg O/ℓ
BOD$_5$ inhabitant equivalent (75g O/I_{Eqb})	233,330

Solid matter capacity

Solid matter inflow	7,920 kg/d
Solid matter concentration	150 mg/ℓ

Dimensioning

Activating basins, 1st and 2nd stage

2 Rectangular basins each stage

	1st Stage	2nd Stage
Useful volume, each basin	4,250 m^3	4,250 m^3
Useful volume, total	8,500 m^3	8,500 m^3
Surface, each basin	1,250 m^2	1,250 m^2
Surface total	2,500 m^2	2,500 m^2
Surface load		
- at 2,200 m^3/h	0.88 m^3/m^{2}h	0.88 m^3/m^{2}h
- at 2,500 m^3/h	1.00 m^3/m^{2}h	1.00 m^3/m^{2}h
- at 3,000 m^3/h	1.2 m^3/m^{2}h	1.2 m^3/m^{2}h
Detention time at 2,200 m^3/h	232 min	232 min

Sludge Thickener

2 thickeners for biological sludge

Surface area	169 m^2 each
Useful volume	507 m^3 each
Surface load	0.29 m^3/m^{2}h
Solid matter surface load	49.3 kg/m$^2\cdot$d

(continues..)

(continues...)

Sludge storage containers
2 rectangular containers

Useful volume	705 m³ each
Detention time	2.1 d

Air Supply

1st Stage

6 Rotary piston blowers, capacity 5,136 m³/h = 31,000 m³/h
1 reserve blower (for 1st & 2nd stage)

Alpha factor	0.8
O₂ utilization (mammoth pumps)	approx. 7.5%
Oxygen requirement	526 kg/h
OC-load	0.72

2nd Stage

2 Rotary piston blowers, capacity 4,000 m³/h = 8,000 m³/h

Alpha factor	0.8
O₂ utilization	approx. 12.7%
Oxygen requirement	227 kg/h
OC-load	1.56
OC-load, 1st & 2nd stage together	1.03

Sludge store (for biological sludge)

2 Rotary piston blowers, capacity 1,000 m³/h = 2,000 m³/h
1 reserve blower

Oxygen requirement	44 kg/h

Electrical energy requirement

1st Stage (ventilation)	13,248 kWh/d
2nd Stage (ventilation)	3,840 kWh/d
Sludge Thickening & Aeration	1,056 kWh/d
Electrical energy	0.5 kWh/m³ wastewater or 1.35 kWh/kg BSB₅
Demulgator	approx. 5 g/m³ wastewater
Operation	3 men

Dewatering and Incineration Plant

Incineration plant data

Calculated data	normal capacity	max. capacity
Thoughput of bark (50% dry subst.)	550	3,000 kg/h
Throughout of biol. sludge (19% d.s.)	3,310	3,310 kg/h
Throughout of mech. sludge (25% d.s.)	650	650 kg/h
Additional fuel (crude oil)	480	240 kg/h

(continued..)

(continues...)

Combustion air	10,390	14,500 m³/h
Exhaust Gas	15,700	22,100 m³/h
Steam production (13 atm., saturated)	5,950	8,700 kg/h
Energy requirement	3,840	4,776 Kwh/d
Ash Production	304	353 kg/h

Investment costs

Dewatering and Incineration	9,600,000 Sfr.

Working stock consumption (Nov. 1975)

Electrical energy	4,900 kWh/d
Quartz sand	900 kg/d
Crude oil	6.0 tons/d
Flocculator (polyelectrolyte)	0.5 kg/d

Steam production (Nov. 1975)

Saturated steam, 13 atm.	6.5 tons/h
Operation (four shifts)	11 men

REFERENCES

Eckenfelder, W.W. and E.B. Barnhardt (1960). Paper presented to Amer. Inst. of Chem. Engrs., Atlanta, Georgia.

Hunken, K.H. (1960). *Untersuchungen über den Reinigungsverlauf und den Sauerstoffverbrauch bei der Abwasserreinigung durch das Belebtschlammverfahren.* Kommissionsverlag R. Oldenburg, Munchen.

Imhoff K. *Taschenbuch der Stadtentwasserung.*

Munz W. *Abwasser.* 4th Edition, Lehrmittelverlag Juventus, Zurich.

Nogai, R.J. (1972). Selecting Wastewater Aeration Equipment. *Chemical Engineering,* April 17.

Sawyer, C.N. (1958). *An Evaluation of High Rate Digestion, Biological Treatment of Sewage and Industrial Wastes.* Eds., J. McCabe and W.W. Eckenfelder. Vol. 2. Reinhold Pub. Gov., New York, N.Y.

Verband Schweizerische Abwasserfachleute (1962). Erfahrungen beim Betrieb von chemischen, biologischen und Schlammaufbereitungsanlagen unter besonderer Berucksichtigung der Probleme bei industriellen Abwassern. *Verbands-Bericht,* Zurich, 74/1.

Wastewater Renovation in Paper Reprocessing

A. HAMZA

High Institute of Public Health, Alexandria University, Egypt

ABSTRACT

The escalating prices of imported pulp in Egypt and the concominant increased demand for paper have contributed to the recent expansion in wastepaper reprocessing. The ALBA plant in Alexandria is liable to strict pollution control as its effluent is discharged into an important fishing lake at the city entrance. This paper evaluates the feasibility of wastewater renovation and reuse at the plant. Despite the increased cost of the renovation system, potential savings are expected due to the waiving of the surcharge cost proposed by the local authorities.

INTRODUCTION

The inflationary trends of recent years have greatly increased the price of imported pulp, which represents a major portion of the raw material used by the paper industry in Egypt. Concomitantly, the population growth and the increase in the amount of paper per capita have contributed to the recent trend of expanding wastepaper reprocessing.

The ALBA paper-reprocessing plant in Alexandria liable is to strict pollution control since its effluent is discharged into an important fishing lake at the city entrance (Lake Maruit).

Development of practical technology for internal pollution control could ultimately lead to closing the water cycle in the plant and alleviation of the water supply and pollution problems. Additional benefits are increased process efficiency, raw materials and energy savings, as well as lower investment costs compared to the costs of external treatment.

Therefore, this study was undertaken to characterize process effluents and to determine the feasibility of wastewater renovation and reuse in a closed system at the ALBA paper-reprocessing plant.

LITERATURE REVIEW

Interest in the application of water recycling in the paper industry has been stimulated by recent public concern and improvements in waste treatment technology. Stelmakh (1974) described the improvements made in the treatment system of a cylinder board mill to enable reuse of treated water for felt and wirewashing. Fibers were recovered on filter savealls and the water was then clarified using alum and polyelectrolyte at 240 mg/l and 10 mg/l, respectively. The recycled water contained 58 mg/l suspended residue and 192 mg/l (COD). Streebin, Reed and Law (1976) described the results of a full-scale project undertaken to determine the feasibility of complete water reuse in a paper reprocessing mill producing roofing felt. They concluded that one hundred percent water reuse was technically and economically feasible in an organic felt mill and that the product quality was not adversely affected by water reuse. Kimura and Izumisawa (1976) described the application of carbon adsorption to resulted to BOD and suspended residue values of the treatment of board mill wastewater after clarification and sand filtration. This treatment reduced the COD by 80% and less than 10 mg/l. Thomas (1972) demonstrated a "closed loop" system for the manufacture of tissue paper. By closing up the white water system, the effluent from the mill was reduced by 4900 m^3/d while fiber loss was reduced from 2.3% to less than 1%. Davis and others (1973) examined a system in which the process effluent was clarified by coagulation and sand filtration. Problems encountered in this system included clogging of the sand filter, slime built-up in the holding tank and residual color remaining in the filtrate.

MANUFACTURING PROCESSES AND WATER USE AT THE ALBA PLANT

Wastepaper and residual cuts of paper from processing operations represent about 80% of the raw materials, while rice straw comprises the other 20%.

The process involves the cutting of straw to an average size of 5 cm and separation of foreign matter in cyclons, followed by semi-chemical digestion using 8-10% sodium hydroxide solution. The pulp yield is about 65-70% of the raw materials by weight. The black liquor from the digestion process contains a mixture of cellulose, hemi-cellulose, lignin and residual alkaline solution. This liquor is directly discharged to the plant sewer system.

The straw pulp produced is mixed with the wastepaper and subjected to a series of screening and homogenizing processes before feeding to the papermaking machine, where the slurry at a fixed consistency of 1% is fed at a constant rate to an endless wire screen to reduce the water content to 80%. The paper layer moves on a felt blanket through a series of pressing rolls and then through the drying section of the machine where its moisture content is reduced to about 10%.

The flow rate of the process water averages 5500 m^3/d or 183 m^3/ton paper. About 10% of the process water is taken from the city supply, while 90% is untreated water taken from the nearby Mahmodia canal.

The canal water is passed to a large settling tank (350 m^3) for plain sedimentation before use as process water for the semichemical digestion of the straw, as well as in the collar gangs, dump chest, beater and screens.

Most of the water drained from the paper machine is reused in the damp chest, beaters and screens. Reuse of this water results in a significant reduction in the waste load. Details of the various processes at the ALBA plant are illustrated in Fig. 1.

EXPERIMENTAL PROCEDURES

Characteristics of Wastewater

Laboratory analyses were made to establish the physical and chemical characteristics of process water and wastewater in seven areas. Sampling points are numerically identified in Fig. 1.

All analyses were performed according to the *Standard Methods* (1971).

Laboratory Experiments

Tests for the renovation of the plant effluent were conducted by coagulation using a six-gang flow stirrer with an adjustable speed ranging from 0-120 rpm. One - liter

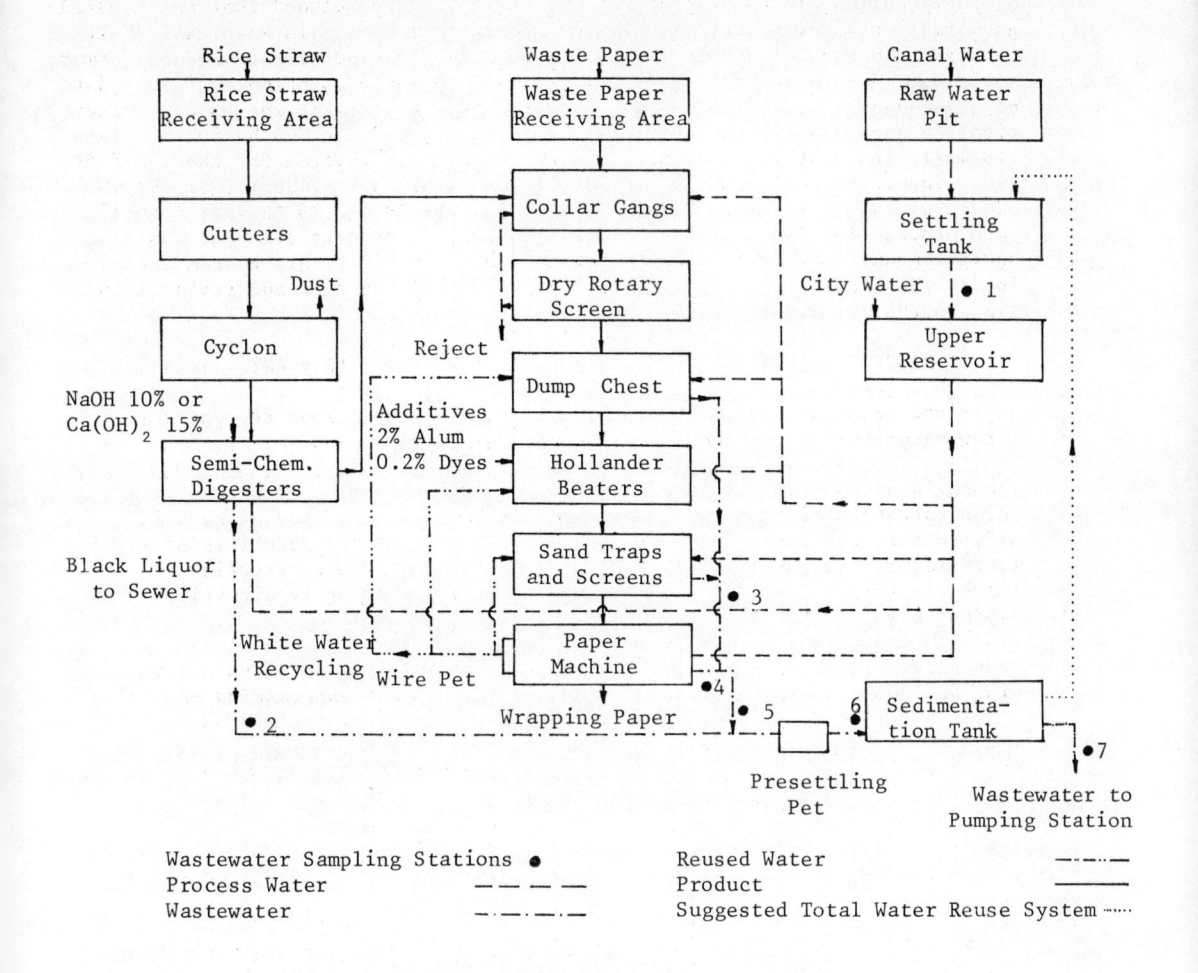

Fig. 1. Wastepaper Reprocessing at the ALBA Plant

samples were used in all experiments. The chemicals were added during flash mixing at 120 rpm. After a flocculation period of 10 min at 30 rpm, the samples were allowed to settle for 30 min. The results of a preliminary investigation indicated that the above conditions were suitable for the flocculation and settling of paper-reprocessing waste.

The coagulation study was performed according to the following scheme:

a) The effect of $Al_2(SO_4)$ and $FeCl_3$, using doses ranging from 50 to 200 mg/l at a pH range of 5.0 - 6.5, and the effect of $FeSO_4$, using the same dose range and pH 9.0 - 11, were investigated.

b) The combined effect of polyelectrolyte (Hercules anionic No 819.2), using doses of 1 to 5 mg/l, and the coagulants $Al_2(SO_4)_3$ and $FeCl_3$, using doses of 100 to 200 mg/l, was examined.

Settling Column Study

The settleability study was conducted using a column 200 cm high and 45 cm in inner diameter. Taps were located at 30 cm deep intervals. A variable speed stirrer was used for flash mixing at 140 rpm, while flocculation was carried out at 35 rpm for 5 min. Extending the flocculation time in the settling column beyond 5 min resulted in adverse effects on floc formation.

Samples were drawn from six taps at selected time intervals up to 120 min. The one-minute period required to take a set of samples and the slight drop in the surface level due to the withdrawal of samples were ignored in labelling the settling characteristic curves.

The study was conducted according to the following plan:

a) The effect of $Al_2(SO_4)_3$, $FeCl_3$ and $FeSO_4$ was measured using the same conditions as in part (a) of the laboratory experiments. The removal of total organic carbon (TOC) was the selected criterion for this set of experiments.

b) Based on the results of the above experiments, the setleability study was repeated using $Al_2(SO_4)_3$ and $FeCl_3$ at doses of 150 and 200 mg/l and a pH of 5.5 settling characteristic paths were established, based on the removal of the suspended residue.

RESULTS AND DISCUSSION

The characteristics of the process water and wastewater effluents are shown in Table 1. The influent water withdrawn from the Mahmodia canal is characterized by high concentrations of COD, BOD, residues and chlorides. The canal serves as a convenient basin for the uncontrolled discharge of industrial effluents in this area.

All plant effluents contain substantial levels of COD, BOD, settleable solids, chlorides and residues. Straw pulping effluent is extremely polluted. However, it is not advisable to segregate this waste as its volume represents less than 5% of the total effluent.

The results of stations 6 and 7 show that the removal percentages of COD, total residue and settleable solids in the existing sedimentation tank are 22.2%, 20.6% and 12.3%, respectively.

The data of Table 2 demonstrate the concomitant increase in percentage removal of COD, total residue and suspended residue with an increase in the coagulant dose. Within the range of the pH tested in this study, it appears that good flocculation occurred at pH ranges of 5.0 - 5.5 for $Al_2(SO_4)_3$ and $FeCl_3$, and 10.0 - 11.0 for $FeSO_4$. These ranges are not entirely explicit, but few inconsistencies emerged as a result of variations in the initial concentration of the tested samples. In general, the results indicate that $Al_2(SO_4)_3$ and $FeCl_3$ were more effective coagulants than $FeSO_4$.

It is evident from the data of Table 3 that increasing the dose of the anionic polyelectrolyte to 2 mg/l may slightly improve the removal efficiency. However, it is expected that polymer costs will offset this slight improvement. Excessive addition of polyelectrolyte resulted in adverse effects on the flocculation process. It is postulated that excess polyelectrolyte might have occupied the adsorption sites and hindered the bridging, resulting in dispersion rather than flocculation. In addition, it is observed that the active anionics used in this study function well at pH levels above 7.0, although all tests were carried out at pH 5.5.

An impression of the change in the settling characteristics under different flocculation conditions may be gained from Figs. 2, 3 and 4. The curves show a distinct increase in TOC concentration with depth at a given time, and the slope becomes proportionately steeper by increasing the settling time. The results of the settling study reveal approximate agreement on the suitability of $Al_2(SO_4)_3$ and $FeCl_3$ for the treatment of paper-reprocessing wastewater.

Table 1 Characteristics of Process Effluents at the Alba Paper Reprocessing Plant

Sampling Location		DO Mg/ℓ	pH	COD Mg/ℓ	BOD Mg/ℓ	Cl Mg/ℓ	Alkalinity Mg/ℓ	Settleable Solids ml/ℓ	Total Residue Mg/ℓ	Volatile Residue Mg/ℓ	Filtrable Residue Mg/ℓ
1. Process water (river water)	$\bar{X}$	8.2	7.9	92	65	175	253	0.0	418	150	275
	SD	1.3	0.2	25	26	29	25	0.0	135	78	54
2. Straw pulping effluent	$\bar{X}$	3.4	9.4	1748	575	192	453	34.8	1879	1090	985
	SD	1.8	0.3	440	198	92	56	7.6	698	452	295
3. Screening effluent	$\bar{X}$	3.7	7.4	1500	520	255	400	28.0	1601	884	518
	SD	2.1	0.2	326	136	106	215	10.3	211	211	122
4. Paper machine effluent	$\bar{X}$	2.5	8.2	1535	1128	463	341	40.0	1424	700	635
	SD	1.0	0.38	1393	1069	223	113	12.5	346	343	237
5. Paper processing effluent	$\bar{X}$	1.3	7.3	1286	565	268	320	38.6	1375	419	513
	SD	0.8	0.2	159	115	156	46	11.2	586	224	306
6. Final effluent before settling	$\bar{X}$	3.1	7.4	1335	760	262	308	39.1	1585	831	688
	SD	0.9	0.2	268	182	138	41	18	688	314	370
7. Final effluent after settlings	$\bar{X}$	3.8	7.4	1053	489	267	276	34.2	1258	808	415
	SD	1.1	0.2	202	65	149	49	7.3	238	115	335

$\bar{X}$ = Average of five samples
SD = Standard Deviation

Table 2 Effect of Coagulant Dose on Paper-Reprocessing Wastewater
(Percent Removal)

I. $Al_2(SO_4)_3$

DOSE	50 mg/ℓ			100 mg/ℓ			150 mg/ℓ			200 mg/ℓ		
pH	5.0	5.5	6.0	6.5	5.0	5.5	6.0	6.5	5.0	5.5	6.0	6.5	5.0	5.5	6.0	6.5
COD	31	34	29	27	51	61	57	52	82	81	78	75	87	92	88	88
TR	28	30	29	26	50	58	53	48	71	75	72	71	81	84	78	80
SR	30	35	33	36	54	65	62	57	73	77	78	74	88	90	83	84

II. $FeCl_3$

DOSE	50 mg/ℓ				100 mg/ℓ				150 mg/ℓ				200 mg/ℓ			
pH	5.0	5.5	6.0	6.5	5.0	5.5	6.0	6.5	5.0	5.5	6.0	6.5	5.0	5.5	6.0	6.5
COD	17	18	14	12	44	42	38	45	63	65	54	58	71	69	62	60
TR	20	22	16	15	46	50	43	39	58	56	42	47	65	63	55	57
SR	27	26	22	21	45	49	54	49	67	64	53	55	73	70	63	65

III. $FeSO_4$

DOSE	50 mg/ℓ		100 mg/ℓ		150 mg/ℓ		200 mg/ℓ	
pH	9.0	10.0	11.0	9.0	10.0	11.0	9.0	10.0	11.0	9.0	10.0	11.0
COD	29	34	33	38	36	35	52	58	62	63	67	68
TR	36	38	41	33	41	43	45	49	58	65	71	69
SR	42	44	50	43	49	53	61	64	64	63	67	73

[1] The average values are given of two samples with characteristics as follows: COD, 985-1150 mg/ℓ; total residue (TR), 1322-1585 mg/ℓ; and suspended residue (SR) 733-1130 mg/ℓ.

Table 3 Effect on Paper-Reprocessing Wastewater of the Addition of Polymeric Flocculant (Percent Removal at pH = 5.5)[1]

I. $Al_2(SO_4)_3$ + Polymer[2]

$Al_2(SO_4)_3$ Polymer	Initial Charac- teristics	100 % Removal				Initial Charac- teristics	150 % Removal				Initial Charac- teristics	200 % Removal			
		1	2	3	4	5		1	2	3	4	5		1	2	3	4	5
COD	1220	50	49	51	52	50	1135	79	82	76	73	75	1035	93	88	86	86	82
TR	1418	48	47	45	43	40	1615	71	73	68	65	64	1510	86	84	81	79	80
SR	927	60	58	55	55	52	1153	78	84	79	74	77	984	92	88	85	84	85

II. $FeCl_3$ + Polymer

$FeCl_3$ Polymer	Initial Charac- teristics	100 % Removal				Initial Charac- teristics	150 % Removal				Initial Charac- teristics	200 % Removal			
		1	2	3	4	5		1	2	3	4	5		1	2	3	4	5
COD	1115	43	46	38	35	32	1050	69	63	60	58	60	980	78	83	83	76	76
TR	1367	44	47	42	40	40	1415	60	62	63	65	63	1245	68	73	76	75	76
SR	844	52	53	50	54	45	997	66	71	73	73	71	812	77	79	82	80	81

[1] The average values of two samples are given

[2] Hercules anionic polymer No. 819.2. and $Al_2(SO_4)_3$ were added in mg/l.

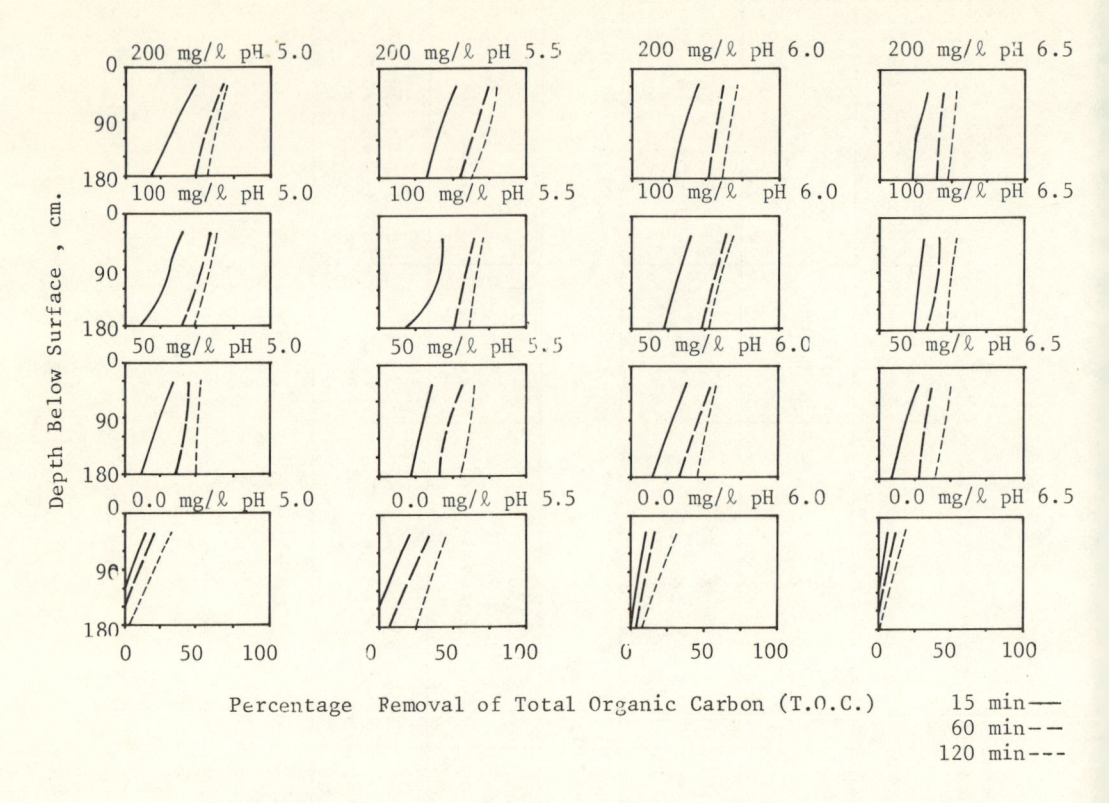

Fig. 2. Effect of $Al_2(SO_4)_3$ on the Settling of Paper-Reprocessing Wastewater

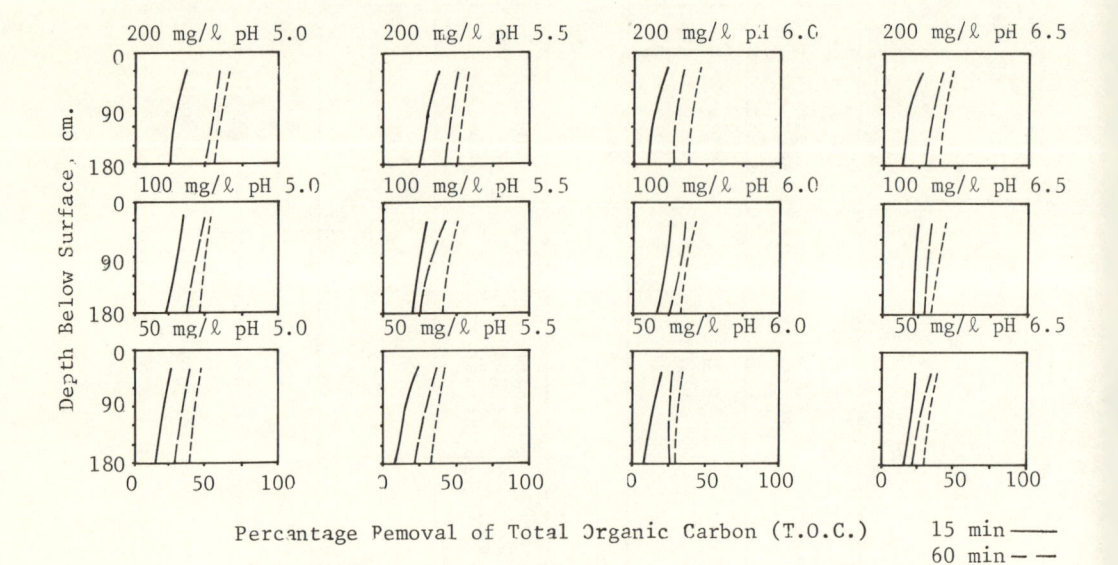

Fig. 3. Effect of $FeCl_3$ on the Settling of Paper-Reprocessing Wastewater

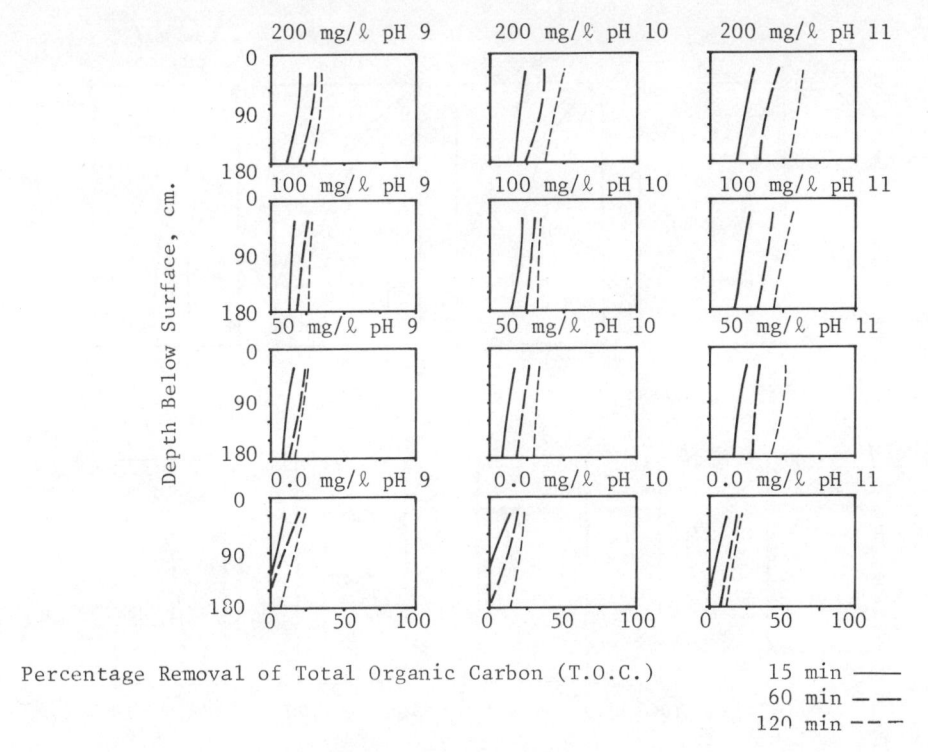

Fig. 4. Effect of FeSO$_4$ on the Settling of Paper Reprocessing Wastewater

The sedimentation performances of Al$_2$(SO$_4$)$_3$ and FeCl$_3$ were further evaluated by comparing their suspended residue removal curves, as shown in Fig. 5. The better performance of Al$_2$(SO$_4$)$_3$ was attributed to the use of pH = 5.5. Stuart and Russel (1969) indicated that the optimum pH for FeCl$_3$ was 3.9. However, such a low pH would be unfavorable for the recycled water.

Based on the results of this study, it is proposed to develop a simple system for water renovation and reuse without major capital expenditure. The new system comprises the following features:

 a) pH adjustment to 5.5 and the addition of 150-200 mg/l Al$_2$(SO$_4$)$_3$ to the existing presettling tank ahead of the final sedimentation tank.

 b) The current detention time in the sedimentation tank is 57 min. This detention time is sufficient for the removal of 85 - 90% suspended residue, as shown in Fig. 5. It is anticipated that this simple treatment may upgrade the effluent quality above that of the process water withdrawn from the canal.

 c) The renovated water can be recycled to the process water tank for further settling before use in the plant. This system can effectively close the water cycle without any adverse effects on the equipment and the quality of the paper produced.

The Alexandria Governorate is initiating a strict abatement program which will be implemented soon. This plan calls for imposing a surcharge cost on ALBA at an annual rate of 30,000 Egyptian Liras (L.E.) to permit the plant to discharge its effluent to

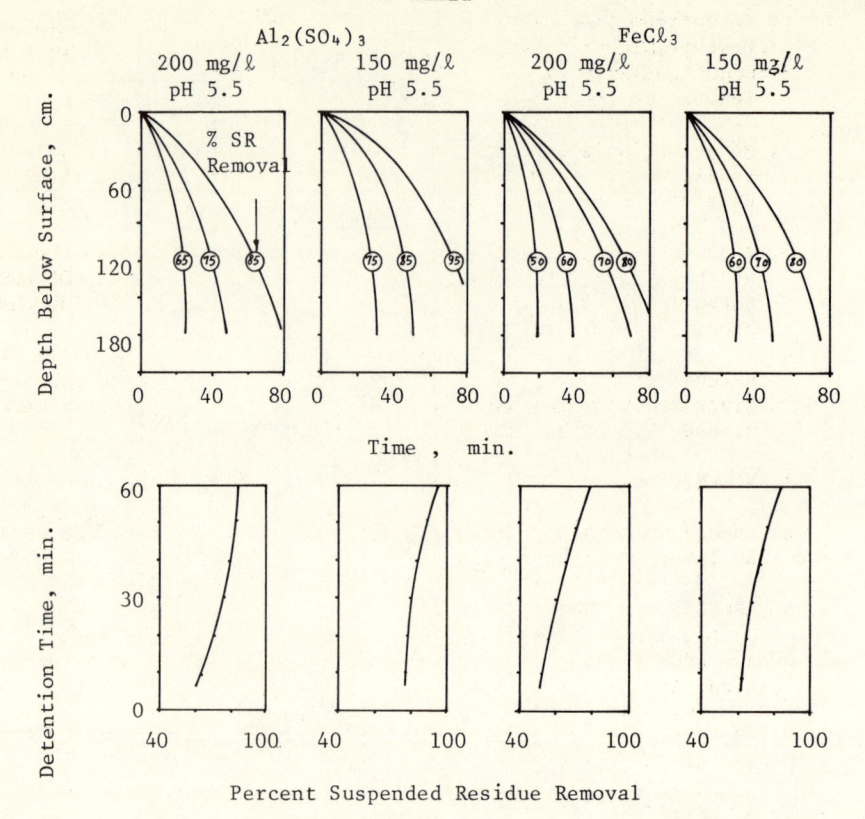

Fig. 5. Effect of Different Coagulants on Settling Characteristic Curves of Paper-Reprocessing Wastewater

the public sewer system. On the other hand, the cost of the new renovation system is estimated as follows:

   a. Annual operating costs (chemical and manpower) 24,000 L.E.
   b. Annual depreciation of capital expenditure (5%) 1,000 L.E.

Therefore, closing the water cycle at ALBA can result in a potential saving of 5,000 L.E./year, when the new abatement program is enforced.

CONCLUSIONS

Based on the results of this study, the following conclusions may be made:

   1. Process effluents generated at ALBA are generally heavily polluted. Plain sedimentation is inadequate for treating the combined effluent.

2. An optimum flocculation can occur at pH 5.5 using 200 mg/l $Al_2(SO_4)_3$. However, lower doses of $Al_2(SO_4)_3$ may be used, depending on the appropriate quality of the recycled water and the economics of the proposed renovation system.

3. No synergistic effect is expected by adding anionic polyelectrolyte to $Al_2(SO_4)_3$ at pH 5.5. Excessive polyelectrolyte may result in dispersion rather than flocculation.

4. A new renovation and reuse system is proposed to close the water cycle at the plant. The annual cost of the new system is 25,000 L.E., and potential saving is expected due to the waiving of the surcharge cost proposed by the local authorities.

5. Extensive recycling will not affect the equipment and paper quality, since the recycled water will meet or exceed the process water quality already in use at ALBA.

ACKNOWLEDGEMENT

This study was funded by US EPA grant NO 3-542-4. Experimental work was conducted at the HIPH and SERC laboratories of Alexandria.

REFERENCES

APHA - AWWA - WPCF (1971). *Standard Methods for Examination of Water and Wastewater*. 13th ed. American Public Health Association, New York, N.Y.

Davis, W., R. Kraiman, J. Parker, and C. Thurburg (1973). Recycling Fine Paper Mill Effluent by Means of Pressure Filtration. *TAPPI, 56,* (1), 89.

Kimurage, R. and K. Izumizawa (1976). Approach to Entirely Closed Board Mill by Activated Carbon Adsorbtion. *TAPPI, 30,* (1), 20.

Stelmakh, B. (1974). Improved System for the Treatment of Wastewater at the Lvov Board Mill. *Lisove Gospod. Prom, 6,* 21.

Streebin, E.G. Reid and P. Law (1976). Water Reuse in a Paper Processing Plant. *TAPPI, 59,* (5), 105.

Stuart E. and F. Russel (1969). Coagulation of Pulping Waste for the Removal of Color. *J. Water Poll. Control Fed., 41,* (2), 222.

Thomas, A. (1972). Mill White Water Process and Reuse System. *Amer. Paper Ind., 54,* (3), 38.

DISCUSSIONS

Leentvar, J. (Holland) : Did you increase the mixing intensity when you increased the coagulant aid dose to 5 mg/ℓ? Such a high dose of polyelectrolyte may result in the flocs becoming so large that they settle during flocculation.

Hamza, A. (Egypt) : We discovered that increasing the dose of anionic polyelectrolyte to 2 mg/ℓ led to a slight improvement in removal efficiency, primarily due to the unusually low pH, we think. Good flocculation was found to occur at a pH of 5.5. Excessive polyelectrolyte, however, may result in dispersion rather than flocculation.

Gür, A. (Turkey) : In the final effluent after settling you seem to have quite a large amount of dissolved oxygen, 3.8 mg/ℓ. Which method did you use to measure your dissolved oxygen?

Hamza, A. (Egypt) : We used the Winkler method. Samples were taken from the site for analysis in the lab, 10 minutes away. 3.8 mg/ℓ is an average value of 5 samples. Actually, when compared with the value of 3.1 mg/ℓ obtained before settling, it shows only a slight increase. This is probably due to exposure to the sun when the samples were being taken from the settling tank.

Çetin, C. I. (Turkey) : It has been said that Lake Maruit is very important in Egypt with regard to fish production. Have any other studies, similar to those carried out on the wastewater from the Alba paper-processing plant, been conducted on the wastewater discharged from other industrial plants situated around the lake?

Hamza, A. (Egypt) : With the help of a 2 million dollar grant from the U.S. Environmental Protection Agency, an initial study is being carried out of the whole area. There is also a long-term project for the actual abatement of pollution from industry in this area. The cost, which will be in the range of 60 million dollars, will be met by the USEPA and the Egyptian government. We at the Institute are surveying all the industrial plants in the region, for example, starch, textile and edible oil plants. We are very much aware of the importance of Lake Maruit for fish production, and we are very concerned that this has now decreased to 20% of the production 15 years ago and that there is now also toxic accumulation in the tissue of the fish.

Müezzinoğlu, A. (Turkey): Should averages be used for design purposes, when there is, for example, such a variation in the data for the same hour on succeeding days? Why didn't you use standard deviations or probabilistic representations instead?

Hamza, A. (Egypt) : Discrete combined samples were taken. First, a combined sample was taken over a period of 24 hours. Five sets were then obtained over a period of 3 months. Variations in the results are unavoidable; indeed, we are interested in showing that too many variations exist in the process. In my paper, for example, I suggested for design purposes the use of 150-200 mg/ℓ alum. However, this will again depend on how much variation is encountered during processing. This is something we cannot help in terms of design, but it will be taken into consideration.

Akçın, G. (Turkey) : When taking samples, did you use special equipment?

Hamza, A. (Egypt) : We have an automatic sampling system, by which we can collect samples on an hourly basis and cool them in 24 or 30 discrete bottles in the field. There is nothing new in this method.

Bunning, W. (Holland) : When using ferrous sulphate as flocculant, did you observe any oxidation into ferric compounds because of the relatively high level of dissolved oxygen in the water?

Hamza, A. (Egypt) : We have not yet reached the point of testing whether in practice there is flocculation with ferric rather than ferrous compounds.

Conditioning and Disposal of Industrial Sludges by Direct Slurry Freezing Process

M. Z. ALI KHAN

Department of Civil Engineering, King Abdulaziz University, Jeddah, Saudi Arabia

ABSTRACT

This paper presents a unique process for the conditioning and disposal of pulp and paper sludge. The sludge is slurry frozen (not solid frozen), by direct contact with the refrigerant, butane. The process, by the addition of small quantities of the coagulants such as alum prior to freezing, results in 3 to 4 times faster dewatering rates on sand-drying beds as well as accelerated vacuum filtration rates as compared to those of the unconditioned sludge. The filter cake moisture content is only 40%. The supernatant and filtrate qualities are much better.

The refrigerant is recovered and reused, thus making the process economical. The overall disposal costs are very close to or in some cases less than those of the other conventional disposal methods.

INTRODUCTION

The surveys (Warrick, 1947) of the pulp and paper mills indicate that the wastes generated, i.e. side streams, from different processes involved in producing pulp and paper, are generally quite high in sulfates, sulfites, solids, etc. The BOD tends to be high, especially with wastes containing sulfites. These side streams are usually treated physically, chemicalyy or biologically (Ng and others, 1978; Barton and Byrd, 1975; Peterson, 1975; Voelkel, 1974; Zimmerman, 1958) to produce effluents which meet water quality standards for discharge into the receiving water body. This treatment results in residue called "sludge", which contains a lot of chemicals, especially sulfates and sulfites, in colloidal suspension and dissolved forms, making this sludge less amenable to settling, thickening, conditioning, dewatering etc. before final discharge. For final discharge of the sludge, it is very important to reduce its volume by removing the associated water by physical or mechanical means. Lower volumes mean lower disposal costs.

Most of the sludge treatment processes (Vogler and Rudolf, 1951;Brecht, 1965;Coogan, 1965;Follet and Gehm, 1966;Andrews, 1967;Swets and others, 1974) presently used, have their difficulties and their drawbacks, particularly with regard to certain types of sludges. To overcome or reduce these problems, a direct freezing process (Khan, 1974;Khan, Randall and Stephens, 1976) has been developed for the conditioning and disposal of the waste activated sludge from sewage treatment works.

It is the purpose of this research to attempt to evaluate the feasibility of the direct slurry freezing process for conditioning, dewatering and disposal of sludge obtained from the treatment of pulp and paper waste streams.

SAMPLING and TESTING

The sludge samples were collected from the outlet of the return sludge line from the secondary clarifier of the activated sludge process utilized for treating the side-streams of the Wesvaco Pulp and Paper Mill at Covington, Virginia, U.S.A. The samples were transported directly to the laboratory, located at Blacksburgh, Virginia, and stored by refrigeration for a few hours before testing and analysis. The analysis, before and after direct slurry freezing, was conducted according to Standard Methods (1971). A Becman 215 analyzer was used for total carbon measurements and the pH was measured using a Leads and Northrup probe.

FREEZING PROCESS DESCRIPTION

The experimental sludge at a certain solids concentration is allowed to flow from a sludge tank "ST" (surrounded by ice) through a flow regulator "FR" into the freezer "F", where liquid butane, flowing through the heat exchanger "H", and flow regulator "R", comes in contact with the sludge (intimate contact is achieved by mechanical mixing). After heat exchange between butane and sludge, the slurry frozen sludge is pumped through the condenser Coil "C_1" to the heat exchanger "H_1", from where it flows to heat exchanger "H_2" and finally to the butane stripper "BS" for recovery of dissolved butane from the supernatant for reuse, if necessary. The hot butane vapors from the freezer "F" are compressed by vacuum-pressure pump "VP", condensed on the Coil "C_1" and recovered as liquid for reuse in the freezer. The slurry frozen sludge in the heat exchange "H_1", condenser Coil "C_1" and heat exchanger "H_2" liquifies the butane vapors. The schematic representation is given in Fig. 1.

PROCESS THEORY

Liquid butane has a boiling point lower than the freezing point of water. Butane extracts the heat of vaporization from the sludge, thus inducing freezing and forming ice-slurry. If the freezer operates under a vacuum of 3 cm of mercury, the butane boiling point is lowered to about -1.5^0C to insure better freezing. The sludge freezing point, due to some dissolved solids, can be as low as -0.5^0C. The butane vapors, when condensed on the previously frozen sludge, are liquified for reuse and thaw the frozen sludge. Hence, most of the energy is recovered and reused.

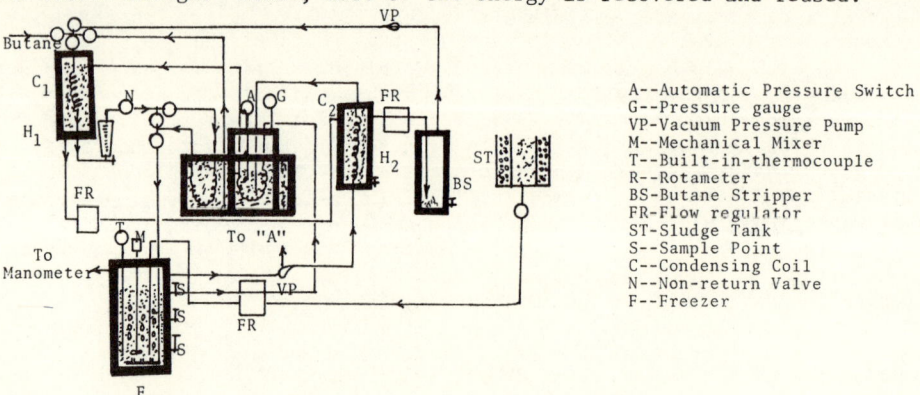

Fig. 1: Butane Recovery and Continuous Flow Apparatus.

A--Automatic Pressure Switch
G--Pressure gauge
VP-Vacuum Pressure Pump
M--Mechanical Mixer
T--Built-in-thermocouple
R--Rotameter
BS-Butane Stripper
FR-Flow regulator
ST-Sludge Tank
S--Sample Point
C--Condensing Coil
N--Non-return Valve
F--Freezer

EXPERIMENTAL DESIGN

The process described in Fig. 1 is for a continuous flow process and was developed by Khan (1974, 1976). For the purpose of this study, only batch studies were conducted by putting about 2 liters of pulp and paper sludge having total solids of about 1.33% in the freezer. The liquid butane flow rate was about 10 mℓ/min. and the butane contact time or residence time was 5-6 hours. A mechanical mixer was used to prevent the solid from freezing. At the end of 6 hours the slurry frozen sludge was taken out of the freezer and allowed to thaw at room temperature; different parameters were then tested to check the improvement in the sludge characteristics.

The following set of experiments were conducted:

1) direct slurry freezing without chemicals.
2) direct slurry freezing using Al^{+++} coagulant dosages of 100 to 1500 p/p/m.
3) direct solid freezing without chemicals.
4) direct solid freezing using Al^{+++} dosage of 100 to 1500 mg/ℓ
5) original sludge with chemicals but without any freezing.

The effectiveness of direct freezing was determined by the specific resistance test as described by Coakley and Jones (1972); the sludge settling characteristics, by calculating the interfacial settling velocity and sludge volume index (S.V.I) as presented by Mohlman (1934); gravity drainage, on sand beds (Nebiker, 1967 and 1969); and the quality of the filtrate and supernatant, by T.O.C. analyzer.

EXPERIMENTAL RESULTS

Results indicate that the addition of chemicals to the sludge, without freezing, shows a reduction in specific resistance from 13×10^{-13} m/kg to about 7×10^{-13} at a chemical dosage of 1000 p/p/m of Al^{+++}, but solid freezing and slurry freezing reduced this specific resistance to $<1 \times 10^{-13}$ m/kg (see Fig. 2).

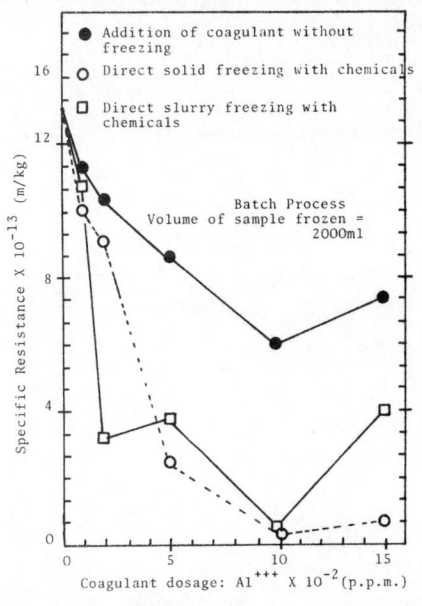

Fig. 2: Effect of coagulant dosage on the specific resistance of the pulp and paper waste sludge.

The cake moisture, (15 min of vacuum filtration) after slurry and solid freezing was about 30%, but the unfrozen sludge still had more than 60% moisture (see Fig.3). The rate of solids production (lbs/ft^2/hr) in the case of solid freezing was twice as much as in the case of slurry freezing (see Fig. 4). The optimum conditioning effect seemed to be at about 1000 ppm dosage of alum. There was practically improvement by slurry freezing without the addition of any chemicals.

The direct slurry freezing resulted in about 50% volume reduction after 30 min settling, thus producing an improvement in the thickening and settling process. The original sludge without any chemical addition or freezing did not settle at all, whereas solid freezing gave results very close to slurry freezing (see Fig. 5).

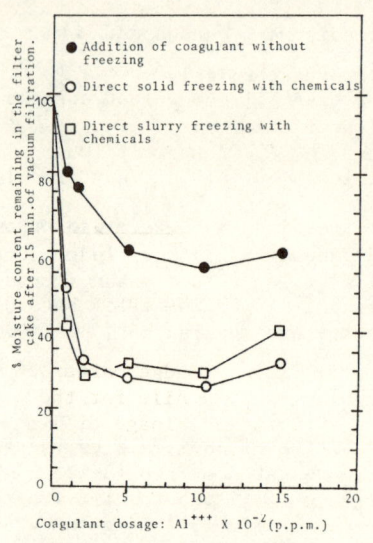

Fig. 3: Effect of coagulant dosage on the filter cake quality of the pulp and paper waste sludge.

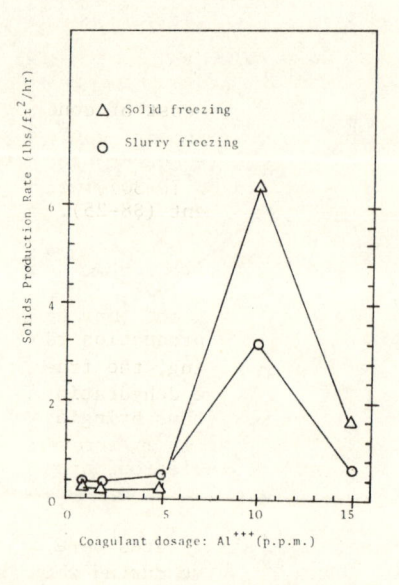

Fig. 4: Variation in rate of solids production with coagulant dosage using vacuum system.

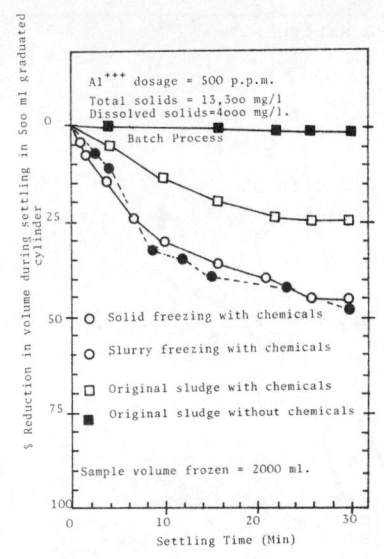

Fig. 5: Settleability of pulp and paper waste sludge after direct freezing

The gravity drainage of directly frozen sludge, after 24 hours, indicated that the % moisture content left was about 92.5%, while for the unfrozen sludge the same value was about 97.5%. After the gravity drainage of 1 day, any further moisture content reduction occurring was considered due to air-drying.

Air drying for 7 days resulted in further reduction of moisture content to about 77% and 86% for the directly frozen slurry and the unfrozen sludges, respectively. The supernatant and filtrate were slightly higher in dissolved solids as compared to the original sludge.

ECONOMICS of the PROCESS

Khan (1974, 1976), while utilizing this process for the waste activated sludge from sewage treatment works, indicated that the cost of conditioning and disposal of sludge, without the cost of dewatering equipment, would be about $ 20-30 per ton of dry solids. This cost is quite close to the other processes, such as chemical conditioning ($8-25), direct solid freezing ($ 10-30), wet air oxidation ($ 30-35), anaerobic digestion ($ 15-20) and heat treatment ($8-25).

DISCUSSION of RESULTS

The improved settling rates (after freezing and thawing of sludge), reduction in specific resistance and the better solids production rate can be attributed to the fact that during the process of slow freezing, the freezing ice exerted a pressure of 12 atm (Clements and others, 1950). The dehydration, during slow freezing, allows the internal cell water to move out, thus bringing the internal proteins so close as to form S-S bonds, thus resulting in an irreversible reaction. This would result in cellular precipitation and flocculation, upon thawing of the frozen sludge.

In the case of drainage, the initial faster rates were due to the filtration of free water, after which there was a lag time during which the sludge particles compacted together and then released the water held in between them, resulting in a secondary increase in drainage but at rates slower than the initial rates. In the

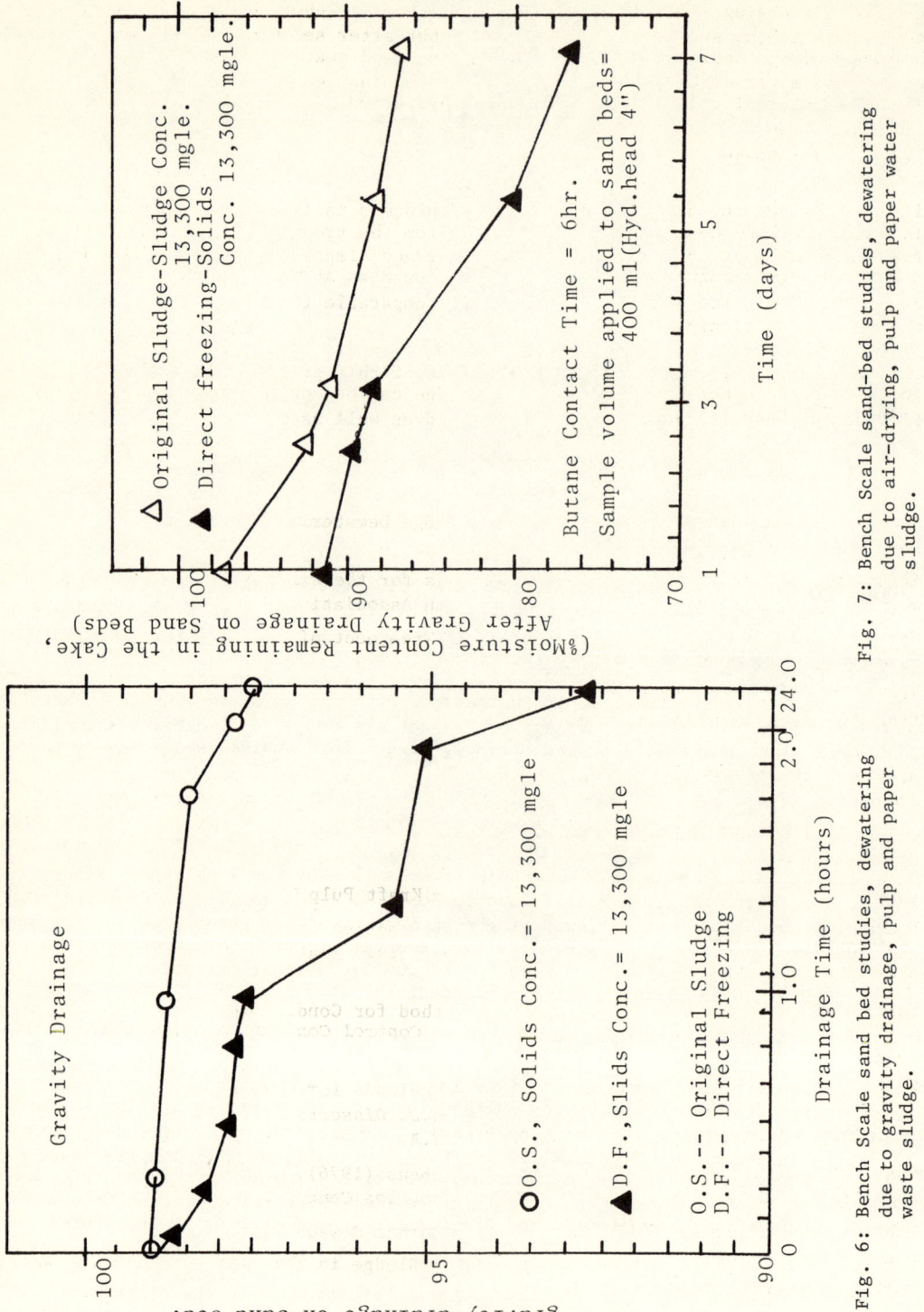

Fig. 6: Bench Scale sand bed studies, dewatering due to gravity drainage, pulp and paper waste sludge.

Fig. 7: Bench Scale sand-bed studies, dewatering due to air-drying, pulp and paper water sludge.

case of air drying, initial moisture reduction was higher, because the moisture could move to the surface for evaporation, but after seven days there was no further appreciable improvement. The higher dissolved content in the filtrate and supernatant, after direct slurry freezing and thawing, was probably due to the release of internal cellular salts during dehydration.

SUMMARY

It appears that the direct slurry freezing process is feasible for the conditionining of biological waste sludges generated from the treatment of the waste streams from pulp and paper mills. The conditioning and disposal cost (not including the dewatering equipment cost), using chemicals such as Al^{+++}, is about $20-30 per ton of dry solids in the sludge. This cost is comparable to or less than the cost of the other conditioning and disposal methods.

One (Warrick, 1947), of the major advantages of this process, which the other freezing processes do not have, is that the sludge can be conditioned on a continuous flow basis. And it can be used out in cold as well as hot climates.

REFERENCES

Andrews, G. and others (1967). Effluent Sludge Dewatering Practiced by Two Pulp and Paper Mills of Mead Corpn. *Tappi, 50,* 99A.

APHA, AWWA and WPCF (1971). *Standard Methods for the Examination of Water and Wastewater.* 13th Edition. American Public Health Association, Washington, D.C.

Barton, C.A., and J.F. Byrd (1975). Joint Treatment of Pulping and Municipal Waste. *Water Pollution Control Federation,* 998, May, 1973.

Brecht, W. and others (1965). Studies on the Dewatering of Water - water Sludges from Paper Mills. *Papier 18,* 741.

Coakley, P. and B.R.S. Jones (1972). Vacuum Sludge Filtration (I). Interpretation of Results by the Concept of Specific Resistance. *Sewage and Industrial Waste, 28,* 963.

Clements, G.S. and others (1950). Sludge Dewatering by Freezing with Added Chemicals. *Journal Inst. of Sewage Purif.,* Part 4, 318.

Coogan (1965). Incineration of Sludges from Kraft Pulp Mill Effluents. *Tappi,* 44A.

Follet, R. and H.W. Gehm (1966). Manual of Practice for Sludge Handling in the Pulp and Paper Industry. *Technical Bulletin,* No. *190.* National Council for Stream Improvement, New York.

Katz, W.J. and others (1967). Freezing Method for Conditioning Activated Sludge. *16th Southern Water Resources and Pollution Control Conference,* Duke Univ., April, 1967.

Khan, M. Z. Ali (1974). Principles and Techniques for Conditioning of Waste Activated Sludge by Direct Slurry Freezing. *Ph.D. Dissertation, Virginia Polytechnic and State Univ., Blacksburgh, Virginia, U.S.A.*

Khan, M.Z. Ali, C.W. Randall, and N.T. Stephens (1976). Direct Slurry Freezing of Waste Activated Sludge. *Virginia Water Resources Center Bulletin,* No. *94.*

Mohlman, F.W. (1934). The Sludge Index, *Journal Sewage Works, 6,* 119.

Nebiker, J.H. (1967). Drying of Wastewater Sludge in the Open Air. *Journal Water Pollution Control Federation, 39,* 608.

Nebiker, J.H. and others (1969). An Investigation of Sludge Dewatering Rates. *Journal Water Pollution Control Federation, 41,* R255.

Ng., K.S., J.C. Mueller, and C.C. Walden (1978). Ozone Treatment of Kraft Mill Waste *Water Pollution Control Federation,* 1742.

Peterson, R.R. (1975). Design for Criteria for High Purity Oxygen Treatment for Kraft Mill Effluents. *Water Pollution Control Federation,* 2137. Sept., 1975.

Swets, D.H., L. Pratt and E.E. Metcalf (1974). Thermal Conditioning in Kalamazoo, Michigan. *Water Pollution Control Federation,* 575, March. 1974.

Voelkel K.G., and R. W. Deering. Joint Treatment of Municipal and Pulping Effluents. *Water Pollution Control Federation,* 634. April. 1974.

Vogler, J.F., and W. Rudolf (1951). White Water Sludge Drainage Factors. *Sewage Industrial Wastes,* 23, 699.

Warrick, L.F. (1947). Pulp and Paper Industry Wastes. *Ind. Eng. Chem.* 39, 670.

Zimmerman, F.J. (1958). New Waste Disposal Process. *Chemical Engineering,* 65, 117.

Possibilities in Reducing the Effluent from Chemical Pulping Operations

H. FLECKSEDER

Universitatsassistent am Institut fur Wasserversogung, Abwasserreinigung und Gewasserschutz, TU Wien, Wien, Austria

ABSTRACT

Chemical pulping is an operation which yields the raw material for papermaking and to some extent for the chemical industry. Despite immense efforts on the process engineering side, the pollution loads from modern pulp mitts can have a severe effect on the ecology of receiving waters, especially in acidic environment. This paper deals with the amount of pollution discharged from modern pulp mitts and the link between internal and external treatment.

GENERAL ASPECTS

Introduction

Chemical pulping is the exposing of cellulosic fibres from fibrous raw material and the cleaning of these fibres under the influence of chemicals, pressure and temperature. Depending on the type of cooking liquor one can differentiate sulfite-pulping, sulfate -(or kraft) pulping and NSSC-(neutal sulfite semichemical(pulping. Newer processes are being developed. The cleaning of the cellulosic fibres is achieved by intensive washing with water which removes dissolved matter and by centrifuging which separates unwanted solid matter (knots, shives, inorganics). A chemical cleaning step (bleaching) is applied in case the pulp requires additional properties (brightness, viscosity, content of α-cellulose). From all this, it can be concluded that water constitutes a key raw material in pulping.

The amount of pollutants discharged from this industry depends on the process technology applied in pulping. This discharged matter can have the following effects in aquatic ecosystems:

- primary and secondary oxygen depletion by microbial growth and the shifting of large quantities of hetherotrophs;

- change in the light climate by lignins and the derivatives of lignins, with a shift in primary productivity and life in the water;

- acute as well, as chronic toxic effects, associated mainly with kraft pulping than with sulfite pulping.

Raw Materials

The raw material for pulping is normally wood, but in certain areas reed and straw are also used. Prerequisites for large investments in a pulp mill are safe usage of raw material, reliability of supply, as well as knowledge of the cost of the raw material and its storing properties. From the water pollution control point of view it should be noted that most wood has to be debarked before subjecting it to cooking and that the chemical composition of wood or any other raw material gives some insight into the pollution potential.

Groups and classes of chemicals present in wood are

- α-cellulose, a monopolymer from d-glucose
- hemicelluloses
- lignin
- extractibles
- inorganic components

The distribution of these components depends on the type of wood (or raw material), the place where and the conditions under which it grows. For Scandinavian wood (dry weight base unextracted) the figures given in Table 1, show that cellulose, extrac-

Table 1 : Distribution of Chemicals in Scandinavian Wood

	cellulose	hemi-celluloses	lignin	Extractives	in-Organics
pine	44%	25%	28%	2.6%	0.4%
birch	40%	38%	20%	1.7%	0.3%

tives and inorganics are present in roughly equal amounts in hardwoods as well as in softwoods. However, softwoods are richer in lignins than hardwoods, whereas with hemicelluloses, it is just the reverse. Chemical pulping is more or less equivalent to dissolving hemicelluloses, lignins, extractives and inorganics, and as the breakdown of hemicelluloses yields sugars and other carbohydrates, spent cooking liquors from pulping hardwood will have a higher specific BOD-load than when pulping softwood, based on the premise that the specific COD-load is equal, i.e. equal amounts of wood substance have been dissolved in pulping.

Basic Principles of Pulping

When cooking, e.g., chipped wodd, the main purpose of cooking is to dissolve the middle lamella which glues cellulosic fibres together. In full cooking - depending on the type and quality of pulp desired - only 40 to 60% of the dry weight of raw material put into the process becomes a manufactured product, whereas the remainder is, at least at first, waste. In mechanical pulping, however, close to 100% of the raw material is manufacturable, but the qualities of the cooked and the mechanically separated pulps are different. In between fully cooked and mechanically separated fibres, there are the "semicooked" pulps with a yield in cooking of from 60 to 90%.

The chipping or segregating of the raw material into smaller pieces before cooking is mandatory. The product (i.e. the cooked pulp) can be categorized not only according to the yield, but also by the Kappa number (which is a measure of delignification), by the brightness (which represents the spectral reflection relative to absolute white) and, for pulps which are determined for chemical industry, also by the degree of finish, which is a measure of the amount of α-cellulose present in the pulp.

Cooking processes used at present most widely are *sulfite-pulping* and *kraft-pulping*. Sulfite-pulping was more predominantly used in Europe than kraft-pulping up to 20 years ago due to the high degree of brightness achieved without bleaching and the relative ease in bleaching compared with pulps from sulfate (kraft) cooking. However, types of wood with a high resin content could not be cooked by Ca-sulfite-pulping.

In batch cooking, the process runs as follows: The debarked and chipped wood is fed

Chemical Pulping Operations

into the digester; the cooking liquor is added; heat is applied according to certain rules and maintained over a certain period of time; after cooking, the cooking liquor is discharged and the pulp taken out from the digester and a new cooking cycle starts again. When the spent cooking liquor and the cooked pulp are still under pressure in the digester and they are removed, they are said to blow out; therefore, the emptying is also called "a blow". In sulfite-pulping, the cooking liquor is composed of 7-10 kg't of pulp of methanol, 30-130 kg/t of acetic acid, 5-16 kg/t of furfural, 150-250 kg/t of sulfonic and aldonic acids, 200-400 kg/t of sugars and 600-800 kg/t of ligninsulfonates and other minor fractions. A quick glance at these compounds tells us that

- per ton of product there is a large amount of pollution in the spent liquor;
- both decomposable and resistant organic matter are present;
- in sulfite-pulping, there is also a fair amount of volatile organic matter (methanol, acetic acid,...).

In sulfate (kraft)-pulping, the cooking liquor is composed of 5 kg/t of methanol, 100-200 kg/t of acetate, 40 kg/t of formiate, 100 kg/t of lactate, in the order of 250 kg/t of sugar acids and lactones, 20-100 kg/t of tall oil and 400-600 kg/t of alkaline lignins. A glance at these compounds agains tells us that

- per ton of product there is a huge amount of pollution in the spent liquor;
- both decomposible and resistant organic matter are present;
- in sulfate-pulping, the volatile fraction is much smaller than in sulfite-pulping.

From this, it must be concluded that the holding back of the spent liquor from the receiving waters is of prime importance in water pollution control at pulp mills. To this end, process equipment has been designed which allows liquor recovery rates of 98 to 99.5% for sulfite-pulping and 99.5% for kraft-pulping. The separated spent liquor is concentrated by evaporation (where volatile compounds can give rise to polluting discharges in the condensate) and is either burnt by utilizing its heat content (electricity, steam) or sold as a raw material.

After cooking, cellolosic fibres may have a color appear which makes them unfit for various uses (paper-making). Also, the residue from the hemicelluloses and lignins may have to be removed before the product can undergo further processing (rayon fibre production, speciality products). In order to undertake this, the cooked pulp is subjected to a chemical cleaning (bleaching). *Bleaching* involves both a pre-bleaching step, which consists of splitting the lignin with oxidants (e.g. dissolved oxygen, chlorine or chlorine compounds) and dissolving the split lignin and carbohydrates with an alkali (e.g. NaOH) and a final bleaching step, in which again oxidants (e.g. ClO_2 or NaOCl) and alkalis (e.g. NaOH) are used successively to get rid of unwanted hemicelluloses and lignins. The tendency in pulp mill design at present is to take the organic matter dissolved in bleaching to the washing department after cooking with the aim of incorporating all pollution in the concentrated spent liquor and disposing of it by burning. These are internal measures for water pollution control at pulp mills.

POLLUTION LOADS ASSOCIATED with VARIOUS MODIFICATIONS of THE PULPING PROCESS

Sulfite-Pulping

1. *Sulfite-pulping without recovery:* In the old Ca-based sulfite-pulping process without recovery (see Fig. 1), there is in general no usage of the spent liquor, i.e. also no cycling of chemicals. As an order of magnitude for softwood pulping, 400 kg BOD_5/t of pulp and 1600 kg COD/t (out of which 720 kg COD_b/t is biodegradable and 880 kg COD_{nb}/t is refractory) originate from cooking and 20 kg BOD_5/t and 100 kg COD/t (36 kg COD_b/t and 64 kg COD_{nb}/t) from bleaching. Such mills are completely out-

moded and should be replaced by mills with a soluble base and liquor recovery.

2. *Sulfite-pulping with a soluble base and liquor recovery:* The schematic representation of the process given in Fig. 2 and as can be seen, the cooking liquor is recirculated in Fig. 3, the specific pollution load is presented for an operation without neutralisation of spent liquor. By neutralising the spent liquor prior to evaporation, organic acids are held back by 70-90%, giving rise to a reduction in BOD_5 of 50-80%. Mill superintendents, however, are not completely in favor of such a solution as losses in efficiency in evaporation as well as difficulties in burning and recovery have been reported.

Specific pollution loads associated with these solutions, where there is 95%-99% of liquor recovery and no neutralisation, are 60-80 kg BOD_5/t, 175-240 kg COD/t and 75-110 kg COD_{nb}/t; whereas with 97%-99% liquor recovery and neutralization of the spent liquor, the figures are 40-50 kg BOD_5/t, 135-170 kg COD/t and 75-90 kg COD_{nb}/t

3. *Linking an O_2-prebleaching stage with the previous measures:* On the way to the "pollution free pulp mill", O_2- delignification plays an important role. The most advanced step reporded at present for producing a bleached sulfite paper pulp comes from Norway, where a Cl_2-NaOH-NaOCl-ClO_2 bleaching sequence is at present converted into an O_2 $(Mg(OH)_2)$-Cl_2-NaOCl sequence. Thus, it is possible to reduce the dischar of overall chlorine from 67 kg/t to 23 kg/t, and it is felt that the organically bound chlorine is reduced by the same ratio (say by two thirds). Furthermore, it is assumed that with 99% liquor recovery, spent liquor neutralization and prebleaching by oxygen, 25 kg BOD_5/t, 70 kg COD/t and 30 kg COD_{nb}/t will be the figures obtained However, spills are not included in these figures, and the lower the continuously discharged load, the higher the effect of the load in spills.

In speciality pulp production, preoxide bleaching is already applied at Schwäbische Zellstoff AG at Ehingen (FRG) and will soon be applied at Chemiefaser Lenzing AG in Austria.

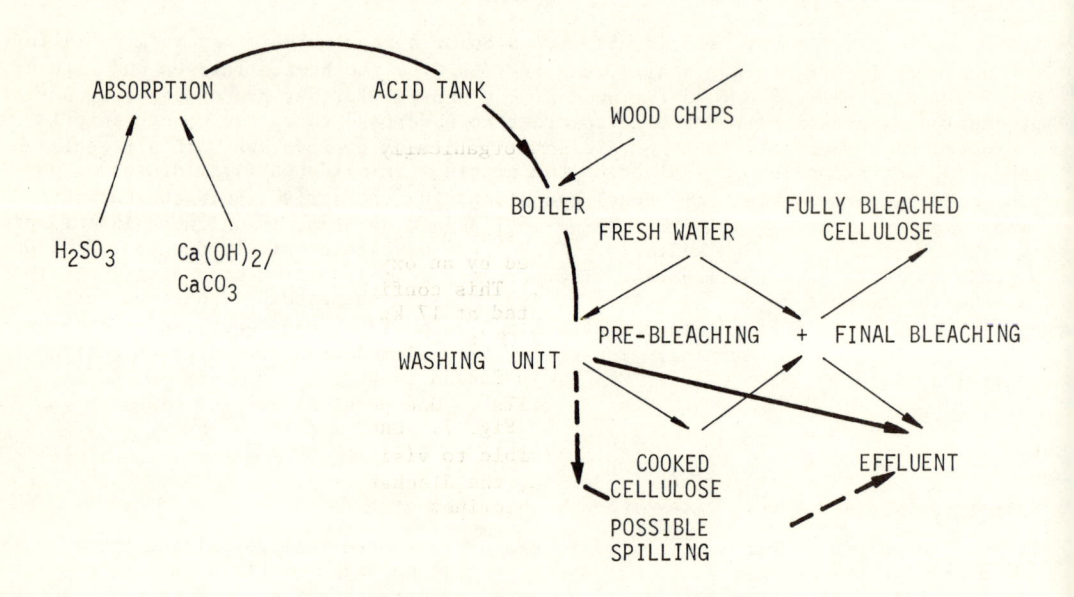

Fig. 1: Schematic Representation of the Ca-sulphite pulping process without liquid recovery.

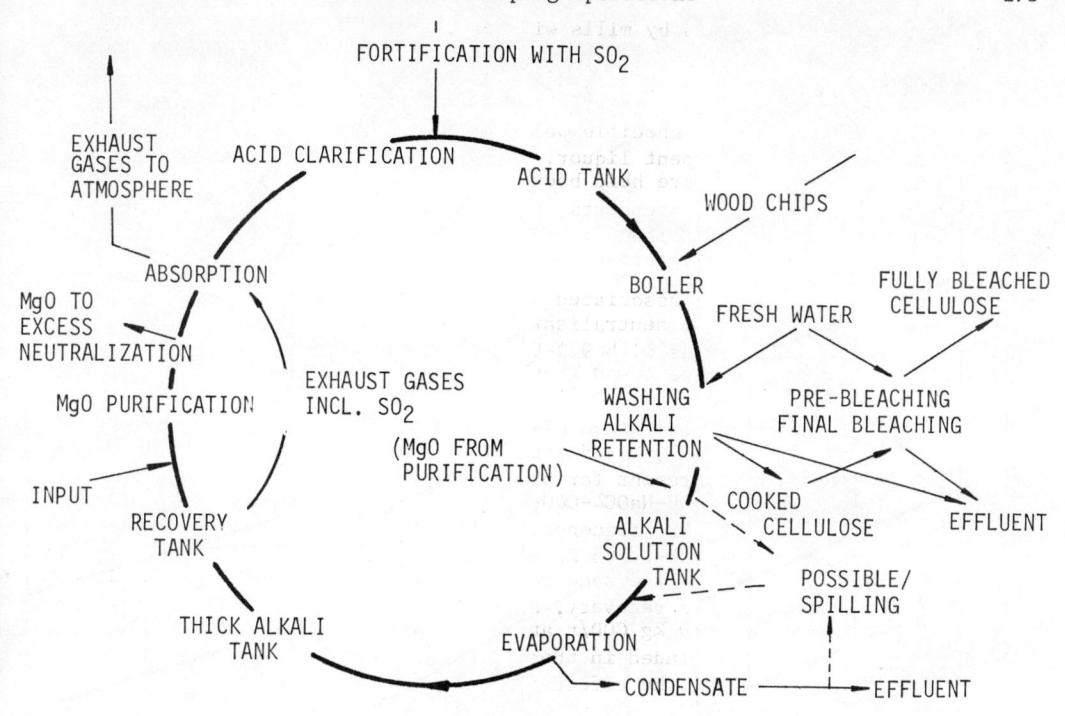

Fig. 2: Schematic representation of the Mg-sulfite pulping process with liquid recovery.

Sulfate-Pulping

1. *Sulfate with a conventional bleach plant:* The process is shown in Fig. 5. The base is a quite modern mill which has been in operation for three years. The Na_2SO_4 - loss in the washed pulp is 8 kg/t, and the methanol is stripped off from the condensate. The bleach plant is of the Cl_2 (with some ClO_2) $-NaOH-NaOCl$ $-ClO_2-$ $NaOH-ClO_2$-type. The load of pollution discharged amounts to 25 kg BOD_5/t, 150 kg COD/t, 105 kg COD_{nb}/t and roughly 14-15 kg organically bound Cl/t. Most of this comes from bleaching.

2. *Sulfate with an O_2 - prebleaching stage*: Cooking, liquor recovery, evaporation and the treatment of condensates is the same as in the previous case. However, the closed screening is immediately followed by an oxygen bleach, the wash-water of which goes to the cooking liquor recovery. This configuration is shown in Fig. 6. The load of pollution discharged is estimated at 17 kg BOD_5/t, 60 kg COD/t, 30 kg COD_{nb}/t and roughly 3 kg organically bound Cl/t.

3. *Sulfate in closed cycle:* Developments in Canada as well as in Scandinavia are directed towards "pollution-free kraft mills". One possibility is the Rapson-Reeve-process from Canada, which is presented in Fig. 7. Such a mill has been in operation since the end of 1976, but it was not possible to visit it by mid-1978. According to the inventor of this process, Prof. Rapson, the discharge is 3 kg BOD_5/t, 15 kg COD/t, 10 kg COD_{nb}/t, with no organically bound chlorine.

EXTERNAL TREATMENT

Introduction

External treatment, commonly known as wastewater treatment, is needed in all those

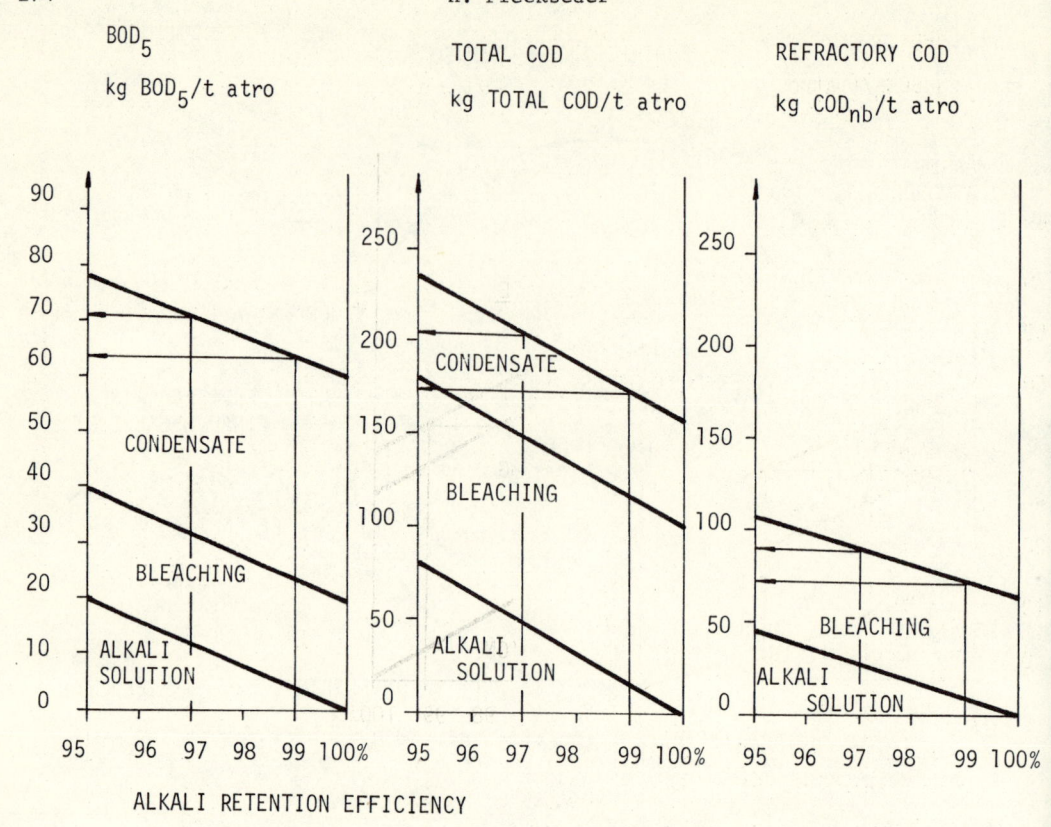

Fig. 3: Specific pollution load per ton for Mg-sulfite, with liquor recovery but no neutralization.

cases where water is used in a production process as an essential raw material and where the water leaves the process enriched with parts of other raw materials. Although not yet applied, it is possible to conceive of external treatment also as a means of internal treatment as soon as no wastewater leaves the process, but internal removal of wastewater constituents is also required.

Removal of Settleable Solids

At modern pulp mills, most of the solids are already retained within the mill, and relative to the BOD_5- load the solids load is small. As long as the suspended solids have to be utilized, sedimentation should be applied. Whenever this utilization is not required, the removal of settleable solids can be omitted and incorporated into the activated sludge process where the settleable solids serve as growth surfaces for microorganisms.

Removal of Biodegrable Matter

The following formulation can be looked upon as a criterion for "biological treatment or its equivalent":

> A degree of treatment equivalent to the one obtainable by biological treatment exists when the specific pollution load after biological treatment and the method of treatment compared with it have the same dissolved load in BOD_5. Fractionation into a dissolved and a solid fraction should be done by a membrane filter with a pore size of 0.45 μm.

Chemical Pulping Operations 275

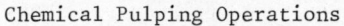

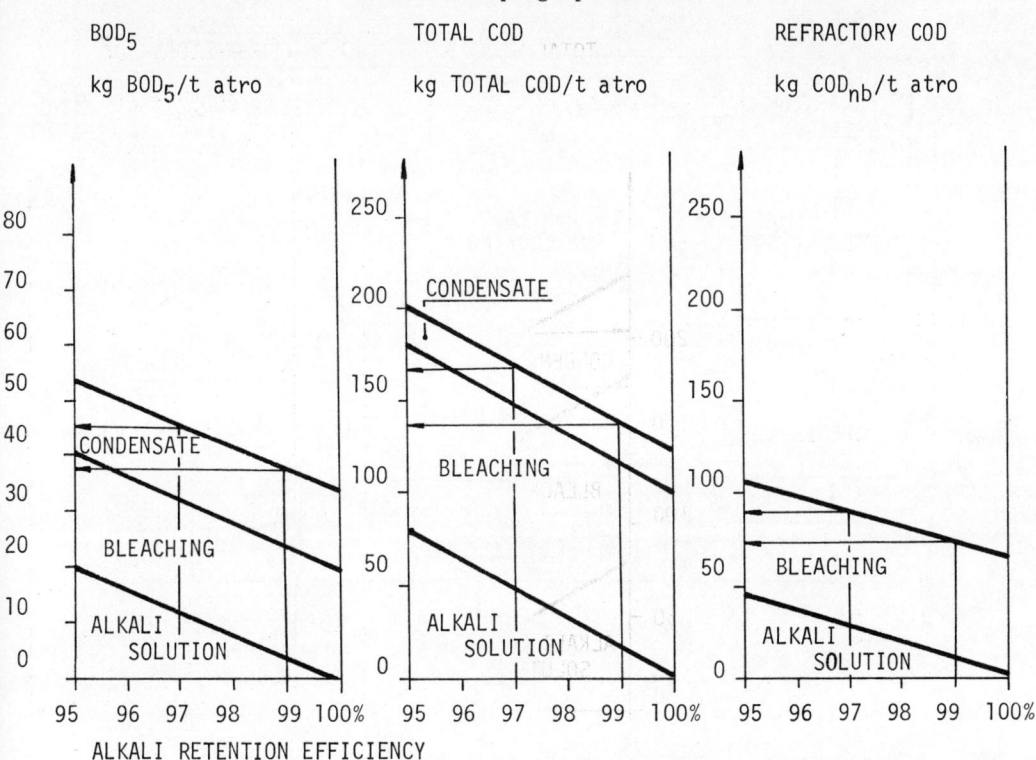

Fig. 4: Specific pollution load per ton for Mg-sulfite with liquor recovery and spent liquor neutralization

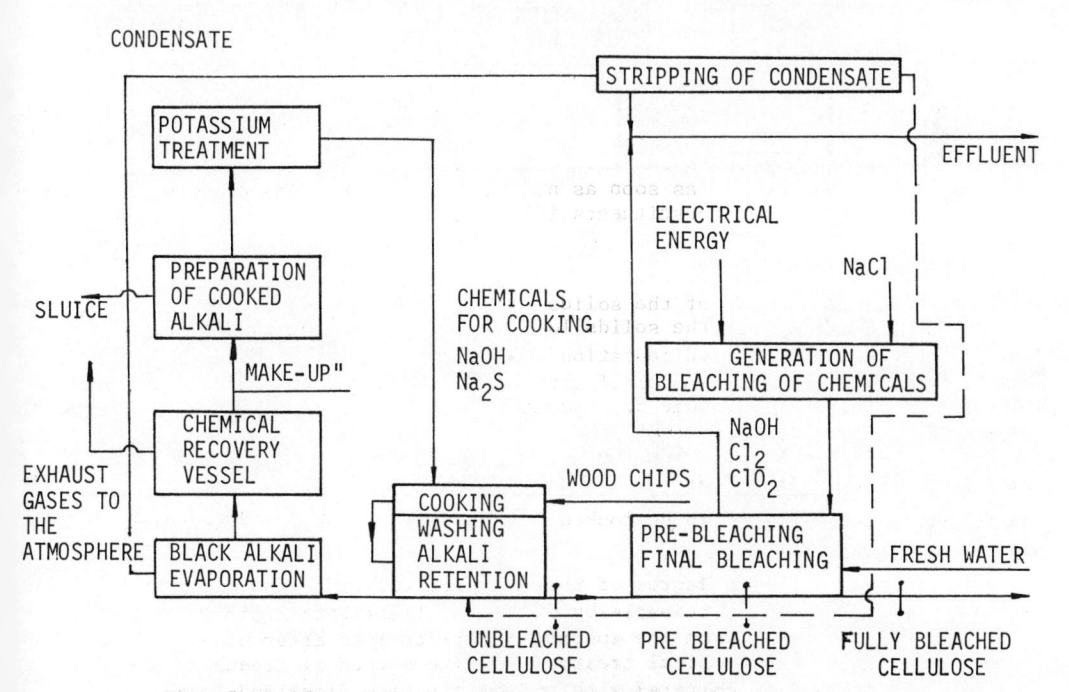

Fig. 5: Sulfate process with a conventional bleach plant

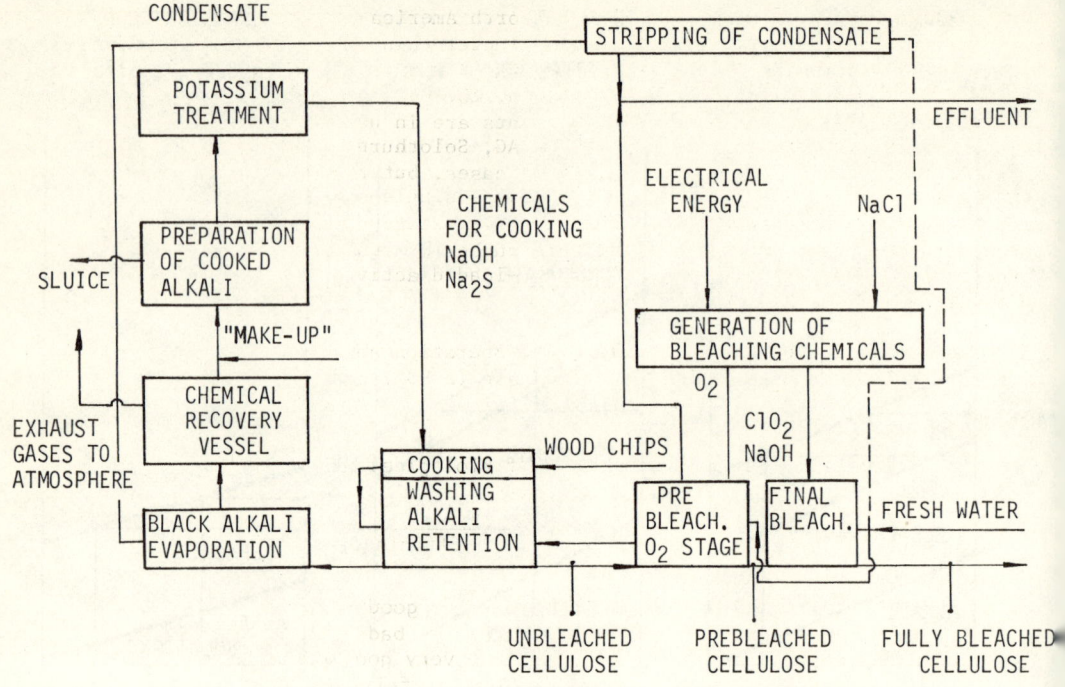

Fig. 6: Sulfate process with an oxygen pre-bleaching stage

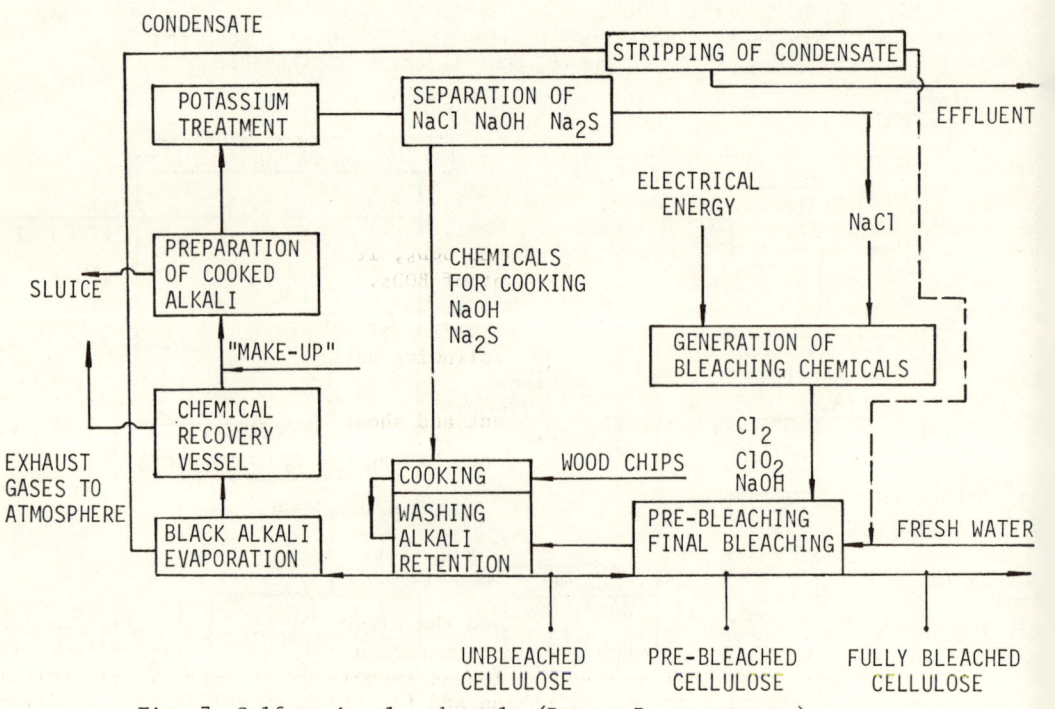

Fig. 7: Sulfate in closed cycle (Rapson-Reeve process)

Aerated lagoons are in wide use all over North America and to some extent also in Scandinavia; their detention time must be longer than or at least 5 days. Highly loaded activated sludge plants (with a sludge age of 2 to 5 days) are also in use in the FRG, Japan, Poland, the Soviet Union, the U.S. and undoubtedly also in other countries. Two-stage activated sludge plants are in use in Switzerland as well as in the GDR (System of Cellulose Attisholz AG, Solothurn/CH). Plastic media trickling filters have also been applied in some cases, but the degree of treatment obtainable has to be increased by a second stage. A single-stage system which offers the highest degree of treatment obtainable, namely extended aeration, has not yet been applied to any great extent. For quite a number of reasons which will be presented later, extended aeration systems, i.e. low-loaded activated sludge systems, should be favored.

The main prerequisite, before putting into operation any biological treatment sequence, is at least a high liquor recovery rate ($\geq 98\%$) and a reliable utilization of the concentrated spent liquor.

Table 2: Removal Efficiency of Biological Processes

Constituent	Efficiency
toxicity towards fish	good*
color	bad
BOD$_5$	very good
COD	fair
gettleable solids	very good

*holds for low-loaded systems

The efficiency of biological processes to remove certain wastewater constituents are given in Table 2.

According to more recent investigations from Scandinavia, not only acute toxic but also mutagenous compounds can be removed by biological systems (Hultberg, 1978). The removal of COD depends on the ratio of COD to BOD$_5$; it can be assumed that the COD removed (COD$_b$) is 1.5 - 1.8 times the amount of BOD$_5$.

Extended aeration plants with a volumetric loading of less than 0.5 kg BOD$_5$/m^3·d and a sludge age of 10 days and more hold the following advantages over more highly loaded systems:

- very high process stability under transient and shock loadings
- smallest amount of nutrients required by any biological system
- no primary sedimentation needed
- rare occurrence of bulking sludge
- very simple operation.

The somewhat greater requirement in space and the higher investment in the aeration tank are compensated by omitting primary sedimentation and a surge tank. The energy cost will be higher than with more highly loaded systems; other expenses in operation like those for nutrients, for neutralisation and for fighting foaming will be lower.

The special advantage of extended aeration is the very reliable operation and the more or less constant effluent quality obtainable even over a wide spread of loading (e.g. 0.2-0.8 kg $BOD_5/m^3 \cdot d$), provided adequate care has been taken in design regarding oxygen transfer as well as final sedimentation capacity. This, in short, implies that extended aeration can be put into operation now and remains a working tool even when the discharge from the mill is reduced by further internal measures.

Table 3: Treatment of a Sulfide Mill Bleach Plant Effluent in Laboratory Scale

Laboratory Plant		a	b	c	d	e	f	g	h
MLSS	(kg/m^3)	2.6	2.6	3.2	3.7	5.5	6.1	4.2	7.4
Vol. BOD_5-loading	($kg/m^3 \cdot d$)	0.18	0.29	0.31	0.39	0.46	0.68	0.73	1.31
F/M-Ratio (BOD_5)	($kg/kg \cdot d$)	0.07	0.11	0.10	0.11	0.08	0.11	0.17	0.18
η-BOD_5	(%)	95	93	95	93	95	92	91	89
Influent BOD_5	(mg/ℓ)	197	197	321	197	321	321	197	321
Effluent BOD_5	(mg/ℓ)	11	14	16	15	16	25	18	34
Vol. COD.-loading	($kg/m^3 \cdot d$)	0.94	1.50	1.47	1.97	2.25	3.22	3.72	6.20
F/M -Ratio (COD)	($kg/kg \cdot d$)	0.36	0.58	0.46	0.54	0.39	0.53	0.89	0.83
η-COD	(%)	39	37	34	33	34	32	35	31
Influent COD	(mg/ℓ)	1000	1000	1520	1000	1520	1520	1000	1520
Effluent COD	(mg/ℓ)	610	630	1000	680	1000	1040	660	1050
(Carbon-OU/MLSS)	($kg/kg \cdot d$)	0.068	0.098	0.068	0.131	0.088	0.105	0.142	0.145
Excess Sld/MLSS	($kg/kg \cdot d$)	0.054	0.086	0.074	0.038	0.050	0.062	0.100	0.107
I_{SV}	(mg/ℓ)	38	39	29	47	33	37	58	63
Duration after Adaptation	(d)	63	63	77	63	77	77	63	77

(e.g. 0.2-0.8 kg $BOD_5/m^3 \cdot d$), provided adequate care has been taken in design regarding oxygen transfer as well as final sedimentation capacity. This, in short, implies that extended aeration can be put into operation now and remains a working tool even when the discharge from the mill is reduced by further internal measures.

From Table 3 the following conclusions can be drawn:

- BOD_5 (as an indicator of quickly biodegradable pollution) can be removed from a bleach plant effluent at volumetric loadings from 0.18-1.31 kg $BOD_5/m^3 \cdot d$ with more than 90% efficiency.

- The mixed liquor in the laboratory scale plants (T at room temperature) had excellent settling properties.

- Variations in volumetric loading from 1:7 and sludge loading from 1:2.5 could be

applied with ease. I_{SV} increased with an increase in loading.
- The COD, the BOD_5, as well as the ratio COD/BOD_5, of the influent can vary greatly. This is caused by the wood processed and by the control on bleaching. The treatment efficiency, therefore, can only be measured safely by the effluent-BOD_5 relative to the influent and/or by a continuous determination of oxygen uptake within the process. The effluent-BOD_5, however, should not be equated with the overall effluent quality.

The excess sludge from the operation should, after thickening and mechanical dewatering, be burnt together with bark and other organic residues as the most favorable solution.

Removal of Refractory Compounds

Color and refractory compounds can be further reduced by physical-chemical methods. They help to reduce

toxicity towards fish	well
color	very well
BOD_5	badly
COD	well
settleable solids	very well

Thus, biological and physical-chemical methods in pulp mill wastewater treatment overlap. Unsolved technical problems as well as high cost, however, will hinder a quick and widespread application of physical-chemical methods (e.g. precipitation and flocculation, adsorption, ion exchange, reverse osmosis, etc.).

THE LINK BETWEEN INTERNAL and EXTERNAL MEASURES

Present-day technology does not permit solution of the questions of water pollution control at pulp mills *either* by internal measure *or* by external measures *alone*. Therefore, all internal measures a mill superintendent thinks can be applied should be and implemented. Water pollution control will in most instances dictate a further reduction in BOD_5, COD_b and toxicity, and this can very well be obtained by biological treatment. Therefore we should apply internal as well as external measures and must find a common language for the designers and operators of mills, as well as for our colleaques within administration.

COST ASPECTS

Process-Related Measures

Cost estimating in process design is not an easy task, especially if it is to be a precise estimate. Beyond this, however, we also have in pulp mill design the question of which equipment should be classified as process-related and which as pollution control-related. As I am not a pulp mill designer, I shall not go into this question. However, what is quite important is that a large capacity (say 150.000 t/a) will be in terms of investment as well as the running cost per ton of product, 1.3 to 1.5 times less expensive than a mill of, say 50.000 t/a, for fixed costs of raw material. This, therefore, makes large mills very attractive. In the 1973 OECD study, it was estimated that 40% of the investments in a new mill were related to environmental protection (water, air), and my estimate is that 25-30% are related to water.

External Measures

Cost estimating in this field is not an easy task, either, but in nearly all cases the cost can be charged to environmental protection (exception: reuse of sedimented fibres).

In Central Europe, the two-stage activated sludge plant at Attisholz/CH was the first to be designed and constructed. Both for wastewater treatment and sludge disposal, together with burning bark, a figure of 50-60 SFr/t of product has been quoted. This is an extremely high figure quoted by everyone in pulp manufacture.

We had to undertake cost estimates for three Mg-based sulfite mills in Austria, which at the time of our estimation still worked on a Ca-base. The internal measures foreseen varied from mill to mill and the external treatment was incorporated into a regional wastewater treatment scheme (combined treatment with municipal wastewater).

The cost estimates read as follows:

Table 4: Cost Estimates for Biological Treatment of Mg-Based Mill Effluents

Mill	A	B	C	
Production	180,000	90,000	72,000	t abs·dry/a
Primary Treatment	yes	yes	no	
BOD_5-load	48	68	58	kg/t abs·dry
COD-load	180	190	200	kg/t abs·dry
Flow (estimate)	140	200	220	m^3/t abs·dry
Investment (1980)	1320	2050	1500	AS/t abs·dry/a
Capital Cost (10%/a)	132	205	150	AS/t abs·dry
Operating Cost (1980)	90	141	150	AS/t abs·dry
Total Cost (1980)	222	346	300	AS/t abs·dry

In addition, the investments for the plant at Mill A may be broken down as follows:

Biological part, 46% (aeration tanks 19%, final tanks 15%, sludge handling 12%)
Primary sedimentation including sludge handling, 10%;
Commonly used parts, 30% (lot area, roads, connecting pipes, operators building etc.)
Interceptor sewers, 14%.

Again, what seems important is the economy of scale which means that the cost of biological treatment and sludge handling is at a mill of 50,000 t/a 1.3 to 1.6 times as expensive as at a mill of 150,000 t/a.

The cost associated with physical-chemical treatment is hard to estimate. As an order of magnitude, it is as high as the one for biological treatment and associated sludge handling, but in certain cases it will be much higher (e.g. adsorption with γ-Al_2O_3).

REFERENCES
General
Aticelca (1973). Harmonizing Pulp and Paper Industry with Environment., *Proceedings of the 15th EUCEPA Conference, Rom.*

Dalpke, H.-L. (1977). Convenor of the Discussion "Die biologische Behandlung von Zellstoff- und Papierfabriksabwässern". In *Das Papier, 31*. Jg., Heft 10A.

v.d. Emde, W., H. Fleckseder, L. Huber and K. Viehl (1973). Zellstoffabwässer - Anfall und Reinigung. *Wasser, Abwasser, Gewässer, 13*. Wien.

Environment Canada (1974). *The Basic Technology of the Pulp and Paper Industry and its Waste Reduction Practices*.

Environment Canada (1976). *Proceedings of Seminars on Water Pollution Abatement Technology in the Pulp and Paper Industry*.

Environment Canada (1976). *Review of Color Removal Technology in the Pulp and Paper Industry*.

Eucepa/Ozepa (1977). *Manuskripte und Diskussionen zur 17. Eucepa-Konferenz*, Wien.

OECD (1973). *Pollution by the Pulp and Paper Industry: Present Situation and Trends*. Paris.

SSVL (1974). *The SSVL Environment Care Project*. Technical Summary. Stockholm.

Special Publications Related to Selected Aspects:

Arbeitsunterlagen der Arbeitsgruppe 24 (Zellstoff) des BM d. Inneren (BRD) zur Festlegung von Mindestanforderungen gemäß Pargraph 7a (1) Wasserhaushaltsgesetz (1978).

Christensen, P.K. (1978). *Bleichen von Sulfitzellstoff mit O_2 und $Mg(OH)_2$ - Vom Laboratorium zum Ausführungsprojekt*. Paper presented at SPCI-78.

Franzreb, J.P. (1976). Einige innerbetriebliche Maßnahmen zur Verminderung der Vorfluterbelastung, dargestellt am Beispiel einer Calciumbisulfitfabrik. *Das Papier, 30*. Jg., Heft 9.

Hornke, R. (1977). In Eucepa/Ozepa, *Manuskripte und Diskussion zur 17. Eucepa-Konferenz* zum Thema "Technischer Stand der abwasserfrei arbeitenden Sulfitzellstoffabrik".

Hultberg, H. (1978). Ecological Aspects on Water Pollution Caused by Pulp and Paper Production. IAWPR, 9th International Conference, *Post Conference Seminar on Water Pollution Control in Pulp and Paper Industries*.

Märki, E. (1979). Industrieabwasserprobleme in der Schweiz. *Vortrag am 14. ÖWWV-Seminar in Raach (NÖ) zum Thema "Industrieabwasserbehandlung - Neuere Entwicklungen"*.

Norrström, H (1978). Bleaching of Pulp and Water Pollution. IAWPR, 9th International Conference, *Post Conference Seminar PS 4 on Water Pollution Control in Pulp and Paper Industries*.

Rapson, H. (1978). The Closed Cycle Bleached Kraft Pulp Mill. IAWPR, 9th International Conference, *Post Conference Seminar on Water Pollution Control in Pulp and Paper Industries*.

Solbach, H. (1977). Erfahrungen mit der Abwasserbehandlung in einer integrierten Zellstoff -und Papierfabrik. In H.L. Dalpke, *Allgemeine Übersichten*.

Teder and S. A/Norden (1978). Prolonged Pulping and Oxygen Bleaching - A Way to Pollution Reduction. IAWPR, 9th International Conference, *Post Conference Seminar on Water Pollution Control in Pulp and Paper Industries*.

Ullrich, H. (1978). Purification of Pulp Bleaching Waste Water with Aluminium Oxide. *Progress in Water Technology, 10*, Nr. 5.

DISCUSSIONS

Hernandez, J. W. (U.S.) : Are the lignins still in suspension the source of the colour in the paper mill effluent? Have you tried colour removal with activated carbon?

Fleckseder, H. (Austria): Yes. The colour is due predominantly to lignins of all kinds. We didn't try activated carbon since it is expensive and only applicable to kraft mill effluents. For sulphite mill effluents, γ-aluminium oxide is more amenable to adsorption. As Prof. Hanisch mentioned, one mill in Germany already runs on that. The overall cost per ton of material produced, however, is two or three times greater than that for biological treatment.

Velioğlu, S. G. (Turkey): Why is an air pollution problem encountered when a calcium-based sulphite is used and not when a magnesium-based sulphite is used?

Fleckseder, H. (Austria): In calcium-sulphite pulping, $CaSO_3$ is formed, when burning in the incinerator; in magnesium-sulphite pulping $MgSO_3$ is formed. For $CaSO_3$ to be split by heat, a temperature of roughly 1350°C is required, and at present no steel exists which will withstand such a high temperature. This means that, since the $CaSO_3$ is not partly split, sulphite will be discharged. For a temperature of 850°C, at which $MgSO_3$ can be split, incinerator construction materials exist. With magnesium-based pulping there is, therefore, no noxious discharge into the atmosphere.

A Case Study of the Treatment of Wastewater from Paper Mills

B. BAYSAL, F. ŞENGÜL, A. MÜEZZINOGLU and A. SAMSUNLU

Environmental Engineering Department, Civil Engineering Faculty, Ege University, Izmir, Turkey

ABSTRACT

In this article a case study for designing a wastewater treatment plant is presented. Steps for this design are discussed; generalizations, as far as possible, are made and evaluated.

The polluting industry in this study is a paper mill which also has a small scrap pulper. It creates problems in the environment and causes difficulties in treatment typical of the pulp and paper industry.

Alternative schemes for the treatment plant, as well as information on each unit and selection criteria are given in the paper. Finally, the selected alternative is discussed, and some diagrams showing the design of the units in this alternative are presented.

INTRODUCTION

Pulp and Paper Production in Turkey, Existing Situation and Trends

In Turkey, the majority of the pulp and paper mills belong to a Government Enterprise, SEKA, and a few plants are privately owned. The oldest and biggest plant is in Izmit, but there are other important plants scattered in Anatolia. At some of these plants, sulfite pulp is manufactured and others are for mechanical and even for scrap pulp production. The larger plants usually produce more than one type of pulp. There are a number of plants still in the construction and project stage.

Local pulp manufacture is limited by the forestry resources of the country. The percentage of imported pulp used in the total paper production is, however, appreciably high. A logical approach in limiting the pulp imports is to increase the recycling rate, in other words, to reuse increasing amounts of waste paper. The paper mill in our case has a special unit for producing pulp from waste paper. Naturally, the repeated use of recycled paper pulp is controlled by the quality of paper required. And undoubtedly, it affects the environmental impact of the wastewater discharged from the industry.

The quality of paper produced, as well as the raw materials utilized during production, determines the quality of the wastewater discharged. The major indicators of quality for paper mill wastewater are its suspended, colloidal and dissolved solids content. Typical of the industry, the solids concentration is very high, consisting mainly of fibers and some paper additives. That is why the first units in paper wastewater treatment plants are screens, sieves, and presettlers. Generally, these mechanical treatment units are followed by chemical coagulation units.

Chemical coagulation and flox settling units, may, depending on the expected quality of the final effluent, be followed by a biological treatment unit. However, it should always be kept in mind that biological treatment of paper mill wastes is a difficult process for the following reasons:

- Biodegradability of cellulose and lignin fibers is slow.

- Nitrogen and phosphorus concentrations are not enough for treatment, and unless these are added in sufficient amounts, the right microorganisms for biological decomposition of paper mill wastewaters may not be cultivated.

- Usually there is not enough seed in the wastewater to start biological growth.

The wastewater is strongly alkaline, and unless pH adjustment is made, biological treatment is not possible.

However, generally after chemical treatment, BOD_5:N:P ratios and pH levels become more favorable for biological treatment.

In this presentation, a paper mill located near Izmir has been extensively studied for effluent characteristics using laboratory treatment models; and test results are evaluated to propose alternative treatment systems. The major alternatives are discussed and design characteristics are presented.

The plant studied consists of a paper machine with a high percentage of wash and cooling water recycling. A flotation unit collects most of the cellulose fibers in all wash and cooling waters, after treatment with suitable additions of casein, alum, and polyelectrolyte. The remaining water is recycled back into the system. Water recycling is complete except for a small amount of make-up water. Obviously the amount of wastewater from the flotation system, which is also the effluent quantity, is equal to the amount of feed-up water.

Theory Related to Treatability of
Pulp and Paper Mill Waste

The major contaminants found in the paper industry's wastewater are cellulose and lignin. Both are present in the paper industry wastes to an extent determined by the type of raw materials and processes.

Cellulose is one of nature's most common polysaccharides. Its molecule is polyglucose linked by beta 1-4 bonds. The number of glucose molecules making up a cellulose molecule is in the order of 200,000 to 1,000.000. Due to the high molecular weight created by this high number of glucose units, cellulose is insoluble in water.

The beta 1-4 linkage creates a resistance in the cellulose molecule against biodegration. On the other hand, the insolubility of the cellulose chain causes all the biodegradation activity to take place on the water - substrate interface.

Cellulose digestion is a slow process retarded by the intermediate products like cellobiose, which controls the overall decomposition rate. Digestion is carried out by a group of microorganisms which poliferate under acidic conditions. This process is exothermic and the heat dissipated, as well as the nutrients produced, is substantial enough for symbiosis with sulfate-reducing bacteria.

Lignin is the name of a group of complex polycyclic aromatics synthesized by plants. The exact formula of the lignin differs according to the type of plant, but they all contain a series of cross-linked phenyl propane monomers with some phenolic and methoxy groups attached to the phenyl rings.

In wastewater from the paper industry, lignin is generally in the form of lignin sulfonate. This chemical is formed by sulfonation of the complex molecule during the sulfite process. This compound has a brownish-black color and imparts color to the waste. Fortunately, lignin sulfonate is more biodegradable than lignin.

Only a few obligate aerobe bacteria found in natural waters and soil can decompose lignin molecules. This causes the oxidative decomposition processes to be successful to a limited extent, but anaerobic lignin decomposition is totally impossible. That means, in a water environment lignin may decompose as long as oxygen is present. As the particles containing lignin settle to the bottom of water bodies where anaerobic conditions prevail, the cellulosic portion degrades while the lignins accumulate. This process is responsible for the build-up of organics in the sediment and is thought to yield coal and shale deposits at the bottom in the long run (Dugan, 1972).

MATERIALS and METHOD

Three test groups are applied on wastewater taken from a scrap mill, paper machine and mixed outlet canal under different production conditions in our case study.

- The main physical and chemical parameters indicating the quality of the wastewater.
- A thorough study of the parameters showing the possibility and efficiency of chemical treatability of the wastewater.
- A thorough study of the parameters showing the possibility and efficiency of biological treatment of the wastewater. The main parameters studied in Group A are BOD, COD, turbidity, pH, total solids, dissolved solids, suspended solids, volatile solids, and nutrients, such as total Kjeldahl nitrogen and phosphate concentrations.

Test results show that only a very small percentage of the total BOD and solids load can be removed by direct mechanical treatment, i.e. using screens, grit chambers and presettlers. To increase the removal efficiency for these loads, artificial flocculation methods can generally be applied: (a) chemical coagulation and (b) biological processes. In *chemical coagulation*, solid particles, which are all negatively charged and therefore resist flocculation, are brought together by high-valence positive ions such as Fe^{+++} or Al^{+++}. These ions are supplied with the aqueous solutions of chemical compounds commercially available in the form of $FeCl_3$, $Fe_2(SO_4)_3$, and alum. These high-valence positive ions bind negatively charged colloids and form flocs, provided the pH, temperature, co-ions and other conditions are suitable. Coagulation is improved by the addition of very small doses of coagulation aids, such as certain polyelectrolytes forming long chain polymers in a water solution, which contain radicals at certain parts of the molecules to bind the flocs already present in the water.

Jar tests with the addition of coagulants and coagulation aids at several concentrations are applied. pH and temperature conditions are adjusted according to the requirements of the type of coagulant. Sedimentation rates are determined for each run. BOD and COD inputs and outputs after each coagulation test are also determined.

Biological treatment processes include cultivation of suitable species of microorganisms by using the carbonaceous pollutants as substrate. The microorganism flocs thus formed are settled and pollutants are removed from the water. Biological processes are subject to limitations in the following respects:

- Pollutants in the water must be easily decomposable by enzymatic action.

- Environmental conditions like pH, temperature, toxicity, availability of nutrients like nitrogen and phosphorus must be at suitable ranges for the biological growth of microorganisms. For instance, a $BOD_5:N:P$ ratio of 100:5:1 shows the minimum amount of N and P to be provided in order to start biodegradation.

These limiting factors must all be tested and the ranges required must be provided by the addition of certain chemicals or other wastes.

In pulp and paper manufacturing wastewater containing substantial amounts of cellulose and lignin is discharged. As was mentioned before, these two are among the hardest materials to decompose by enzymatic processes. Another unfavorable condition for paper mill wastewater is the poor ratio of $BOD_5:N:P$. Unless these two nutrients are added to the wastewater, or BOD_5 is removed by non-biological means, this ratio limits the growth of microorganisms according to the Liebig's law of minima.

Taking all these factors into consideration, artificially aerated laboratory models are utilized to test biological treatment possibilities.

Receiving Waters

All of the process wastewater is directly discharged into a river flowing nearby. The point of discharge is very close to the point where the river enters the sea. The high solids content and occasionally highly colored alkaline wastewater form a large bed of wet sludge, as the river is quite sluggish at that part. This sludge bed, especially at the edges, acts as if the river were facultative lagoon. It is as unsightly and disturbing as a natural waste treatment plant at the point of discharge. However, it is not a treatment plant but part of a living river, which with it all the nutrients of the rich Ege soil it is flowing through.

The river, therefore, used to be a good water source for fisheries in the past. But now that the plant wastes are discharged and the water is alkaline, toxic even after dilution and very shallow at the point of discharge, the fish cannot swim to and fro between the sea and the upper parts of the river to lay eggs; they are now almost completely exterminated. That means the ecological balance is impeded; natural cycles are destroyed. The water quality of the river just before the point of discharge was determined and the following figures found:

pH	:	7.5
BOD_5	:	10 mg/ℓ
COD	:	20 mg/ℓ
Total Suspended Solids	:	4 mg/ℓ
Total Dissolved Solids	:	656 mg/ℓ
Turbidity	:	60 Jtu
Nitrogen	:	0.84 mg/ℓ
Phosphorus	:	0.5 mg/ℓ

Flow Rates

The quantities of effluents discharged from the scrap pulp unit, from the paper machine and their mixture in the outlet canal were separately measured. Flow rate measurements were made using a propeller, the revolutions of which were automatically counted, totalled and recorded. The measurement was timed with a stop watch; the wet cross sectional area was measured; and the effluent flow rate at the point of measurement was calculated utilizing the appropriate formula. The flow rate measurement points are schematically shown in the drawing in Fig. 1.

Determined flow rates are:

Scrap pulp unit : 200 ℓ/sec.
Mixed Effluent Canal: 100 ℓ/sec.

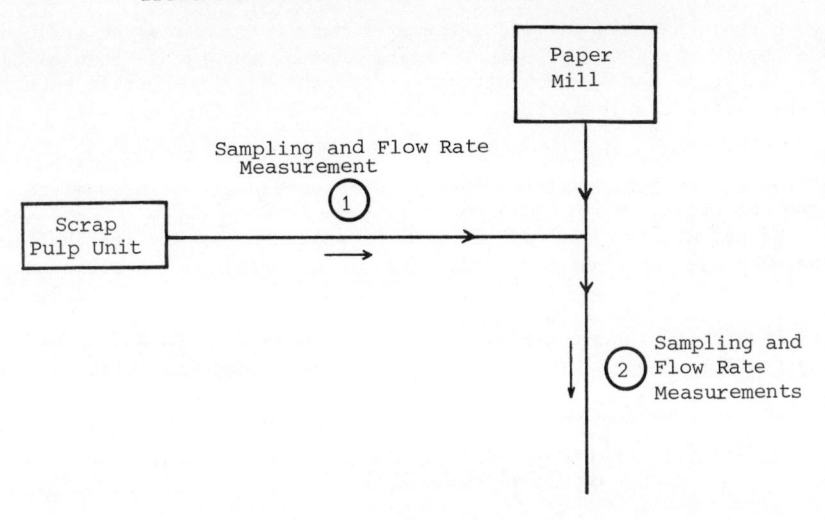

Fig. 1: Schematic representation of flow rate measurement and sampling points

Experimental Results

By laboratory tests the COD, BOD5, total, suspended and dissolved solids, total Kjeldahl Nitrogen, phosphate, pH and temperature were determined. Sampling points are shown in Fig. 1. Integrated and composited point samples were taken from the scrap pulp unit, paper machine and mixed effluent canal. The ranges of the parameters were

- pH in the range of 8-9, with an average of 8.5.

- temperature range between (17^0C - 20^0C), most of the values being around 20^0C.

- total Kjeldahl Nitrogen concentration (0-18 mg/ℓ) with most of the samples containing very little nitrogen;

- phosphate concentration (as phosphorus) ranging between (1.5-8.4 mg/ℓ), the average value being in the order of 2-3 mg/ℓ.

- total solids concentration ranging between the limits 1146-2960 mg/ℓ, the average figure being in the order of 1500-1700 mg/ℓ.

- total suspended solids ranging in a very broad field 108-1412 mg/ℓ

- COD concentration in the 114-1500 mg/ℓ range the average value being 800 mg/ℓ.

- BOD$_5$ values varying between 180-800 mg/ℓ.

Samples taken from the main canal when the scrap pulp unit was in full operation contained BOD$_5$ values as high as 800 mg/ℓ. Other samples gave BOD$_5$ values in the order of 200-300 mg/ℓ.

- Chlorides 5 were in the range of (83-120 mg/ℓ).

- Turbidities were in the range of (200-600 Jackson Turbidity Units)

- Alkalinities were in the range of (200-400 mg/ℓ).

Deep well water used for wash and cooling water was also analyzed for certain parameters:

```
pH             :   7.0
NO3-N          :  10 mg/ℓ
Chlorides      :  89 mg/ℓ
Total Hardness: 230 mg/ℓ
```

This proves that the chlorides in wastewater mainly come from the influent water, which can be classified as brackish. Also, most of the total Kjeldahl Nitrogen comes from well water, in the form of nitrates. And the high proportion of dissolved solids in the wastewater is partly due to the high total hardness values of the influent.

Jar test results and their variations for each sample are also submitted in Fig. 2. Tests were conducted for alum concentrations of 10-20-25-50 mg/ℓ with the addition of 1 mg/ℓ of polyelectrolyte (polyacrylamide solution in ethyl alcohol and water mixture) for each jar. The pH level was fixed in the range 6-8 and the temperature at 20^0C.

Jar tests and the subsequent sedimentation processes were rated for each variable by following the COD changes in each case. Results of these tests are shown in Tables 1 and 2.

Table 1: Results of Coagulation (Jar) Test on Paper Mill Wastewater Sample dated 11 Nov., 1978.

Influent COD (mg/ℓ)	Effluent COD (mg/ℓ)	COD Reduction (%)	Alum Dosage (ppm)
591.2	216.2	63	10
591.2	77.5	87	25
591.2	20.0	96.6	500
591.2	-	-	-

Table 2: Results of Coagulation (Jar) Test on Paper Mill Wastewater Sample dated 8 December, 1978.

Influent COD (mg/ℓ)	Effluent COD (mg/ℓ)	COD Reduction (%)	Alum Dosage (ppm)
835	-	-	0
835	228.48	72.6	10
835	150.96	82.0	20
835	89.36	89.2	25
835	224.40	73.0	50
835	20.0	97.6	500

Coagulant additives and their effects on pollutant removal rates were followed by timing the sedimentation process. Results obtained for several alum doses are plotted in Fig. 3. From this graph it can be observed that an inflection point in the sediment volume is obtained within 4-5 minutes, but to be safe a 10-minute period is required for satisfactorily cleared waters. This graph also shows that even without the addition of coagulants and coagulation aids in the laboratory experiment setup, a fairly good flocculation is possible, provided enough time is available. This is because alum and polyelectolytes are already present in paper machine wastewater,

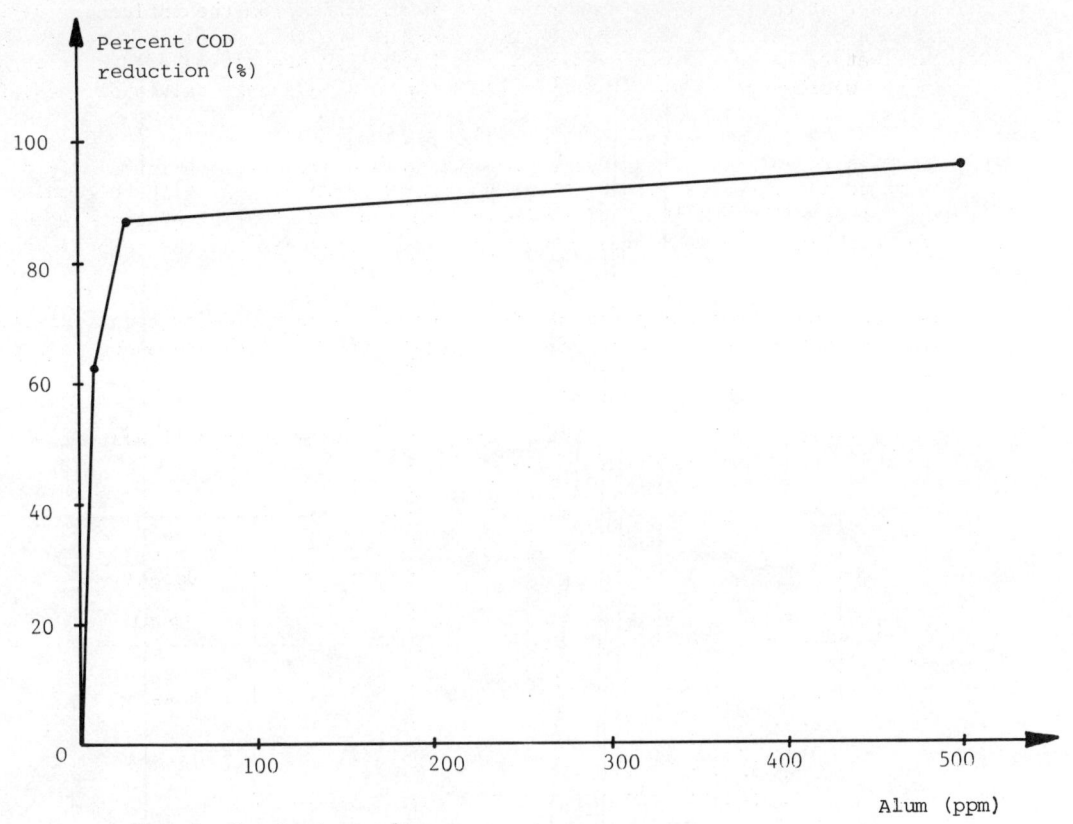

Fig. 2: Jar test results

having been added right before the flotation unit, which aims at returning cellulose fibers and large volumes of wash and cooling water back into the system. But for a safe estimation of the coagulant dose, 10 ppm alum solution is advocated for the design of a chemical treatment plant. This dose might of course be increased or decreased very easily, to answer the demands of various types of wastewater.

Various characteristics of waste before and after chemical coagulation with 10 ppm alum + 1 ppm polyelectrolyte dosage are shown in Table 3.

Table 3: Analysis Results of Final Effluent Before and After Chemical Coagulation Test

Parameter	Before Coagulation	After Coagulation
BOD_5 (mg/ℓ)	600	60
COD (mg/ℓ)	624	60
Total Kjeldahl N (mg/ℓ)	5.46	5.32
Phosphate (mg/ℓ)	12.5	2.50
Turbidity (JTU)	650	100
pH	7.5	7.1
Total Solids (mg/ℓ)	6060	1380
Total Suspended Solids (mg/ℓ)	4320	660
Total Dissolved Solids (mg/ℓ)	1740	720
Volatile Solids (mg/ℓ)	3600	760
Settling Solids (mg/ℓ)	-	100

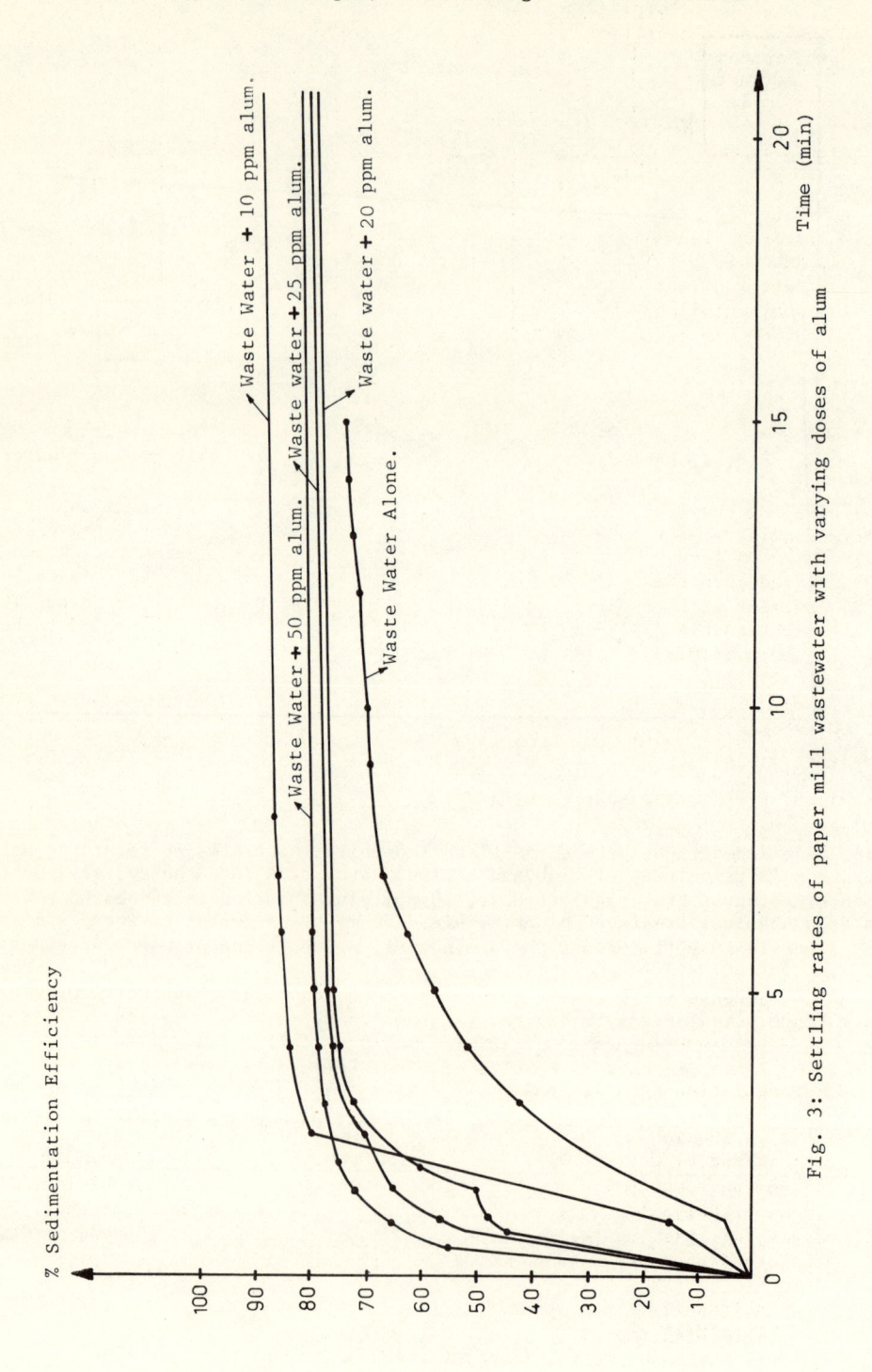

Fig. 3: Settling rates of paper mill wastewater with varying doses of alum

```
                     Treated of Wastewater from Paper Mills
```

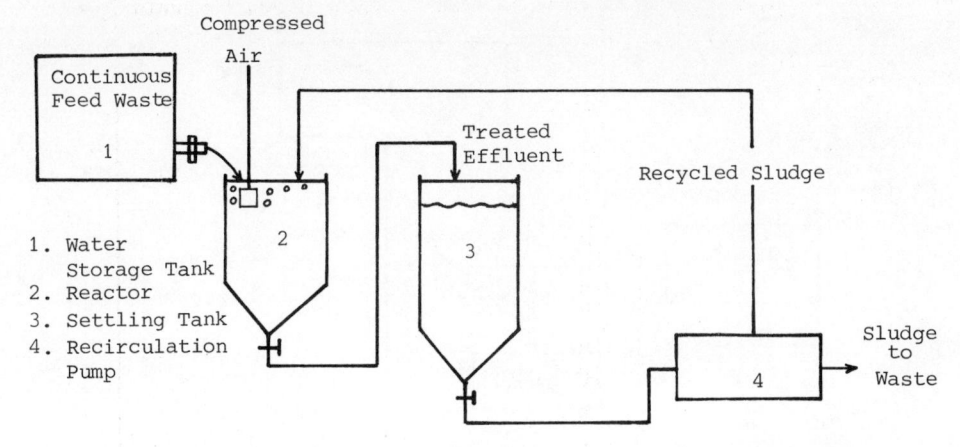

Fig. 4: Laboratory model for biological treatability of wastewater

Along with the chemical coagulation tests, the final effluent was tested for biological treatability. For this test a laboratory biological treatment model was utilized with continuous feed, oxygenation and sludge return. In the first run, mixed canal water was fed into the model, seeded with soil inoculum and aerated (Fig. 4). It was found impossible to grow activated sludge with direct wastewater feed. This is quite understandable as the pH was high and the BOD_5:N:P ratio was found to be unfavorable (100:0.55:0.75 in this case).

In the second run, the pH was adjusted to 7, and nitrogen, phosphorus seed and glycine were added. The wastewater in the feed tank was used to acclimatize the soil+ sewage plant effluent seed used for inocculum. But still an effective growth of expected protozoa and bacteria could not be obtained. This finding runs parallel with the values of BOD_5, BOD_9, BOD_{21} test results obtained from the same wastewater samples. This proven resistance to biodegradation is due to the difficulty of decomposing cellulose and lignin molecules by enzymatic action.

In the third run a Zahn-Wellens test for determining the rate of biodegradation was carried out on a static system. A 1.5-liter neutralized batch of mixed canal wastewater was aerated continuously after the addition of nutrient, mineral, trace element and yeast extract solutions of a calculated strength according to (OECD, 1978). This test showed that an effective activated sludge was obtained only after 5 days of operation. The test system was inoculated with 5 mℓ of filtered effluent from a sewage treatment plant (a oxidation ditch treating only municipal waste).

However, coagulation test effluents that were analyzed for pH, nitrogen and phosphorus showed that the BOD_5:N:P ratio became more favorable for biological treatment (100:9:4.1) as the BOD_5 load decreased to a near- acceptable value (60 mg/ℓ). Therefore, if demanded by Government Regulations, this value could be decreased to conform to any standards required of effluents or receiving water bodies by biological treatment following chemical coagulation. That is why the following wastewater treatment flow - chart for this case was proposed. A treatment plant was designed in accordance with an alternative set forth in this flow chart.

ALTERNATIVES FOR THE DESIGN OF A PAPER MILL WASTEWATER TREATMENT and ACCORDING TO THE SELECTED ONE

Under the light of analysis and model test results obtained for the wastewaters from the paper mill in our case study, an attempt was made to design an economical wastewater plant with maximum ease of operation. Two such systems were proposed as possible alternatives. Flow diagrams for these alternatives are given in Figs. 5 and 6.

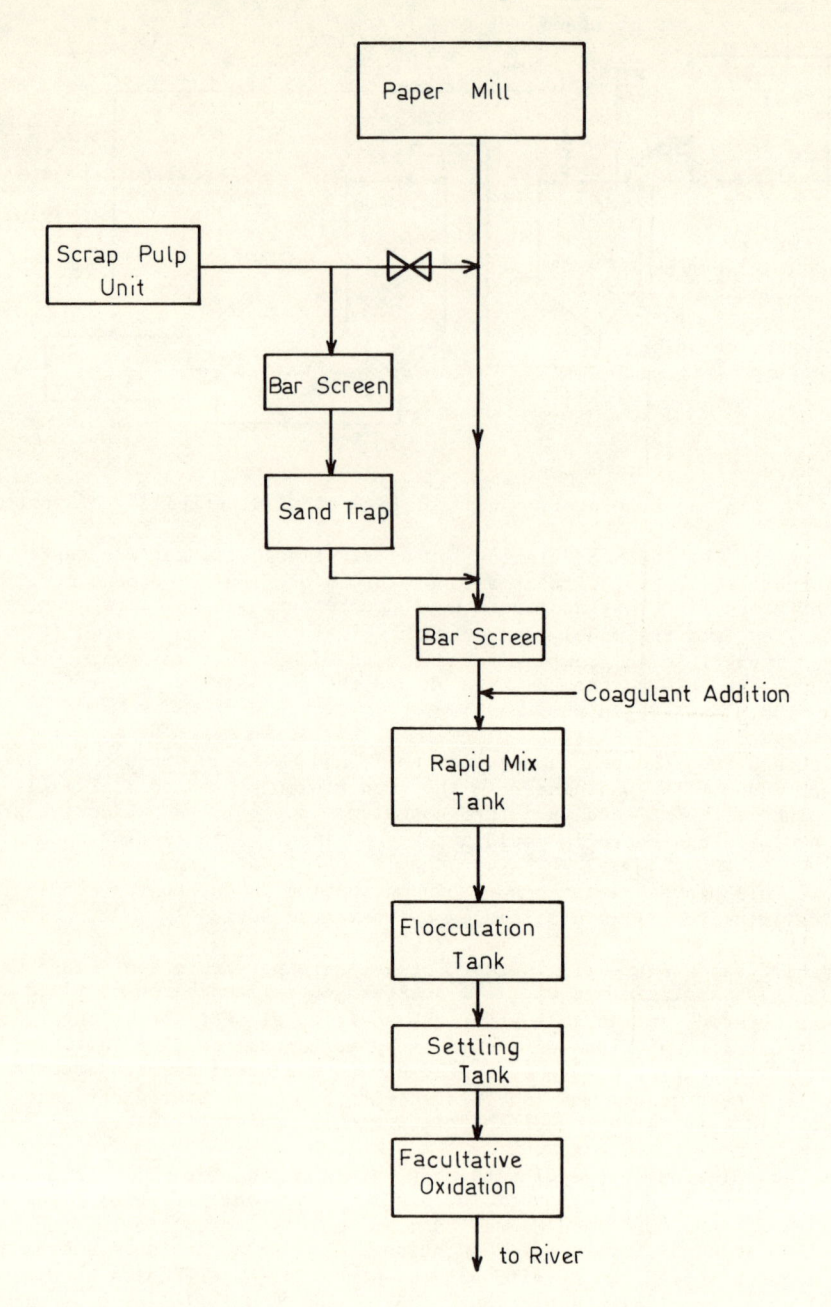

Fig. 5: Proposed alternative plant flow diagram

Treatment of Wastewater from Paper Mills

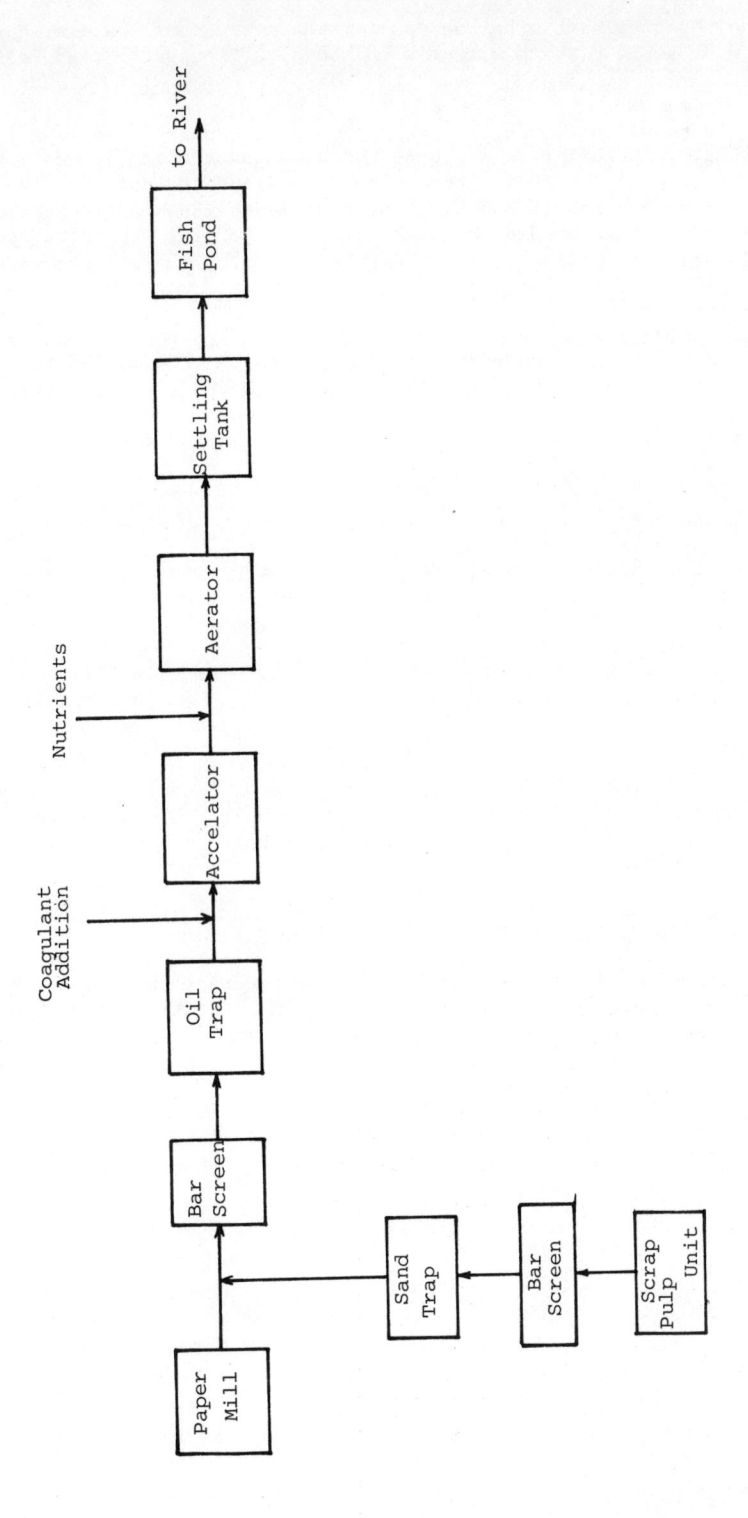

Fig. 6: Alternative - 2.

wastewater line together with a bar screen for the mixed canal waters, coagulant addition followed by rapid mix and slow mix chambers, a floc settler and a facultative pond.

In the second alternative (Fig. 6), the scrap pulp unit waste water was passed through a grit removal unit, and after mixing paper machine wastewater, it was screened to remove coarse fibers, etc. In this alternative, coagulant addition, mixing, flocculation and settling took place in one piece of equipment, known as an accelerator, was patented, which meant that we did not have to design for it. The cleared water coming from the accelerator is then aerated and settled again, which meant biological treatment.

The second alternative, proposed as a complete solution to the wastewater problem is generally applied in the rich industrialized countries. It is expensive both to construct and to operate, requiring large sums of money in the form of foreign currency. However, the first alternative, which could be realized by the plant's own personel with very little expenditure on fancy equipment, also showed maximum ease in operation and maintenance.

These are the reasons why only the first alternative was favored and units for it were designed. Design of the units and system was based on the existing plant operation and analysis results under the worst production and operating conditions. This meant maximum utilization of scrap pulp in the paper machine, which is about 40% for the time being.

Design characteristics of the favored alternative are summarized below:

- Flow rates for the design were 20 ℓ/sec for scrap pulp and 100 ℓ/sec for mixed canal wastewater.

- Two parallel grit removal chambers were considered necessary for the coarse particles (0.1 mm diameter), entering the scrap pulp unit. It was planned to put the chambers into use alternatively so that the one not in operation would be manually shoveled out and cleaned.

The mixed canal water was screened through bars of 2 cm width, which had to be manually cleaned.

- 10 mg/ℓ alum and 1 mg/ℓ polyelectrolyte solutions were to be added for coagulation. Both compounds were already being used in the plant operation as secondary raw materials; therefore, were present in their storerooms.

- For rapid mix after coagulant addition, a volume of 45 seconds detention time was designed for. The rapid mixer was a 100-120 rpm propeller.

- For effective flocculation, a slow mix tank with a detention time of 30 minutes and 1-3 rpm paddle mixer was planned. The settling of the flows would take place effectively in a 2-hour detention tank.

This tank could also be used as an oil skimmer for the floatables coming from the paper machine with the cooling water.

- Laboratory analyses and tests showed that following chemical addition, the effluent BOD_5 would be 60 mg/ℓ. In order to decrease that concentration to a more acceptable figure of 25 mg/ℓ, a 3-day detention time oxidation pond was designed. 5 meters deep, it was a facultative pond with mechanical aeration.

- Part of the final effluent to be discharged from the treatment plant will be passed into a small fish pond, to demonstrate the efficiency of the treatment.

CONCLUSIONS

Industrial waste treatment plant design in an important task for the environmental engineer. In order to design a successfully operating plant with realistic financial requirements, operational ease and minimal foreign currency demands, engineering skills must be utilized in accordance with the country's pollutional, legal and economic situation.

Sophisticated equipment might give him self-confidence and guarantee a given effluent quality, but if not in accordance with the country's pollution abatement regulations and financial situation, might lead to the idea of wastewater treatment being totally abandoned.

Depending on the climatic conditions prevailing in a large part of our country, simple waste treatment plants might safely be advocated. The use, especially of mechanically aerated facultative lagoons would be a good step towards the cheap treatment of industrial wastewater. Whenever necessary, however, chemical pretreatment might be included, again with the minimum use of mechanical equipment, steel tanks and connections.

This case study has shown that a thorough physical and chemical examination of industrial wastewater, followed by laboratory model tests indicating settling rates, biological and chemical treatability and rates, is of the greatest importance for the design of the most economical treatment plant.

REFERENCES

Dugan R. (1972). *Biochemical Ecology of Water Pollution*. Plenum Press, New York.

OECD (1978). *OECD - Ring Test Program on Detecting Biodegradability of Chemicals in Water*. Berlin.

DISCUSSIONS

Hernandez, J. W. (U.S.) : You indicated in your talk that the pH was adjusted to 7. Will you in the proposed plant have pH control?

Baysal, B. (Turkey) : We will certainly have pH control for the proposed plant. This is only a preliminary report. We are still conducting tests on the wastewaters from paper mills.

Hernandez, J. W. (U.S.) : What are the dewatering characteristics of the sludge after sedimentation?

Baysal, B. (Turkey) : These are still being studied. The project has not been completed yet. Several sludge disposal alternatives are at present being investigated, in collaboration with the plant authorities. For example, the sludge after drying could be burnt, or it could be thickened and used in compost plants.

Akçın, G. (Turkey) : Did you consider using any coagulants other than alum in the jar test? Were polyelectrolytes used at all?

Baysal, B. (Turkey) : Ferric sulphate and ferric aluminium sulphate were also tested for coagulation, but alum was found to be the most successful. Another reason for preferring alum as the coagulant in chemical treatment was that the plant already had in its stock, since alum solutions are utilized for plant operation. With regard to polyelectrolytes, we used in our laboratory studies one particular type, which again was already available in the plant.

Velioğlu, S. G. (Turkey): Why was it necessary in the jar tests to fix the pH at 6-8, considering that you had quite a good alkalinity level ranging from 200 kg to 400 mg/ℓ? Such a level would seem to me to be sufficient to provide buffer capacity.

Baysal, B. (Turkey) : As can be seen from the tables, the pH levels could vary considerably. We occasionally adjusted the levels in order to have steady conditions during the jar tests.

Velioğlu, S. G. (Turkey): How did you fix the pH at 6-8?

Müezzinoğlu, A. (Turkey): The pH was already at this level. For biological testing, the alkali in the solution must be neutralized, by the addition of a dilute acidic solution, of course.

Gür, A. (Turkey) : Since you are proposing to have a grit removal chamber, you must collect a considerable amount of grit.

Baysal, B. (Turkey) : This was a problem. The wastewater channels of the plant were full of grit. We tested the grit to establish particle size and settling velocity and studies are continuing. Since a grit removal channel is clearly inadequate, we may decide to use aerated grit removal.

Leentvaar, J. (Holland) : Our institute found that nitrifying bacteria were inhibited by a small concentration of alum in the effluent from the chemical treatment plant. In your case, too, the difficulties you had in obtaining a good sludge growth after coagulation-flocculation may have been due to the fact that the special bacteria needed to break down the cellulose compounds were also inhibited by the alum residue in the wastewater.

Baysal, B. (Turkey) : Our experiments showed, however, that even before coagulation-flocculation the biodegradability of the wastes was very slight or completely impossible. This may have been caused by a deficiency in nutrients.

The Biological Treatment of Sewage from Hide Glue Factories

KH. KRAUTH and K. F. STAAB

Institut fur Siedlungswasserbau, Wassergute-und Abfallwirtschaft, Universitat Stuttgart

ABSTRACT

During several periods in different seasons the Institut für Siedlungswasserbau, Wassergüte-und Abfallwirtschaft investigated the sewage of a hide glue factory. 97% of the sewage flow of about 440 m^3 per ton of glue originated in the liming and in the washing process. The pollution load ranged between 400 and 2700 kg BOD_5 per ton of glue, dependent on the kind of raw material.

Generally the sewage from hide glue factories is characterized by high pH-values (12 to 13) and very low phosphorous contents compared to the BOD. This sewage can be treated by the activated sludge process to a degree corresponding with the effluent standards that are normally required in the Federal Republic of Germany. In any case the undissolved lime is to be removed in a primary sedimentation tank.

In the case of aerobic stabilization the settled, highly alkaline sewage can be let directly into the activated sludge tanks. Higher loaded processes require an neutralization step and additional removal of the precipitated $CaCO_3$.

The flue gas neutralization is characterized not only by the saving of chemicals but essentially by the low amount of sludge produced. An overacidification of the sewage is impossible so that an automatic regulation for saving operational costs is possible by minimal measuring efforts. By adding phosphate into the influent of the aeration tank the efficiency of the biological process can be increased.

THE ORIGIN OF THE SEWAGE

Places of Origin

Hide glue is made by wastes of animal hide, delivered from leather factories and slaughter houses. Those wastes differ widely as to the degree of contamination caused by flesh, grease, blood, and hair. In the hide glue factory the raw material is conditioned with lime. The liming process is followed by an intensive washing of the raw material to remove the lime and the hide substances dissolved during liming. The washing procedure can take 24 to 48 hours, depending on the number and kind of washing machines.

After washing the hide wastes that are disintegrated and released from undesired components, the glue is extracted by melting in a boiler. The course of manufacture and the places of sewage origin are schematically shown in Fig. 1. It can be seen that about 7% of the sewage originates in the liming process and about 90% in

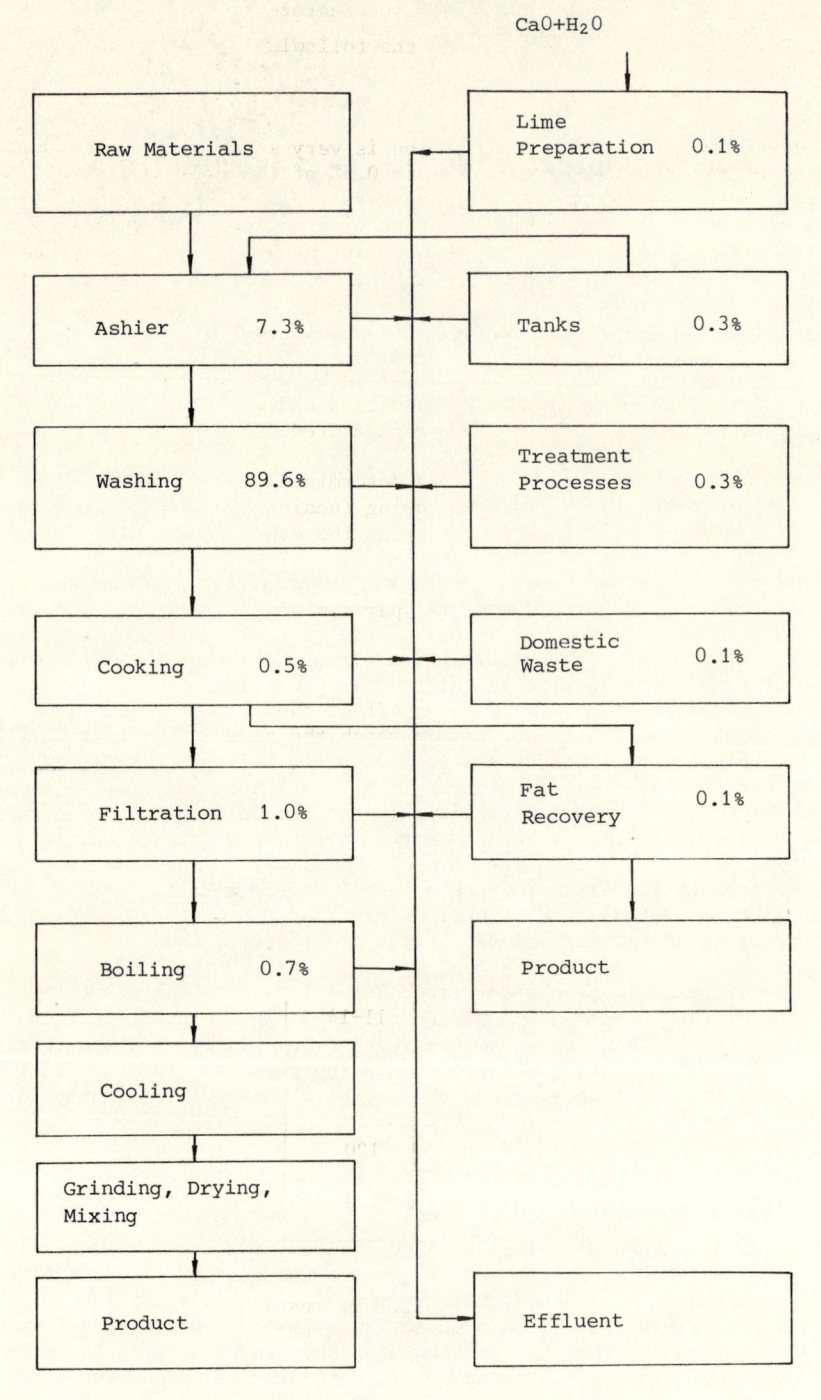

```
                                    CaO+H₂O
                                       ↓
┌─────────────────────┐        ┌─────────────────────┐
│                     │        │ Lime                │
│  Raw Materials      │        │ Preparation   0.1%  │
│                     │        │                     │
└──────────┬──────────┘        └─────────────────────┘
           ↓
┌─────────────────────┐        ┌─────────────────────┐
│                     │        │                     │
│  Ashier      7.3%   │───────▶│ Tanks         0.3%  │
│                     │        │                     │
└──────────┬──────────┘        └─────────────────────┘
           ↓
┌─────────────────────┐        ┌─────────────────────┐
│                     │        │ Treatment           │
│  Washing     89.6%  │───────▶│ Processes     0.3%  │
│                     │        │                     │
└──────────┬──────────┘        └─────────────────────┘
           ↓
┌─────────────────────┐        ┌─────────────────────┐
│                     │        │ Domestic            │
│  Cooking     0.5%   │───────▶│ Waste         0.1%  │
│                     │        │                     │
└──────────┬──────────┘        └─────────────────────┘
           ↓
┌─────────────────────┐        ┌─────────────────────┐
│                     │        │ Fat                 │
│  Filtration  1.0%   │───────▶│ Recovery      0.1%  │
│                     │        │                     │
└──────────┬──────────┘        └─────────────────────┘
           ↓
┌─────────────────────┐        ┌─────────────────────┐
│                     │        │                     │
│  Boiling     0.7%   │───────▶│ Product             │
│                     │        │                     │
└──────────┬──────────┘        └─────────────────────┘
           ↓
┌─────────────────────┐
│                     │
│  Cooling            │
│                     │
└──────────┬──────────┘
           ↓
┌─────────────────────┐
│ Grinding, Drying,   │
│ Mixing              │
│                     │
└──────────┬──────────┘
           ↓
┌─────────────────────┐        ┌─────────────────────┐
│                     │        │                     │
│  Product            │        │ Effluent            │
│                     │        │                     │
└─────────────────────┘        └─────────────────────┘
```

Fig. 1: Shematic representation of the wastewater sources.

the laundry. During the liming process and the following washing step all components of the hide that are unfit for the glue production are eliminated and thus they are comprised in the sewage.

The sewage flow from the glue boiling station is very small compared to that originated in the preceding steps. It amounts to 0.5% of the total flow. This sewage, like that from the following filter and evaporating plant, is mainly composed of the cleaning water that is used in those production units. The rest of the sewage originates in the slaked lime preparation, the boiler house, the sanitary equipment, etc.

Sewage Flow

The sewage flow during production days fluctuated between 1800 and 7000 m^3/d. Cleaning activities during Saturdays without production caused a flow from 1250 to 3200 m^3/d. On Sundays only 180 to 1050 m^3/d were measured.

The hydrography curve of the sewage flow is determined by the working rhythm of the factory. The minimal flow was registered during feeding the washing-machines with fresh material from the liming process, whereas the maximum flow corresponded to the beginning of the washing procedure.

The mean sewage flow was found to be 440 m^3 per ton of glue produced.

Sewage Components

The impurities of the raw material that can effect the quality of the hide glue, by liming the hide wastes, and thereby they reach the sewage. The main aspect of this procedure is the conversion of the collogen into glutine. Thereby the protoplasmic connective tissue cells, that intersperse all the collogenous tissue of the derma are loosened and removed by the strong alkaline swelling of the tissue. Together with the sewage, those wastes reach the effluent. The mean values of BOD, nitrogen and phosphorus that were measured in the sewage are presented in Table 1.

TABLO 1: Variation of BOD$_5$, N and P Mean Concentration During One Day

Time-Interval	8-11	11-14	14-17	17-20	20-8
BOD$_5$, mg/ℓ	586	823	705	316	467
Total N, mg/ℓ	103	120	136	54	80
Total P, mg/ℓ	1.2	1.3	1.4	0.7	0.9

The permenganate consumption of the sewage based on results obtained on intervals of 3 hours varied between 170 and 5900 mg/ℓ. The fluctuations were much more distinct and irregular compared to those of the BOD.

The pH-value varied between 11.2 and 12.8. 92% of all values fell into the range between 12.3 and 12.8.

The total alkalinity (m-value) fluctuated between 11.3 and 194.5 mval/ℓ. 61% of the values were found to be in the range between 30 and 50 mval/ℓ.

Pollution Load of the Sewage

The highest BOD_5 - concentrations in the effluent of the primary sedimentation were measured during the filling of the washers. Because of the minimum sewage flow, however, only a low corresponding BOD_5 - was calculated. The maximum BOD - load occured at the beginning of the washing process and decreased during the washing process.

The loads fluctuated between 1500 and 3500 kg BOD_5/d with a mean value of 2300 kg/d. On Saturdays they ranged between 700 and 2600 kg/d, on Sundays between 260 and 600 kg/d.

The daily phosphorus loads concerning the working days varied between 3 and 8 kg P/d with a mean of 5 kg/d.

Ratio of Pollution Loads

The data that refer only to pollutant concentrations are not sufficient for the characterization of a sewage. The ratio of their loads are just as well important.

The average nitrogen load was 18% of the BOD_5 - load. This ratio is similar to that of municipal sewage.

The average phosphorus load was 0.2% compared to the BOD_5 - load. This means that the phosphorus fraction in the sewage of hide factories is very low.

BOD_5 - load in Terms of a Production Unit

Accurate values relating analytical results to a production unit can only be fixed if long term data are available. In this connexion it must be considered that there is a time difference between the washing of the material, i.e. the main sewage flow, and the extraction of the glue by melting, and that seasonal fluctuations of temperature with liming can influence those relationships as well as differences in the composition of the raw materials. In this sense the following data can only represent the magnitudes of possible values. They permit however, an estimation of the relations in similar applications.

In the effluent of the primary sedimentation a BOD_5 - load of 220 kg per ton of glue produced was determined. This value can fluctuate distincly according to the characteristics of the raw material. If handglue leather is worked up, the BOD_5 - load in the raw sewage is about 400 to 500 kg BOD_5 per ton of glue compared to about 2700 kg BOD_5 per ton of glue. In the case of machine made glue leather the yield is about 180 kg BOD_5 compared to 60 to 80 kg BOD_5 per ton of material if other materials are used.

CHOICE OF THE TREATMENT PROCESS

One basic supposition for the successful biological treatment is the extensive removal of sludge of the sewage by mechanical pretreatment. The high pH-value can be buffered by a neutralization step or in the biological stage itself if the pH-values are adequately considered with dimensioning.

To examine the biodegradability of the sewage and to derive design parameters, comprehensive experiments in the laboratory and semitechnical scale were made.

By feeding mechanically cleaned, not neutralized sewage with a pH-value of about 13, the neutralization of the sewage by the CO_2 - production of the activated sludge and the biological purification of the sewage could be accomplished if the BOD_5 - sludge loading was kept below 0.1 kg/(kg MLVSS . d). By that however the precipitated $CaCO_3$, was accumulated in the activated sludge unit. The excess sludge showed a $CaCO_3$ portion of more than 50%. When the admissible loading was exceeded, the activated sludge was irreversibly damaged as the amount of CO_2 produced was not sufficient for neutralization and the pH-value in the aeration tanks increased excessively.

The neutralization of the sewage by injecting CO_2 resulted in a quick and improved settling of the solids and the carbonate precipitated during neutralization.

The activated sludge showed a higher organic portion when feeding neutralized sewage. In case of influent concentrations of 740 mg/ℓ and BOD_5 - sludge loadings below 0.1 kg/(kg MLVSS · d) BOD_5 - values in the effluent between 2 and 3 mg/ℓ were achieved. The rise of the sludge loading to 0.5 kg/(kg MLVSS · d) led to a deterioration of the effluent quality with corresponding BOD_5 - values of about 25 mg/ℓ. An increase of both the sludge loading and the biodegradability was accomplished by adding phosphate. The low phosphorus concentration in the sewage of the glue factory (P : BOD_5 = 2 : 1000) was the limiting factor for sludge growth and hence affected the efficiency of the process. In the case of a mean sludge loading of 0.25 kg BOD_5/(kg MLVSS · d) and highly polluted influent, the average BOD_5 - concentration in the effluent without adding phosphorus was 47 mg/ℓ. Values below 25 mg/ℓ were not found. On the other hand, however, the additional feed of phosphate led to a mean BOD_5 - concentration in the effluent of 16 mg/ℓ at a mean sludge loading of 0.58 kg BOD_5/(kg MLVSS · d).

DESIGN PARAMETERS

a) Treatment with a Dosage of Phosphate

The following design parameters are recommended for an activated sludge plant treating the sewage of hide glue factories with additional dosage of phosphate:

The neutralization step should lead to a pH-value of 7 or at least 8. The minimum detention time in the sedimentation tank following the neutralization should be 30 minutes. The dimensioning of the aeration tank should conform to the following data: BOD_5 -sludge - loading 0.4 kg/(kg MLVSS · d), mixed liquor suspended solids 6 g/ℓ, organic fraction in the sludge 50%, phosphate dosage related to the daily BOD_5 - load: 1.5 kg P/100 kg BOD_5. When the secondary sedimentation is dimensioned the high sludge concentration in the activated sludge tank must be considered. The sludge volume load should not exceed 0.3 m^3/ (m^2 · h).

The combined treatment of the sludges coming from the primary, intermediate and secondary sedimentation in thickener and filter process in supposed to be the most suitable technical solution.

In order to equalize fluctuations of flow and concentration it is useful to combine the primary sedimentation tank with a buffer tank. A storage capacity of 50 to 100 % of the daily flow is recommended. The neutralization of the sewage has to be done by injecting flue gas into a serpentine ditch. The CO_2 should be put into the water by large bubble aeration. As intensive foaming must be taken into account a high freeboard is necessary. Because of the danger of foaming a large bubble aeration has to be installed. To meet the peaks without difficulties it is useful to use a step aeration process.

b) Treatment Without Additional Dosage of Phosphate

In the case of treatment plants without adding phosphate the following design parameters are recommended.

Following the preliminary elimination of the settleable solids in the primary sedimentation and the buffer tanks, the activated sludge plant should be dimensioned with a sludge loading of 0.05 kg BOD_5/ (kg MLVSS · d) and 3.5 g/ℓ mixed liquor suspended solids. The activated sludge plant should be build as an oxidation ditch with aeration by cage rotors.

The final sedimentation is to be designed with the parameters usally applied.

PRACTICAL EXPERIENCES WITH BOTH PLANTS

The low loaded activated sludge plant without neutralization and without dosage of phosphorus didn't cause any operational problems and led to an excellent effluent quality with BOD_5 - concentrations below 10 mg/ℓ and a permanganate consumption below 25 mg/ℓ. The efficiency was very stable.

The operation of the higher loaded activated sludge plant was accompanied by considerable problems. The effluent quality fluctuated distinctly. BOD_5 - concentrations below 10 mg/ℓ were measured. In the other side, however, the BOD_5 also reached values in the magnitude of 50 mg/ℓ.

DISCUSSIONS

Tabasaran, O. (Germany) : Could you give more information about the German regulations governing the quality of wastewater to be discharged into rivers and lakes?

Krauth, K. (Germany) : The quality requirements are undergoing a great change in Germany on account of the new legislation. At the beginning of 1979, the Federal Republic issued the so-called minimum regulations for the discharge of wastewater. Up to now these have applied to communal treatment plants. The requirements, however, are less stringent for treatment plants built before 1978 than for those still under construction. Furthermore, every branch of industry, for example, the hide-glue industry, has to fulfill special conditions in compliance with the generally approved regulations of the technology used in the treatment of the effluent. The following values from the hide-glue industry would meet with the quality requirements for a neutral discharge: a BOD of 20 mg/ℓ for a 24-hour period, a BOD of 30 mg/ℓ for a 2 -hour period and a COD of under 120 mg/ℓ.

Samsunlu, A. (Turkey) : The sludge loading during the treatment of the effluent from the hide-glue industry is very low, especially when phosphate is not added. A great amount of space must be required. Couldn't the loading be increased without the addition of phosphate?

Krauth, K. (Germany) : No. Higher loadings are only possible with phosphate addition. With high sludge loadings, a larger cellular mass is formed, for which phosphate is required.

Coagulation-Flocculation of Beet-Sugar Wastewater

J. LEENTVAAR, H. M. M. KOPPERS and W. G. BUNING

Agricultural University, Department of Water Pollution Control, Wageningen, The Netherlands

ABSTRACT

The coagulation-flocculation of wastewater from a beet-sugar mill was studied. Special attention was paid to the dose and type of coagulant, to the effect of coagulant aids and to the amount of sludge produced. The strength of the sludge floc was examined simply in order to design a transportation pipe. The experiments were carried out on a batch scale and in continuous-flow laboratory and pilot-plant experiments. It was shown that batch experiments in rectangular vessels describe the continuous coagulation-flocculation process quite well. The settling velocity of the sludge was determined in a vertical settling cylinder. This simple technique proved to be successful in designing the sedimentation tank. Economic aspects were briefly surveyed.

INTRODUCTION

This study was carried out on the wastewater of sugar mill in Groningen, the Netherlands, having a discharge of 35 m^3/min. The wastewater originates mainly from hydraulic transportation or "fluming" of the sugar beets. Besides soil and sugar by-products, the wastewater also consists of some excess condensate containing high ammonia and some other fermented sugar products.

The present treatment system consists of a sedimentation basin for the settling of coarse particulates, a cooling pond and an aerated lagoon (Fig. 1). The aim of this study was to investigate the feasibility of coagulation-flocculation of the effluent from the aerated lagoon for in-plant reuse or discharge into a surface water recipient.

The study performed during one working-season included:

- A series of jar tests with three inorganic coagulants to establish the required dose, pH-range, possible use of coagulant aids and the amount of sludge produced.

- Comparison of the batch experiments with the continuous-flow laboratory experiments (60 l/h) and pilot-plant studies (6 m^3/h).

- Determination of the design criteria for a coagulation-flocculation plant treating sugar-mill wastewater.

EXPERIMENTAL SET-UP

Experiment were carried out on a batch scale, in a laboratory-scale pilot plant and in a pilot plant with a flow of 6 m^3/h. The batch coagulation-flocculation tests were

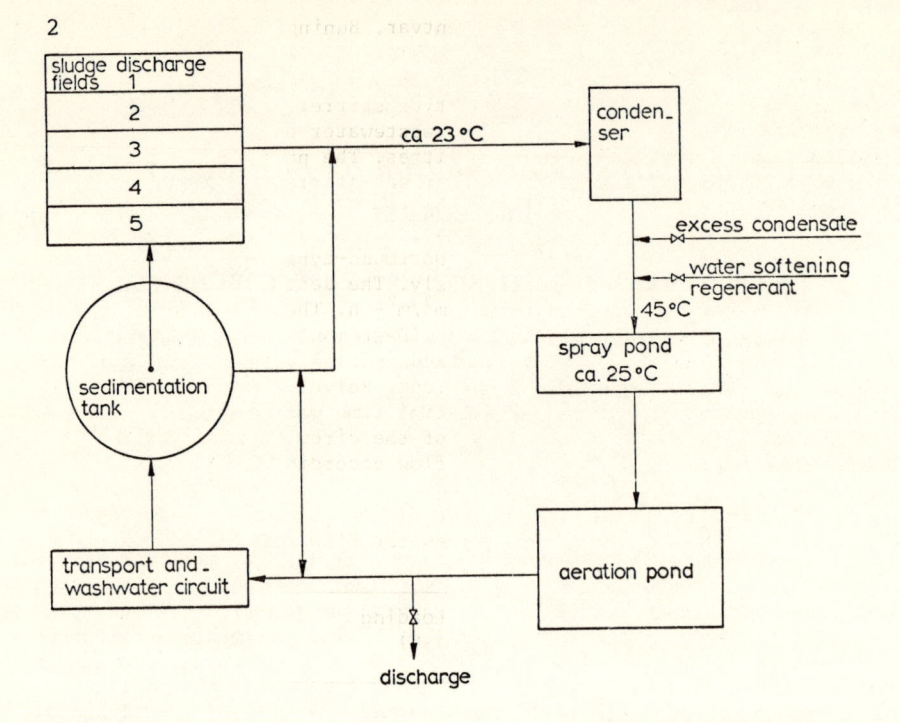

Fig. 1. Simplified Water Flow Diagram of the Sugar Mill

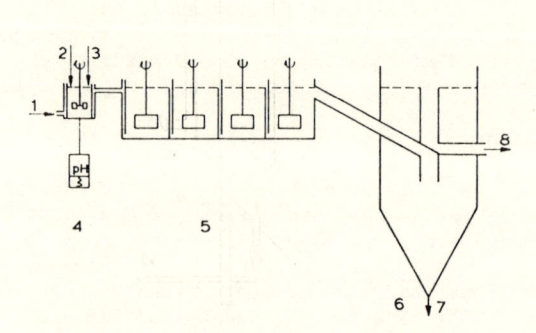

Fig. 2. The Laboratory-Scale Pilot Plant (60 l/h)
1. Influent. 2. Coagulant Dose. 3. pH-Adjustment. 4. Mixing Vessel. 5. Flocculator. 6. Sedimentation Tank. 7. Sludge Discharge. 8. Effluent.

carried out as described previously by Leentvar, Buning and Koppers (1978). The laboratory-scale flocculator (Fig. 2) had a flow of 0.06 m^3/h and consisted of a rapid mix vessel for coagulant feed and for pH-adjustement. This vessel had a volume of 1 litre and was stirred with a turbine-type stirrer at 365 rpm, which gave a G value of 150 s^{-1}. After this initial mixing, the wastewater entered a plug flow flocculator, with 4 reactors and a total volume of 20 litres. The pH-electrode was placed in the first flocculation compartment. Each reactor was stirred by a flat blade stirrer at 63, 50, 22 and 10 rpm. For technical reasons the G values were measured at the end of the experimental period; they amounted to 60, 32, 14 and 5 s^{-1}, respectively. Finally, the wastewater was allowed to settle in a Dortmund-type sedimentation tank, from which the settled sludge was drawn off continuously. The detention time in the tank was 1 hour; the surface loading amounted to 0.5 m^3/m^2·h. The second pilot plant was a sludge blanket flocculator of the circulator-type (Degrémont Holland B.V.). In the circulator (Fig. 3), wastewater and coagulant were mixed initially in the influent pipe. Flocculation occurred in a conical pipe inside the tank. Polyelectrolytes were added at the overflow of the conical pipe. The flocculation time was governed by the influent flow which ranged from 4 to 8 m^3/h. The volume of the circulator was 5 m^3. The surface loading was also determined by the wastewater flow according to the table below.

Table 1   Relation between Wastewater Flow and Surface Loading

Wastewater Flow (m^3/h)	Surface Loading (m^3/m^2-h)	Detention Time (h)
4	2.2	1.25
5	2.7	1.00
6	3.3	0.83
7	3.8	0.71
8	4.4	0.63

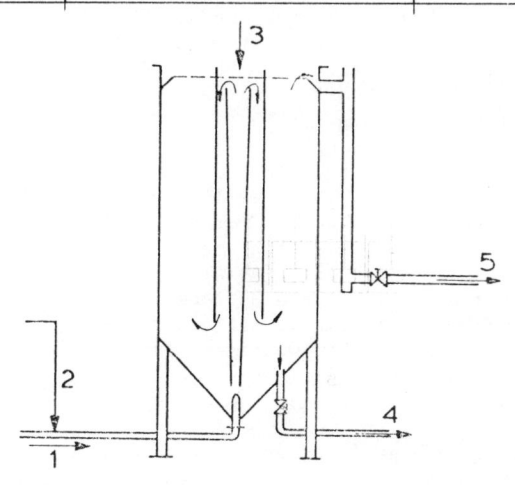

Fig. 3. The Sludge Blanket Flocculator (4-8 m^3/h).
1. Influent. 2. Coagulant Dose. 3. Coagulant Aid Dose.
4. Sludge Discharge. 5. Effluent.

To estimate the settling velocity of the sludge floc, some experiments were carried out in a settling cylinder, with sampling points at different heights. The volume of this cylinder was 0.2 m^3 and its diameter was 0.30 m.

Ferric chloride (40 % w/w solution), aluminium sulphate (Al$_2$(SO$_4$)$_3 \cdot$ 18 H$_2$O) and hydrated lime (Ca(OH)$_2$) were used as coagulants. Superfloc A 100, N 100 and C 100, anionic, nonionic and cationic synthetic polymers, respectively, (American Cyanamid Company, Wayne, New Jersey) were used as coagulant aids.

RESULTS

Batch experiments

Batch experiments were carried out in order to determine the ratio of coagulant dose to flocculation effect, the optimal flocculation pH, the effect of the addition of coagulant aids and the amount of sludge produced.

Experiments with ferric chloride as a coagulant showed that the pH had little effect on the COD-removal in a pH range from 5.0 to 7.0.

Depending on the coagulant dose, up to 85% COD-elimination was achieved. The experiments showed that the addition of 1-2 mg/l anionic polymer A 100 besides ferric chloride resulted in an extra COD-reduction of about 5% and a reduction in the volume of sludge produced of about 30%. The addition of a cationic or a nonionic polymer as a coagulant aid had a negligible effect. Several experiments showed that COD-removal was optimal when more than 200 mg Fe/l were added.

When alum (Al$_2$(SO$_4$)$_3 \cdot$ 18H$_2$O) was used as a coagulant, batch experiments showed that the optimal pH for coagulation-flocculation was also higher than 5.0. The maximal COD-reduction of about 92% occurred with doses higher than 100 mg Al/l. Considerable amounts of sludge were produced. This sludge volume could be reduced by adding 1-2 mg/l nonionic polymer N 100 as a coagulant aid, while anionic and cationic polymers were not so effective. In contradiction to previous coagulation-flocculation experiments with waste water of domestic origin (Leentvaar and others, 1978; Minton and Carlson, 1973), the coagulation-flocculation of beet-sugar wastewater with hydrated lime as a coagulant was not determined to any great extent by the pH. A fixed amount of hydrated lime added to the wastewater gave a higher COD-removal percentage, up to 79%, than when a lime dose was added until a fixed pH was reached. The COD-removal was maximal at lime doses above 2500 mg Ca(OH)$_2$/l. The addition of organic polymers together with lime had almost no effect on the COD-removal and the volume of sludge produced. The addition of small quantities of ferric chloride (ca. 10 mg Fe/l) with the lime led to a substantial improvement in COD-removal, but also to an increase in the volume of sludge produced.
Fig. 4 summarizes the results of the batch coagulation-flocculation experiments carried out with the three different coagulants. As nitrogen plays an important role in the cost of discharge of polluted waters into the surface water or into the local sewerage system, the elimination of organic and inorganic nitrogen compounds was examined indicentally both in batch experiments and in the two pilot plants. All data showed a linear relationship between COD and Kjeldahl nitrogen-removal (See Fig. 5). The correlation coefficient of the r, obtained from 17 observations was 0.88.

Table 2 shows that volatile organic acids are not removed by coagulation-flocculation. The concentration of these acids increased during the treatment. The removal percentage based on COD was somewhat higher than the removal percentage based on TOC.

The wastewater of the beet-sugar mill does not have a constant quality. During the working season the total COD(COD$_t$) and the COD of the supernatant after centrifuging at about 2500 g (COD$_c$) increases, as illustrated in Fig. 6. This COD$_c$ is the COD of soluble products such as sugars and volatile organic acids, as well as of some colloidal materials It was assumed that as the wastewater COD increased during the working season,

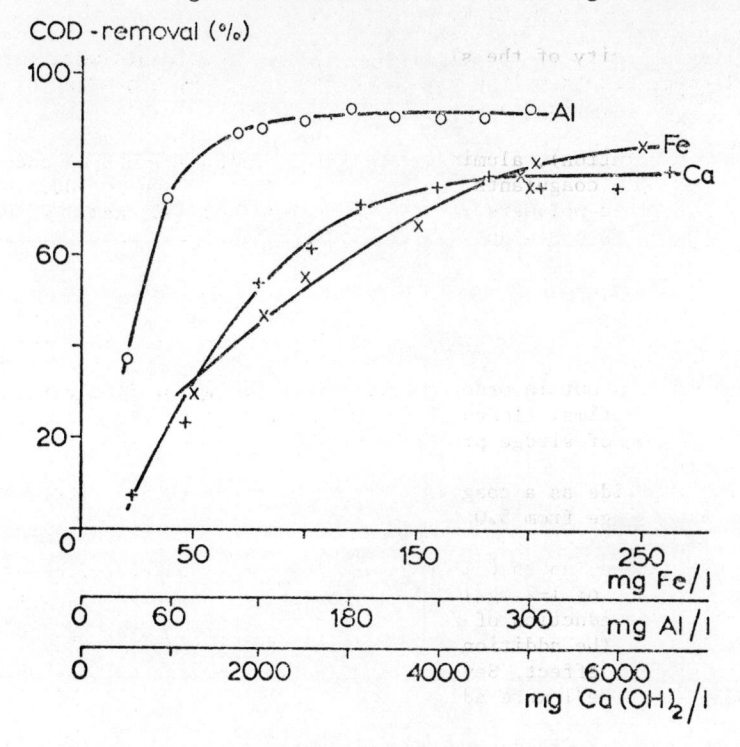

Fig. 4. The Results of Batch Coagulation-Flocculation Experiments Carried out with Ferric Chloride, Alum and Hydrated Lime

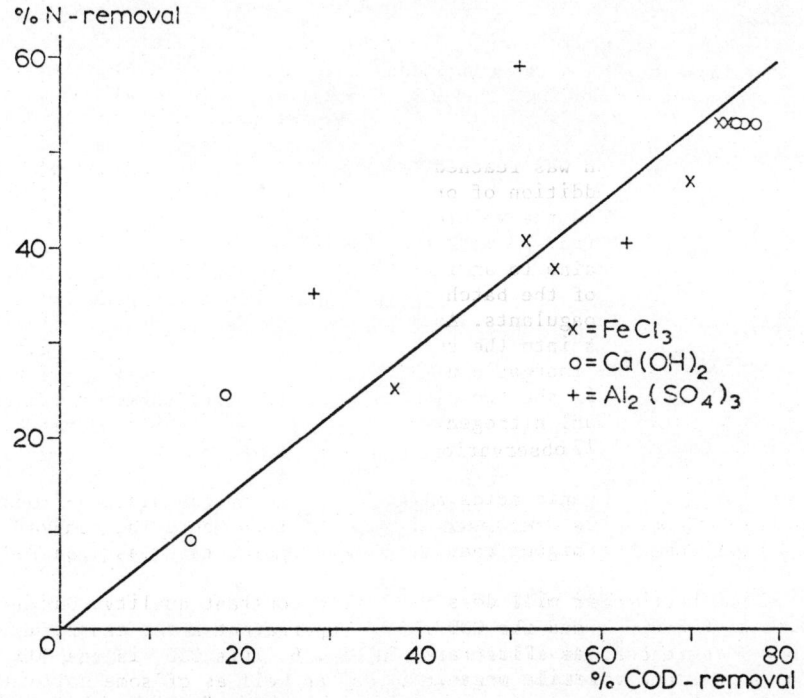

Fig. 5. The Relation between COD-Removal and Nitrogen-Removal with different Coagulants Used

Table 2 The TOC, COD Values and the Concentration of Volatile Organic Acids after Coagulation-Flocculation in a Batch Experiment[1]

Coagulant dose (mg/l)	COD (mg/l)	TOC (mg/l)	Volatile organic acids[2] C_2 (mg/l)	C_3 (mg/l)	iC_4 (mg/l)
0	4610	1230	250	400	5
50 Fe	3960	1070	270	420	5
150 Fe	2120	595	295	465	7
250 Fe	1210	450	290	460	4
30 Al	3640	1135	250	405	5
90 Al	1690	505	315	500	6
150 Al	1400	465	280	440	5
1000 Ca(OH)$_2$	3670	1030	310	470	5
2000 Ca(OH)$_2$	2830	920	310	485	6
3000 Ca(OH)$_2$	2160	635	253	410	6

[1] pH of wastewater : 7.5

[2] C_2 = acetic acid; C_3 = propionic acid; iC_4 = isobutyric acid; C_4, iC_5 and C_5 components < 5 mg/l

the COD-removal percentage would decrease with a fixed coagulant dose. Batch experiments with fixed doses of the three coagulants, however, showed a constant removal efficiency with respect to the COD. The COD_t-removal decreased when there was an extra supply of molasses (COD_c); the COD_t - removal decreased because this soluble product was not involved in coagulation.

Pilot plant experiments

As mentioned previously, two continuous-flow pilot plants were used: a laboratory-scale plant with a wastewater flow of 0.06 m^3/h and a circulator with a flow adjustable from 4 to 8 m^3/. The experiments with the laboratory pilot plant were to determine whether the COD-removal percentages of the jar test corresponded with those of the continuous experiments.

For this reason the surface loading of the sedimentation tank was very low: 0.5 m/h. The experiments with the circulator, together with the experiments in the sedimentation cylinder served to obtain design criteria. For this reason the surface loading and with that, the average detention time of the liquid in the circulator was varied. The experiments were carried out with a large and a small coagulant dose and the addition of a coagulant aid. For coagulation-flocculation with ferric chloride as a coagulant, 2.0 mg/l anionic polymer A 100 was added; with alum, 2.0 mg/l nonion polymer N 100; and with hydrated lime as coagulant, 30 mg Fe/l was added.

A synopsis of the data of the experiments on a batch scale, laboratory scale and sem technical scale is given in Table 3. Other experiments carried out gave the same results.

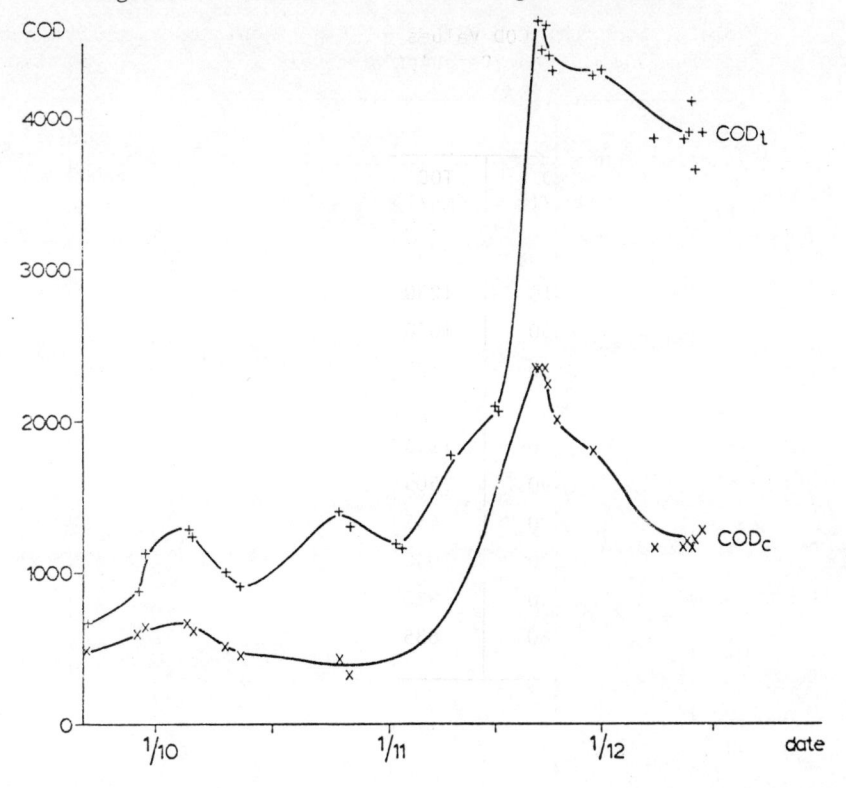

Fig. 6. COD_t and COD_c of the Wastewater During the Working Season

Table 3 illustrates that the removal percentages of both the laboratory-scale flocculator and circulator corresponded well with those obtained in the jar tests for all three coagulants. The correlation coefficient (r) between the laboratory-scale data and the batch experiments data was 0.92 (N = 17) and that between the circulator experimental results and the results of the jar tests was 0.93 (N = 24), in which N is the number of experiments.

Design

As the cost of the sedimentation tank is the most important part of the investment cost, this tank should be designed very carefully. For this reason the settling velocity of the flocs was measured under different conditions in a vertical settling cylinder, as previously described. The wastewater in the cylinder was flocculated by adding a large or small amount of coagulant together with a coagulant aid and by mixing with a perforated plate. During the settling of the flocs, samples were taken at different heights of the cylinder, from which the COD was determined. As the flocculent particles settled, the settling velocity increased in some way or another with time and depth. In the case of the sedimentation cylinder, the efficiency of the sedimentation decreased with the height of the sampling point. From these data the surface loading of an upflow sedimentation tank with a known depth can be determined at

Table 3. Synopsis of the Data of Experiments on Batch Scale Laboratory Scale and Semi-Technical Scale

Coagulant dose[1] (mg/l)	COD_t/COD_c (mg O_2/l)	COD_t-reduction (%) Batch Scale	Laboratory Scale	Semi Technical Scale	Surface loading C (m/h)
50 Fe	1180/495	38	56	39	2.7
50 Fe	1160/515	48	57	{39 / 37	2.7 / 3.8
200 Fe	1340/375	88	85	{92 / 80	2.2 / 3.3
30 Al	2190/795	51	40	64	2.2
30 Al	2070/665	47	45	64	2.2
30 Al	4640/2410	24	30	{23 / 28	2.2 / 2.7
30 Al	4430/2350	13	17	{20 / 21	2.2 / 3.8
60 Al	4390/2240	46	52	{45 / 37	2.2 / 3.8
60 Al	4310/2000	42	56	{51 / 28	2.2 / 3.8
60 Al	4210/1790	41	28	{49 / 52	2.2 / 3.3
60 Al	4320/ –	44	53	42	3.8
1000 Ca(OH)$_2$	3860/1170	9	11	10	3.8
1000 Ca(OH)$_2$	3910/1200	12	11	10	3.8
1000 Ca(OH)$_2$	4100/1165	14	12	14	3.8
2000 Ca(OH)$_2$	3640/1200	20	16	24	3.8
2000 Ca(OH)$_2$	5530/1200	43	47	47	3.8
2000 Ca(OH)$_2$	3890/1270	25	19	19	4.5

[1] The following coagulant aids were used:
    With Fe : 2.0 mg A 100/l
    With Al : 2.0 mg N 100/l
    With Ca : 30 mg Fe/l

a fixed degree of purification. These values are given in Table 4. Table 4 shows that a small coagulant dose led to some flocculent settling and that a high dose led to zone sedimentation (Compare 30 and 90 mg Al/l). Similar settling tests proved that an increase of suspended solids in the wastewater had a positive effect on the settling rate. Hence, the settling rate was lower at the start of the working season than at the end of the season. The tests also showed that the addition of coagulant aids result

Table 4 Relation between Settling Rate and Depth of Sedimentation Tank for Different Coagulants (95% Removal)

Coagulant Dose	Depth of Sedimentation Tank (m)	Settling Rate (m/h)
50 mg Fe/ℓ + 2.0 mg A 100/ℓ	0.25 0.75 1.25 1.75	8.0 8.5 10.0 8.5
200 mg Fe/ℓ + 2.0 mg A 100/ℓ	0.25 0.75 1.25	0.6 3.8 4.3
30 mg Al/ℓ + 2.0 mg N 100/ℓ	0.25 0.75 1.25 1.75	2.7 3.3 4.1 4.1
90 mg Al/ℓ + 2.0 mg N 100/ℓ	0.25 0.75 1.25	1.4 1.6 1.7
1000 mg Ca(OH)$_2$/ℓ + 30 mg Fe/ℓ	0.25 0.75 1.25 1.75	6.5 12.0 13.0 13.0
2000 mg Ca(OH)$_2$/ℓ + 30 Fe/ℓ	0.75 1.25 1.75	2.4 3.2 <4.2

in larger flocs which settled more rapidly. The results of the tests in the sedimentation cylinder agreed reasonably well with those obtained with the circulator, as illustrated in Table 3. Table 3 shows that in the case of coagulation with 60 mg Al/l, a surface loading of 3.8 m/h was too high for a good separation of the flocs and the effluent, because flocs appeared in the effluent and the COD removal percentage decreased.

This result corresponded with those of the column settling test (Table 4) with a dose of 90 mg Al/l, in which the settling velocity at a depth of 1.25 m was 1.7 m/h; if the values for the settling velocity at 0.25, 0.75 and 1.25 m are extrapolated, then the sedimentation rate will not be higher than 3.8 m/h at a depth of 3.5 m.

Sludge

As illustrated in Fig. 7, the amount of sludge produced depends on the COD of the wastewater and the dose and kind of coagulant. The dry solids content of the sludge was 10 g/l with ferric chloride and alum as coagulants, and 25-30 g/l when coagulation-flocculation was with hydrated lime. The volume of the sludge and its dry solids content were measured after 30 min of sedimentation. The addition of coagulant aids and sludge recirculation would have a positive effect on the dry solids content of the sludge.

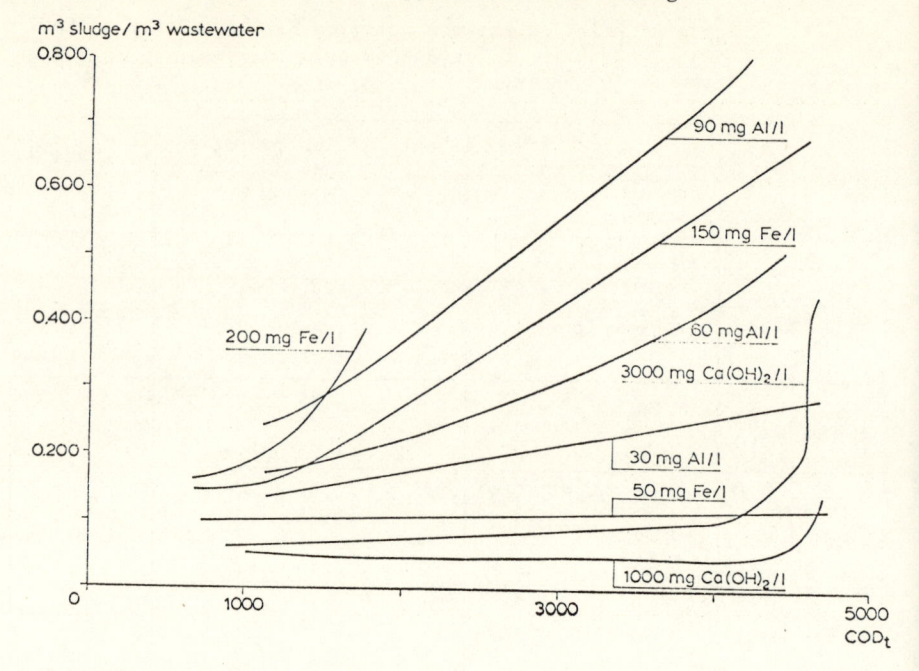

Fig. 7. The Volume of Sludge Produced per m³ Wastewater as a Function of Wastewater, COD and Type of Coagulant

The sludge produced in the physico-chemical treatment had to be transported for about 450 m and was dumped together with solid material from the sugar production on a sewa field. A jar test was set up to gain an impression of the sludge floc.

For this reason, sugar-waste sludge, physico-chemical sludge and a 5:1 mixture of bot sludges were agitated with a velocity gradient of about 200 s^{-1} for 3 min. After 24 the COD_t of the supernatant and the sludge volume were measured and compared with these values before agitation. Table 5 shows that the agitation of iron sludge led to a 22% increase in COD and also to an increase in the volume of sludge. Alum and li sludges were very strongly disintegrated by agitation. Agitation of a 5:1 mixture of sugar-waste sludge with iron or lime sludge resulted in an increase in the COD of the supernatant by about 22% and a decrease in sludge volume of 14%.

Agitation of the sugar-waste sludge increased the COD of the supernatant by 5%. As t flow of the suspended sugar-waste sludge was 480 m³/h and the flow of the physico-chemical sludge was in practice 20 m³/h, the percentage COD increase due to the agitation/transportation of this mixture would be less than 5% provided that agitation wa not higher than 200 s^{-1}.

The G value of 200 s^{-1} corresponds roughly to a pipe diameter of 0.15 m at a flow of 20 m³/h.

Table 5 Influence of Agitation on the Release of COD from Physico-Chemical Sludge and Sugar-Waste Sludge

Sludge type	COD_t supernatant[1] Before agitation (mg/l)	After agitation (mg/l)	Increase (%)	m³ sludge / m³ wastewater Before agitation (m³/m³)	After agitation (m³/m³)	Increase (%)
Sugar-waste sludge	4650	4890	(+5)	0.140	0.145	(+4)
Iron sludge	2020	2460	(+22)	0.420	0.465	(+11)
Mixture[2] (Fe)	4400	5320	(+21)	0.175	0.150	(−14)
Alum Sludge	2310	4900	(+112)	0.410	0.340	(−17)
Mixture (Al)	4530	6169	(+36)	0.145	0.125	(−14)
Lime Sludge	3680	9880	(+168)	0.500	0.425	(−15)
Mixture (Ca)	4660	5790	(+24)	0.175	0.150	(−14)

[1] The COD of the supernatant was measured after 24 hours of sedimentation.
[2] The mixture consisted of a 5:1 mixture of sugar-waste sludge and physico-chemical sludge.

ECONOMIC ASPECTS

Though the capital cost of a physico-chemical wastewater treatment plant is lower than that of a mechanical-biological plant, the recurring costs of a physico-chemical plant are in most cases higher. The recurring costs of the coagulation-flocculation step are mainly due to the cost of the coagulants (See Table 6).

Coagulant efficiency is defined as the total amount of COD removed per unit coagulant and expressed as kg COD/kg coagulant. Table 6 surveys the efficiency and cost of the three coagulants used in this study. This table illustrates that at a certain coagulant dose, coagulant efficiency increased with an increase in the wastewater COD. At a given wastewater COD, the coagulant efficiency decreased with an increase coagulant doses.

In view of the coagulant cost, the following coagulant dose is recommended: 50-100 mg Fe/l, 30-60 mg Al/l or 1500-2000 mg Ca(OH)$_2$/l. The choice of the coagulant dose at an equal cost per kg COD removed depends on the desired effluent quality and/or the sludge-handling and disposal facilities.

DISCUSSION

The wastewater of the CSM sugar mill at Vierverlaten after sedimentation and aeration can be coagulated-flocculated easily using the three coagulants, ferric chloride, alum or hydrated lime, under normal conditions in the sugar mill. The degree of purification depends partly on the ratio of total COD and COD of the supernatant after centrifuging and on the kind and dose of the coagulant. The addition of coagulant aids such as Superfloc A 100 with ferric chloride as coagulant and Superfloc N 100 with alum as

Table 6 A Survey of the Coagulant Efficiency and the Coagulant Cost at Different Wastewater COD Values and with Different Coagulants.[1]

Coagulant Dose (mg/l)	Wastewater COD (mg t_{O_2}/l)	Coagulant Efficiency (kg COD removed/ kg coagulant)	Coagulant Cost/ kg COD removed (f)[2]
Fe 50	<1000	4.0– 7.5	0.18–0.34
50	>2000	14.0–15.0	0.09–0.10
100	<1000	4.0– 6.0	0.23–0.34
200	<1000	2.7– 3.6	0.38–0.51
200	1000–2000	3.6– 8.0	0.17–0.38
250	<1000	2.5– 3.0	0.46–0.55
250	>2000	ca. 13	ca. 0.11
Al 30	<1000	ca. 15	ca. 0.25
30	>2000	15.0–45.0	0.08–0.25
60	<1000	ca. 15	ca. 0.25
60	>2000	15.0–40.0	0.09–0.25
90	<1000	ca. 11	ca. 0.34
90	>2000	11.0–25.0	0.15–0.34
150	<1000	ca. 7.5	va. 0.50
150	>2000	7.5–20.0	0.19–0.50
Ca(OH)$_2$ 1000	<1000	ca. 0.4	ca. 0.39
1000	>2000	0.3– 1.0	0.16–0.52
2000	<1000	ca. 0.35	ca. 0.44
2000	>2000	0.6– 1.0	0.16–0.26
3000	<1000	ca. 0.25	ca. 0.26
3000	>2000	0.5– 0.7	0.22–0.31

[1]
The costs given in this table were calculated using the following prices:
Ferric chloride   (40% w/w FeCl$_3$ solution)   :f. 188.80/tonne
Alum   (98% Al$_2$(SO$_4$)$_3 \cdot$ 18H$_2$O)   :f. 295.00/tonne
Lime   (98% Ca(OH)$_2$:f. 155.00/tonne

[2]
f 1.00 = $0.50

coagulant has a positive effect on the COD-removal, the settling rate of the flocs produced and on the volume of sludge produced. To minimize the volume of sludge produced, the coagulant dose should not be too large. The experiments proved that batch coagulation-flocculation tests in rectangular tanks can describe the continuous-flow experiments very well and are very useful for evaluating the coagulation-flocculation process in practice, for surveying the coagulation-flocculation plant and for determining the optimal coagulant feed. The settling rate of the flocs as determined using the vertical sedimentation tube gives a good indication of the surface loading of the vertical flow settling tank that can be applied. The disadvantage of using alum as a coagulant is the possibility of sulphate reduction coupled with the release of H$_2$S when the oxygen supply is insufficient. The disadvantage of hydrated lime as a coagulant is the high effluent pH together with the severe hardness of the treated water. With respect to the coagulant cost per kg COD removed, the use of ferric chloride as a coagulant, especially in small doses, is the most favourable.

The settling rate of the flocs with small ferric chloride doses is also favourable.
The application of a coagulation-flocculation system also depends on the kind of biological wastewater treatment process applied, because the coagulation-flocculation process removes mainly suspended and colloidal material, while soluble, easily biodegradable compounds remain in solution. Recent experiments have shown that coagulation-flocculation especially in combination with biological pretreatment of wastewater can be successful. (Lettinga and others, 1977).

ACKNOWLEDGEMENTS

The authors wish to thank J. van Driessum, R. de Vletter, E. Wind and J. Baar of the Centrale Suiker Maatschappij (CSM) for their kind cooperation and valuable discussions.

REFERENCES

Leentvaar,J., W.G. Werumeus Buning and H.M.M. Koppers (1978). Physico-Chemical Treatment of Municipal Wastewater. Coagulation-Flocculation. *Water Res.*, *12*, 35.

Lettinga, G. and others (1977). Anaerobic Treatment of Beet-Sugar Factory Wastewater in a 6 m^3 Pilot Plant (Dutch). *H$_2$O*, 526.

Minton, G.R. and D.A. Carlson (1973). Primary Sludges Produced by the Addition of Lime to Raw Wastewater. *Water Res.*, *7*, 1821.

DISCUSSIONS

Mitwally, H. H. (Egypt) : Did you try separating the sludge cake by technological methods before disposing of the slurry. We tried this successfully in a cane-sugar plant.

Leentvaar,J. (Holland) : No, we didn't. As advisors, we were not allowed to make process changes. We had to work with the sewage system already present in the sugar mill.

Sestini, U. (Italy) : Do you have any figures for the specific amount if nitrogen removed?

Leentvaar,J. (Holland) : It will probably be in the order of 5 mg total N/ℓ.

Erdin, E. (Turkey) : Three chemical coagulants were used in your experiments. Have you made a cost-efficiency comparison for each coagulant?

Leentvaar,J. (Holland) : The figures for the coagulant cost per kg of COD removed are expressed in Table 5. The costs are lower at a high influent COD than at a low influent COD because the removal efficiency increases. Bear in mind that the figures given are based on the Dutch price system, where lime costs less and alum costs more than ferric chloride. For alum, for example, the costs for a low dose and a high dose would be 12 ¢ and 25 ¢, respectively, per kg of COD removed.

Erdin, E. (Turkey) : Is sludge disposal a problem in the beet-sugar mill?

Leentvaar,J. (Holland) : The beet-sugar mill is one of the very few plants where it is not a problem. Of the total waste to be disposed of, less than 1% is sludge from the chemical treatment of the wastewater. This chemical sludge can easily be disposed of with the soil from washing the beets.

Yılmaz, M. (Turkey) : On what basis were the cationic and anionic polyelectrolytes selected? Could you also give information on the amounts added in your tests?

Leentvaar, J. (Holland) : Batch experiments were made to study the effect of anionic, cationic and nonionic polymers, respectively. In these tests the effect of polymer doses ranging from .1 to 5 mg/ℓ was investigated, with respect to COD removal and the volume of sludge produced.

Lo, S.N. (Canada) : Why were four reactors in series used in the laboratory-scale pilot plant?

Leentvaar, J. (Holland) : From the theoretical point of view, when a plug flow system is used, several reactors in series have a positive effect on the coagulation-flocculation performance in the case of a downward-tapered mixing intensity. Experiments with domestic sewage, however, have shown that 4 reactors are unnecessary; in fact, 2 reactors or even a single reactor can give just as favorable a removal efficiency.

Purification of Sugar Factory Wastewater in the RT—Lefrancois System

N. TAYGUN, N. ŞENDÖKMEN and G. ÜLKÜ

Seker Enstitüsü, Ankara, Turkey

ABSTRACT

The sugar industry has long been regarded as a consumer of large amounts of water and as a highly polluting industry. Recent developments in the water recirculation and wastewater treatment systems have made this industry a water-producing rather than a consuming industry, and have reduced its pollution charge greatly.

The RT-Lefrancois wastewater treatment system has played an important role in this field. By this method the wastewater can be treated in a short time and reused again as beet-flume and wash water. This is the most economical method used for the partial treatment of the highly polluted wastewater. A solids concentration as high as 20 g per liter is worked with in the reactor, and 90% of BOD_5 removal is achieved in 3-4 hours.

The nutrient requirements are BOD_5: N:P=100:3:0.5 in this system.

INTRODUCTION

In the sugar factory about 15 m^3 of water is required for processing 1 ton of beets. Formerly, the wastewater produced was discharged into a river, often without treatment, and the fresh water required for the process was taken from the aquatic environment. This caused some serious problems with regard to environmental protection and the fresh water supply from the environment. Technological developments made it possible to reuse the water by recirculation. The sugar industry has made great efforts in this field to reach the most economical solutions satisfacotry to the authorities and industry, without damage to the natural environment.

Because of these efforts, the sugar factories have decreased their water requirements from 1500 m^3 to 26 m^3 per 100 tons of processed beets.

Parallel to these efforts, new biological wastewater treatment systems have also been developed. Since 1967, all over the world natural purification of wastewater has been carried out. Water purification by this system takes a very long time (7-8 months) and large areas are needed. Because of these disadvantages, new, more economical methods for the purification of sugar factory wastewater have also been developed and put into practice.

TYPES and AVERAGE AMOUNT of
WASTEWATER USED IN THE FACTORY

The water used in the factory consists of four main water circuits. The water in a given circuit is recirculated within that circuit. Table 1 gives the types and average amount of water in these circuits.

Table 1: Types and Average Amount of Water Used in the Factory

Water Circuits	m³ water per 100 ton of beets
Process water	200-250
Barometric condenser water	400-600
Mechanical cooling water	100-130
Beet-flume and washwater	600-800

As long as the water in these water circuits is clean enough to be recirculated, it is not discharged into the environment. The amount of water lost by evaporation or by droplets is added to the water circuits using the fresh water.

With the beet-flume and wash water, the situation is a little different. In the settling pond this water loses its mud, and to discharge it to this settling pond, water from the same circuit is used. Therefore, the lost amount of water must be added to the circuit by using fresh water (See Fig. 1).

TYPES OF WATER TO BE TREATED
IN THE BIOLOGICAL TREATMENT SYSTEM

The list of the various types of highly polluted waters, their amounts and degree of pollution is given in Table 2 (Schneider, 1968).

Table 2: Type, Amount and Degree of Pollution of Water to be Treated

Types of Water	Amount m³/100 ton beet	Degree of Pollution as PC mg/ℓ	BOD$_5$ mg O$_2$/ℓ
Cloth wash water	2.5	25,000	5,000
Filter wash water	1.5	30,000	6,000
Cleaning water	2	10,000	3,500
Sanitary water	3	500	300
Carbonation-mud discharge water	5	15,000	3,000
Beet mud discharge	10	2,000	800
Boiler purge water	2	-	-

These highly polluted wastewaters amount to 26 m³ per ton of processed beets. As seen from their degree of pollution, they can not be discharged to the aquatic environment, without purification. Purification of these polluted waters is important for the following two reasons:

1. To keep the environment clean,
2. To cut down the fresh water needs of the factory

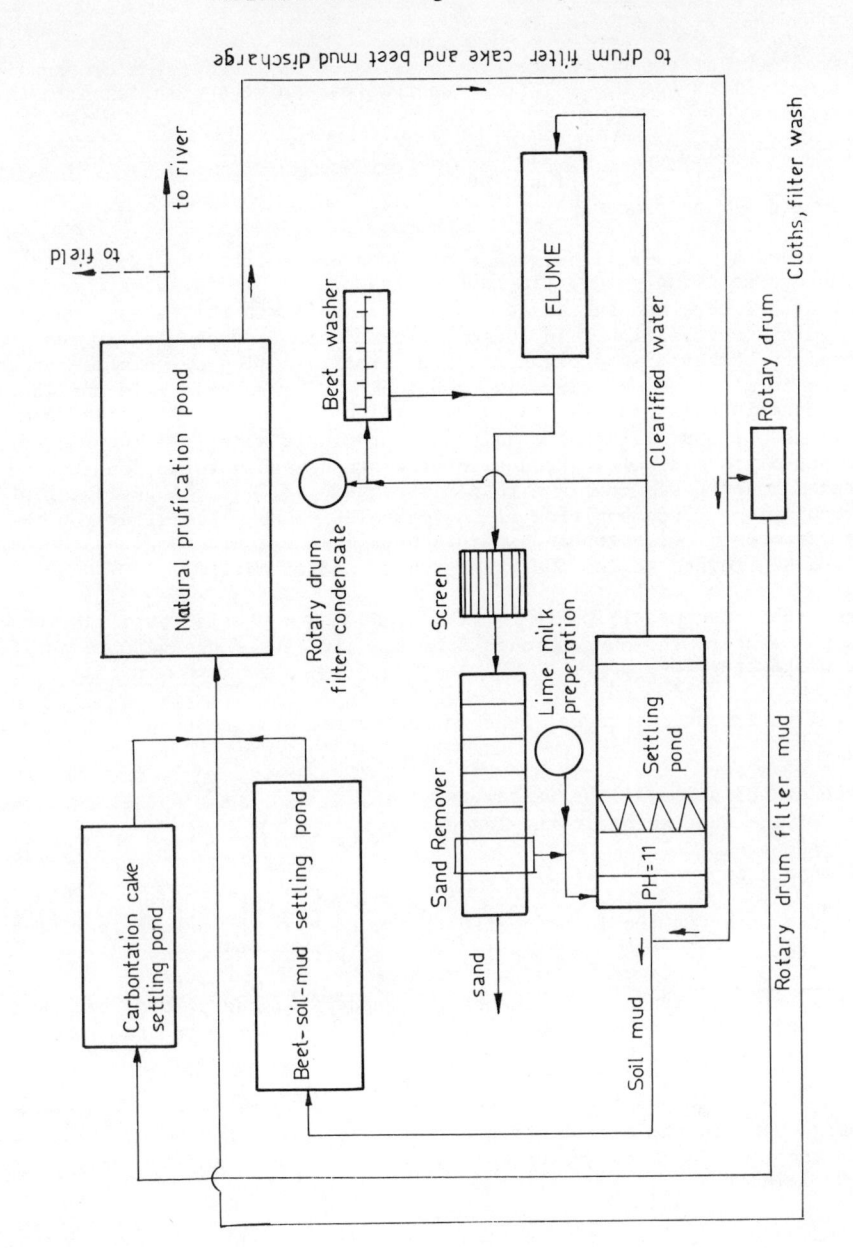

Fig. 1: Natural purification scheme of factory wastewater

Biological Treatment of the Wastewaters

Biological treatment of the wastewaters is carried out by means of microorganisms. Biological treatment of the sugar factory wastewater can be accomplished in three ways:

1. Natural purification
2. Accelerated natural purification, and
3. The Activated sludge process.

Natural purification: Non-recirculatable polluted water from the factory is collected in the mud-settling ponds. In these ponds settleable solids are separated from the polluted water, and the wastewater is collected in the lagoons to be subjected to natural purification. In these lagoons, carbohydrates in the wastewater are decomposed by bacteria into organic acids. The formation of volatile organic acids (such as formic acid, acetic acid, propionic acid, butyric acid and lactic acid) takes place in this process. The composition of the organic acids formed in this stage is as follows: 45% acetic acid, 35% propionic acid, 10% butyric acid, 5% lactic acid and 5% formic acid. Because of the organic acid formation, the pH-value of the wastewater decreases to 5 and then rises again to 7 at the beginning of the methane fermentation. Decomposition of the proteins takes place after the pH-value of the water reaches 7. Nitrogen and sulfur-containing organic compounds decompose, giving NH_3 and H_2S to the medium under anaerobic conditions.

After the anaerobic decomposition is completed, having a partial purification effect on the wastewater, the oxygen taken from the air causes the aerobic purification of the wastewater to start.

In the natural purification process, the following factors are taken into consideration:

1. The degree of pollution of the wastewater
2. The depth of the wastewater in the lagoons
3. Total surface area
4. Local climate (temperature, wind)

The depth of the water in the lagoons affects the time of the natural purification. The shallower the depth, the less time is taken to purify the water.

In the natural purification process the oxygen needed for the aerobic treatment is taken from the air. Therefore, more water surface means more air from the aerobic process.

Temperature plays an important role in the bacterial growth. A rise in the temperature causes a rise in the purification effect. To obtain the most satisfactory results, the temperature of the water must not be below +5°C and the depth of water must be about 1 meter.

In Fig. 1 the natural purification system is shown. In this system waste water purification takes a long time, but there is no need for nutrient salts, competent personnel and electrical supplies.

Accelerated Natural Purification (Oxydation Ponds): In this system the wastewater in the oxidation ponds is aerated mechanically. By doing so, vile odor formation is prevented. It takes from 20 days to 2 months to purify the wastewater in these oxidation ponds. The depth of the water in the ponds can be between 1 and 3 meters. Hence, the area used for this purpose is less than ½ to ¼ of the area used in the natural purification system (Schlanitz, 1972). For the aeration of the water, compressed air or surface aerators can be used.

Activated Sludge Process: Biological treatment of the wastewater can occur in a short time by using activated sludge. The microorganisms in the activated sludge need sufficient amounts of carbon, nitrogen and phosphate for their growth. In the sugar factory wastewater, there are enough carbon compounds, but there are not sufficient nitrogen and phosphate salts for their growth. These salts must be added externally to the wastewater.

In the activated sludge process, to achieve satisfactory treatment, the following ratio of values must be kept constant: $BOD_5/N/P = 100/5/1$.

Since the sugar factory wastewater contains a great amount of sucrose (sugar), the direct aerobic treatment of the wastewater is not possible. Sphaerotilus natans growth prevents the attainment of a good settleable (flocked) activated sludge. Hence, before the aerobic treatment of the wastewater, an anaerobic treatment process must take place for the decomposition of the sugar into the organic acids.

Treatment wastewater in the RT-Lefrancois System: This system was developed by the sugar factory at Tirlemont, Belgium and applied by the Dormagen and Wevelighofen factories in Germany. Since 1975 these factories treated their wastewaters using this system.

In the Dormagen sugar factory, the pH of the beet-flume and wash water was kept constant at 11 by adding solid CaO. This water was separated first from the sand in it, by the use of sand removers, and then transferred to the settling pond to remove the beet-mud from the wastewater. The beet-mud taken from this pond was sent to the beet-mud settling pond. Into the same pond the carbonation cake-mud was also pumped. This is shown in Fig. 1. The water leaving the pond is then passed into the RT-Lefrancois purification reactor.

The advantages of this system were as follows:

- In the fermenter (reactor), air, activated sludge and water were mixed very thoroughly.
- Because of this good mixing, oxygen dissolved very rapidly in the water.
- Microorganisms that get sufficient oxygen in this way decomposed more organic compounds.
- Even highly polluted water ($BOD_5 \sim 10,000$ mg O_2/ℓ) can be treated in this way.
- The system could be operated with a high activated sludge concentration in the reactor (11-16 g activated sludge per liter).
- Aeration time is very short, 3-4 hours.
- BOD_5 removal efficiency was 90%.
- The energy requirement need is between 0.55-0.75 kWh/kg BOD_5
- The nutrient (N and P) requirements are low.
- The area required for the fermenter is small (500 m^2)
- Effluent leaving the clarifier can be reused as flume and wash water back in the factory.

The beet flume and wash water recirculated in the beet-flume and wash water circuits gets very polluted in a short time. Its BOD_5 value reaches to 5,000-10,000 mg/ℓ in one week.

This polluted water caused some difficulty in transporting the beets in the factory. To treat this water and to make it reusable, two procedures could be applied:

1. Discarding some part of it and adding fresh water to dilute the main water body to a pollution level which allows it to be used again in the factory, or

2. Treating some part of it and adding the same treated portion to the main water circuits.

The latter procedure is more useful since no discharge of the polluted water and no fresh water supply are required. To make this procedure possible, a new purification system had to be developed. The RT-Lefrancois wastewater treatment system was used. The water gained from this treatment system is not pure enough to discharge into a river, but is good enough to be used as the flume and wash water in the factory.

In 1975 from Sept. 29 to Nov. 14, the following study was carried out using the RT-Lefrancois system in the Dormagen Sugar factory.

The plan of the system is shown in Fig. 2. The RT-Lefrancois system has the following parts:

(a) Fermenter. This has a volume of 240 m^3. It consists of two cylindrical tanks, one inside the other. The outside tank has a diameter of 8.7 and a height of 8.1. The inside tank has a diameter of 8.1 m and a height of 3 m. Between these two tanks there is a hole 30 cm long all around. In the middle of this hole there is an air pipe-line going around the inside tank. The air enters the fermenter through this air pipe-line and recirculates the wastewater and activated sludge mixture thoroughly in the fermenter (See Fig. 2). The temperature in the fermenter is kept around 20°C. The influent, before being passed into the fermenter, is mixed with N and P nutrient salts, and the recycled activated sludge enters the fermenter center at the bottom. The effluent leaves the fermenter from the inside tank at the half height.

This outlet position can be adjusted to 4-5 m. The detention time can also be adjusted to between 2-3 hours. The activated sludge concentration in the fermenter can be kept between 12-15 g solids per liter.

(b) Clarifier. This has a conical bottom portion to collect the settled activated sludge, and has a usable volume of 500 m^3. The effluent leaving the fermenter is passed into a cylindrical column at the center of the clarifier. The settled activated sludge is taken from the conical bottom of the clarifier to circulate back to the fermenter. The excess sludge is returned to the mud settling pond. The clarified water leaves the clarifier from the top level water-groove.

The level of the activated sludge in the conical part of the clarifier is controlled very often to prevent excess sludge formation in the clarifier because the excess sludge may decompose anaerobically in the clarifier, causing some pollution in the effluent.

(c) Air Compressor. This has a capacity to pump 10,000 m^3 air in an hour. And the pressure of the air is approximately 0.5 kg/cm^2.

The air taken from the compressor is used in the fermenter to obtain a homogenized mixing and to supply enough oxygen for bacterial growth.

(d) Preheater. This is a heat exchanger. It is used to keep the water temperature in the fermenter around 20°C. The condensate water from the factory is used to preheat the wastewater in the preheater.

To reach the satisfactory growth rate of the bacteria in the activated sludge, nutrient salts containing N and P are used. For the N and P sources, urea and ammonium phosphate, respectively, are used.

The bacterial composition can be formulated generally as follows:

$$C_{106} H_{180} D_{45} N_{16} P$$

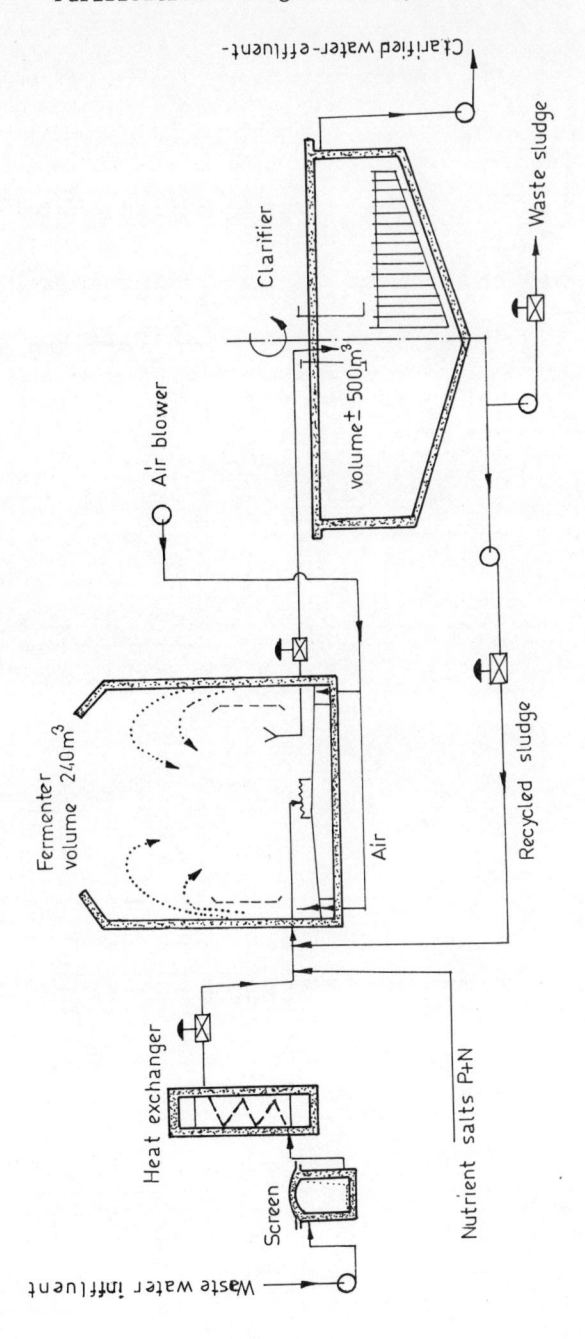

Fig. 2: RT-Lefrancois biological wastewater treatment system

Table 3: Data Obtained from the RT-Lefrançois Water Purification System in Dormagen Sugar Factory (Campaign 1975)

DATE	Time	INFLUENT m³/h	pH	COD	BOD₅*	FERMENTER Tem. °C	pH	SS**g/ℓ	EFFLUENT pH	COD	BOD₅	RECYCLE SLUDGE m³/h	SS g/ℓ	WASTE SLUDGE m³/h	pH	REMOVAL EFFICIENCY %COD	%BOD₅	AMOUNT OF WASTE REMOVED COD kg/h	BOD₅ kg/h
11/10/75	7.30	–	5.0	7772	4827	–	–	–	5.3	6323	3927	–	–	–	–	18	18	14	9
12/10/75	8.30	–	5.2	–	–	–	–	–	6.3	–	–	–	–	–	–	–	–	–	–
13/10/75	9.00	10	5.3	7447	4626	–	–	–	–	2286	1731	–	–	–	–	69	62	51	29
14/10/75	13.00	25	5.3	6664	4137	–	–	–	7.5	1960	1607	–	–	–	–	70	73	141	91
15/10/75	8.30	30	6.2	6078	3775	–	–	–	7.7	1487	890	–	–	–	–	75	76	206	130
16/10/75	10.00	60	6.4	5949	3695	26	–	11.1	7.8	1358	813	75	–	–	–	77	78	275	173
17/10/75	8.00	60	6.4	6047	3756	26	–	13.5	8.0	1435	859	75	–	7	–	76	77	277	174
18/10/75	9.00	70	6.6	6111	3759	25	–	–	8.0	1273	826	84	–	2	–	79	78	290	178
19/10/75	8.00	70	6.5	6150	3819	22	–	–	7.9	1650	988	84	–	5	–	73	74	270	170
20/10/75	8.00	60	6.5	6275	3897	21	–	–	7.7	1979	1118	74	–	9	–	68	71	258	167
21/10/75	8.00	60	6.6	6340	3937	23.5	–	–	7.7	1460	874	70	–	6	–	77	78	293	184
22/10/75	8.00	60	6.4	6340	3937	23	–	–	7.8	1378	825	70	–	15	–	78	79	298	187
23/10/75	8.00	60	5.9	6507	4041	23	–	–	7.7	1225	561	75	–	–	–	81	86	317	209
24/10/75	8.00	60	5.6	6741	4186	23.5	–	–	7.9	1221	560	75	–	6	–	82	86	331	217
25/10/75	7.30	60	6.8	7197	4470	23.5	–	–	7.8	1481	886	75	–	–	–	79	80	336	215
26/10/75	10.00	60	5.5	7386	4587	24.5	–	–	7.8	1784	1007	75	–	–	–	76	78	336	215
27/10/75	7.20	60	6.5	7216	4481	24	–	–	7.7	1810	1022	75	–	–	–	75	77	324	207
28/10/75	7.45	60	6.8	7250	4503	22	–	–	7.8	1920	1084	75	–	9	–	73	76	320	205
29/10/75	7.45	60	6.8	7425	4612	21	7.5	–	7.7	2227	1258	75	–	–	7.1	70	73	312	201
30/10/75	7.45	60	6.9	7620	4733	21	7.5	–	7.6	2833	2038	75	–	–	7.1	63	57	287	162
31/10/75	7.30	60	7.1	7840	4869	–	7.5	17.5	7.7	2799	1581	66	23.6	–	7.0	64	67	302	197
1/11/75	7.45	60	7.1	8100	5031	23	7.1	18.3	7.7	2371	1339	68	26.1	6.9	6.9	71	73	344	221
2/11/75	10.00	60	7.1	8389	5210	22	7.1	18.5	7.6	2826	2033	68	27.6	6.9	6.9	66	61	334	191
3/11/75	7.30	60	7.2	8350	5186	22	7.1	18.0	7.7	2874	2067	63	24.2	6.9	6.9	65	60	328	187
4/11/75	7.30	60	7.1	7873	4890	23	7.1	18.1	7.7	2497	1410	75	27.8	7.0	7.0	68	71	322	208
5/11/75	7.30	60	7.1	6647	4128	23	7.1	17.8	7.7	1930	1091	73	26.1	–	6.8	71	73	283	182
6/11/75	7.45	60	7.1	6819	4235	22	7.8	18.7	7.8	1524	912	75	28.6	–	7.2	78	78	317	199
7/11/75	9.00	60	7.2	7655	4754	23	7.8	19.6	7.9	1428	855	75	30.0	10	7.2	81	82	373	234
8/11/75	7.30	60	7.2	7591	4715	23	7.8	19.8	7.9	1447	866	75	27.6	10	7.3	81	82	368	231
9/11/75	8.30	60	7.2	7398	4595	22.5	7.8	19.0	7.9	1505	901	76	30.5	10	7.4	80	80	353	221
10/11/75	7.40	60	7.1	7400	4596	22	7.9	18.0	7.6	1512	905	76	28.2	10	7.5	80	80	353	221
11/11/75	7.10	60	7.2	7273	4517	22	7.8	18.0	7.9	1531	917	76	27.2	10	7.4	79	80	341	216
12/11/75	6.40	60	7.3	7273	4517	22.5	7.7	17.4	7.9	1684	1008	72	26.7	10	7.2	77	78	335	210
13/11/75	7.30	60	7.3	7409	4602	22	7.9	16.1	8.0	1679	1005	70	25.4	10	7.4	77	78	344	216
14/11/75	7.30	60	7.3	7950	4938	21.5	8.0	–	8.0	1679	1005	72	–	10	7.6	79	80	376	236
15/11/75	7.40	60	–	8268	5135	–	7.9	–	–	1831	1034	–	–	10	7.7	78	80	386	246

*BOD₅ Calculated values (Reinefeld, 1975)
**SS Solid substance

and the relation between the BOD_5 pollution value and the N, P values for the sugar factory wastewater can be expressed as follows:

$$BOD_5/N/P = 100/5/1$$

This means that for the decomposition of 1000 g of BOD_5, 50 g of nitrogen and 10 g of phosphorus are required. This ratio is much more economical in the RT-Lefrancois purification system, as seen below:

$$BOD_5/N/P = 100/3/0.5$$

By controlling the BOD_5 values of the influent, the required amount of N and P salts can be added by a dosing pump in the influent pipe-line.

The activated sludge used in the fermenter can be supplied by the domestic water treatment plant. For the RT-Lefrancois purification plant in the Dormagen sugar factory, 25 m^3 of activated sludge from the nearby domestic water treatment plant was taken.

On the first day, 25 m^3 of activated sludge and 10 m^3/h of wastewater, with the required amount of nutrient salts added were aerated at $20°C$ in the fermenter. To acclimatize the bacterial growth to the new medium, the wastewater supply to the fermenter was adjusted in the following four days, as noted below:

1st day, 10 m^3/h: 2nd day, 20 m^3/h; 3rd day, 40 m^3/h; and and 4th day, 60 m^3/h.

In the meantime, the solids concentration in the fermenter was controlled. In one week the solids concentration reached 15 g/ℓ in the fermenter, so normal working conditions in the fermenter were obtained.

After reaching these conditions, the normal continuous wastewater purification took place in the fermenter, and the resulting excess activated sludge was removed from the system.

By fermentation of 100 kg of BOD_5 pollution, about 60 kg of activated sludge-solid substances was obtained. Since the activated sludge concentration increased as a result of this process, the activated sludge level in the clarifier conical part had to be kept around 2 m high. The activated sludge concentration in the fermenter should not exceed 15-20 g solids/liter to achieve a satisfactory purification effect.

Analysis: In order to check the working conditions of the system, and to make the necessary adjustments, the necessary analyses were conducted and the results shown in Table 3 were obtained. In Fig. 3, the purification process run is shown graphically.

For a comparison, the technical data given for this system by the RT-Tierlemont Company is also shown in Table 4.

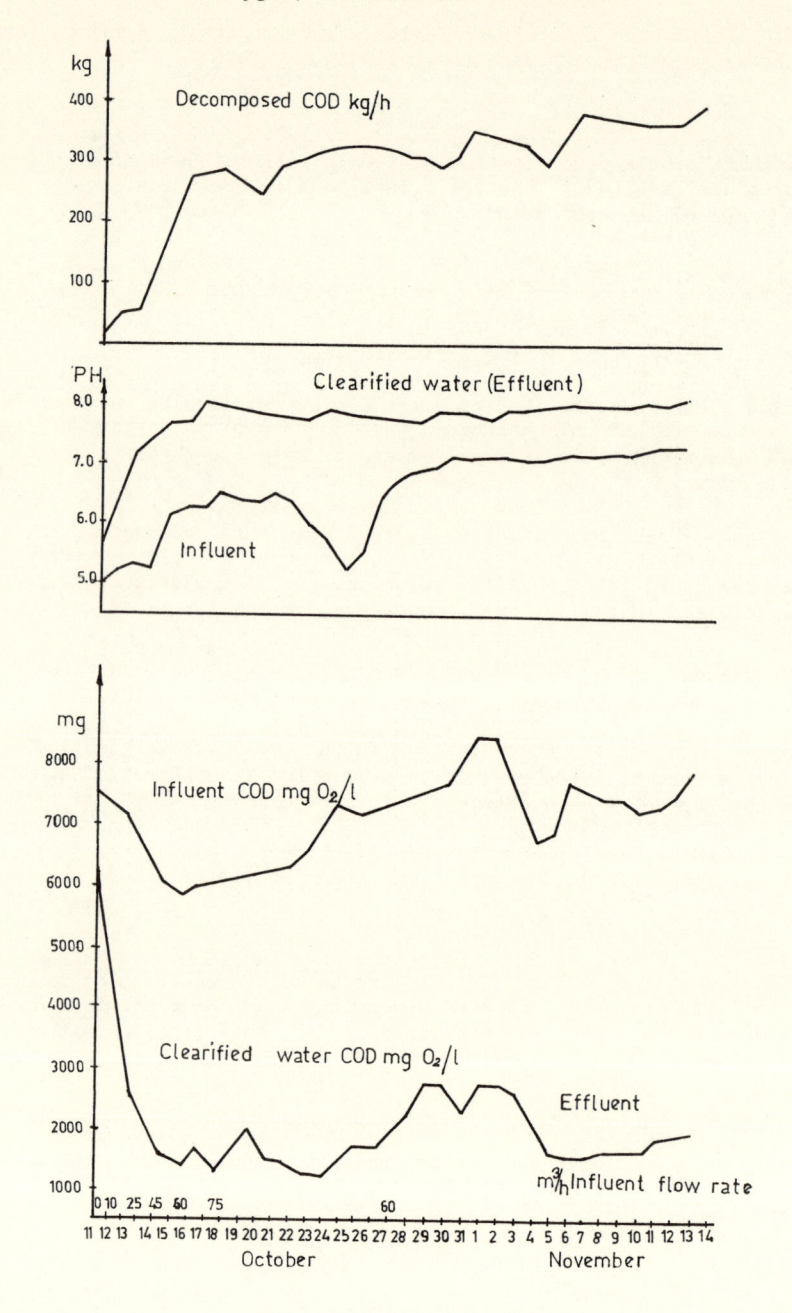

Fig. 3: Graphs showing purification process of wastewater in the Dormagen Sugar Factory

Table 4: Technical Data Given for the RT-Lefrancois Biological
Wastewater Treatment Plant by the RT-Tirlemont Company

```
Influent flow-rate                            :  60-70 m³/h
Recycled activated sludge flow-rate           :  75-90 m³/h
Duration time in the fermenter                :   3-4 h
Daily total COD-value of the influent         :  9000-11,000 kg COD/d
Daily total BOD₅-value of the influent        :  7000- 9,000 kg BOD₅/d
Daily total COD removal (clarified sample)    :  7100- 9,000 kg COD/d
Daily total COD removal (centrifuged sample)  :  8000-10,000 kg COD/d
Daily total BOD₅ removal (clarified sample)   :  5900- 8,100 kg BOD₅/d
Daily total BOD₅ removal (centrifuged sample) :  6300- 8,700 kg BOD₅/d
```

Average removal efficiency.

For COD-values (clarified sample)   : 80%
For COD-values (centrifuged sample) : 91%
For BOD₅-values (clarified sample)   : 90%
For BOD₅-values (centrifuged sample) : 96%

Energy requirements.

For COD -values (clarified sample)   : 0.60 kWh/kg COD
For COD -values (centrifuged sample) : 0.55 kWh/kg COD
For BOD₅-values (clarified sample)   : 0.70 kWh/kg BOD₅
For BOD₅-values (centrifuged sample) : 0.65 kWh/kg BOD₅

Amount of Nutrient Salts

18 g of P_2O_5/kg BOD₅

40 g of N/kg BOD₅

Sludge Load

1.5 - 2.5 kg BOD₅/kg solid/d

Fermenter Volume Load

15-35 kg BOD₅/m³ · d

Activated sludge concentration in fermenter (Biomass) 10-15 kg Solids /m³
Solids formation in fermenter 60-70 kg/100 kg BOD₅

REFERENCES

Reinefeld, E. and H.P. Hoffman-Walbeck (1975). Analytische Untersuchungen an Zuckerfabrikabwassern, insbesondere uber den CSB-Wert von Stapelteichwassern. *Zucker, 28,* Nr. 4. 165.

Schlanitz F. (1972). Gedanken zur totalen Losung des Abwasserproblems in Zuckerfabriken. *Zucker, 25,* Nr. 13, 427.

Scheider, F. (1968). *Technologie des Zuckers*. Germany.

Simonart, A. J. P. Dubois and R. Peck (1976). Die Verwendung ammoniakhaltiger Wasser fur die biologische Abwasserreinigung RT-Lefrancois. *Zuckerindustrie, 26,* Nr. 3, 189.

DISCUSSIONS

Buning, W. (Holland) : In the Netherlands problems with bulking sludge have been encountered when beet sugar wastewater is treated aerobically. What were the characteristics of the sludge in your fermenter?

Taygun, N. (Turkey) : The sludge in the fermenter was completely normal, as were the flocs and the settling rate. No mould, in other words, no growth of sphaeritilus natans was encountered. The amount of oxygen in the fermenter was low, around 0.2 - 0.3 mg/ℓ.

Case Study of Rice Starch Waste Treatment

S. G. SAAD

Department of Environmental Health, High Institute of Public Health,
Alexandria University, Egypt

ABSTRACT

Waste from the starch and glucose industry is characterized by heavy pollutional organic loads. This research covers the various treatment methods which can be used to recover the organic colloids for further processing as animal feed. Coagulation with $FeSO_4$ and polyelectrolytes followed by carbon adsorption under various coagulating and adsorption conditions were studied with the aim of reducing the waste characteristics to the permissible limits for wastes discharged to lakes.

INTRODUCTION

Starch processing is a fifty-year old industry in Egypt. This industry is a rather unique process, as there are relatively very few starch production processes around the world which start from low-grade rice as a basic raw material.

Corn starch is the most common and widespread product. Nemerow (1971) characterized the wastes from starch processing according to their high polluting loads and heavy contents of suspended colloidal gluten, soluble organic matter and starches. Recovery systems were developed, which proved of economic advantage to the plants using them and also helped to reduce the stream pollution, as described by Hatfied (1953). The problem of recovery or prevention of wastes from a starch-processing plant is an old one, regardless of the starting raw material.

Bisset (1976) conducted a study to determine the effects of potato starch wastes on trout at different concentrations. The major toxic agents were found to be the high organic loads, cleaning and disinfecting compounds, ammonia and hydrogen sulfide generated during decomposition of the wastes. Biological treatment rendered these wastes non-toxic.

The Egyptian Starch and Yeast Company in Alexandria has one of its processing plants at Moharrem-Bey. Though this plant is an old one with rather outdated technology, it produces 400 per month of rice starch and 600 per month of glucose from part of the processed starch.

The plant water supply is taken partly from the Mahmoudia Canal. About 1200 m^3/d is withdrawn for cooling purposes. The processing water is taken from the city water at a rate of 800 m^3/d.

The steeping wastewater and the primary concentration process waste coming out of the centrifuges are the main sources of pollution in the starch production process. Their organic loads are very high and capable of raising the final effluent to a high pollutional level although they comprise less than 30% of the total waste. The remaining 70% comes from the cooling water and in-plant cleaning operations, which cause a relatively small amount of pollution. The glucose-processing plant is a completely closed cycle, with no liquid wastes other than those from cleaning operations.

In this research the final effluent and seperate process waters are subjected to various treatment methods, and recommendations are made regarding the reduction of the pollutional load which enters Lake Maruit from the Moharrem Bey plant, thus contributing to the deterioration in the water quality of the lake.

As there is a shortage of animal feed, the solids seperated from those wastes which are rich in gluten (plant protein) and starches as well as glucose, can be further processed to give a reasonably cheap cattle feed, by adding fibrous matter seperated from the waste treated in the adjacent paper re-processing plant.

The available area for the construction of a complete treatment unit is limited; consequently, the proposed treatment facilities have to use the minimum space required.

DESCRIPTION OF THE STARCH PROCESSING

With low-grade rice, vibrating screens are used to remove most of the foreign matter. The clean broken seeds are placed in caustic soda solution (0.8-10 Be^0) for a period of 10-12 hours to dissolve the proteinaceous matter. The effluent from this process is characterized by high pH values and high contents of suspended and soluble matter. This is termed the steeping process, after which the swollen starch seeds are mechanically reduced in size and water is added to form a slurry. This is further screened, and the rejected larger starch particles are returned to the mill for further grinding. The milk of starch produced from this screening process is then concentrated by means of centrifugal force and the wastewater discharged to the sewer. The thicker slurry thus produced is concentrated still further in a horizontal or vertical basket centrifuge, where a multi-layer cake is formed.

After removal of the cake, the starch layer is separated from the ash and precipitated proteins, to be further dried on trays in a tunnel drier.

MATERIALS AND METHODS

Sampling

24-hour discrete samples were collected every hour to determine the characteristics of the final effluent. These samples were therefore taken from the drain discharging the final wastewater from the whole processing plant. All samples were used promtly after collection as storing drastically alters their chemical characteristics. Representative samples from the batch steeping and continuous concentration processes were also collected to evaluate the amount of seperated solids rich in gluten and starches to be furthe used in the processing of cattle feed.

Characterization of Samples

The raw samples before treatment were characterized by pH, turbidity and COD, as well as total and volatile residues. Tests were carried out according to the procedures listed in Standard Methods (1971). Total organic carbon (TOC) analyses were performed to confirm the results obtained from the best treatment conditions. A Beckmen 915A TOC Analyser was used for all TOC analyses. The discrete 24-hour samples were characterized by temperature, pH, settleable solids, total, volatile, fixed, suspended and dissolved residues. Chlorides, sulfates, total alkalinity and hardness, total phosphates, all nitrogen parameters, as well as COD, BOD and dissolved oxygen contents were tested. TOC analyses were performed to determine the carbon contents of the waste.

Determination of Suspended Residue (S.R.)

In this type of waste, determination of suspended residues following the standard procedure was rather difficult due to its high colloidal concentration.

A modification of this procedure was carried out by centrifuging a 10 ml sample at 10,000 rpm for 3 min or until a clear supernatant was obtained. The suspended solids in a compact layer from were separated at the bottom of the centrifuge tube, thus allowing an easy separation of the supernatant liquid; the decanted solids were washed and transferred with distilled water to a preweighed crucible, where they were left to dry at 105°C, and their amount was determined in mg/l.

Treatment of Final Waste

1. a) The effect of $FeSO_4$ locally processed at the Kafr El-Zayat Industrial and Financial company was tested. The raw waste pH was in the range of 10-11 which is known to be the best for coagulation with $FeSO_4$, as indicated by Fair, Ieyer and Okun (1968). Doses applied were 100, 200, 300, 400, 500 and 600 mg/l. These were tried only on the final wastewater to be discarded.

   b) Acidification at different pH levels from 2-7 was carried out by the addition of a dilute solution (10%) of sulfuric acid. The acid was added while stirring till the inserted combined electrodes of the pH meter indicated the proper pH. Acidification was carried out in a one-liter beaker for each pH level. The Jar Tester, method was used, the six beakers being stirred for 3 min at 20 rpm and then left to decant for half an hour. Flocculation time and speed were optimized after trial and error procedures to promote the formation of large flocs rather than their cleavage at higher speeds or during a prolonged stiring time.

   c) Polymer Coagulation

   The following list of polymers were added to the final waste after acidification and reduction of the pH to 3. This was found to be the best value for maximum precipitation of colloids to occur.

Type of Polymer	Commercial number
Anionic	Hercules 816.2
	Hercules 831.2
	Hercules 819.2
Cationic	Hercules 844
	Hercules 829.3
	Hercules 859
Nonionic	Hercules 824.3

   Polymer solutions were prepared according to the specifications of the manufacturing firm, by dissolving 1 g of the polymer in 1 ℓ of tap water. This was placed in a beaker, put on a magnetic stirrer and maintained at a sufficiently high revolving speed to creat a vortex. The polymer was powdered slowly at the shoulder of the vortex. Stirring was continued until complete dissolution of the polymer had taken place. The polymer solution was left to age for 24 hours before use and discarded one week after preparation.

d) Carbon adsorption following pH adjustment and separation of the precipitated solids was also carried out to remove part of the soluble organics not removed by precipitation and or polymer coagulation. A 0.1% dose of very finely pulverized active pow carbon was added and stirred with the samples for half an hour at 25^0C and a speed of 30 rpm., after which it was separated from the samples by filtration. Chemical analyses were performed on the supernatant filtrate.

Treatment of Steeping Waste

This waste had a high concentration of suspended solids. Samples were acidified at different pH levels from 2-6. The supernatant, after operation of the Jar Tester, was left to settle. A volume of 10 ml of the supernatant was withdrawn&centrifuged at 10,000 rpm for 3 min to asses the effectiveness of this technique in eliminating most of the remaining suspended solids. To 500 ml of the supernatant, a dose of 3 mg/l nonionic polymer (Hercules 824.3) was added to study the effect of polymer addition on further coagulation of the suspended and dissolved organic solids.

Treatment of Concentration Waste

The concent of soluble and colloidal organic matter in this waste was lower than that in the steeping waste. The effect of acidification at different pH levels followed by polymer (nonionic H. 824.3) addition was investigated.

Coagulation Tests

A six-paddle Jar Tester was used to evaluate the effect of various pH level adjustments on the removal of suspended and colloidal matter from the separate and final wastes. A rapid mixing of the waste at 100 rpm for 1 min was followed by slow mixing at 20 rpm for 2 min. The treated wastewater was left to decant for half an hour. The supernatant was withdrawn for chemical analysis and evaluation of the organic matter removal efficiency.

Testing Procedures for the Precipitated Wastes

The effect of the various precipitating and coagulating agents was tested in a specially designed plexiglass column, 8 ft high and 1½ ft in diameter, fitted with a variable speed stirrer and side openings for the with-drawal of samples at six points, 1 ft apart.

The column was also fitted with a drain valve to discharge the sedimented solids at the bottom and was filled by means of a submerged pump placed in a tank where the starch samples were collected. The samples were poured from an opening at the top of the column to ensure uniform mixing of the solids in the samples.

The settling characteristics were tested using the treatment conditions based on the Jar Test results. The speeds of flash mixing and flocculation were similar to those used in the Jar Tester. Higher speeds were not tried, to minimize the changes in the flocculated particles formed directly after pH adjustment. Suspended residue measurements were the describing parameter for the collected samples at different heights and settling times.

RESULTS AND DISCUSSIONS

Analysis of the final wastewater discrete samples (Fig. 1) shows that the waste contained a high concentration of organic matter. The suspended residue contents were rather high and difficult to settle due to their colloidal nature. The peaks of the total residue plot indicate the batch discharge of the steeping waste. The increase in the TOC, COD

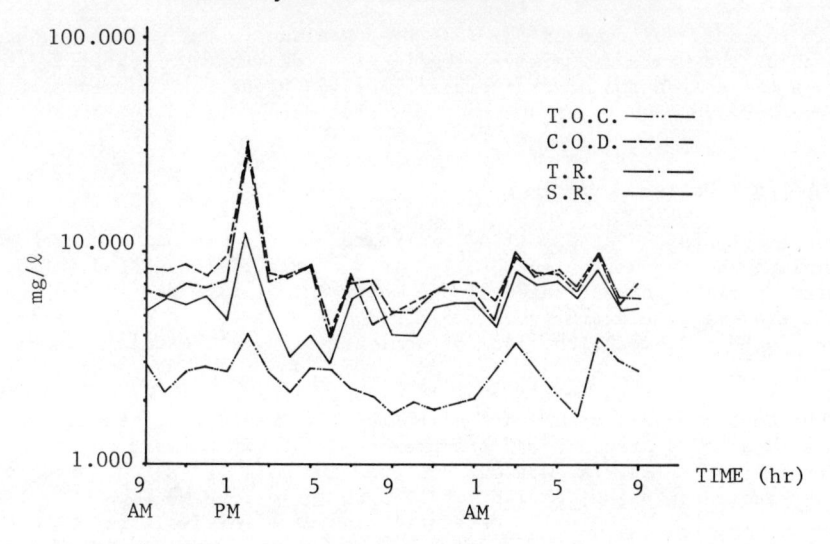

Fig. 1: Typical characteristics of wastewater from a starch plant.

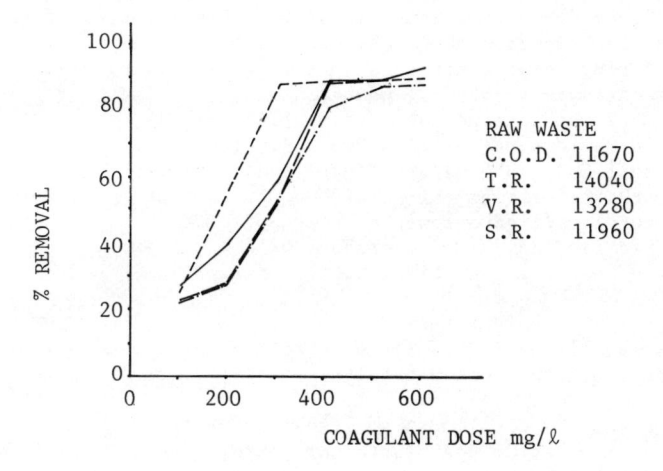

Fig. 2: Effect of FeSO₄ on final waste.

and volatile residues observed at the same time was due to the organic nature of the waste. The pH of the waste was always alkaline at a pH range of 9.5-11.0, if measured immediately after collection of the samples. Acidic values were obtained if the pH measurements were delayed overnight due to the active decomposition of glucose and starches to acids.

Treatment of Final Waste

The studies of Helmers and others (1951) indicated that a minimum nitrogen content of 7% and a minimum phosphorus content of 1.2% in the volatile suspended solids were needed to support an efficient activated sludge unit. The final waste in the present study did not on the average maintain those two minimum limits, being a nutrient-deficient waste. Based on this finding, the idea of activated sludge treatment was excluded.

Coagulation with $FeSO_4$ at the various applied doses is shown in Fig. 2. At a dose of 600 mg/l, the best removals of COD, total residue, volatile residue and suspended residue were achieved, but only drawback was the presence of brown coloration indicating an excess of soluble iron salts. The optimum dose was 400 mg/l, at which the coloration was minimum. The remaining values of 1080 mg/l, 2500 mg/l, 1940 mg/l as COD, total residue and volatile residue, respectively, were still too high for discharge into the heavily polluted Lake Maruit. This procedure, although cheap, did not provide the necessary level of treatment.

Acidification of the waste at different pH levels from 2-7 gave reasonably good removal percentages, as shown in Fig. 3. The best removals of all the indicated parameters were achieved at pH 3. The average values were determined on the basis of three different sets of data. At higher initial raw waste concentrations, better removals were obtained by acidification to any pH, the maximum removal being at pH 3.

At the most favourable pH, the waste was found on the average to have a COD of 1078 mg/l and a total residue of 1362 mg/l, more than 90% of which was volatile residue. This waste was still concentrated and could neither be discharged nor reused in further processes in the plant. Its use in cooling purposes might lead to precipitation and the accumulation of organic matter in the condensers. Further coagulation and dissolved solid removals were needed. Coagulation with Alum or $FeCl_3$ was not feasible as the waste was acidified to a pH below that considered best for coagulation with these two coagulants. The different polyelectrolytes applied gave varying degrees of removals. The nonionic polyelectrolyte (H.824.3), when applied, gave the best removal percentages A final COD of 300 mg/l was attained at a dose of 1 mg/l. The total residue, volatile residue and suspended residue values were 900 mg/l, 1,400 mg/l and 30 mg/l respectively compared with 1200 mg/l COD, 1355 mg/l of total residue and 350 mg/l of suspended residue, the values characterizing the acidified waste at pH 3. Anionic polyelectrolyte (H-816.2) at a dose of 1 mg/l gave lower percentage removals than the other two anionic polyelectrolytes. This finding is in accordance with the results of Novak and O'Brien (1975), who indicated that anionic polyelectrolytes functioned better at a near neutral pH range.

Cationic polymers generally performed better than the anionic polyelectrolytes but still with lower efficiency than the nonionic polyelectrolytes, as shown in Fig. 4.

In practically all cases of polyelectrolyte treatment, higher doses than the indicated optimum dose did not give better coagulation or improve the removal efficiencies. According to the theory of polymer bridging as stated by Stumm and O'Melia (1968), optimum destabilization occurs when only a fraction of the available adsorption sites on the surface of the colloidal particles are covered. Dosages which are sufficiently large to saturate the colloidal surfaces produce restabilized colloids, since there will be no sites available for the formation of polymer bridging. Prolonged stirring can destabilize the aggregates formed, by breaking the polymer-colloidal surface bonds and by folding back the extended segments on the surface of the particles, thus increasing their volume

Case Study of Rice Starch Waste Treatment

Table 1 Summary of Polyelectrolyte Performance at pH 3 on the Final Waste

Polymer No	Type	Best dose mg/ℓ	COD R[1]	Total Residue R	Volatile Residue R	Suspended Residue R	Total Solids R
H 844	Cationic	1	96.0	90.5	93.6	98.7	98.8
H 829.3	Cationic	3	94.0	90.1	89.2	98.0	98.0
H 859	Cationic	3	93.2	90.0	98.3	89.0	98.5
H 824.3	Nonionic	1	98.0	94.0	98.0	97.0	99.5
H 816.2	Anionic	1	94.9	94.4	93.0	95.2	98.6
H 831.2	Anionic	2	93.1	94.2	92.0	92.6	98.3
H 819.2	Anionic	2	97.2	95.3	93.0	96.1	99.2

[1] R = Removal percentage.

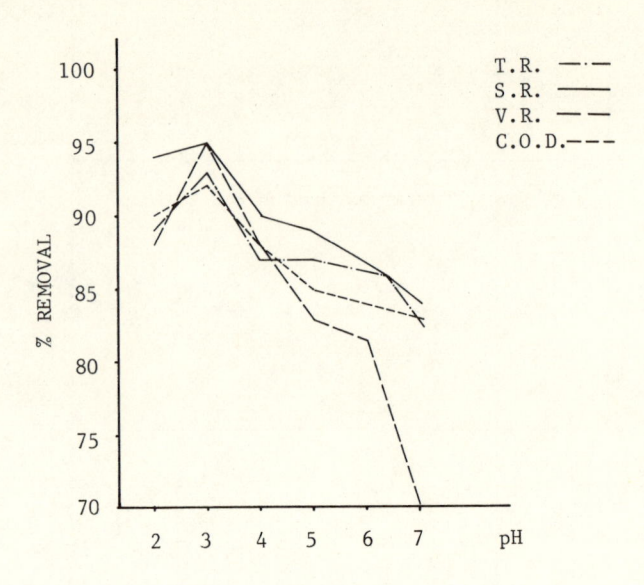

Fig. 3. Effect of Acidification on Final Waste

to weight ratio, which will enhance their buoyancy rather than cause settling.

In the present study, a speed of 20 rpm for two minutes was found to give the optimum floc formation. A prolonged flocculation time led to the destruction of the aggregates. This was also supported by the findings of Bugg, King and Randall (1970).

Carbon Adsorption

A 0.1% dose of very finely pulverized active carbon was added at different pH levels. Best removal efficiencies were obtained at pH 3, at which maximum removals by acidification were achieved. This treatment was carried out by adding the carbon dose in thick slurry form to the final waste after acidification and decantation. The mixing of the decanted waste for half an hour at pH 3 resulted in 99% COD, 95.5% total residue and 98.2% volatile residue removals (fig. 5), starting from respective initial values of 10,000 mg/l, 10,920 mg/l and 9,800 mg/l. The suspended residue, 5960 mg/l, was completely removed by passing the waste through a filter press after carbon addition. Carbon adsorption and the filter press technique are not an innovation for this plant as the same techniques are already in use in glucose processing.

After this treatment, the effluent, with a final COD of 100 mg/l and TOC of 35 mg/l,

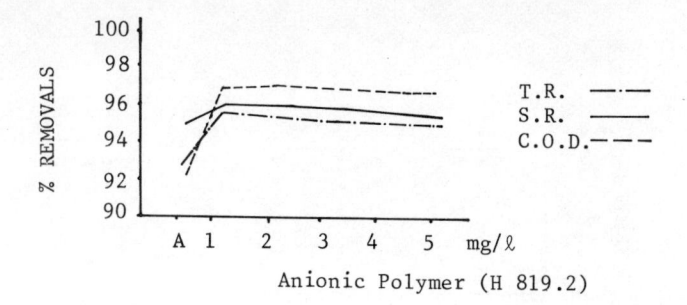

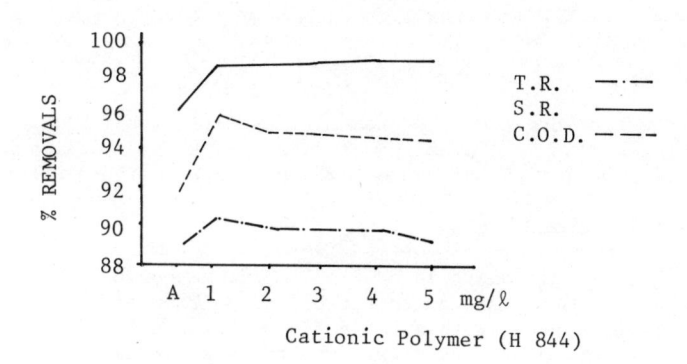

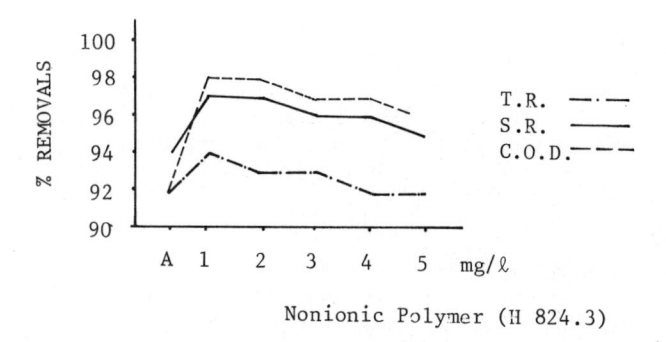

Fig. 4. Effect of Polymer Addition on Final Waste

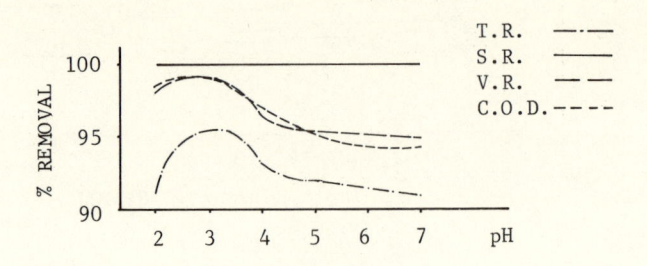

Fig. 5. Effect of Carbon Adsorption of Acidified Final Waste

can be safely discharged into the lake with minimal pollutional hazards. This waste can also be recycled for cooling and cleaning operations with practically no restrictions, its quality comparing well with the Mahmoudia Canal water withdrawn for cooling.

Steeping Wastewater Treatment

The rate of discharge of this waste is about 60 m^3/d. Although it comprises 3% of the final waste, due to its very high concentration it contributes greatly to the final waste loads. The steeping process is a batch process, the waste being dumped three times a day at the beginning of each working shift. The average waste characteristics were 46,250 mg/l COD, 37,760 mg/l of total residue, 35, 280 mg/l of volatile residue, 32,800 mg/l of suspended residue and 10,000 t.U as turbidity. Decantation of the waste for I h resulted in 9%, 35.5%, 34.2%, 12.2% and 2.5% removals, respectively. Centrifugation of the raw waste for 10 min at 10,000 rpm led to 45%, 50.5%, 40.9%, 69.2% and 65% removals, respectively. The effect of acidification at different pH levels from 2-6 followed by either centrifugation or nonionic poly-electrolyte coagulation are shown in Figs. 6, 7 and 8.

The highest removal percentages, namely, 79% COD, 84% turbidity, 70.6% total residue, 80.7% volatile residue and 83.5% suspended residue were achieved at pH 4. Centrifugation of the accified wastes at the different pH values from 2-6 also gave the highest removals at pH 4. The centrifugation time of 5 min was found to be optimum, as increasing the time for longer periods up to 10 min did not drastically change the removal percentages at the different pH values. Besides a 99.5% COD removal, turbidity, total residue, volatile residue and suspended residue removals of 98.9%, 99.3% and 99.1%, respectively, were obtained by centrifugation following acidification at pH 4.

Nonionic polyelectrolyte addition at a dose of 3 mg/l was found to give the optimum removals if applied to the heavily concentrated steeping waste. Higher doses gave lower removals and led to overdosing of the waste and consequently less flocculation and more resuspension of parts of the precipitated colloids. Nevertheless, centrifugation would be the recommended procedure to follow after acidifying the waste to pH 4. This treatment would require construction of a tank where 20 m^3 of the steeping water per shift

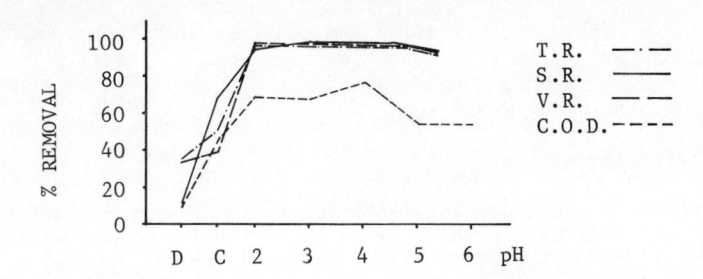

Fig. 6. Effect of Acidification on the Steeping Waste

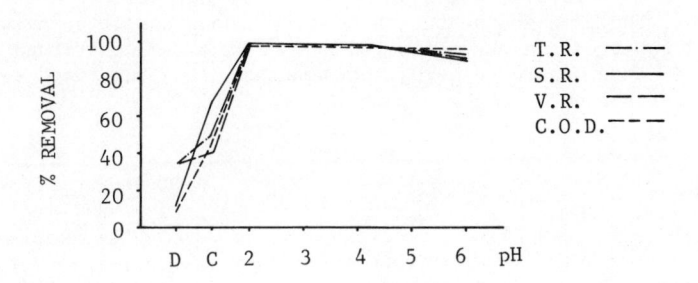

Fig. 7. Effect of Acidification and Centrifugation on the Steeping Waste

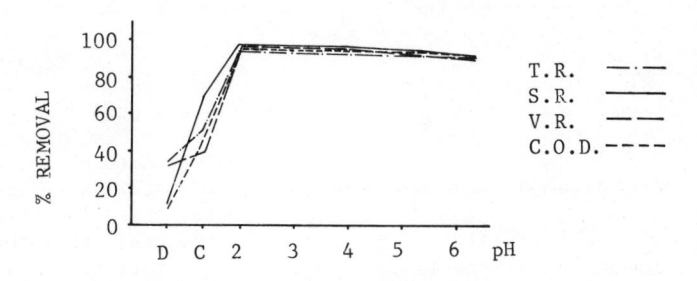

Fig. 8. Effect of Polymer Addition on Acidified Steeping Waste

could be acidified. The supernatant could then be centrifuged and discharged to the sewer.

The separated solids could be further dried and combined with those obtained after the proposed treatment of waste from the reused paper-processing plant. The economics of such a step would require further consideration. At any rate if this segregation procedure were followed, the final waste load would be drastically reduced.

Concentration Waste Treatment

This wastewater, which is continuously discharged to the sewer, is less concentrated than the steeping waste. It has on the average the following characteristics: 11,250 mg/l COD, 11,740 mg/l of total residue, 9980 mg/l of volatile residue, 7000 mg/l of suspended residue and 3375 t.U. as turbidity. Decantation of this waste for 1 h without any additions gave removals of 59%, 55%, 53%, 30% and 70%, respectively. The supernatant was still too polluted to be discharged to the sewer. Acidification to different pH levels from 2-6 gave the various removals shown in Fig. 9. Maximum removals were obtained at pH 3 as follows: 78% COD, 76% total residue, 74% volatile residue, 93% suspended residue and 97% turbidity. The supernatant, when coagulated with nonion polymer (H 824.3) at pH 3, gave reasonably good removals (Fig. 10) at a dose of 1 mg/ The decanted waste, when withdrawn for analysis, had the following characteristics: 380 mg/l COD, 810 mg/l of total residue, 460 mg/l of volatile residue, 50 mg/l of suspended residue, 30 t.U as turbidity and 30 mg/l TOC. Higher polymer doses gave lower removal percentages. The same sedimentation tank for the batch steeping waste after i discharge was used for treating the collected wastes from the centrifugation process. The collected solids were separated using the same technique and added to the steeping wastes solids for further use.

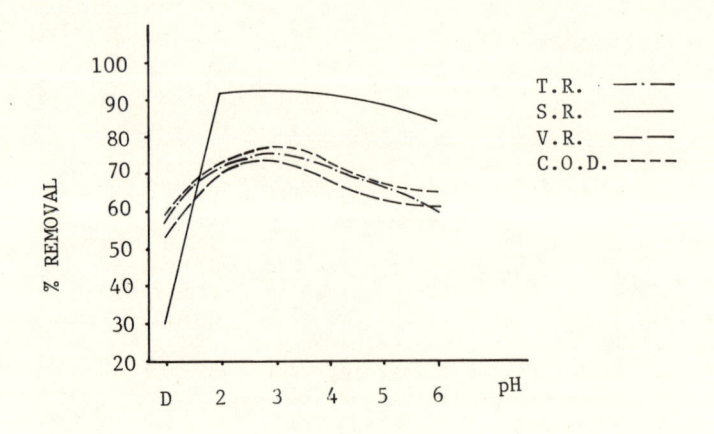

Fig. 9. Effect of Acidification on Concentration Wastes

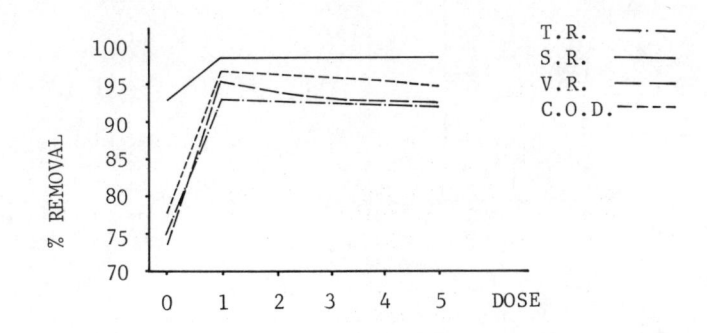

Fig. 10. Effect of Nonionic Polyelectrolyte on Acidified Concentration Wastes

Settling Characteristics of the Treated Final Wastes

The effect on the settling curves of $FeSO_4$, acidification and polymer addition after acidification are shown in Fig. 11. When $FeSO_4$ coagulation was carried out followed by sedimentation, 90% of the suspended residue was removed in 44 min in the sedimentation column at a surface flow rate of 1500 gpd/ft^2. Acidification of the waste to pH 3 gave the same results. Upon nonionic polymer addition, 95% suspended residue was removed in 36 min at a flow rate of 1650 gpd/ft^2. Thus, a more efficient solids removal was achieved, along with a maximum gain of residues to be utilized for animal feed processing.

SUMMARY AND CONCLUSIONS

Lake Maruit, the final recipient of the waste from the starch and glucose industry, which is only one of various industries in Moharrem Bey, has reached an alarming level of pollution. The Alexandria Governorate authorites are starting to enforce waste treatment measures in all plants in the complex.

The Egyptian Starch and Yeast Company can proceed with treating its wastes following one of the two possible alternatives discussed in this research. The first alternative is to segregate the cooling water system from the processing waste. The cooling water can be discharged along with the treated effluents to the sewer without any further treatment. For the heavily polluted steeping waste, acidification to pH 4 followed by centrifugation will produce an effluent that can be either discharged to the sewer or reused in cleaning operations.

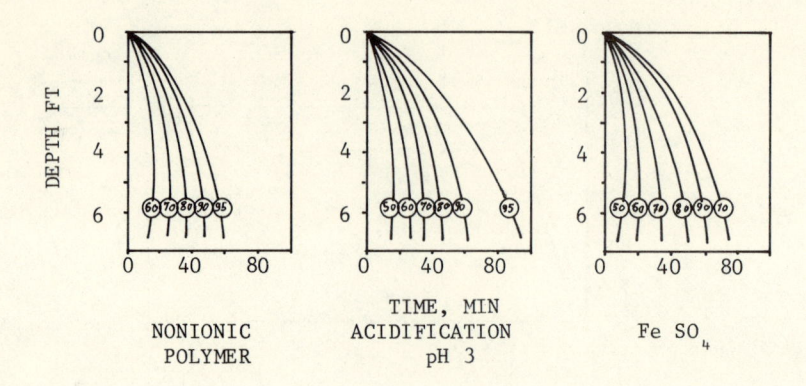

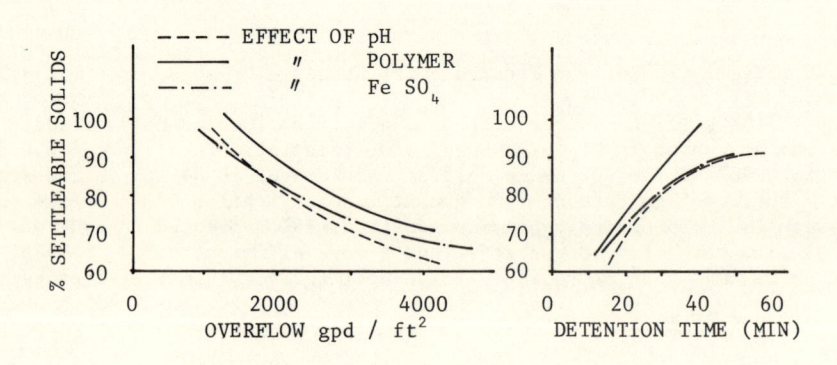

Fig. 11. Settling Characteristic Curves of the Final Waste

The pollutional level of the concentration wastes, if acidified to pH 3 and if 1 mg/l nonionic polymer is added, will be drastically reduced so that the final effluent will have 380 mg/l COD, 810 mg/l of total residue, 460 mg/l of volatile residue and 50 mg/l of suspended residue. This effluent will be further diluted to half its concentration values when discharged with the cooling water. All cleaning operation wastes can be treated with this waste.

The second alternative is to treat the final effluent of the plant without any modifications. Acidification to pH 3 followed by the addition of 1 mg/l nonionic polymer and 0.1% active carbon will produce a final effluent of 100 mg/l COD, 490 mg/l of total residue, 80 mg/l of volatile residue and 30 mg/l TOC.

The final decision will be based on the economics of each alternative, which will be investigated in the future stages of this research.

The Jar Tester technique proved to be a good dependable tool for screening the flocculation capabilities of different polyelectrolytes as it can save both time and money when applying research results to the plant scale.

ACKNOWLEDGEMENT

This research was funded by USEPA grant No.3-542-4. The author would like to extend her appreciation to Prof. Dr. Alexan Y. Salem and his staff at The Sanitary Engineering Center and to the Environmental Health Dept. Staff at The High Institute of Public Health Alexandria University, Egypt, for their enthusiastic and invaluable help. Special thanks are due to Mrs. Nariman Sohil for her assistance in the laboratory analyses.

REFERENCES

AWWA-APHA-WPCF (1971). *Stardard Methods for The Examination of Water and Wastewater*. 13th Ed. American Public Health Association, New York.

Bisset, D.W. (1976). The Toxicity of Food Processing Effluents to Fish. Presented at 6th Environmental Food Processing Industries Conference, Pasific grove, California.

Bugg, H.M., P.H. King and C.W. Randall (1970). Polyelectrolyte Conditioning of Alum Sludges. *J. Am. Water Works Assoc.*, 62, 792.

Fair, G.M., J.C. Geyer and D.A. Okun (1968). *Water and Wastewater Engineering*. John Wiley & Sons, Inc., New York.

Hatfield, W.D. (1953). Corn starch processes. In W. Rudolfs (Ed.), *Industrial Wastes; their Disposal and Treatment*. Reinhold Publishing Corp., New York.

Helmers, E.N., J.D. Frame, A.F. Greenberg and C.N. Sawyer (1951). Nutritional Requirements in the Biological Stabilization of Industrial Wastes. *Sewage and Industrial Wastes*, 23:7, 834.

Nemerow, N.L. (1971) *Liquid Wastes of Industry; Theories, Practice and Treatment*. Addison Wesley Publishing Co., Inc., Menlo Park, California.

Novak, J.N. and J.H.Ò. Brien (1975). Polymer Conditioning of Chemical Sludges. *J. Water Poll. Control Fed.*, 47: 10, 2397.

Stumm, W. and C.R. O'Melia (1968). Stoichiometry of Coagulation. *J. Am. Water Works Assoc.*, 60, 514.

DISCUSSIONS

Müezzinoğlu, A. (Turkey): Didn't you consider biological treatment of the starch wastewater, since starch is mostly a soluble and biodegradable material?

Saad, S. G. (Egypt) : We thought of this. However, two main facts have to be taken into consideration. The company has no space whatsoever for the construction of the necessary basins. Secondly, for the running of a biological degradation system, a C:N:P ratio is essential. In starch wastes there is very little N and P. Moreover, in experiments done on a larger basin we ended up with an excessive growth of sphaeritulus natans, which causes the sludge to float rather than settle.

Müezzinoğlu, A. (Turkey): Due to the abundance of ferrous and sulphate ions, the coagulants and polyelectrolytes probably create toxicity in the sludge. Wouldn't this prohibit the use of this sludge as animal feed?

Saad, S. G. (Egypt) : That is why we decided against using the sulphate with the polyelectrolyte. Excess ferrous sulphate was produced, which on exposure to the air became oxidized, giving a brown coloration and making it impossible to discharge the final waste to the sewers.

Rüffer, H. (Germany) : The use of acids normally requires a neutralizing material like calcium hydroxide, which results in an additional content of dissolved inorganics - for example, calcium sulphate of sulphuric acid is used. Since your plant is near the sea, this method may be acceptable. In many cases, however, where plants are far from the sea, an additional inorganic concentration tends to be avoided. Rivers are often used for other purposes, such as producing drinking water. I would like to draw your attention to this point.

Saad, S. G. (Egypt) : Unlike your situation in Germany, the methods at my disposal are very limited. The use of heat, for example, to destabilize the colloids is not possible because of the presence of gluten. To reduce the pH and destabilize the solution, the only way for me is to use cheap sulphuric acid. If the segregation process is used in treating the effluent, the pH of about 30% of the process water can first be diluted. Acidification of the total process water to a pH of 3 will produce an effluent which, when mixed with the cooling water (120 m^3), will be sufficiently diluted to be discharged without further neutralization. Egyptian law allows the discharge of wastewater having a pH of 6-9.

Rüffer, H. (Germany) : I agree that in your case you might be right to use acids.

Fundamental Studies on the Treatment of Fish-Processing Wastewater by the Activated Sludge Process

T. OKUBO, J. E. ISHIHARA and J. MATSUMOTO

Department of Civil Engineering, Tohoku University, Aoba, Sendai 980, Japan

ABSTRACT

The objective of this study is an evaluation of substrate removal, sludge production and oxygen requirement with the aim of establishing the basic design criteria for the activated sludge process. The BOD removal rate was that of a first order reaction. The removal rate constant was 1.2×10^{-2} ℓ/mg·d. The fraction of BOD converted to new cells was 0.33. The apparent sludge accumulation did not occur when the BOD loading value was less than 0.12 ℓ/day. The effluent COD concentration was expressed in terms of a power function of the influent COD concentration. Biological nitrification occurred when the hydraulic retention time was longer than 12 hours.

INTRODUCTION

Fish-processing operations are largely seasonal in nature and the wastewater is far from uniform with respect to both quantity and quality. The wastewater of a fish-processing plant usually contains a large amount of protein, oil and grease. Therefore, a floatation system has been used for the separation of protein, oil and grease, and then a biological treatment system has been employed.

There are many papers which deal with the characterization of wastewater from various types of fish-processing plants and with the biological degradation of fish-processing wastewater (Eckenfelder and Ford, 1970; Levenspiel, 1972). These studies have given much information on the biological treatment of fish-processing wastewater. However, they have provided little information on the process kinetics of activated sludge.

The purpose of this paper is an evaluation of substrate removal, sludge production and oxygen requirement with the aim of establishing the basic design criteria for the activated sludge process (Maksumoto and others, 1965; Riddle and Murphy, 1972; Pearson and others, 1970).

PRACTICE of TREATMENT of FISH-PROCESSING WASTEWATER in JAPAN

There are a number of fish-processing plants around the fishing ports in Japan. Most of them are small in size and lie scattered in cities. Therefore, it is difficult to expect satisfactory wastewater treatment of individual plants. In order to overcome the above-mentioned difficulties, national and local governments have encouraged fish-processing industries to move their plants to a newly developed industrial area, to collect raw (or pretreated) wastewater from individual factories in this area and to treat it at a treatment plant constructed and operated by the industrial association. Two typical examples are shown in Figs. 1 and 2.

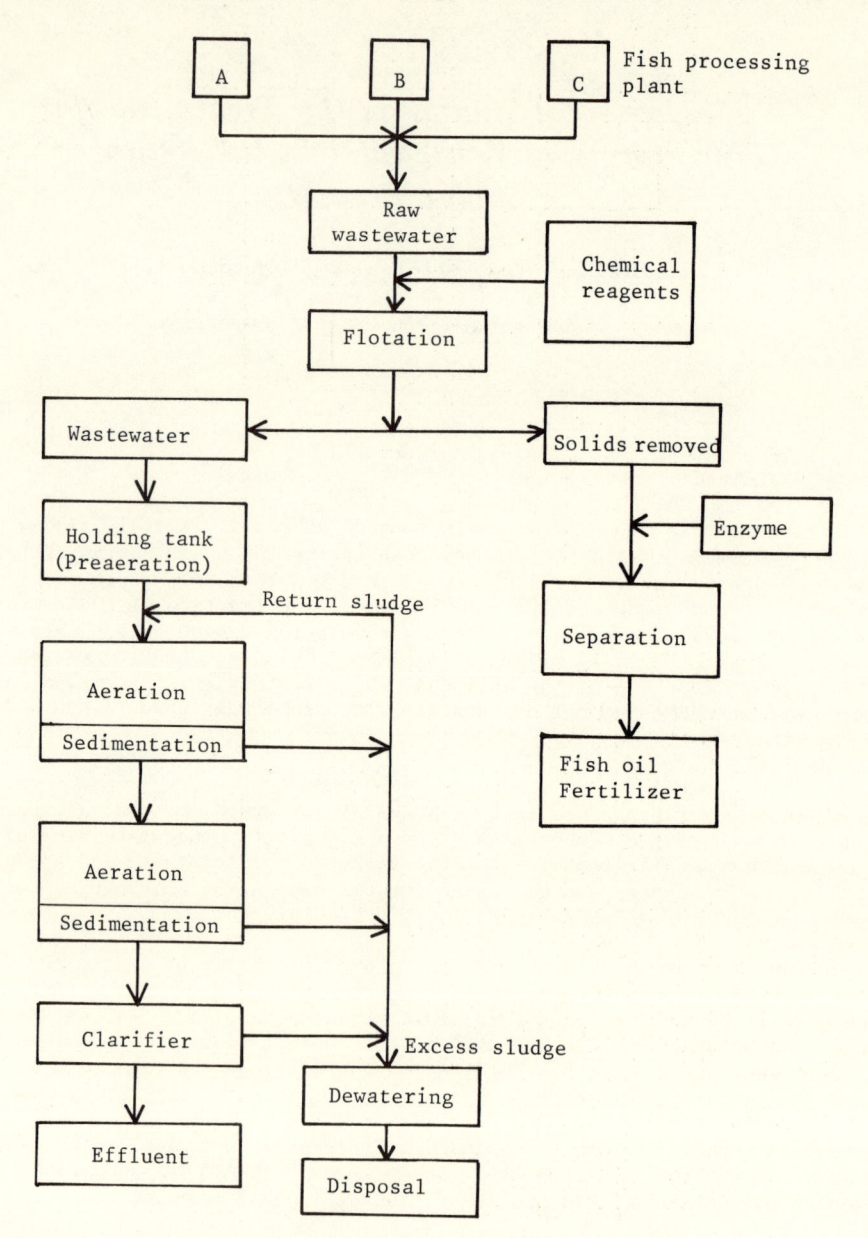

Fig. 1: Shiogama treatment plant

Besides a seasonal variation, the hourly variation in fish-processing wastewater flow and strength is great. It is, therefore, desirable to prepare an equialization tank in which the fluctuations in flow and strength of wastewater can be ironed out. The hydraulic retention time of the activated sludge process should be as long as possible for the stable operation of the process to take place.

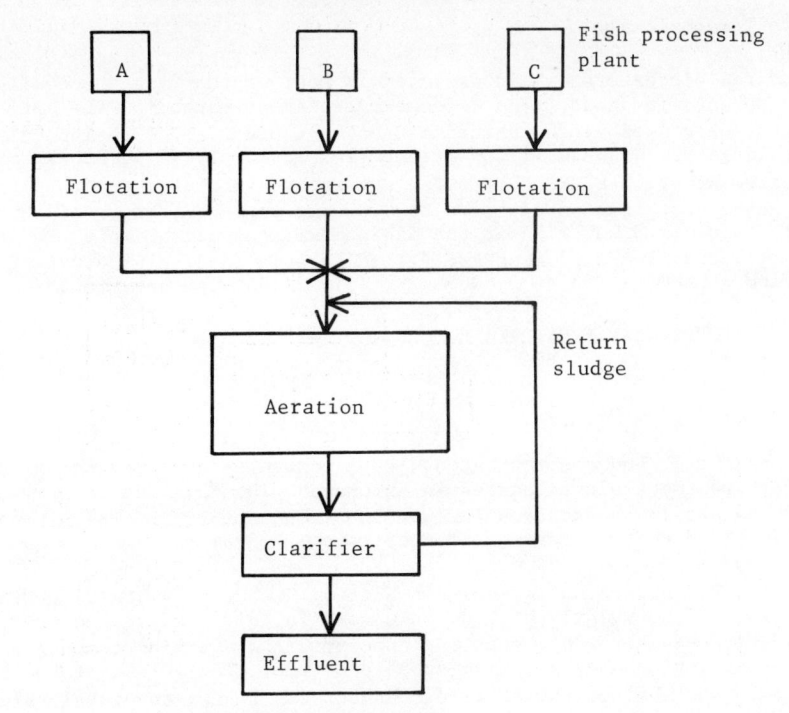

```
                                    Fish processing
   ┌───┐         ┌───┐         ┌───┐  plant
   │ A │         │ B │         │ C │
   └─┬─┘         └─┬─┘         └─┬─┘
     ▼             ▼             ▼
 ┌─────────┐ ┌─────────┐ ┌─────────┐
 │Flotation│ │Flotation│ │Flotation│
 └────┬────┘ └────┬────┘ └────┬────┘
      └───────────┼───────────┘
                  ▼
            ┌─────────┐           Return
            │ Aeration│◄───────── sludge
            └────┬────┘
                 ▼
            ┌─────────┐
            │Clarifier│
            └────┬────┘
                 ▼
            ┌─────────┐
            │ Effluent│
            └─────────┘
```

Fig. 2: Ishinomaki treatment plant

In Table 1, the quality of raw wastewater in the Shiogama treatment plant is presented. Fig. 3 shows the relationship between the BOD and COD of the diluted raw wastewater used in this experiment. The BOD/COD ratio of 0.80 for fish-processing, when compared to a BOD/COD ratio of 0.50 for domestic wastes (Hunter and Heukelekian, 1965), indicates the high degree of degradability of the first.

Primary Treatment

The raw wastewater from fish-processing plants contains a large amount of protein, oil and grease. First, the coarse solids are removed from the wastewater by screens and then the pH value is lowered to 4.8 - 5.1 to coagulate the protein. Subsequently, the wastewater is introduced into a flotation tank, in which flocculated particles are lifted to the surface, with the help of air and sodium polyacrylate. The improvement in by-product recovery techniques has made it possible to recover fish oil and fertilizer.

At the Shiogama treatment plant, 73% of the BOD, 80% of the COD and 97% of the grease were removed by the flotation process. The remaining BOD concentration was about 1000 mg/ℓ. Then, the effluent from the flotation tank was introduced to a preparation tank, which raised the pH value of the wastewater to 6.0-6.5.

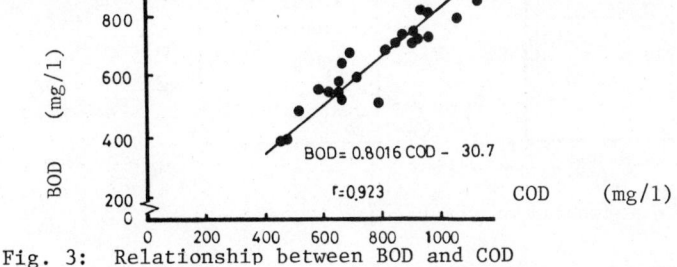

Fig. 3: Relationship between BOD and COD

$BOD = 0.8016\, COD - 30.7$

$r = 0.923$

Secondary Treatment

The activated sludge process is employed in most of the fish-processing wastewater treatment plants in Japan. The two-stage treatment method of the activated sludge process has been used at the Shiogama treatment plant. The design values of the activated sludge process are (1) an organic loading of 0.178 kg BOD/kg SS·d; (2) an air requirement of 60 m^3/kg BOD; (3) a sufficient volume of return sludge to maintain an MLSS concentration of 4,000 to 5,000 mg/ℓ; (4) 23 hours of aeration. The wastewater flow is 3,000 m^3/d at the Ishinomaki treatment plant. The design values of the activated sludge process are (1) an organic loading of 0.094 kg BOD/kg SS·d; (2) an MLSS concentration of 3,500 mg/ℓ; (3) 7.2 days of aeration.

EXPERIMENTAL MATERIALS, APPARATUS and PROCEDURE

Materials and Apparatus

The raw wastewater and the activated sludge were obtained from the fish-processing wastewater treatment plant located in Shiogama. The raw wastewater was diluted with tap water to obtain the desired influent COD concentration. The activated sludge was acclimatized to temperature and substrate.

The experimental apparatus is shown in Fig. 4. Laboratory-scale continuous flow tank reactors made of polyvinyl chloride were used in this study. Each reactor consisted of an aeration chamber and a settling chamber. The aeration chamber was separated from the settling chamber by an adjustable baffle. The volume of the aeration chamber was 6.5 ℓ and that of the settling chamber was 1.5 ℓ; the total volume of the reactor was 8 ℓ. The culture temperature was maintained at 20± 2 C^0 by putting the reactor in a water bath.

Table 1: Water Quality of Fish-Processing Wastewater in the Shiogama Treatment Plant (1976)

Date	Sept. 13	Sept. 22	Oct. 12	Nov. 1	Nov. 18	Dec. 1
COD, mg/ℓ	1,060	1,610	4,010	2,190	2,790	7,360
BOD, mg/ℓ	662	1,430	2,870	-	2,610	6,810
Grease, mg/ℓ	-	598	636	607	761	-
Kjel-N, mg/ℓ	53.1	174	374	185	336	575

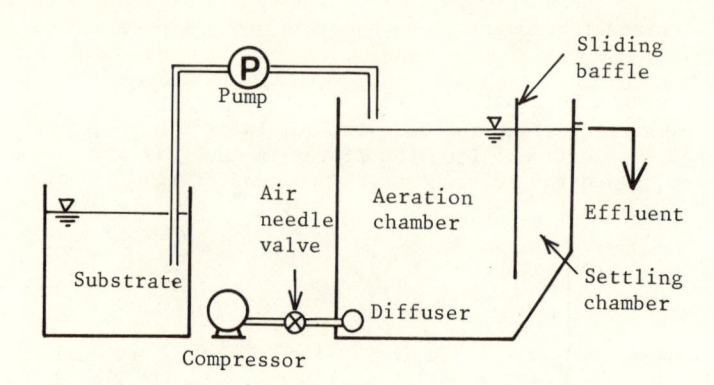

Fig. 4: Experimental apparatus

Procedure

The substrate (diluted raw wastewater) was continuosly fed into the aeration chamber at a flow rate necessary to obtain the desired hydraulic retention time. The MLSS concentrations were maintained at 2,000 to 2,500 mg/ℓ. In the case of excessive microbial growth, a certain volume of mixed liquor was withdrawn from the aeration chamber, and the same volume of effluent from the settling chamber was introduced into the aeration chamber to dilute the remaining mixed liquor. The continuous flow system was maintained until a steady state condition prevailed. This was assumed once the effluent COD was stabilized.

The effect of the influent COD concentration on the specific rate of substrate utilization was studied by using the influent COD concentration and hydraulic retention time, as shown in Table 2, so as to impose a nearly constant organic loading on the reactor.

Table 2: Effect of Influent COD Concentration on Treatability

	Run 1-1	Run 1-2	Run 1-3	Run 1-4
COD loading (mg/mgSS·d)	0.55	0.69	0.56	0.60
COD (mg/ℓ) influent	680	879	2,440	5,370
effluent	99	121	216	261
MLSS (mg/ℓ)	1,744	2,561	2,202	2,201
Hydraulic Retention Time (day)	0.71	0.50	1.98	4.09

In order to study the effect of pH on the activated sludge performance, two reactors were used. The pH of the influent to one of the reactors was lowered with HCℓ to about 4.5, while the other reactor was used as a control. The influent COD and hydraulic retention time were kept constant as far as possible. Further, the coefficients necessary for establishing basic design criteria were estimated by imposing different organic loadings on the reactor. The BOD loadings were in the range of 0.07 to 1.34 mg BOD/mg SS·d.

EFFECT of INFLUENT COD CONCENTRATION on TREATABILITY

The steady state data in this experiment are shown in Table 2. The relationship between the specific rate of substrate utilization and influent COD is shown in Fig. 5. The specific rates of substrate utilization were nearly constant. The influent COD concentration had no effect on the specific rate of substrate utilization. When the COD loadings were in the range of 0.55 to 0.69 mg COD/mg SS·d, the effluent COD could be estimated by the following equation as shown in Fig. 5:

$$S_e = 4.94 \, S_o^{0.48} \tag{1}$$

EFFECT of pH on TREATABILITY

Experimental results are shown in Table 3. The percentage removal of five-day BOD was not affected by the pH value, which was in the range of 4 to 8. The percentage removals of COD, grease and total Kjeldahl nitrogen in Run 2-2 were lower than in Run 2-1. The total Kjeldahl nitrogen removal was especially inhibited by the low pH.

Fig. 5: Effect of influent COD on treatability

$$S_e = 4.94 S_o^{0.48}$$

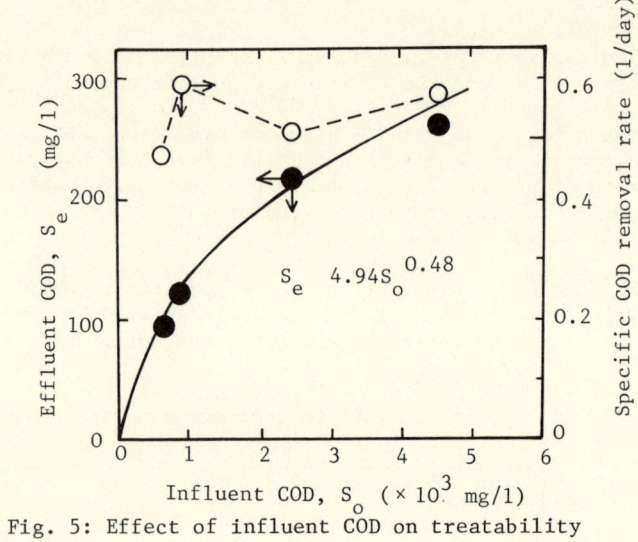

When the pH of the mixed liquor was less than 4, the percentage removal of total Kjeldahl nitrogen came to be less than 5%. The microbial respiration rate and the dehydrogenase activity of cells (Ford and others, 1966) were not affected by the pH which was in the range of 4 to 5, but it was observed that a low pH affected the microbial growth.

Table 3: Effect of pH on Treatability

	Run 2-1	Run 2-2
pH	7.5	4.4
COD (mg/ℓ)		
influent	684	664
effluent	48.7	108
BOD (mg/ℓ)		
influent	470	560
effluent	25.6	36.9
Grease (mg/ℓ)		
influent	399	222
effluent	57	84
Kjel-N (mg/ℓ)		
influent	78.2	73.7
effluent	15.3	51.4

Fig. 6: BOD removal rate (S_o: BOD_5, $k : 1.20 \times 10^{-2}$)

Fig. 7: COD removal rate (S_o: COD_{cr})

PROCESS KINETICS

The steady state data are shown in Table 4.

Table 4: Experimental Result

	3-1	3-2	3-3	3-4	3-5	3-6
Soluble COD (mg/ℓ)						
influent	577	646	577	489	489	646
effluent	161	121	98.5	52.2	55.2	47.2
Soluble BOD (mg/ℓ)						
influent	516	607	516	495	495	607
effluent	81.3	31.9	30.1	4.5	4.2	2.8
Grease (mg/ℓ)						
influent	–	116	–	145	145	116
effluent	–	18.5	–	19.1	17.7	12.3
Kjel-N (mg/ℓ)						
influent	60.4	80.2	60.4	63.9	63.9	80.2
effluent	44.0	48.2	21.2	13.4	15.1	14.9
MLSS (mg/ℓ)	1,280	2,561	1,744	2,253	2,156	1,991
Hydraulic Retention Time (day)	0.32	0.50	0.71	1.06	1.95	5.01

Substrate Removal

A completely mixed system was used in this study. The soluble BOD in the effluent was equal to that in the aeration chamber. A material balance under steady state conditions results in the following relationship:

$$QS_o - QS_e = \left(-\frac{dS}{dt}\right)V \tag{2}$$

Assuming $\frac{dS}{dt}$ to be a function of the substrate remaining according to first order removal kinetics;

$$-\frac{dS}{dt} = kXS_e, \tag{3}$$

the following relationship results;

$$\frac{S_o - S_e}{XT} = kS_e \tag{4}$$

The removal rate constant, k, was determined in Fig. 6. The BOD removal rate was approximately that of a first order reaction and the removal rate constant was 1.2x 10^{-2} ℓ/mgd. The relationship between the effluent COD and the specific rate of substrate utilisation is shown in Fig. 7. The COD removal rate was not a function of the COD remaining according to first order removal kinetics. The residual COD was more than 40 mg/ℓ and the percentage of non-degradable COD was 7% at the hydraulic retention time of 5 days. The relationship between the effluent COD and effluent BOD is shown in Fig. 8.

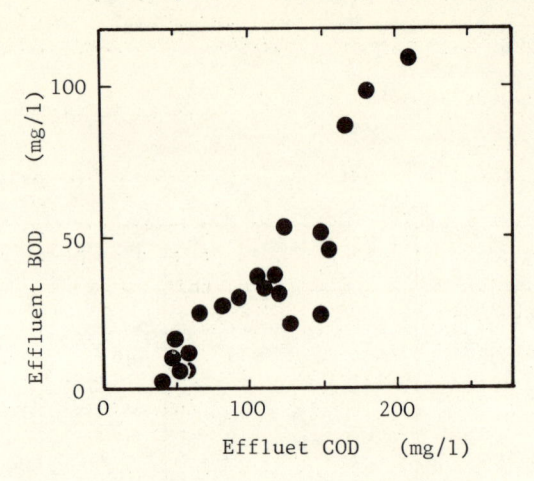

Fig. 8: Relationship between BOD and COD

Sludge Production

Sludge accumulation in the activated sludge system can be estimated by using the material balance expression.

$$\Delta X = QX_o + a(S_o - S_e)Q - (bXV + QX_e) \tag{5}$$

$$\frac{\Delta X_a}{X} = \frac{a(S_o - S_e)}{XT} - b \tag{6}$$

where:

$$\Delta X = \frac{\Delta X - QX_o + QX_e}{V} \tag{7}$$

The sludge production was determined by using Eq.(6). "a" and "b" values were taken as the slope and intercept values respectively, as shown in Fig. 9. The fraction of BOD converted to new cells was 0.33 and the fraction per day of SS oxidized was 0.04. The fraction of COD converted to new cells was 0.35. It is suggested that the apparent sludge accumulation in the aeration chamber does not occur when the BOD loading is less than 0.12 mgBOD/mg SS·d.

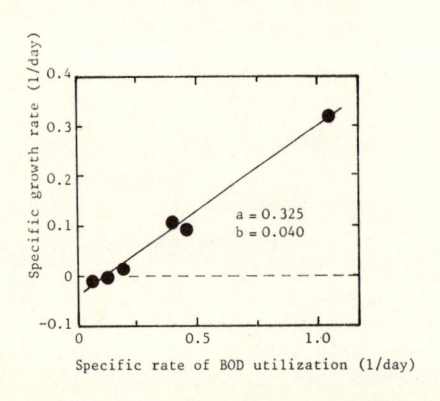

Fig. 9: Sludge production

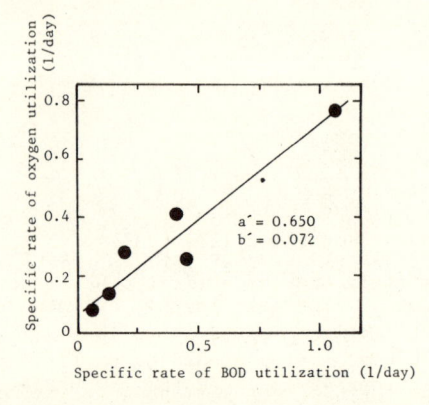

Fig. 10: Oxygen Requirement

Oxygen Requirement

In a biological system the total oxygen requirements are related to the oxygen consumed to supply energy for synthesis, oxygen for endogenous respiration and oxygen for chemical oxidation. This relation can be expressed as follows:

$$R_r = a'(S_o - S_e) Q + b'XV + K_c Q \qquad (8)$$

The oxygen for chemical oxidation is neglected as this demand is usually satisfied prior to testing. Then Eq.(8) can be rearranged as;

$$\frac{R}{X} = \frac{a'(S_o - S_e)}{XT} + b' \qquad (9)$$

The "a'" and "b'" values were taken as the slope and intercept values respectively, as shown in Fig. 10. The fraction per day of SS oxidized was in the range of 0.065 to 0.071, which was determined by dividing the endogenous respiration rate by the SS concentration, which agreed quite well with the "b'" value in Fig. 10.

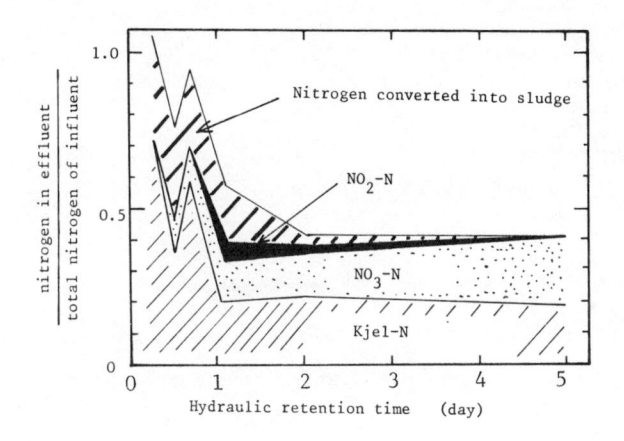

Fig. 11: Relationship between nitrogen removal and hydraulic retention time

Removal of Nitrogen

The residual Kjeldahl nitrogen was more than 10 mg/ℓ when the effluent BOD was less than 5 mg/ℓ, as shown in Table 4. The relationship between nitrogen removal and hydraulic retention time is shown in Fig. 11. Biological nitrification occurred when the hydraulic retention time was longer than 12 hours. The amount of nitrogen consumed for the synthesis of cells was determined by the sludge accumulation. The amount of total nitrogen in the effluent was nearly equal to that in the influent only when the hydraulic retention time was 7.7 hours. It seems that the denitrification occurred in the settling chamber when the hydraulic retention time was longer than 1 day.

CONCLUSIONS

1. The influent COD concentration had no effect on the specific rate of substrate utilization. The effluent COD was expressed in terms of a power function of the influent COD.

2. The total Kjeldahl nitrogen removal was especially inhibited by the low pH.
3. The BOD removal rate was that of a first order reaction and the removal rate constant was 1.2×10^{-2} ℓ/mg·d.
4. The fraction of BOD converted to new cells was 0.33 and the fraction per day of SS oxidized was 0.04. The apparent sludge accumulation did not occur when the BOD loading was less than 0.12 mg BOD/mgSS·d.
5. Biological nitrification occurred when the hydraulic retention time was longer than 12 hours.

NOTATION

a : fraction of BOD converted to new cells (-)
a' : fraction of BOD used for oxidation (-)
b : fraction per day of SS oxidized (T^{-1})
b' : fraction per day of SS oxidized (oxygen basis) (T^{-1})
k : removal rate constant ($M^{-1}L^2T$)
K_c : chemical oxygen demand coefficient ($ML^{-2}T^{-2}$)
Q : flow rate (L^3T^{-1})
R_r : oxygen utilization per day ($ML^{-2}T^{-3}$)
S_o : influent BOD or COD ($ML^{-2}T^{-2}$)
S_e : effluent BOD or COD ($ML^{-2}T^{-2}$)
T : hydraulic retention time (T)
V : volume of aeration chamber (L^3)
X : mixed liquor suspended solids ($ML^{-2}T^{-2}$)
X_o : influent SS ($ML^{-2}T^{-2}$)
X_e : effluent SS ($ML^{-2}T^{-2}$)
ΔX : sludge accumulation per day (MLT^{-3})
ΔX_a : $(\Delta X - QX_o + QX_e)/V$ ($ML^{-2}T^{-3}$)

ACKNOWLEDGEMENT

The authors gratefully acknowledge the assistance afforded by Mr. K. Kondo. Thanks are also due to Mr. A. Endo.

REFERENCES

Eckenfelder, W.W. and D.L. Ford (1970). *Water Pollution Control*, Pemberton Press, Austin.

Ford, D.L., J.T. Yang and W.W. Eckenfelder (1966). Dehydrogenase Enzyme as a Parameter of Activated Sludge Activities. *Proc. 21th Ind. Waste Conf*. Purdue Univ., *1*, 534.

Hunter, J.V. and H. Heukelekian (1965). The Composition of Domestic Sewage Fractions. *J.W.P.C.F.*, 37, 1142.

Levenspiel, O. (1972). *Chemical Reaction Engineering*. Second edition. John Wiley & Sons, New York.

Matsumoto, J., I. Endo, F. Nakamura and S. Wagatsuma (1965). Experimental Studies on the Treatment of Wastes from Fish Products Manufacturing Factories, (in Japanese). *J. Japan Sew. Wrks. Ass.*, 2, 26.

Matsumoto, J. (1975). Technology of Fish-Processing Wastewater Treatment (in Japanese). *Environmental Creation*, 5, 35.

Pearson, B.F., M.J. Chun, R.H.F. Young and N.C. Burbank (1970). Biological Degradation of Tuna Waste. *Proc. 25th Ind. Waste Conf.*, Purdue Univ., *2*, 766.

Riddle, M.J. and K.L. Murphy (1972). An Effluent Study of a Fresh Water -Fish Processing Plant. *Proc. 27th Ind. Waste Conf.*, Purdue Univ., *2*, 777.

DISCUSSIONS

Samsunlu, A. (Turkey) : How were you able to adjust the MLSS concentrations, given that in your model aeration and final settling take place in the same tank?

Okubo, T. (Japan) : The MLSS concentrations, which were maintained at 2000-2500 mg/ℓ, were measured every day. In the case of excessive microbial growth, a certain volume of mixed liquor was withdrawn from the aeration chamber and replaced by the same volume of effluent from the settling chamber.

Rüffer, H. (Germany) : What do you do with the floating fatty matter after flotation and separation?

Okubo, T. (Japan) : The floating fatty matter is discharged into the by-product recovery system to recover fish oil and fertilizer. However, in summer the scum spoils and as a result, the by-product has to be disposed of at a low price. It is necessary for us to develop a recovery system from which by-products of a consistently high quality and commercial value may be obtained.

Evaluation of the Treatability of Industrial Wastewaters in Izmir by Bacterial Respiration Measurements

A. SAMSUNLU

Environmental Engineering Department, Faculty of Civil Engineering, Ege University, Izmir, Turkey

ABSTRACT

Industrial development has not only brought prosperity to humanity but has created many dangerous problems as well. Industrial wastes are among these problems. These vary according to the production methods used and the goods produced. Some industrial wastes can undergo biodegradation; some are subject to partial decomposition; some are toxic to bacteria and cannot degrade biologically.

In this work, specific industrial effluents are studied for their degree of biodegradability. The method used in this study consists of examining the oxygen consumption and respiration activities during the biological degradation process carried out on each effluent, e.g., brewery, textile industry, metal-processing and raw toxic tannery effluents. Results are evaluated and discussed.

INTRODUCTION

The pollution of surface waters has reached tremendous dimensions at the present time. This is caused partly by domestic and partly by industrial wastewater from various sectors. Industrial production is the cause of many important pollution problems. The determination and classification of wastewater characteristics is mandatory for pollution abatement and for the design, construction and operation of waste treatment facilities.

The different technologies and processes used in industry create a broad spectrum of wastewater types. Therefore, technicians must consider the parameters defining the various wastewater characteristics and, if necessary, must try to find new and better methods and techniques for determining the degree of pollution.

The effects of industrial wastewaters on bacterial life, which is very important in biological treatment, can be conveniently determined by measuring the respiration rates and changes in these rates at different dilutions in the laboratory. The adaptability of the microorganisms to the wastewater slugs and efficiency of treatment at various adaptation rates depending on the degree of concentration of pollutants is another important aspect of respiration kinetics.

In this paper, the degree of treatability of selected industrial wastewaters, taken from the Izmir Area, for which respiration activity measurements have been taken, is discussed.

OXYGEN CONSUMPTION and RESPIRATION RATES

Oxygen consumption is an indicator of respiration activity. In order to find out the oxygen consumption and aeration rates, Hixson and Gaden (1950) proposed the following basic relationship:

$$\frac{dc}{dt} = k \ (C_s - C) \tag{1}$$

for zero oxygen consumption, where

$C = $ O_2 concentration in water (mg/ℓ)

$C_s = $ equilibrium O_2 concentration in water (mg/ℓ)

$k = $ reaeration coefficient (h^{-1}).

Assuming that no oxygen enters the reactor with the influent and recycled waters,

$$\frac{dc}{dt} = k' \ (C_s' - C) - OV \tag{2}$$

where

$C = $ oxygen concentration in reactor (mg/ℓ)

$C_s' = $ saturation concentration of oxygen in water under operating conditions

$k' = $ reaeration coefficient under operating conditions (h^{-1})

$OV = $ respiration rates of microorganisms under operating conditions (mg/ℓ-h)

The O_2 concentration in the reactor reaches a fixed level after a certain period of reaeration. This is called the relative or virtual saturation value, C_s^*. Under these conditions C and C_s^* are approximately equal and Eg. (2) becomes

$$\frac{dc}{dt} = k' \ (C_s' - C^*) - OV \tag{3}$$

The saturation value under the operating conditions is as follows:

$$C_s' = C_s^* - \frac{OV}{k'} \tag{4}$$

C_s' may be determined from Eq. (4) by calculating k' and measuring C_s^* and OV values in laboratory set-ups.

For the reduction of pollutants in wastewaters, microorganisms, must consume them as substrate while also consuming oxygen. The oxygen present in the medium, therefore, is the sum of the oxygen consumed under operating conditions (OV) and residual oxygen (OC Reserve). Respiration under operating conditions is the sum of respiration for growth (substrate respiration) and endogenous respiration of the microorganisms. Industrial wastewaters are composed of many different constituents. Microorganisms that multiply in waste waters show different (substrate) respiration activities for each different constituent. As a result, oxygen consumption rates (mgO_2/ℓ-h) are different for each different pollutant.

Oxygen consumption rates and the total solids content (amount of activated sludge), which indicate the microbial growth rates, are interrelated. The total oxygen consumption rate per unit of total solids content (mgO_2/g-TS-h) is defined as the specific respiration rate, OVS. This parameter is related to the total respiration under operating conditions as follows:

$$\genfrac{}{}{0pt}{}{\text{Respiration under}}{\text{operating conditions}} = \genfrac{}{}{0pt}{}{\text{Amount of}}{\text{Total Solids}} \cdot \genfrac{}{}{0pt}{}{\text{Specific}}{\text{Respiration}}$$

By measuring and observing the changes in respiration activity (OV) and specific(OVS) we can determine the biological treatability of industrial wastewaters.

LABORATORY TEST SET-UP

The set-up used for the measurement of respiration activities exists in the Municipal Infra-Structure Institute of Stuttgart Technical University. Tests were carried out in the Institute's laboratories.

The operation of the measurement system is shown in Fig. 1. The observed parameter in the system is dissolved oxygen change per unit time. DO values are transmitted to a recorder via membrane electrodes placed in each reaction vessel which has been filled with different dilutions of industrial waste waters. The reaction vessels are mixed at a constant speed, thermostatted and kept at 20^0C. The contents of the vessels are inoculated with cultured microorganisms taken from a laboratory biological treatment model unit, a flow diagram of which is given in Fig. 2.

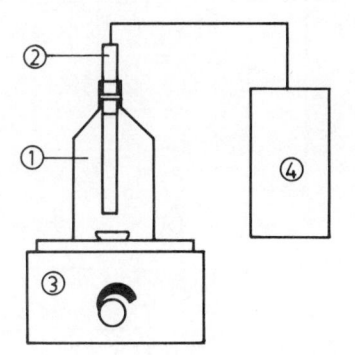

1. Reaction Vessel
2. DO electrode
3. Magnetic mixer
4. Reactor

Fig. 1: Laboratory set-up for measurement of respiration activities

MODEL TESTS

A fixed volume of 65 mℓ inoculum material taken from the aeration unit in Fig. 2 is poured into the reaction vessel in Fig. 1. For each run the total solids content is determined.

Different dilutions of industrial wastewaters taken from breweries, tanneries, textile and metal-plating plants in the Izmir area are prepared, by adding 1, 2, 5, 10, 15, 20, 30, 40, 50 mℓ of each wastewater to the reaction vessels containing the inoculums, and then by filling the vessels with clarified municipal wastewaters taken from the discharge lines of pretreatment units. The dilution plan is shown in Table 1.

The dissolved oxygen reduction in each reaction vessel is continuously recorded. The DO values show different reduction rates for each industrial wastewater at each dif-

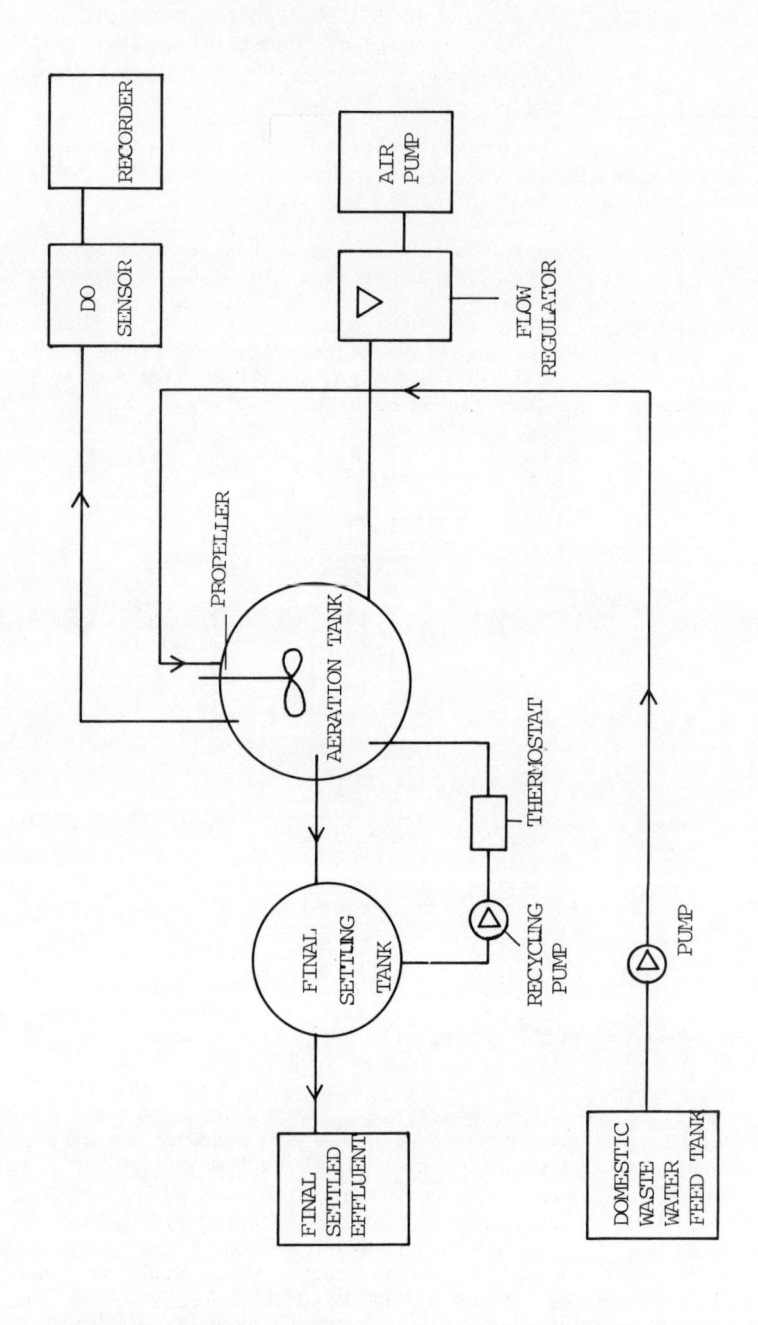

Fig. 2: Laboratory biological treatment model unit - flow diagram

ferent dilution. The reduction lines found by connecting the reduced DO values have a specific slope, which is defined as the respiration activity under operating conditions, OV(mg O_2/ℓ-h).

Table 1: Dilution Ratios of Industrial Wastewaters

Volume of Industrial Wastewaters	Volume of Activated Sludge	Volume of Pretreated Domestic Wastewater
1	65	49
2	65	48
5	65	45
10	65	40
15	65	35
20	65	30
30	65	20
40	65	10
50	65	0

Fig. 3 summarizes the respiration activity changes at different dilutions in graphical form. In Fig. 4 the respiration activity values (OV) and plots obtained from these values for brewery, textile, tannery, and metal-processing wastewaters tested at various dilutions are shown.

Table 2 shows the total solids contents and their changes for each industry (TS). The relationship between the total solids contents in activated sludge and the respiration activity has been made use of in calculating the specific respiration activity (OVS) (mg O_2/g-TS-h). In plotting the curves in Fig. 5, logarithmic (OVS) values are used.

Table 2: Total Solids Contents of Industrial Wastewaters Tested

Industry	TS(g/ℓ)
Efes Pilsen Brewery	2.285
Kula Textile Mill	2.576
Izmir Metal Plating	2.740
Izmir Leather Manufacturing Plant	2.671

DISCUSSION of RESULTS

Untreated wastewaters taken from four industries in the Izmir Area: brewery, tannery, textile production and metal-plating industries, are tested for their biological treatability. The raw effluent water from each industry is mixed with screened and settled (pretreated) municipal wastewater at different ratios, inoculated and tested, using the respirometric dilution method. Graphical representations are made of the volumetric dilution ratio, raw industrial sample activated sludge, versus 1. Specific respiration activity, OVS, and 2. respiration activity, OV. From these graphical results it is to be noted that for metal-plating wastewater, with an increasing volume of industrial effluents OVS values decrease. and especially at ratios exceeding 30% the decrease in OVS becomes substantial. An increase in industrial wastewater volume resulting in increased dilution ratios causes decreasing OVS ratios as a

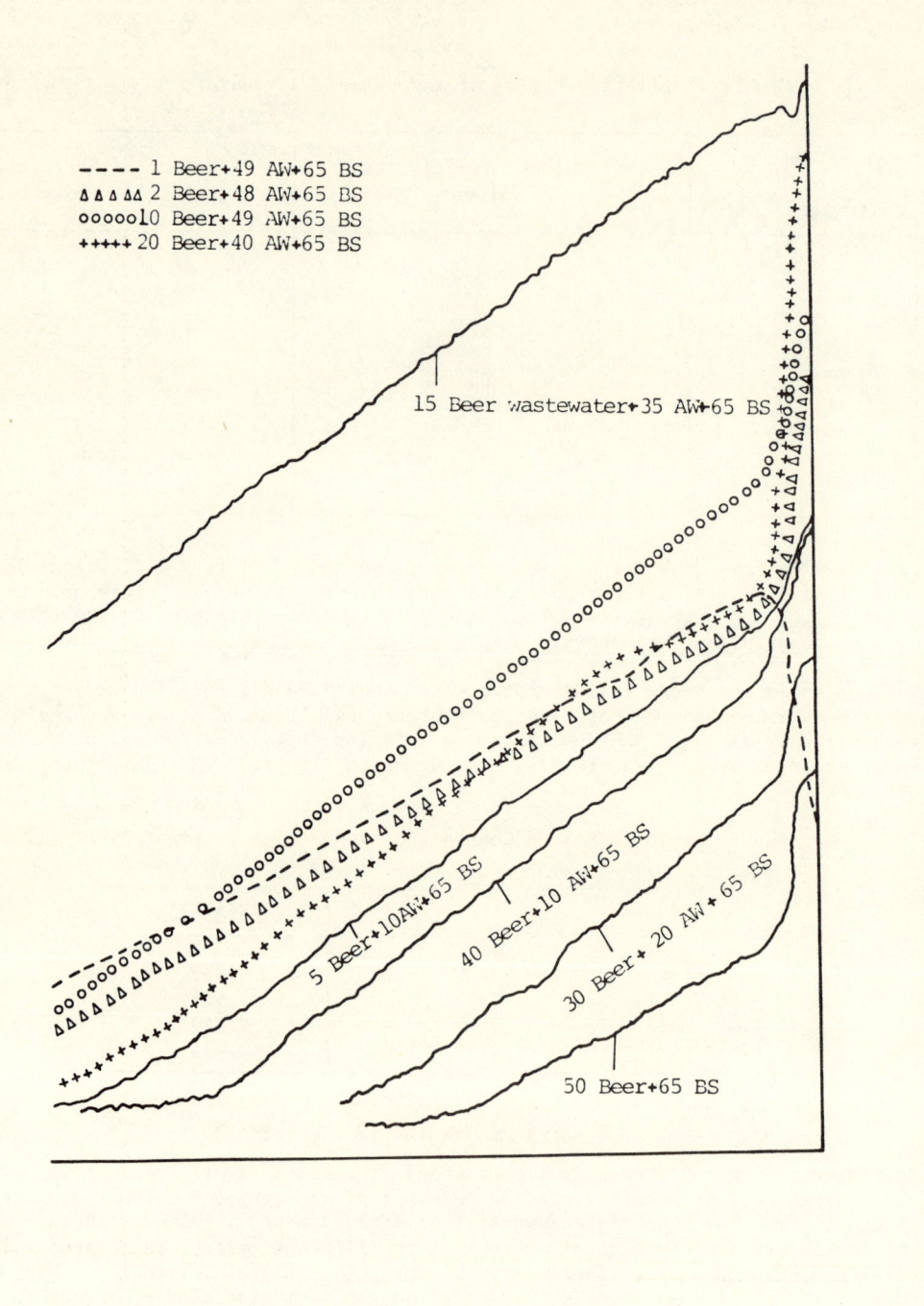

Fig. 3: Recorded respiration activities for brewery wastes at various dilutions

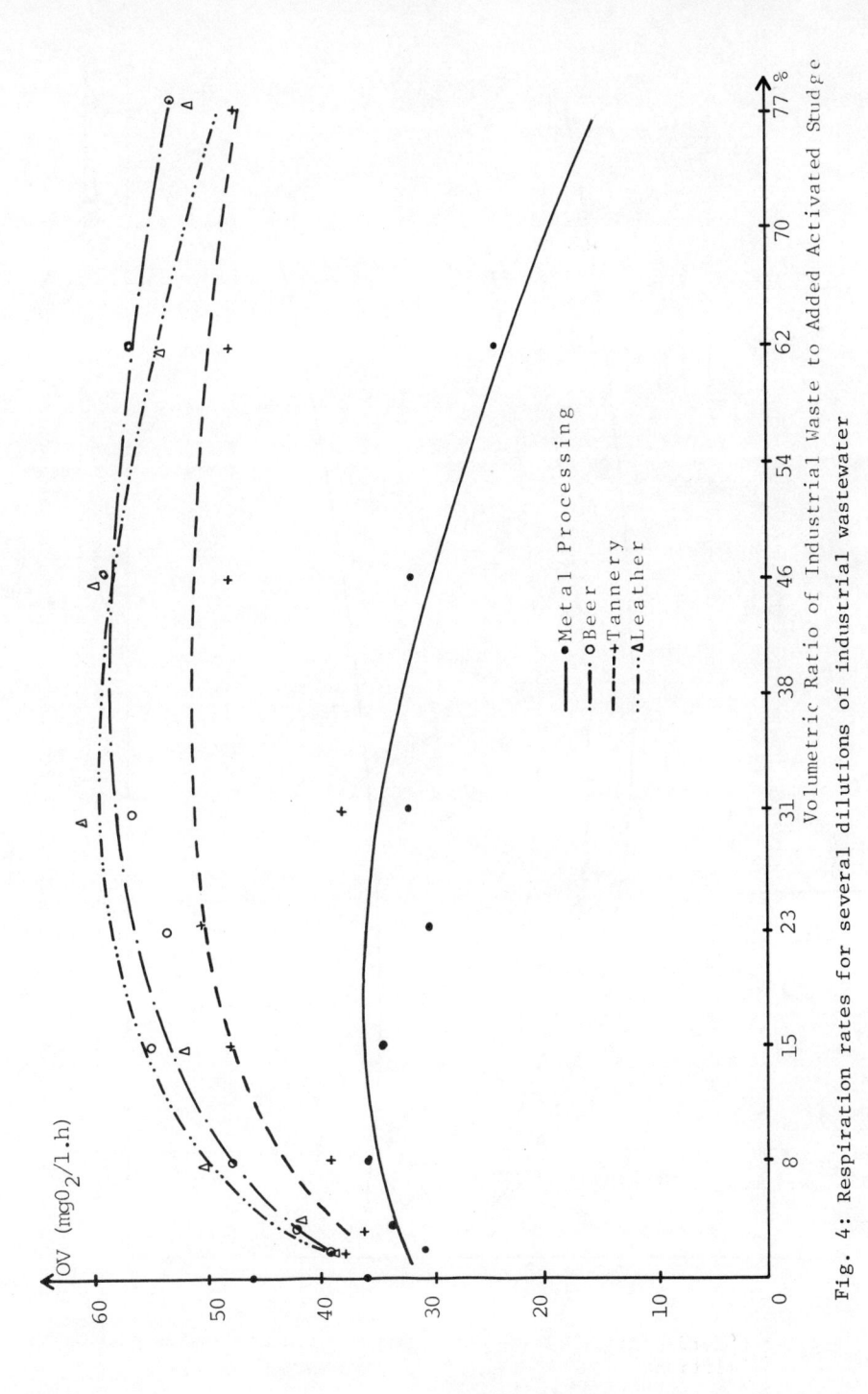

Fig. 4: Respiration rates for several dilutions of industrial wastewater

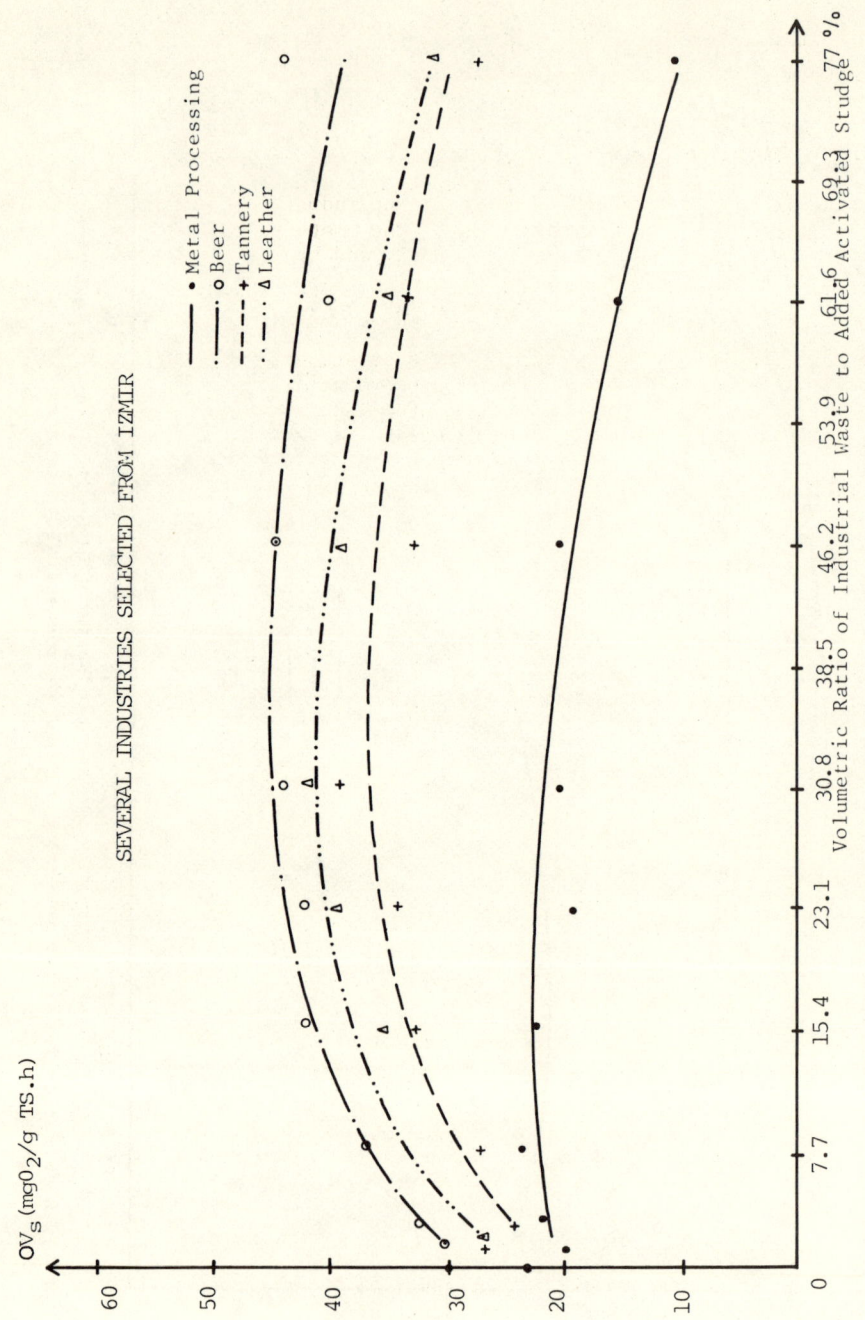

Fig. 5: Specific respiration values for several dilutions of industrial wastewater

rule. Typical of each industry is the rate of decrease in OVS values. This may be observed in wastewaters from other industrial sectors, like breweries and textile plants. Test runs and graphical evaluations with these wastewaters, however, do not show as sharp decreases as the metal-plating wastes. On the other hand, graphical representations of the respiration activity, OV, with respect to several percentage dilutions show that with an increased percentage of industrial wastewater a decrease in respiration activity results. For tannery wastes a rapid decrease in OV is noted for an increased percentage of mixtures, while with metal-plating wastewater a decrease in OV starts only after a volumetric ratio of 23.1% is reached.

From Fig. 4, it may be noted that the different industrial sectors might be classified according to their OV decrease rates. The most rapid OV decrease is seen in metal-plating wastewater, while textile, tannery and brewery wastewaters are listed in the order of decreasing OV reduction rates with respect to increased percentage of mixing.

From the test results, it may be concluded that brewery wastes may easily be treated biologically after mixing with municipal wastewater, even at high mixing ratios. Tannery and textile mill wastewaters, on the other hand, may be digested by bacterial activity rather easily without a major decrease in the respiration rates already obtained in the municipal wastewater treatment. Metal-plating waste, however, cannot be biologically treated alone, and is only treatable when extremely diluted with municipal wastewaters. At lower dilution ratios, treatability is reduced as bacterial life is endangered due to toxicity resulting in decreased respiration activity.

By continuing such tests with various industrial effluents, optimum dilution ratios can be determined for a mixed treatment plant. Also information on the ease of treatment and toxicity can be obtained by this method. The method is applicable to all industrial wastewaters and gives important knowledge on treatability without the necessity of making elaborate laboratory analyses, which increase the cost of design appreciably.

REFERENCES

Grahl, S. (1974). Auswirkungen von Giften aus Industrieabwassern auf die Kläranlage, Seminarbeit, Institut für Siedlungswasserbau und Wassergütewirtschaft der Universität Stuttgart, Stuttgart.

Hixson, A.W. and E.L. Gaden (1950). Oxygen Transfer in Submerged Fermentation, *Ind. and Eng. Che., 42,* 1792.

Hospodka, J. (1966). Oxygen Adsorption Rate-Controlled Feeding of Substrate into Aerobic Microbial Cultures, *8,* 123.

Kaiser, R. (1967). Ermittlung der Savestoffzufuhr von Abwasserlüftern unter Betriebsbedingungen. Heft 1, Technische Hochschule, Braunsweig. Direktor: O. Prof. Dipl.H. Habekost.

Samsunlu, A. (1977). Endüstri Pis Sularının Bakterilerin Aktivitelerine Etkisi. *TÜBİTAK VI. Bilim Kongresi Çevre Araştırmaları Grubu Tebliğleri,* Ankara, 17-21 Ekim, 1977.

Sekulov, I. and B. Dieter (1970). Untersuchungen zur Schnellen Bestimmung der Aktivität von Belebtschlämmen, *3,* 1, 18.

Sekulov, I. and O. Tabasaran (1976). Ergebnisse Orientierender Untersuchungen zur Reinigung von Pyrolyse-Gaswaschwässern, Stuttgart.

Straten, G. and H. Witte. Vorschlag zur Beurtellung der Biologishen Abbaubarkeit von Abwässern Unterschiedlicker Herkunft. *Zeitschrift für Wasser und Abwasser Forschung,* 9, Jahrgang, 38.

DISCUSSIONS

Taygun, N. (Turkey) : Great variations are observed in the nutrients existing in industrial wastewater, depending on their type. In the industrial wastewater you used, there was a nitrogen and phosphate deficiency. Did you compensate for this by adding nitrogen and phosphate from some external source?

Samsunlu, A. (Turkey) : The section of my model plant comprising of the aeration and final settling tanks (Fig. 2) was operated with domestic wastewater, although the actual plant was operated with only industrial wastewater. From this section I took the activated sludge containing the cultured microorganisms and added specific amounts of industrial wastewater without changing its structure in any way. I then examined the nutrients in the activated sludge taken from the domestic wastewater, which was held constant at 65 ml. I was particularly interested in the degradability of the wastewaters when mixed.

Sarıkaya, H. Z. (Turkey): Instead of taking a volumetric mixture of domestic and industrial wastewaters, wouldn't it have been better to take their BOD and COD ratios? Since the pollution load of the industrial wastewater used will vary, the volumetric ratios may give erroneous results.

Samsunlu, A. (Turkey) : In another section of my work the BOD and COD of the wastewaters were measured, but the ratios obtained have not been presented here.

Sarıkaya, H.Z. (Turkey) : How do you explain the reduction in respiration rate which occurs with an increase in the proportion of industrial wastewater, as shown in the graphs? Can it be due to a nutrient deficiency?

Samsunlu, A. (Turkey) : I do not think so. The proportion of activated sludge to which the industrial wastewater was added was always 50% or more. In the first three experiments carried out, the nutrient level was quite high. To my mind, I attribute the decrease in respiration rate more to the fact that the industrial wastewater was not readily biodegradable since the bacteria could not adapt to the environment. This is supported by other studies I conducted on industrial wastewater, including that from the pharmaceutical industry, phenol and brewery wastewater.

Sarıkaya, H.Z. (Turkey) : Why did you choose your method of measuring the respiration rates, rather than the Warburg respirometer?

Samsunlu, A. (Turkey) : You know that there are at least 13 or 14 different methods, chemical and others, for measuring the amount of oxygen. The method I chose was readily available at Stuttgart University, where it has been successfully used for the last few years. It was developed there in order to overcome the difficulties envisaged with the Warburg method with regard to the recording of results and the length of time involved.

Tabasaran, O. (Germany) : Could you give further information on the adaptation phenomenon?

Samsunlu, A. (Turkey) : From the investigations which I conducted on phenol, pharmaceutical, tannery and brewery wastewaters, I have observed that when adaptation of the microorganisms takes place, there is a marked increase in respiration activity and a subsequent increase in treatment efficiency. Adaptation can be achieved when industrial wastewater is fed at a constant amount to the municipal wastewater treatment plant. If, however, there is a shock feeding, the bacteria are greatly affected, and the respiration activity, including the endogeneous respiration, falls to the minimum level.

Velioğlu, S. G. (Turkey): Were you able to prevent any oxygen change when transferring your samples from Izmir to Stuttgart?

Samsunlu, A. (Turkey): Yes. The samples were transferred to the laboratories there and refrigerated within 2 or 3 hours.

Velioğlu, S. G. (Turkey): Did you make any pH measurements in the course of your investigations? I would have thought that acidity would be present in the waste discharged from the metal industry.

Samsunlu, A. (Turkey): The pH of one sample was taken, but we did not take any pH values related to dilution. In any case the dilution level was held constant.

Velioğlu, S. G. (Turkey): In other words, after a composite sample was made, the pH value was not measured. Can you give any figures for the BOD of the wastewater from the brewery - for example, 100 or 1000?

Samsunlu, A. (Turkey): As you know, samples from a brewery and measurements have to be taken at different times. I do not have the results here now. As far as I remember, the BOD value varies between 700 and 1200 for a period of five days.

Velioğlu, S. G. (Turkey): Couldn't the reduction in respiration rate with the increase in the amount of brewery wastewater be attributed to the excessive organic loading - a BOD level of between 700 and 1200?

Samsunlu, A. (Turkey): Certainly. We are working on a method of increasing the total solids content in the aeration tank when a shock feeding is given to the microorganisms.

Pollution Control in the Shuaiba Industrial Area in Kuwait

E. AL-KHATIB* and I. I. ESEN**

*Pollution Control Center, Shuaiba Area Authority, Kuwait
**Department of Civil Engineering, Kuwait University, Kuwait

ABSTRACT

Most large-scale industries in Kuwait are located in the Shuaiba Industrial Area, which lies 50 km south of Kuwait City. Some of these industries are major sources of pollution. Air and water pollution in the area is continuously monitored and controlled by the Air and Water Pollution Control Center of the Shuaiba Area Authority, which is the main regulatory body responsible for administration, general planning and development of the Shuaiba Industrial Area. The pollution control programs developed in the area are a good example of the application of modern technology in the rapidly developing Middle-Eastern countries. This paper discusses the details of the air and water pollution control programs administered by the Shuaiba Area Authority.

INTRODUCTION

The Shuaiba region of Kuwait, situated 50 km south of Kuwait City, is heavily industrialized. At present, there are two oil refineries, two power production and water desalination plants, one major petrochemical plant and a cement factory. In addition to these, an appreciable number of small-scale plants are in operation. Also, some other major and minor industrial plants are either at the planning, design or construction stages.

The Shuaiba Industrial Area is administered by the Shuaiba Area Authority, which is the governmental regulatory agency responsible for administration, general planning and development of the area. The Shuaiba Area Authority has full responsibility for environmental pollution control over the entire region. In this respect, water and air quality is continuously monitored by the Air and Water Pollution Control Center of the Shuaiba Area Authority. In addition, large-scale pollution studies are contracted to various government agencies and to foreign and domestic consulting companies.

In this paper, a general outline of the industrial area will be given, and pollution control programs administered by the Shuaiba Area Authority will be reviewed.

GEOGRAPHY OF THE AREA

Kuwait is situated at the northwestern corner of the Arabian Gulf. It is bounded to the north by Iraq and to the west by Saudi Arabia, occupying an area of 15.000 km^2. It lies between the latitudes 28^{0}N and 30^{0}N and between the longitudes 46^{0}E and 48^{0}E.

The topography of Kuwait is generally flat to gently undulating, occasionally broken by low hills and shallow depressions. The ground surface rises gradually from the east towards the south-west. The surface soil is mainly sand with some fluvial deposits and low vegetation.

The climate of Kuwait can be described as being of a dry hot desert type. The region has mild winters and very hot summers. The barometric pressure varies widely, being highest in winter and lowest in summer.

In winter, the prevailing winds are NW with a frequency of about 14 - 17 days per month and SE with a frequency of 7 - 8 days per month. In spring, the NW winds tend to decrease to 8 - 11 days per month with S or SE winds occurring at about the same frequency. Strong winds blow from the NW and SE. In summer, the NW wind (Shamal) is strong during the daytime, and light at nights. The Shamal winds prevail for nearly 40 days, mostly during the period from June 8 to July 18. During the second half of the summer, winds are more variable. In autumn, NW winds occur for about 10 days in October and 14 days in November.

Winters in Kuwait are cool. The mean January temperature is 12.7^0C. The mean daily minimum temperature falls to 7.9^0C in January with a mean daily maximum of 18.5^0C. Temperatures in spring are very changeable, becoming high during the influence of tropical air masses. Summers are very hot, especially in the moths of July and August. Maximum temperatures are observed in July with a daily maximum of 48^0C and a minimum of 28.8^0C. The autumn temperatures are slightly less than those in summer, the daily maximum being 35.5^0C and the daily minimum 19.5^0C.

THE SHUAIBA INDUSTRIAL AREA

The Shuaiba Industrial Area is situated 50 km south of Kuwait City on the desert coastline of Arabian Gulf, as shown in Fig. 1. This site was originally allocated in 1964 and was mainly selected due to sufficient depth of water near the coastline and its proximity to sources of energy, crude oil and natural gas.

The Shuaiba Industrial Area amounts to 8.4 km^2, of which 60% has been utilized for industry and associated activities. A further area of 5 km^2 was allocated in 1974.

The major industries operating in the area are as follows:

1. The Shuaiba South Power and Water Production Station, which is operated by the Ministry of Electricity and Water, comprises six units for water distillation, each having a distillation capacity of 15.000 m^3/d, together with six steam units having a total power production capacity of 804 MW. Planned expansions show that the eventual water distillation capacity of the plant will be 136.000 m^3/d. The basic feedstocks are seawater, natural gas and light fuel oil. The date of commissioning was 1970.

2. The Shuaiba North Power and Water Production Station is also owned and operated by the Ministry of Electricity and Water. It comprises a steam turbo alternator capable of generating 400 MW of electric power and water distillation units capable of producing a total of 63.500 m^3/d of distilled water. The basic feedstocks are seawater, natural gas and light fuel oil. The date of commissioning was 1965.

3. Kuwait National Petroleum Company Refinery is the world's first all - hydrogen refinery. It is owned by the Government of Kuwait. This refiner is capable of producing a wide range of petroleum products such as gasoline, kerosene, naphtha, diesel oil, fuel oil and sulphur. The present oil production capacity is about 21.500 m^3/d. The basic feedstocks are crude oil and natural gas. The date of commissioning was 1968.

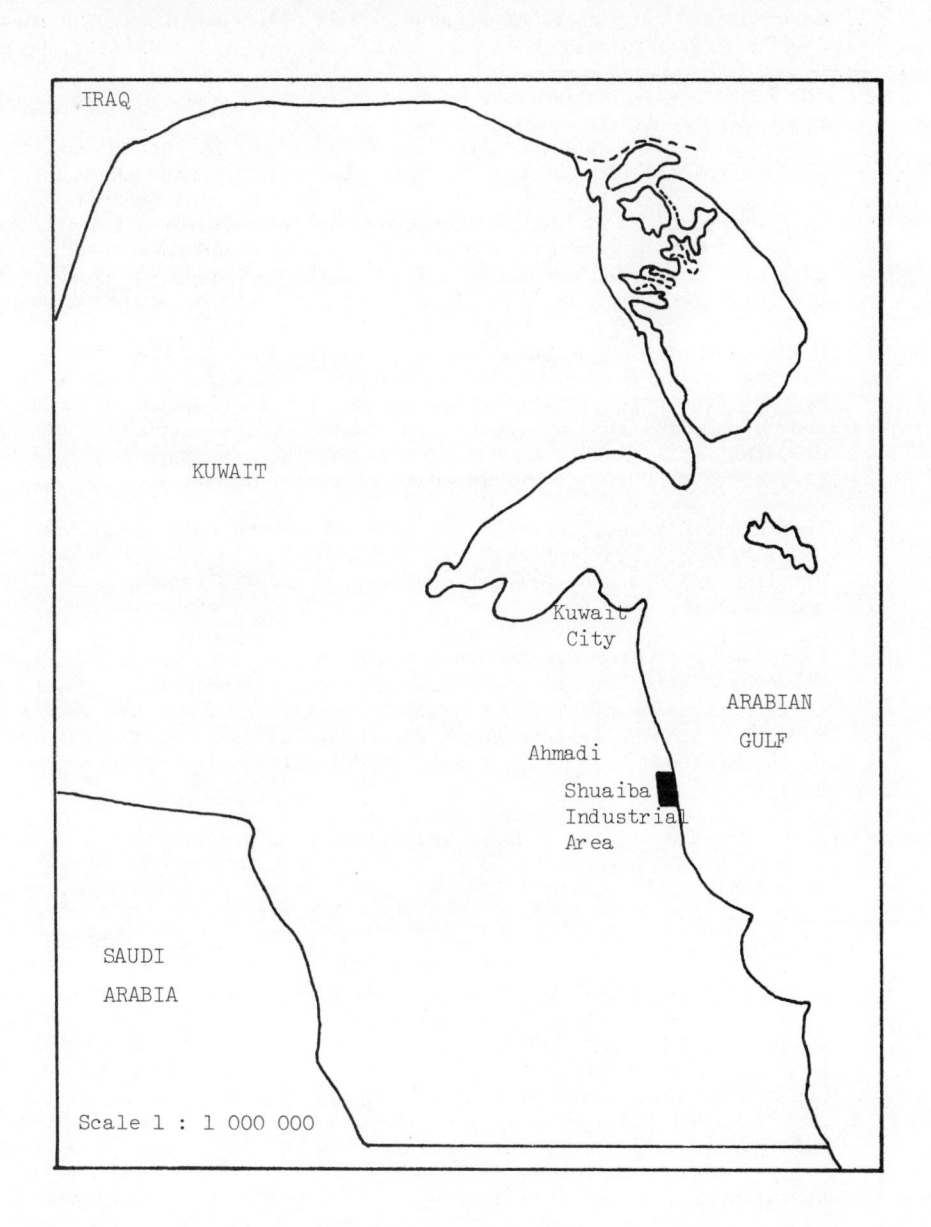

Fig. 1. Geographical Location of Shuaiba Industrial Area

4. Wafra Refinery was previously known as Aminoil Refinery. It is owned by the Government of Kuwait and the major products are naphtha and desulfurized fuel oils. At present, crude oil production is about 17.200 m^3/d.

5. The Petrochemical Industries Company Plant A is a chemical fertilizer plant. It produces liquid ammonia (400 t/d), sulfuric acid (400 t/d), ammonium sulfate (500 t/d) and urea (550 t/d). The basic feedstocks are natural gas and sulfur. The date commissioning was 1966.

6. The Petrochemical Industries Company Plant B is also a chemical fertilizer plant. It produces liquid ammonia (two plants, each with a capacity of 880 t/d) and urea (two plants, each with a capacity 700 t/d). The basic feedstock is natural gas. The date of commissioning was 1971.

7. Kuwait Cement Company produces ordinary Portland cement, moderate sulfate resisting cement (type - II) and sulfate resisting cement (type - U). The production at this stage depends on the importation of clinker and gypsum from abroad. The present capacity is 300,000 t/yr but, more grinding facilities are under construction and the capacity will be increased to 1,000,000 t/yr. The date of commissioning was 1972.

8. Kuwait Oil Company, owned by the Government of Kuwait, produces natural gas, crude oil and refining oil. Other major products are bunker fuels, gasoline and bitumen. Present capacity is about 33,400 m^3/d. The date of commissioning was 1949.

9. Liquefied Petroleum Gas is owned by the Ministries of Oil and Finance. The construction of the plant is almost completed, and at present final testing is being carried out prior to operation. The plant is designed to remove propane, butane and higher hydrocarbons from the gas supplied by the Minageesh oil field and the Kuwait National Petroleum Company Refinery.

The minor industries existing in the Shuaiba Industrial Area are summarized in Table 1.

The two major industrial establishments which are still at the planning stage are as follows:

1. The Petrochemical Industries Company has carried out feasibility studies for a proposed ethylene plant to produce 350.000 tons of ethylene per year, using ethene gas as feedstock obtained from the Liquefied Petroleum Gas project. The ethylene will be used to produce high and low density polyethylene and ethylene glycol.

2. The Petrochemical Industries Company plans the building of an Aromatics Plant to produce benzene and xylene.

The industries operating in the area are served by the following port facilities:

1. Shuaiba Commercial Harbour consists of five berths. Expansion of the northern arm is under construction, and after completion six more berths will be available.

2. Shuaiba Oil Pier consists of four berths to accommodate tankers up to 100,000 DWT.

3. South Ahmadi Oil Pier consists of four berths.

Table 1 Minor Industries within the Shuaiba Industrial Area

Industry	Basic Feedstocks	Major Products & Capacity
Dresser Company		Drilling muds: borite (8 t/h) Bentonite (5 t/h)
Kremenco		Industrial maintenance and cleaning service for process plant equipment
Kuwait Industrial Gases	Ambient air	Liquid nitrogen (200 m^3/h), Liquid oxygen (200 m^3/h), Liquid argon (10 m^3/h)
Kuwait Refrigeration & Oxygen Co. Ltd.	Ambient air	Liquid nitrogen (216 m^3/h), Liquid oxygen (240 m^3/h), Liquid argon (10 m^3/h)
Kuwait Sulfur Company	Sulfur	Grinding and packing of sulfur (5.000 t/annum)
United Fisheries of Kuwait	Shrimps & Fish	Fish processing and freezing, 40 t/d shrimps, 10 t/d fish metal and residue
Packing & Plastic Industries	Polyethylene & Polypropylene	Packing bags of polyethylene and polypropylene (12 million bags per annum)

MAJOR POLLUTANTS IN THE SHUAIBA INDUSTRIAL AREA

In the Shuaiba Industrial Area, the rate of pollution of the environment as a result of various atmospheric emissions and industrial wastewater discharges has increased continuously throughout the last ten years. On the other hand, Kuwait and most other Gulf States are unique in that the sea is their major source of drinking water. Thus, the quality of the Gulf Water is of primary importance. It should also be noted that since the Arabian Gulf is virtually a closed sea, self-purification occurs at a much reduced rate.

Major pollutants in the Shuaiba Industrial Area can be summarized as follows:

1. Atmospheric Pollutants

    a) Sulfur compounds (SO_2, SO_3, H_2SO_4, H_2S, R-SH)

    b) Nitrogen compounds (NH_3, NO_x, $(NH_4)_2SO_4$ dust, urea dust)

    c) Hydrocarbons (C_xH_y)

    d) Heavy and transition metals (compounds of lead and calcium)

    e) Other atmospheric pollutants (oxidants, products of incomplete combustion (CO, C), cement dust)

2. Marine pollutants
   a) Sulfur compounds (SO_4^{2-}, S^{2-})

   b) Nitrogen compounds (NH_4^+, NH_3, urea, NO_3^-, NO_2^{2-})

   c) Hydrocarbons (oil, phenols, mercaptans)

   d) Heavy and transition metals (Pb, Cr, Cu, Hg)

   e) Thermal pollution (cooling water discharges)

   f) Other marine pollutants (CN, detergents, inorganic phosphorous, organic nitrogen compounds)

SHUAIBA AREA AUTHORITY

The Shuaiba Area Authority is the regulatory body responsible for administration, general planning and development of the Shuaiba Industrial Area and Mina Abdulla Industrial Area. Its main functions and responsibilities are laying down conditions for and approval of new industrial projects; monitoring and control of environmental pollution; coordination of direct and indirect services for all the industries in the area in cooperation with the ministries and other governmental agencies; supplying sea water for cooling and natural gas as fuel; serving the industries with port facilities; and construction of roads and drainage systems.

The Shuaiba Area Authority has participated in numerous activities in connection with air and water pollution control problems since 1966, and in 1970 the Air and Water Pollution Control Center was established.

POLLUTION MONITORING IN THE SHUAIBA INDUSTRIAL AREA

The Air and Water Control Center of the Shuaiba Area Authority is responsible for the control and monitoring of pollution in the Shuaiba Industrial Area. In this respect, separate programs are simultaneously administered in the area. In this section, a brief review of these pollution control programs will be presented.

Liquid Effluent Monitoring

The sampling of liquid effluent streams within the Shuaiba Industrial Area and that of the external receiving waters are carried out by instantaneous manual sampling, party automatic on-line sampling, and by taking partly automatic measurements for offshore waters.

Sampling locations and parameters measured are summarized in Tables 2 and 3.

In addition to those programs directly carried out by the Air and Water Pollution Control Center, at present a seperate ocean monitoring program is being introduced with the cooperation of Kuwait Institute for Scientific Research. This program involves the continuous monitoring of water quality parameters such as dissolved oxygen, conductivity and pH, together with oceanographic parameters such as current velocity and direction, wave height, wind speed and direction and temperature. The system used is completely automatic and will continue for 18 months.

Air Quality Monitoring

The scheme for the air quality monitoring of the area is based on three major activities which are the ambient air quality monitoring, monitoring of emissions at the source and routine spot measurements.

Table 2 Sampling Locations and Major Pollutants for Sea Water Systems in the Shuaiba Industrial Area[1]

Sea Water Pollutants	SNPS I	SNPS O	SSPS I	SSPS O	SCWPS I	PIC(A) O	KNPC O	PIC(B) O	KOC O	Marine Monitoring 9 sea locations
Physical parameters										
Temperature	x	x	x	x	x	x	x	x	x	x
Conductivity	x	x	x	x	x				x	
pH	x	x	x	x	x	x	x	x	x	x
Chemical parameters										
Total hydrocarbon	x	x	x	x	x	x	x	x	x	
Total organic carbon	x	x	x	x	x	x	x	x	x	x
Ammoniacal nitrogen	x	x	x	x	x	x	x	x	x	x
Total nitrogen	x	x	x	x	x	x	x	x	x	
Sulphide	x	x	x	x	x	x		x	x	
Residual free chlorine	x	x	x	x						
Urea						x		x	x	
Nitrite/Nitrate	x		x		x	x	x	x	x	x
Phosphate		x		x		x			x	x
Heavy metals										
Plankton count										x

[1] The following abbreviations are used for various industrial establishments: SNPS = Shuaiba North Power Station, SSPS = Shuaiba South Power Station, PIC = Petrochemical Industries Company, KNPC = Kuwait National Petroleum Company, KOC = Kuwait Oil Company, I = Intake, O = Outfall.

Table 3 Sampling Locations and Major Pollutants for the Liquid Effluent Streams within the Shuaiba Industrial Area

Liquid Effluent Pollutants	KNPC API Sep.	PIC(A) East Surf. Drain	PIC(A) West Surf. Drain	PIC(A) Chem. Drain	PIC(B) North Surf. Drain	PIC(B) South Surf. Drain	PIC(B) Chem. Drain	PIC(B) Neut. PIT	KOC API Oil/Water Seperator	KNPC(WAFRA) API Oil/Water Seperator
Physical parameters										
Temperature	x	x	x	x	x	x	x	x	x	x
Conductivity	x	x	x	x	x	x	x	x	x	x
pH	x	x	x	x	x	x	x	x	x	x
Chemical parameters										
Total hydrocarbon	x		x	x	x	x	x	x	x	x
Total organic carbon	x		x	x	x	x	x	x	x	x
Ammoniacal nitrogen	x		x	x	x	x	x	x	x	
Urea			x	x	x	x	x	x		
Nitrite/Nitrate			x	x	x	x	x			
Sulphide	x		x	x	x	x	x	x	x	x
COD	x		x	x	x	x	x	x	x	x
Chloride	x		x	x	x	x	x	x	x	x
Total suspended solids	x		x	x	x	x	x		x	x
Total phosphate	x		x	x	x	x	x		x	x
Phenols	x			x			x			x
Mercaptons	x									x
Total cyanides	x		x	x					x	x
Heavy metals	x	x	x	x	x	x	x	x	x	x

Ambient air quality is monitored by a network of thirteen fixed sites selected at critical locations within the Shuaiba Industrial Area. Average 24 - hour concentrations of NH_3 and SO_2 are measured using gas-train systems. In addition, at three continuous monitoring stations, parameters such as SO_2, NH_3, H_2S, NO_x, hydrocarbons and particulates are continuously measured, and the data collected is transmitted by telephone lines to the computer at the Air and Water Pollution Control Center. Also, at nine locations, dust deposits and particulate material are analyzed calcium chloride, for total ammonia, calcium chloride, silicate, urea and sulfate by using dust bowls and high volume samplers.

Monitoring of emmisions at the source is performed by manual stack tests, and the schedule for this purpose is given in Table 4. At present, continuous monitoring of emissions by a fully automated system is under consideration.

Table 4 Stack Monitoring Schedule

Source of Emission	Measured Parameters	Frequency
KNPC flare stack	Sulfur content, flow rate, temperature	once/week
Sulfur plant incinerator of KNPC	H_2S, SO_2 in tail gas, flow rate temperature	once/week
PIC(B) urea prilling tower	Urea dust concentration, flow rate, temperature	once/2 weeks
PIC(B) ammonia purge stream	NH_3, flow rate, temperature	once/week
KCC vents[1]	Cement dust concentration, flow rate, temperature	once/2 weeks
MEW boiler[2] stacks	SO_2, CO, flow rate, temperature	once/week
PIC(A) sulfuric acid stack	Sulfur compounds, flow rate, temperature	once/week

[1]Kuwait Cement Company
[2]Ministry of Electricity and Water

Routine spot measurement of pollutants such as NH_3, H_2S, SO_2, mercaptans and hydrocarbons is made within the vicinity of the industrial plants for purposes of industrial hygine. Measurements are made three times daily using dragör tubes and other portable equipment.

SPECIFIC PROBLEMS

Besides the environmental effects, pollution directly influences the efficiency of the industrial plants in the Shuaiba Industrial Area. In this section, the details of the specific problems related to water and air pollution will be given.

Problems Related to Water Pollution

The cooling water intakes of the South and North Power Stations are quite close to t[he] outfalls of the various industries. Contamination of the seawater by ammonia causes the following problems in these two power plants:

1. Corrosion of the cupro-nickel tubes in the cooling water exchangers.

2. Great increase in chlorine demand.

3. Deleterious effect on the distilled water due to ammonia being carrie[d] over during the distillation process.

The Shuaiba Industrial Area has huge oil refineries. Thus, it is expected that oil m[ay] find its way to the refinery effluent stream and eventually reach the sea. The probl[ems] caused by oil are:

1. Adverse effects on the efficiency of heat transfer surfaces.

2. Altering of the specifications of the distilled water produced by upsetting power plant operations.

3. Adverse effects on the marine life in the vicinity of Shuaiba offshor[e] waters.

One other major problem related to the pollution of the marine environment is due to thermal pollution resulting from the large amounts of cooling water discharged. Recirculation of hot cooling water greatly reduces the efficiency of the power and water distillation plants.

Problems Related to Air Pollution

Besides public health problems, air pollution is causing serious trouble by accelera[t]ing corrosion. It has also been observed that silver plated contacts are attacked by H_2S and SO_x, and urea deposits on the insulators have caused electric short circuits in the power plants.

AIR AND WATER QUALITY STANDARDS

The standards for air and water quality are summarized in Tables 5, 6 and 7.

Table 5 lists the ambient air quality criteria. The values shown are taken from the 1976 report of the "Pollution Control Project for Shuaiba Industrial Area," prepared by the Cremer and Werner Consulting Company (1967). In this table, the following cri[t]eria are given:

1. Community maximum allowable concentration (CMAC), defined as the invi[]olate level of a pollutant which should never be exceeded outside the site boundary of an industry.

2. Community normal maximum (CNM), defined as the ambient concentration of a pollutant that may only be exceeded on 2% of the occasions for a one - hour sampling period.

3. Long - term average concentration (LTAC), representing the level belo[w] which action to reduce concentrations of pollutants further would not be generally justified.

Table 5 Ambient Air Quality Criteria

Substance	Primary Criteria CMAC (ppm)	Primary Criteria CNMC (ppm)	Secondary Criteria LTAC (ppm)
Ammonia	2.5	0.8	0.13
Carbon monoxide	35.	10.	2.
Chlorine	0.1	0.03	–
Hydrogen sulfide	0.1	0.03	–
Total mercaptans	0.05	0.017	–
Nitrogen oxide	0.5	0.17	0.025
Oxidant (as Ozone)	0.06	0.02	–
Sulfur oxides	0.5	0.17	0.025

Table 6 shows the accepted international standards for particulate emission criteria. It should be noted that the overriding concern of the Shuaiba Area Authority in controlling particulate emissions is to safeguard electric supplies.

Table 6 Particulate Emission Criteria

Substance	Source of Emission	Emission Criteria (mg/m^3)
Cement dust	Cement grinding and crushing	100
Heavy metals, e.g. antimony, lead, mercury		23
Sulfuric acid	Sulfuric acid plant	230
Total solid particulate matter, dust & grit larger than 10 micron	Incinerators, cupolas, process furnaces and heaters, sintering plant	460
Fines and fumes less than 10 micron		115
Urea	Electric power plants, Fertilizer plant	130

The water quality criteria for the liquid effluents for the Arabian Gulf at Shuaiba is shown in Table 7, which lists the desirable environmental values (DEV), threshold hazard values (THV) and the reference discharge loads (RDL) from the total industrial unpolluted sea water coolant (Air and Water Pollution Control Center, 1976).

Table 7 Liquid Effluent Criteria

Parameter	DEV	THV	RDL (t/d)
pH	8.0	5.5 to 9.0	-
DO	5.4 ppm	2.0 ppm	-
COD	2.0 ppm	4.0 ppm	15.0
BOD	2.0 ppm	4.0 ppm	-
Total hydrocarbon	N.D.	0.5 ppm	0
Phenolics	0.05 ppm	0.1 ppm	0.25
Sulfides	0.005 ppm	0.01 ppm	0.025
Ammoniacal nitrogen	0.02 ppm	0.5 ppm	0.025
Total nitrogen	0.5 ppm	1.3 ppm	-
Inorganic phosphate	0.001 ppm	0.02 ppm	0.005 to 0.10
Cyanide	N.D.	0.01 ppm	0
Alkyl mercury	N.D.	0.0001 ppm	0
Total mercury	N.D.	0.0001 ppm	0
Arsenic	0.01 ppm	0.05 ppm	0.05
Cadmium	0.01 ppm	0.03 ppm	0.005
Lead	0.01 ppm	0.05 ppm	0.05
Chromium	0.05 ppm	0.1 ppm	0.25
Copper	0.01 ppm	0.05 ppm	0.005
Zinc	0.05 ppm	0.1 ppm	0.005
Iron	0.05 ppm	0.3 ppm	0.25
Manganese	0.02 ppm	0.1 ppm	-
Nickel	0.002 ppm	0.1 ppm	0.01
Coliform	100 NPN/mℓ	2000 NPN/mℓ	-

CONCLUSIONS

Extensive work is being carried out on the control of air and water pollution in the Shuaiba Industrial Area in Kuwait. The governmental body responsible for the contro of pollution is the Shuaiba Area Authority, and since 1970, the Air and Water Pollution Control Center has been carrying out continuous manual and automatic sampling for this purpose. In addition, large-scale pollution control programs are conducted is cooperation with private companies and other governmental agencies such as the Kuwait Institute for Scientific Research (1977). International standards for air and water quality criteria are rigidly followed.

REFERENCES

Air and Water Pollution Control Center (1976). *Codes of Practice and Environmen Guidelines*. Shuaiba Area Authority, Planning and Development Dept., Kuwait.

Cremer and Werner Consulting Engineers and Scientists (1967). *Pollution Control Project for Shuaiba Industrial Area*. London.

Kuwait Institute for Scientific Research (1977). *Bathymetric Survey of the Shuaiba Offshore Area*. KISR/ESE-R-RT-7801. Kuwait.

Problems Caused by Industrial Waste in Alexandria and their Remedial Action

H. MITWALLY

Department of Environmental Health, High Institute of Public Health,
Alexandria University, Alexandria, Egypt

ABSTRACT

About 35% of the national industries are found in the Alexandria area. The textile, chemical, paper, food products and canning industries are the most important of the various industrial activities. The industrial wastewaters are discharged into the Mediterranean Sea either directly without treatment, or indirectly with or without any treatment. Three main outfalls transfer the mixed industrial wastewater together with either domestic sewage or agricultural drainage or both. This takes place through the collective system of the central city, Abu-Kier drain in the east and Lake Mariot in the west, respectively.

The total amount of wastewater discharged is estimated at six million cubic meters per day; out of this one million is domestic and storm, while a quarter of a million is industrial waste, most of which comes from private sources belonging to the relevant industries. The remaining amount originates from agricultural drainage of the Western Delta area. The total organic load discharged into the sea is estimated at 3,600 tons of oxygen per day. This amount is discharged either directly on to the shore or at a relatively short distance into the open sea; but in both cases this is not more than 1 km from the bathing beaches. Such a short distance is not sufficient for dilution or self-purification; thus, the quality of the recreational and fishing areas is seriously effected.

To control this pollution problem, remedial action ought to be taken so that the treated discharged effluent would conform with the specified standards for recreational beaches. The economical losses as a result of the deterioration in the quality and quantity of fishing and tourism resources are estimated to reach $100 million a year; yet the remedy is just to have firm application of the state law concerning industrial waste treatment. Should treatment take place, this would not cost more than one-tenth of the total losses incurred annually.

SANITARY SURVEY

Introduction

Alexandria city is an elongated strip stretching along the sea shore with a length of more than 30 km and a width ranging from 1 to 5 km. It is bounded on the north by the Mediterranean Sea, by lake Mariot on the south, the Western desert on the west, and by Abu-Kier Bay on the last. It is located uphill with about 80% of its land declining to the north downhill to the sea, and 20% declining to the south downhill to Lake Mariot. According to its unique topographical and geographical situation, Alexandria was planned to dispose of its domestic wastewater either north into the sea or south into the lake.

Present Situation and Research Conducted

The wastewaters from Alexandria are disposed of through three main outfalls, either directly into the sea or indirectly via Lake Mariot as mentioned below.

I. Kait Bay Outfall

The middle part of the city has its collective system directed to a main outfall into the sea which measures 750 m long and 17 m deep. The average daily discharge is 200,000 m^3 per day of domestic sewage. Another 17 small auxiliary outfalls handle the wastewater from the north-eastern part of the city with an average discharge of 50,000 m^3. These small outfalls were designed originally to handle the storm wastewater via overflow weirs directed to the shore. When the discharge increased tremendously as a result of the increase in population, the storm outfalls were resorted to as auxiliary outlets for combined sewage (Sharkawi, 1977).

II. Eastern and Southern Part of the City

A - The collective system of the easternpart of the city is directed to the New Eastern Sewage Treatment Plant with an average daily flow of 60,000 m^3, mostly from industrial sources. The effluent of the treatment plant is discharged into Lake Mariot, which is considered a huge oxidation pond of an area exceeding 6000 acres where self-purification takes place (Saleh, 1974).

B - The industrial area in the southern part of the city disposes of its industrial waste directly into Lake Mariot with an average daily discharge of about 60,000 m^3 through different outlets.

Lake Mariot itself is connected to the sea at the far end from the west. In addition to the industrial waste and the effluent from the eastern sewage treatment plant it receives the agricultural water from the West Bank of Lower Egypt, which ranges from 3 to 4 million m^3 a day. This huge amount of agricultural drainage water-acts as a dilutant on the industrial wastewater received. The action of self-purification in such a large oxidation pond is very obvious. It is worth mentioning here that the BOD of the discharge pumped from the lake to the sea rarely exceeds 30 mg/l. In other words, the total organic load does not exceed 120 tons of oxygen per day (Sharkawi, 1976).

According to the laboratory findings, the effect of industrial wastewater on the Mediterranean Sea at Alexandria from both the Kait Bay and Lake Mariot pumping station outfalls is relatively negligible. The total organic load from Kait Bay and the auxiliary outfalls is about 110 tons per day and it is of a domestic and not an industrial nature. The most important source of industrial wastewater pollution is that discharged into Abu-Kier Bay.

It is worth mentioning here that research conducted during the last ten years on these three major disposal sites has been very thorough.

III. Abu-Kier Bay

The bay can be considered as an extension of the Alexandria sea coast from the East (Abbas, 1971). A study has been carried out to determine the extent and path of pollution in Abu-Kier Bay as a result of the discharging of pumped wastewater from the Abu-Kier drain. The wastewater in the drain consists of irrigation drainage from the agricultural land of Beharia Governorate and industrial waste from the Kafr El-Dawar and Tabia areas.

The drain flows through cultivated areas to the Tabia pumping station on the drain terminal at the south-west corner of Abu-Kier Bay. The main source of pollution in the bay is industrial waste from different establishments discharging their wastewater into the drain. These include textile and weaving, chemical, rayon, die and food-canning factories as well as paper mills.

The amount of discharged wastewater varies from 1½ to 2 million m^3 per day. Such a huge amount of agricultural drainage and industrial waste has detrimental effects on the aquatic life in the bay and the immediate vicinity up to the Edkou area about 20 km from the outfall (El-Samra, 1973). The effect on the resulting pollution can be seen by the naked eye. The different changes in colour of the bay water were so marked that a bounded area like a pollution lake was clearly visible especially during calm and windless days. Some floating waste matter could be seen on the boundaries of the imaginary lake due to the difference in salinity in both the pumped wastewater and the bay water, and the consequent difference in their specific gravity. Most of the time, the lake moves eastward or to the north-east, but when the sea is turbulent it is usually difficult to identify such boundaries as a result of the mutual mixing of both waters. However, it was observed by applying the method of floating bodies, which were tracked from selected sites, that they followed an eastward direction parallel to the shore. This was confirmed by checking the pollutional effects along this path and following the organic decomposed bottom sediments along the shore to El-Meadia and even the Edkou area more than 20 km from the outfall.

RESULTS

Physical, chemical and biological analyses throughout a comlete year revealed the seasonal variation shown in Figs. 1 to 7. Six sampling stations were selected in addition to the outfall sampling station in order to the path of pollution in the bay. The first was about 150 m from the shore in front of the outfall while the last was about 5 km to the east.

Pollution indicators showed the maximum value in spring for the BOD to be 2441 mg/l, while the minimum value was also relatively high, reaching 1179 mg/l in winter. This gives an average of 3900 tons of oxygen per day as the organic load required to cover the BOD. With regard to settleable solids, the amount ranged from 6.6 ml/l in summer to 13.75 in winter, a concentration which gives an average daily amount of 20,000 m^3 as settleable solids. This is in addition to the colloidal but potentially settleable solids which lose their balance as a result of mixing with salty sea water. It was also noticed that a considerable amount of floating oil was observed on the water surface, where the emulsified oil ranged from 7 mg/l in spring to 455 mg/l in winter. Substantial amounts of floating oil and grease were skimmed off by a baffle reaching the bay.

The number of live plankton was infinitesimal near the outfall, while the present ones were dead. However, viable plankton appeared towards the east.

The pollution effect decreased gradually towards the east although it was noticed as organic botom sediments at a distance of more than 20 km. However, the continous addition of organic loads has created a well-defined polluted lake in the bay which extends a few hundred meters westwards and some hundred meters northwards. It stretches eastwards parallel to the sea shore as affected by the water currents and prevailing winds, since there is no efficient mixing of the discharged pollution wastewater and the bay water especially during the calm sea periods.

ECONOMICAL EFFECTS

About $100 million a year is estimated as the total loss in quality and quantity of the fisheries, oyster and mussel beds in the sandy beaches around Alexandria (Mit-

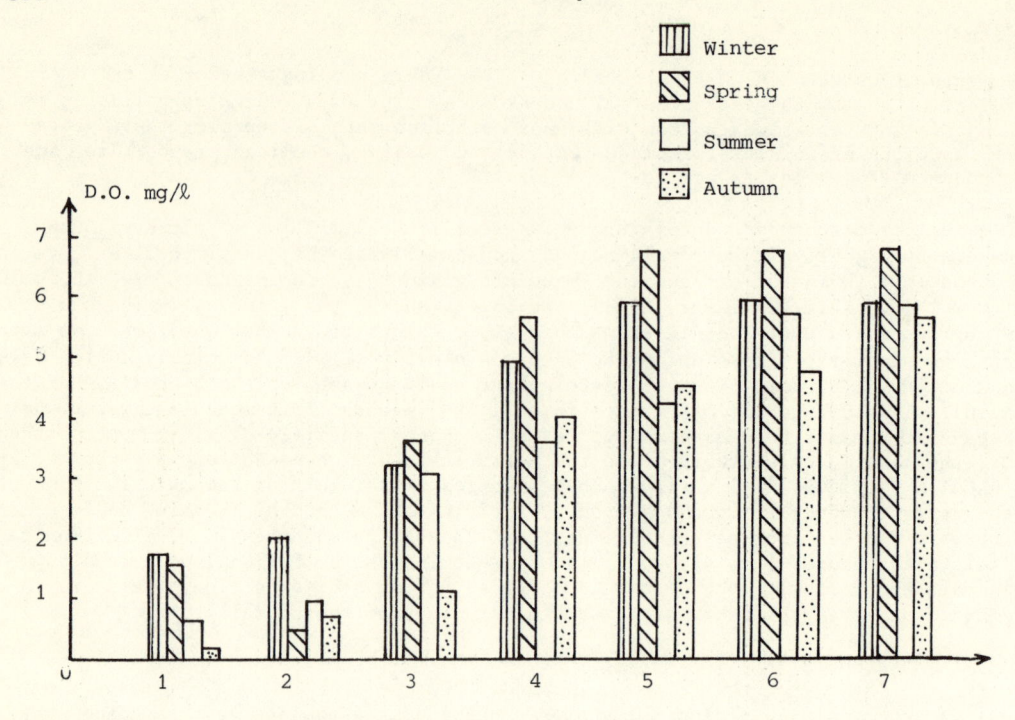

Fig. 1. Mean Levels of Dissolved Oxygen at Different Seasons in Abu-Kier Bay

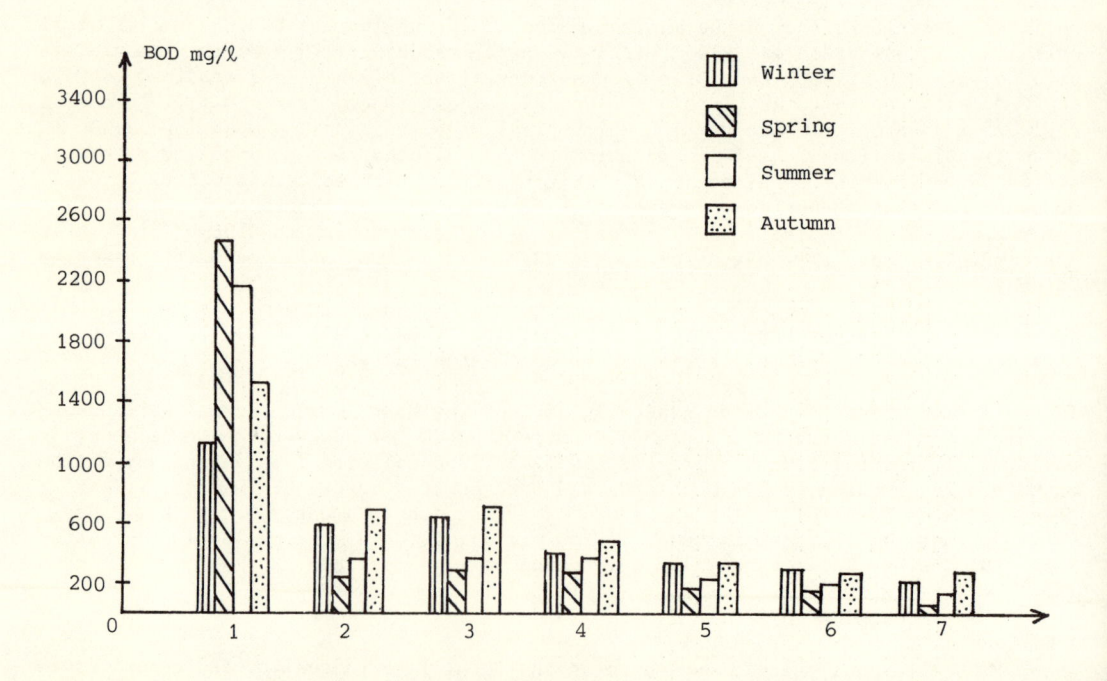

Fig. 2. Mean Levels of Biochemical Oxygen Demand at Different Seasons in Abu-Kier Bay.

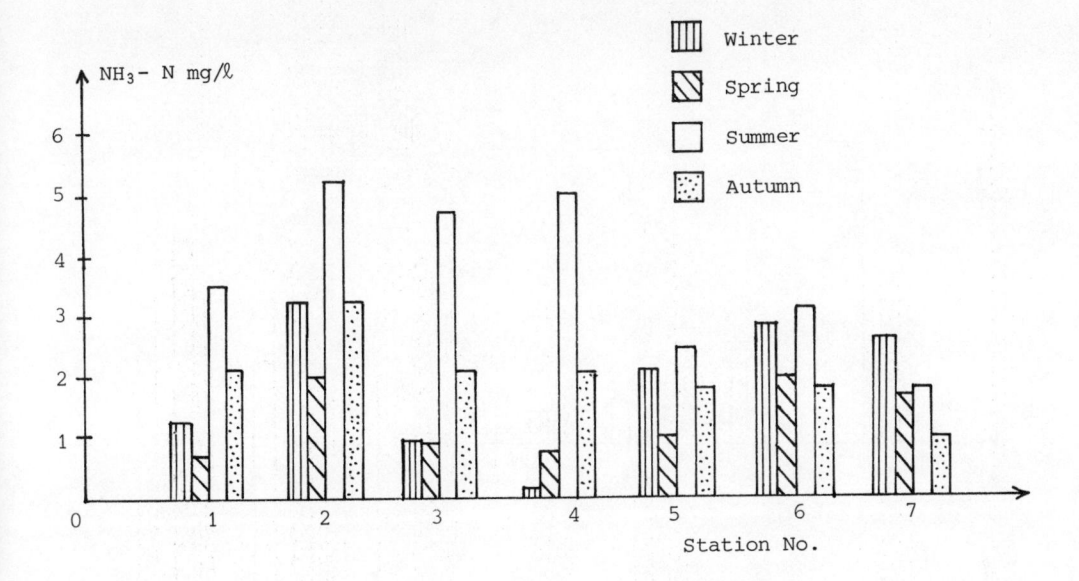

Fig. 3. Mean Levels of Amonia Different Seasons in Abu-Kier Bay

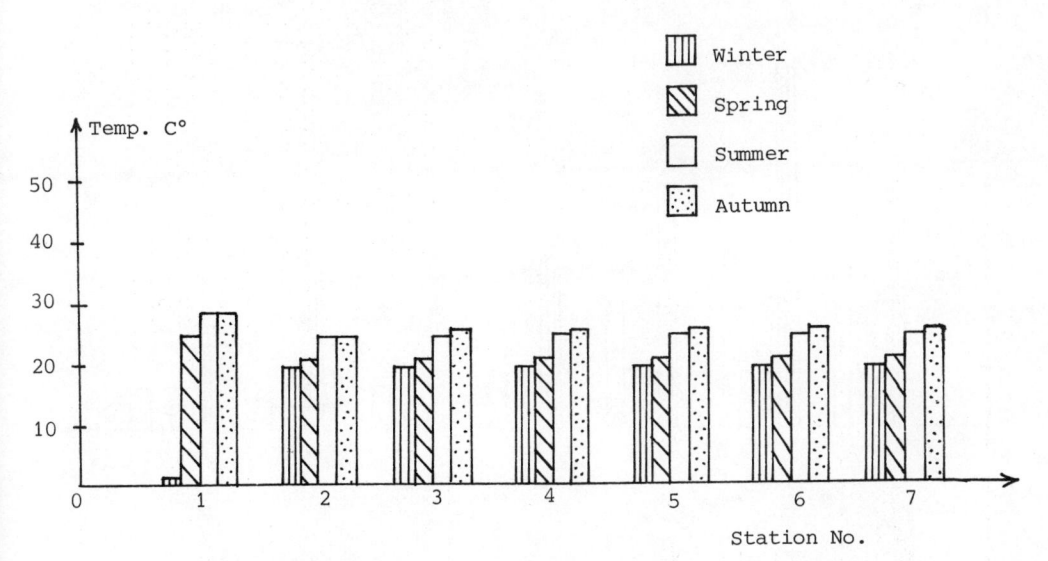

Fig. 4: Mean Temperature at Different Seasons in Abu-Kier Bay

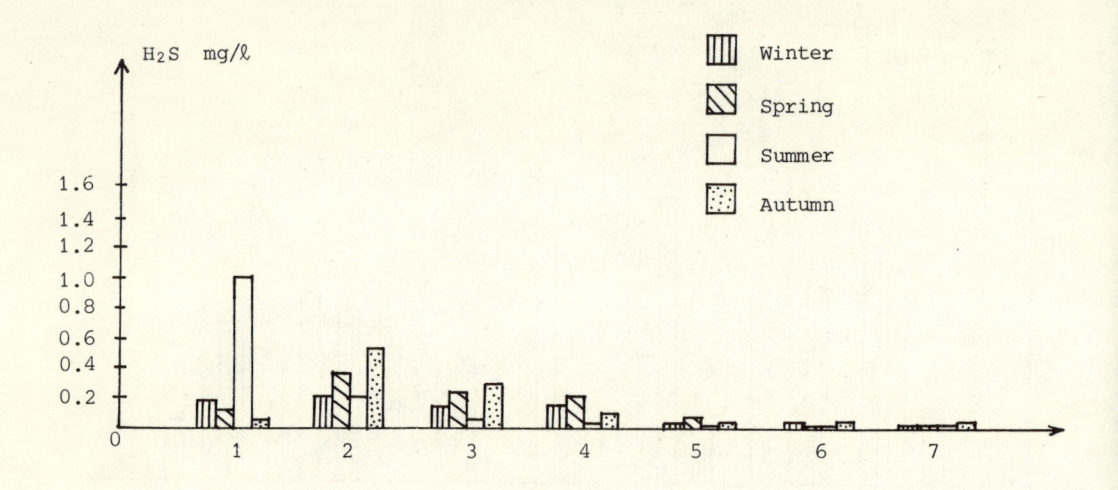

Fig. 5. Mean Levels of Hydrogen Sulfide at Different Seasons in Abu-Kier Bay

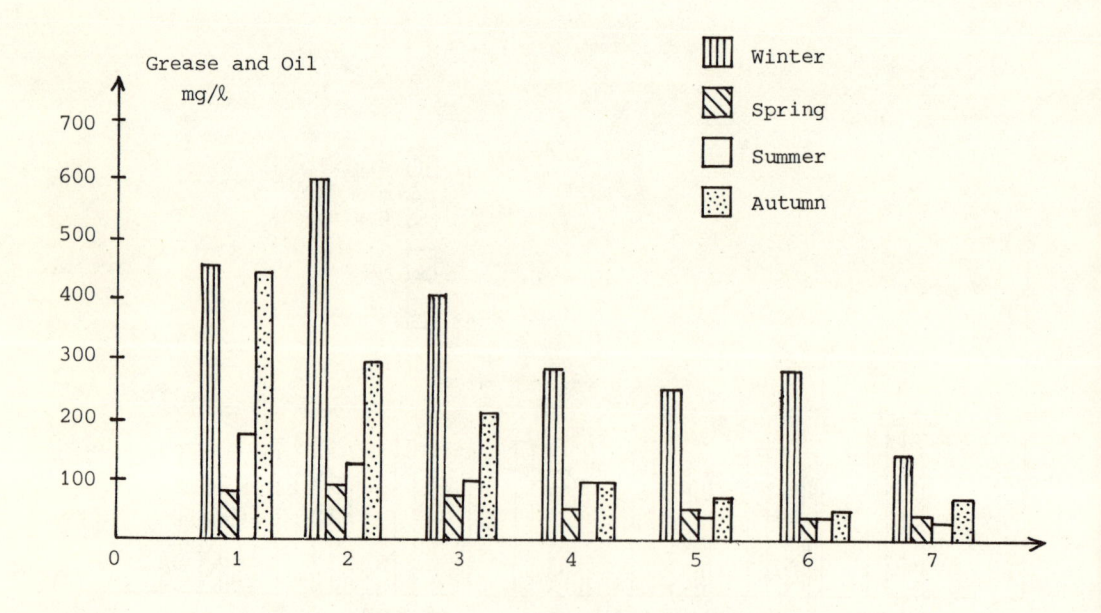

Fig. 6. Mean Levels of Grease and Oil at Different Seasons in Abu-Kier Bay

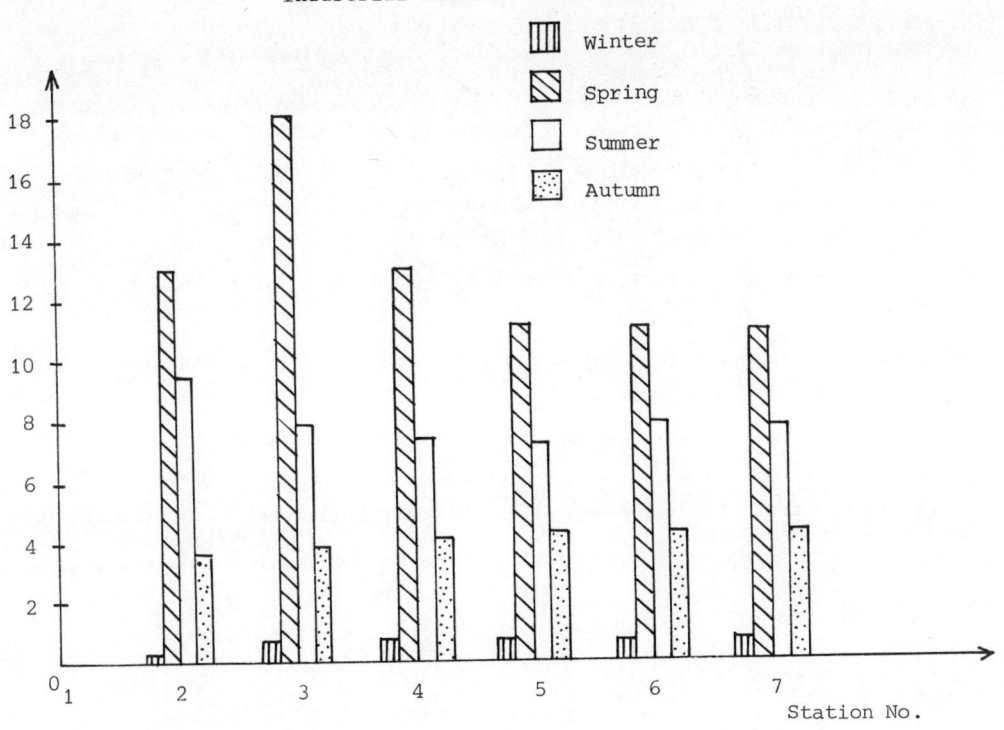

Fig. 7. Mean Plankton Count at Different Seasons in Abu-Kier Bay

wally and Saleh, 1971). Also the effect on recreational returns and tourism investment has been considerable. In addition to these tangible effects, other intangibles such as the effect on public health, and the reduction in the price of neighbouring land should also be considered.

REMEDIAL MEASURES

There has been great reluctance to enforce the law dealing with the disposal of industrial wastes in different water bodies. Sticking to the permissible standards for discharged industrial waste, should they be discharged into water bodies, would not lead to any significant amounts of pollution. The main reason for the leniency in law enforcement is the lack of finance for constructing treatment plants in the industrial establishments belonging to the public sector. However, the cost of having new treatment constructions, either separate for each industry or combined for similar ones, would not amount to more than one-tenth of the economical losses incurred as a result of discharging the raw industrial waste into the sea (Salib, 1979).

Fortunately the State Industrial Authorities are at present seriously considering the subject of treatment and it is expected within the next few years that the pollution of the Mediterranean by industrial wastes in Alexandria will be safely handled.

REFERENCES

Abbas, M. (1971). Studies of Water Pollution of Abu Kier Bay. M.Sc.Thesis, Alexandria University.

Darrel J. J. (1971). Dams and Ecology, *J. Civil Engineering* ASCE., 9.

El-Samra M.A. (1973). Physical and Chemical Studies on Lake Edkou, Abu Kier Bey and the Mixed Water Between them. M.Sc. Thesis, Alexandria University.

Mitwally, H. and F.A. Saleh (1971). Biological Survey on Shellfish in the Local Beaches and Markets Alexandria University, *Bulletin of the Faculty of Engineering, 10.*

Mitwally, H. et al. (1979). Pollutional Status of Abu Kier Bay. Final Report, High Institute of Public Health, Alexandria University.

Saleh, F.A. (1974). Future Planing of Lake Mariot. MPH Thesis, Alexandria University.

Salib A., (1979). Studies on Rakta Paper Mill Wastes. MPH Thesis, Alexandria University.

Sharkawi, F. et al. (1976). Studies on Mariot Lake Pollution Progress Report, Alexandria University.

Sharkawi, F. et al. (1977). Alexandria Sea Coast Pollution. Final Report, Alexandria University.

DISCUSSIONS

Samsunlu, A. (Turkey) : We in Turkey, too, use the method of marine disposal of wastewater. What is your experience in Egypt with marine disposal and your opinion of it? Do you also pretreat the wastewater before discharging it into the sea?

Mitwally, H.H. (Egypt) : We have used marine disposal in Egypt for 70 or 80 years. In the early 1950's a marine outfall was built, 750 m in length and at a depth of 17 m. This was designed to take a flow of 60,000 m^3/d. Right now 1 million m^3/d of wastewater has to be disposed of. Obviously this outfall can no longer cope with such a large flow. Auxiliary outfalls, originally designed to carry rainwater directly into the sea near the beaches and recreational areas, are now used for domestic wastewater, which is not sanitary at all. In the early 60's a treatment plant was built, which takes only 60,000 m^3/d, less than one tenth of the total flow. Its effluent is also discharged indirectly into the sea via a huge oxidation pond called Lake Maruit. At present we are studying a plan for the disposal of 2 million m^3/d of wastewater: 1 million domestic and ¾ million industrial. By the end of this century the total daily discharge of wastewater may reach 3 million m^3. An American company last year submitted their proposal for constructing outfalls 10 km into the sea at a depth of 50 m. This led to heated discussion. A developing country like Egypt with vast areas of arid desert and short of water resources needs this water which is being discharged into the sea. We believe that it would be a much better policy to treat our wastewater for later use in irrigation.

Evaluation of Water Quality in the Kitakami River

M. ONUMA and T. OMURA

Department of Civil Engineering, Iwate University, Morioka, Japan

ABSTRACT

The ferrous ion was evaluated in the Kitakami river by means of linear multiple regression, the group method of data handling and the Kalman filter. Linear multiple regression was suitable for estimation over a long period. Meantime, the group method of data handling was a good method for estimation over a short period. It was found that the Kalman filter was the most suitable method for prediction in the case of the on-line system.

INTRODUCTION

The catchment area of the Kitakami river in northern Japan is predominantly rural with relatively little industry. The aquatic environment has been contaminated with the drainage from a sulphur mine at the upper reach of this river for about 70 years. The Kitakami river has a mean flow rate of 50 m^3/min, receiving the effluent at the rate of 20 m^3/min from the treatment plant. The water quality of this river is characterized by a low pH (about 3), while the mean ferrous ion is 321 mg/ℓ and the mean sulphur is 2200 mg/ℓ.

The treatment plant is located at a distance of about 3 km from the sulphur mine, as shown in Fig. 1, and will be fully operational in 1979. The treatment plant consists of two stages with a hydraulic capacity of 20 m/min, as shown in Fig. 2. In the first stage, biological oxidation from the ferrous to ferric ion, occurs as a result of Thiobacillus Ferrooxidants and in the second stage, neutralization with lime or calcium carbonate takes place.

It is stated that the ferrous ion has a great influence on oxidation from the ferrous to the ferric ion by Thiobacillus Ferrooxidans (Schnaitman and Kortzynski, 1969; Umita, 1978). This means that it is of great significance to predict the ferrous ion in order to control dynamically this plant.

Therefore, the ferrous ion is evaluated by means of linear multiple regression (MR), the group method of data handling (GMDH) and a Kalman filter (KF), using parameters such as water temperature, pH, flow rate, acidity, total iron and ferrous ion, the respective values being taken at the inlet of the treatment plant.

BEHAVIOUR OF TOTAL IRON AND FERROUS ION AT THE INLET

The correlation matrix of water characteristics is shown in Table 1, based on values recorded in 1975 and 1976. Total iron correlated closely with pH and acidity, res-

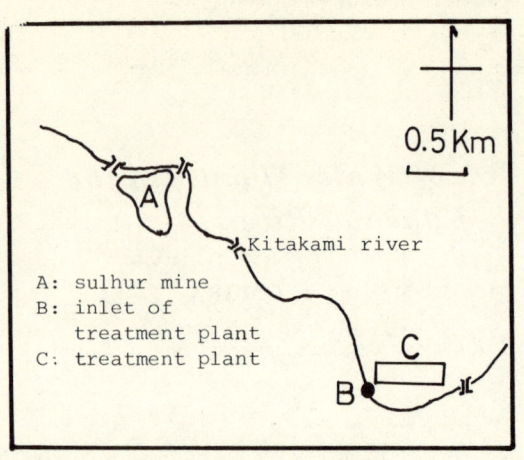

Fig. 1: Location of the inlet

A: sulhur mine
B: inlet of treatment plant
C: treatment plant

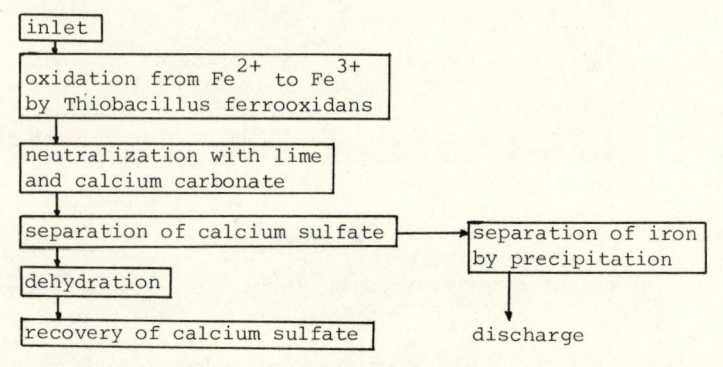

Fig. 2: Flow chart of treatment plant

pectively: (r= -0.817, 0.972). In other words, the pH decreased and acidity increased with an increase in total iron, because the iron was contained in the receiving water as a mixture of ferrous sulphonate and ferric sulphonate. Meanwhile, the ferrous ion did not correlate significantly with other parameters. It did, however, show a slight correlation with water temperature: (r= -0.409). This means that the ratio of the ferrous ion to the total iron changed at the inlet of the treatment plant day by day. The behaviour of the ferrous ion seemed to be influenced by other factors, such as Thiobacillus Ferroxidans at the bottom of the river. The activity of that organism appeared to increase with water temperature, because there was a slight correlation of the ferrous ion with water temperature.

Table 1: Correlation Matrix of Water Qualities

	pH	WT	TF	A8	Q	F2
pH	1.00					
WT	0.20	1.00				
TF	-0.82	0.03	1.00			
A8	-0.85	0.02	0.97	1.00		
Q	0.69	0.21	-0.67	-0.73	1.00	
F2	-0.35	-0.41	0.12	-0.09	0.09	1.00

Evaluation of Water Quality in the Kitakami River 393

Table 2: Partial Regression Coefficient

a_1	a_2	a_3	a_4	a_5	a_6	
3370	−1350	9.40	−0.035	0.523	−0.226	

b_1	b_2	b_3	b_4	b_5	b_6	b_7
1670	−669	3.63	0.079	0.659	−0.130	0.062

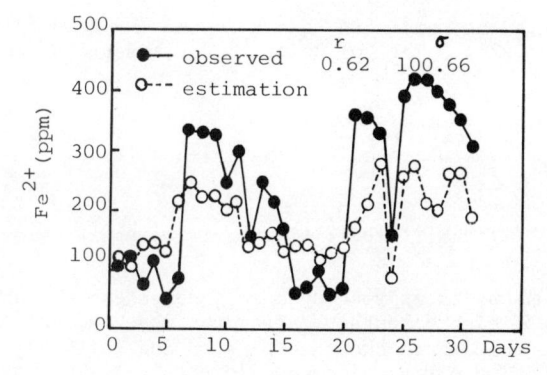

Fig. 3: Estimation by MR(Eq.1) in August, 1975

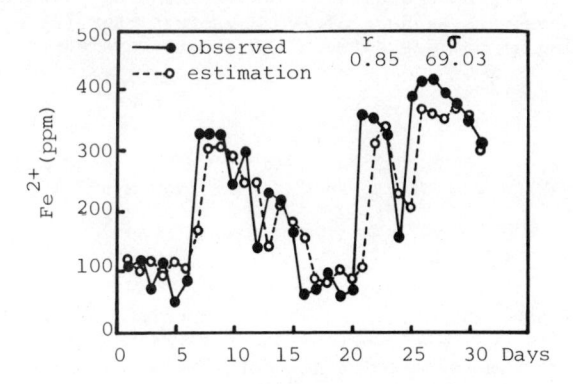

Fig. 4: Estimation by MR (Eq. 2) in August, 1975

EVALUATION OF THE FERROUS ION BY MR AND GMDH

Evaluation by MR

Since the ferrous ion varied depending on the season, it was meaningless to apply the estimated structure of the ferrous ion in winter to that in summer. It was more effective to adopt the method which adjusted the structure to fit the values observed each season.

The criteria used in this study were: a standard deviation of residuals, a good agreement of peak values and a correlation coefficient between the predicted values and the observed values.

In this study, the structure was estimated based on the observed values in summer (July and August, 1975 and 1976). The following linear equations were used to determine the structure:

$$F_e^{2+}(t) = a_1 + a_2 pH(t) + a_3 WT(t) + a_4 TF(t) + a_5 Q(t) + a_6 A8(t) \tag{1}$$

$$F_e^{2+}(t) = b_1 + b_2 pH(t) + b_3 WT(t) + b_4 TF(t) + b_5 F_e^{2+}(t-1) + b_6 Q(t) + b_7 A8(t) \tag{2}$$

Ferrous ion on the previous day is expressed as an independent variable in Eq. (2). The partial regression coefficients, obtained from the observed values, are listed in Table 2. The estimated values are shown in Figs. 3 and 4 (1975), based on Eqs. (1) and (2), respectively. The correlation coefficients between the estimated values and the observed values were 0.62 from Eq. (1) and 0.85 from Eq. (2). On the other hand, the standard deviations of residuals were 100.66 from Eq. (1) and 69.3 from Eq. (2). In fact, it was more desirable to determine the structure of the ferrous ion from Eq. (2). However, because of its structure, Eq. (2) had a weak point, that is to say, a time lag of one day between the observed values and the estimated values.

Evaluation by GMDH

The structure of the ferrous ion evaluated by GMDH may be expressed as the following equation (Ivakhnenko, 1970; Haneda, 1977; Tanaka and Itagaki, 1978):

$$F_e^{2+}(t) = wt(t) + wr(t) \tag{3}$$

where $wt(t)$ is the trend part and $wr(t)$ is the random part. As far as the ferrous ion is concerned, the following four equations were investigated as the expression of the trend part, because the ferrous ion did not show a special correlation with other parameters.

$$wt(t) = c_0 (Q(t))^{c_1} \tag{4}$$

$$wt(t) = c_0 + c_1 pH(t) + c_2 TF(t) + c_3 A8(t) \tag{5}$$

$$wt(t) = c_0 (pH(t))^{c_1} \tag{6}$$

$$wt(t) = c_0 + c_1 pH(t) \tag{7}$$

There was a special reason why water temperature was not found in Eqs. (4),(5),(6), and (7), in spite of a slight correlation of the ferrous ion with water temperature. That is to say, the ferrous ion seemed to be independent of water temperature, be-

Evaluation of Water Quality in the Kitakami River

cause the ferrous ion was estimated each season. The random part of the ferrous ion may be expressed in the following form:

$$wr(t) = f(pH(t), pH(t-1), pH(t-2), WT(t), WT(t-1), WT(t-2), TF(t),$$
$$TF(t-1), TF(t-2), Fe^{2+}(t), Fe^{2+}(t-1), Fe^{2+}(t-2), Q(t),$$
$$Q(t-1), Q(t-2), A8(t), A8(t-1), A8(t-2)) \qquad (8)$$

where (t) represents the observed values on the day, (t-1) those on the previous day and (t-2) those on the two days before. The modified multinominal equations of Kolmogorov-Gaber, as shown in the following, are expressed as a reference function:

$$f(x_i, x_j) = d_0 + d_1 x_i + d_2 x_j + d_3 x_i x_j + d_4 x_i^2 + d_5 x_j^2 \qquad (9)$$

$$f(x_i, x_j) = d_0 + d_1 x_i + d_2 x_j + d_3 x_i x_j \qquad (10)$$

$$f(x_i, x_j) = d_0 + d_1 X_i + d_2 x_j \qquad (11)$$

A conceptual explanation of GMDH is shown in Fig. 5. The first step required the selection of water characteristics. In this step, various water characteristics, considered to be connected with the ferrous ion and observed at the inlet of treatment plant, were selected as input variables in the evaluation by GMDH in this study. A time lag of two days, expressed in Eq. (8), was taken into consideration in this step. In the second step, intermediate variables were determined from the combination of every two variables, taken from all the water characteristics. Thus the number of intermediate variables was $(m\,C_2)$.

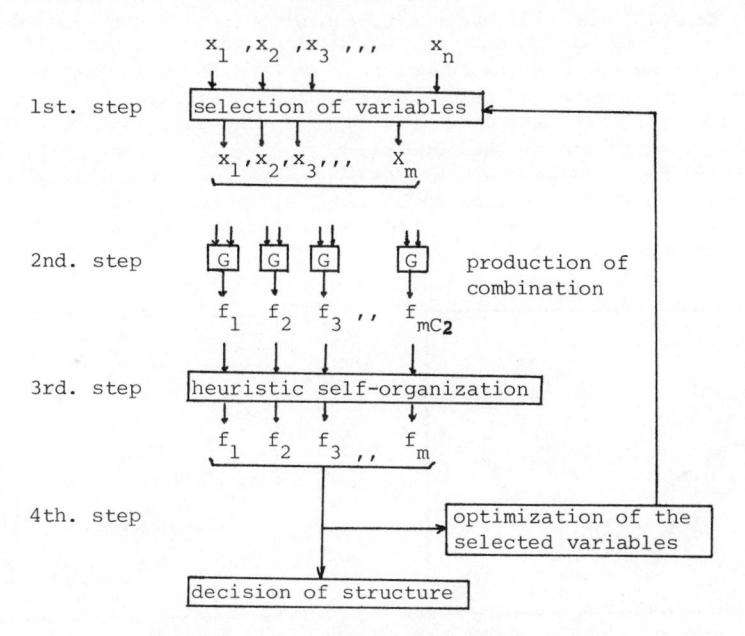

Fig. 5: Conceptual explanation of GMDH

In the third step, the sum of the squares of residuals for the intermediate variable between the estimated values and the observed values was prepared. When the most suitable intermediate variable, which had a minimum sum of squares of residuals, was selected, the computation came to an end; otherwise it returned to the first step. This process was called the fourth step.

The results of the evaluation of the ferrous ion in August (1975) by GMDH are listed in Table 3 and those (1976) are shown in Table 4. The values determined (1975) by GMDH are also shown in Fig. 6. The best agreement of estimated values in August was obtained, based on Eq. (9) as a random part and Eq. (4) as a trend part. The correlation coefficient of 0.850 and the standard deviation of residuals of 73.228 were obtained in 1975. The structure is expressed as follows:

$$wt(t) = 32.989 - 12.245\, pH(t) + 0.011\, TF(t) - 0.229 A8(t)$$

$$wr(t) = 0.249 + 0.085\, x_i + 0.118\, x_j - 0.014\, x_i x_j - 0.182\, x_i^2 + 0.338\, x_j^2 \qquad (12)$$

$$x_i = Fe^{2+}(t-2) - wt(t-2)$$

$$x_j = Fe^{2+}(t-1) - wt(t-1)$$

It seemed that the trend part could be expressed as a function of pH, total iron and acidity on the day. Meanwhile, the random part was expressed as the function which contained a nonlinear term of a higher order. Moreover, it was dependent upon the ferrous ion on the previous day and on the two days before. That meant an autoregression. As far as the evaluation for August (1976) was concerned, it was most suitable to use Eqs. (6) and (10), because the minimum standard deviation of residuals was obtained ($\sigma = 62.454$). However, the maximum correlation coefficient of 0.838 was obtained from the combination of Eqs. (7) and (9). The estimated values, based on Eq. (6) and expressed as follows, are shown in Fig. 7. It was found from Eq. (13), that the trend part was dependent on pH on the day. On the other hand, the random part was expressed by the intermediates (f_1, f_2), which were dependent on the ferrous ion and the flow rate on the previous day, and the total ion and the acidity on the two days before – that is to say, on $Fe^{2+}(t-1)$, $Q(t-1)$, $TF(t-2)$ and $A8(t-2)$. A nonlinear term of a higher order was not contained in Eq. (13).

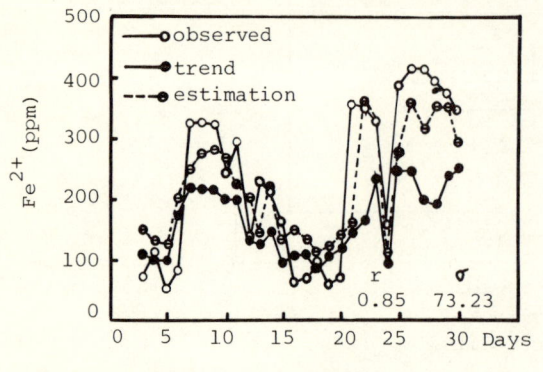

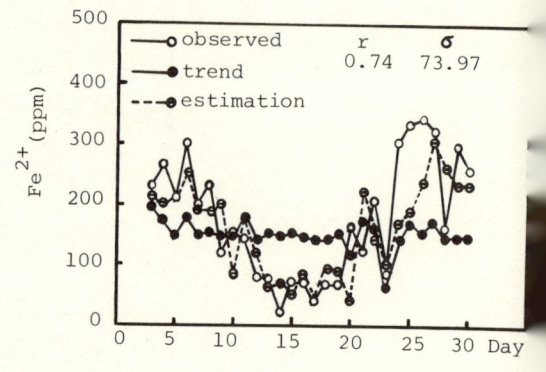

Fig. 6: Estimation by GMDH in August, 1975

Fig. 7: Estimation by GMDH in Aug., 1976

$$wt(t) = 26.367 \, pH(t)^{-3.861}$$

$$wr(t) = 0.104 - 0.017 f_1(t) - 0.0005 f_2(t) + 0.100 f_1(t) f_2(t)$$

$$f_1(t) = 0.025 - 0.076 \, x_i - 0.015 \, x_j + 0.142 \, x_i x_j$$

$$f_2(t) = 0.026 + 0.010 \, x_k + 0.0004 \, x_1 + 0.148 \, x_k x_1 \qquad (13)$$

$$x_i = Fe^{2+}(t-1) - wt(t-1), \quad x_j = TF(t-2),$$

$$x_k = Q(t-1), \qquad x = A8(t-2)$$

It was found in this study that the seasonal structures differed from year to year. In fact, it seemed to be difficult to determine the peculiar structure of ferrous ion for each season, when evaluated by GMDH.

Table 3: Results of the Evaluation of the Ferrous Ion In August, 1975 by GMDH

Eq.	(9)	(10)	(11)
Eq. (4)	r= 0.64 σ=107.42	r= 0.41 σ=154.46	
(5)	r= 0.85 σ= 73.23	r= 0.84 σ= 75.38	r= 0.83 σ= 86.80
(6)	r= 0.75 σ= 95.44	r= 0.75 σ=102.19	
(7)	r= 0.81 σ= 90.38	r= 0.65 σ=122.40	r= 0.61 σ=117.09

r: correlation coefficient
σ: standard deviation of residuals

Table 4: Results of the Evaluation of the Ferrous Ion In August, 1976 by GMDH

Eq.	(9)	(10)	(11)
Eq. (4)	r= 0.33 σ= 94.64	r= 0.74 σ= 72.40	
(5)	r= 0.41 σ=104.28	r= 0.39 σ=103.83	
(6)	r= 0.79 σ= 73.96	r= 0.80 σ= 62.45	
(7)	r= 0.84 σ=105.42	r= 0.49 σ= 86.61	r= 0.48 σ= 86.51

r: correlation coefficient
σ: standard deviation of residuals

PREDICTION BY MR, GMDH and KF

Prediction by MR

The predicted values of the ferrous ion in August (1977) are shown in Fig. 8, based on Eqs. (1) and (2). The ferrous ion on the previous day was expressed as a dependent variable in Eq. (2). The correlation coefficients were 0.87 from Eq. (2) and 0.63 from Eq. (1), while the standard deviations of residuals were 33.79 from Eq.(2) and 52.09 from Eq. (1). It was evident that Eq. (2) had a more suitable structure than Eq. (1) for prediction by MR. The predicted correlation coefficient and standard deviation of residuals were almost in agreement with those estimated in August (1975). That is to say, the predicted values closely approximated the observed values, when the ferrous ion was predicted on the basis of the estimated structure.

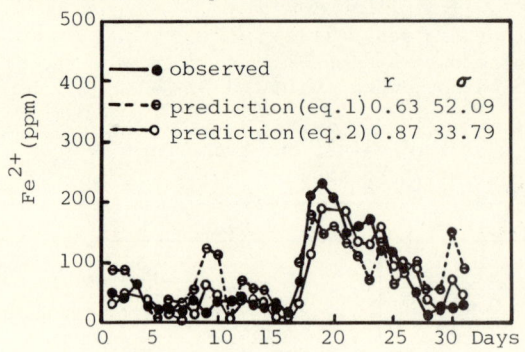

Fig. 8: Prediction by MR in August, 1977

Prediction by GMDH

The predicted values of the ferrous ion in August (1977), based on Eqs. (12) and (13), are shown in Fig. 9. Correlation coefficients were 0.68 from Eq. (12) and 0.31 from Eq. (13), while standard deviations of residuals were 74.68 from Eq. (12) and 0.31 from Eq. (13). It was found that Eq. (12) had a more suitable structure than Eq. (13) in this study. However, GMDH had its own merits and demerits in predicting the ferrous ion. The values predicted by GMDH were not always more suitable than those by MR, as the structure of GMDH changed each season. On the other hand, GMDH had the following advantages over MR.

1- The structure given by GMDH contained smaller numbers of independent variables than that of MR.

2- When the seasonal structures were the same, GMDH was a useful method of prediction because a time lag could be taken into consideration.

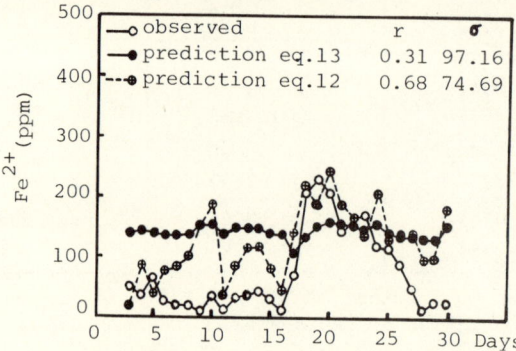

Fig. 9: Prediction by GMDH in August, 1977.

Prediction by KF

Natural phenomena, such as air and water pollution problems, were expressed in the form of partial differential equations. Hino et.al (1977) and Tamura (1978) made a success of applying the Kalman filter theory to the prediction of air and water pollution problems on the assumption that these could be expressed as partial differential equations.

In this study, the ferrous ion was predicted by the Kalman filter theory, based on the assumption by Hino et. al. (1977) and Tamura (1978). The pertinent equations, in discrete form, are given below:

$$X(k+1) = \emptyset(k+1/k)X(k) + N(k) \tag{14}$$

$$Z(k+1) = M(k+1)X(k+1) + V(k+1) \tag{15}$$

where $X(k)$ is a state variable, $\emptyset(k+1/k)$=the state transformation matrix, $N(k)$ = a Gaussian noise, $Z(k+1)$=the observed value, $M(k+1)$=the non-probabilistic matrix and $V(k+1)$=the observed residuals.

A conceptual explanation of KF is shown in Fig. 10. Predicted values are shown in Fig. 11 (1975), Fig. 12 (1976) and Fig. 13 (1977). The correlation coefficients were 0.55 (1975), 0.40 (1976) and 0.73 (1977). Meanwhile, standard deviations of residuals were 128.19 (1975), 118.63 (1976) and 51.44 (1977). The predicted values were in good agreement with the observed values in 1977. KF had its own weak point, namely, a time lag of one observation period. That meant a time lag of one day, in the case of an observation period of one day. This problem could be solved, when the water characteristics were observed continuously. Moreover, KF had the following advantages over other methods.

1- Water characteristics could be predicted in a short time.
2- KF was a good method for prediction in the case of the on-line system, because the algorithms were always the same.

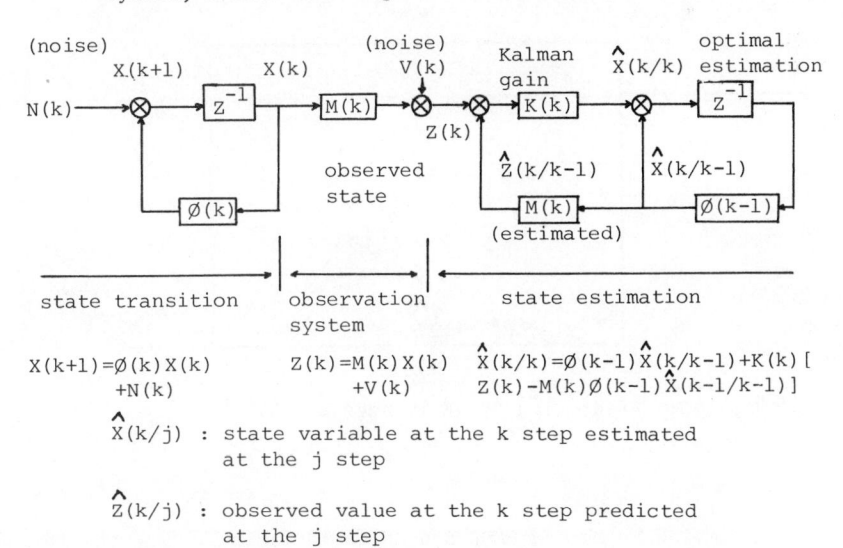

Fig. 10: Conceptional explanation of KF

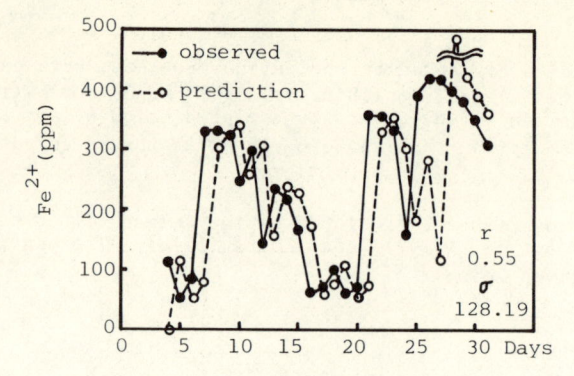

Fig. 11: Prediction by KF in August, 1975

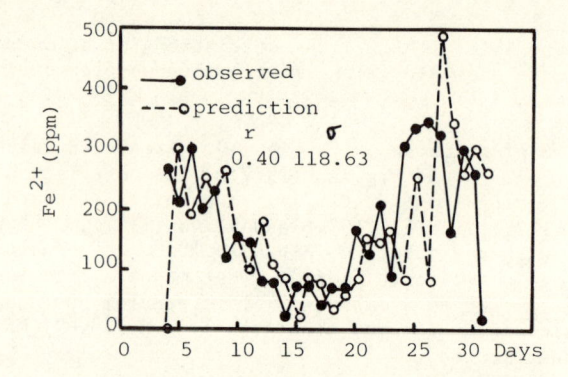

Fig. 12: Prediction by KF in August, 1976

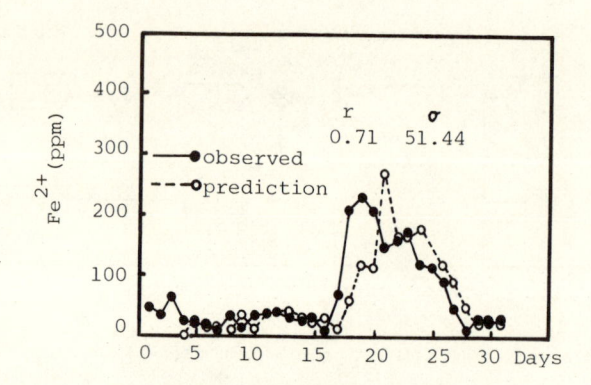

Fig. 13: Prediction by KF in August, 1977

SUMMARY

The results estimated by MR and GMDH are shown in Table 5. The results obtained by MR were estimated on the basis of the observed values in 1975 and 1976. However, the results obtained by GMDH were estimated on the basis of the observed values in 1975 and 1976, respectively. The structure given by MR based on Eq. (2) (six variables) was almost in agreement with the structure given by GMDH. However, MR had the following weak points in comparison with GMDH:

Evaluation of Water Quality in the Kitakami River

1- It was necessary to collect a great deal of data.
2- There were many more variables in the structure.
3- It was difficult to take a time lag into consideration.

Table 5: Result of Estimation by MR and GMDH in August, 1975 and 1976

				r	σ
MR	Eq. (1)	August,	1975	0.62	100.66
	Eq. (2)		1976	0.85	69.03
GMDH	Eq. (12)	August,	1975	0.85	73.23
	Eq. (13)		1976	0.80	62.45

Table 6: Result of Prediction by MR, GMDH and KF in August, 1977

		r	σ
MR	Eq. (1)	0.63	52.09
	Eq. (2)	0.87	33.79
GMDH	Eq. (12)	0.68	74.68
	Eq. (12)	0.31	97.16

Although GMDH had merits it was not suitable for predicting the ferrous ion, because the structure changed each season. In short, MR was more suitable for estimation over a long period. However, GMDH was a good method for estimation over a short period.

The results of prediction by MR, GMDH and KF are shown in Table 6. It was found that prediction by GMDH was inferior to prediction by other methods. In fact, it was difficult to predict the values, as the structure changed from season to season.

On the other hand, MR was a good method for prediction, especially, in the case of the six variables. That is to say, the ferrous ion on the previous day was taken into account.

When the observed values were continuously obtained, KF was the most suitable method for prediction in the case of the on-line system. However, KF had a weak point, a time lag, which could be solved by monitoring the water characteristics.

NOTATION

$A8(t)$	Acidity on the tth day	(mg/ℓ)
a_i, b_i, c_i, d_i	Coefficient	(-)
$Fe^{2+}(t)$	Ferrous ion on the day t	(mg/ℓ)
f_i	Intermediate variable	(-)
$M(k+1)$	Non-probabilistic matrix	(m x m)

N(k)	Gaussian noise	(-)
Q(t)	Flow rate on the tth day	(m^3/min)
r	Correlation coefficient	(-)
TF(t)	Total iron on the day t	(mg/ℓ)
V(k+1)	Observed residuals	(vector m)
WT(t)	Water temperature	(°C)
wr	Random part of ferrous ion	(mg/ℓ)
wt	Trend part of ferrous ion	(mg/ℓ)
X(k)	State variable	(vector n)
x_i, x_j, x_k, x_l	Variable	(-)
Z(k+1)	Observed value	(vector n)
σ	Standard deviation of residuals	(-)
∅(k°1/k)	State transformation matrix	(n x n)

ACKNOWLEDGEMENT

This work was sponsored by the Ministry of Education, Japan, in 1978. Sincere thanks go to the Ministry of Construction for their assistance in observing the water characteristics. Thanks are also due to Associate Professor M. Haneda (Akita Technical College), K. Takahashi and K. Murakami.

REFERENCES

Haneda, M. (1977). Fluctuation Characteristics of Water Quality in the Omono River *J. of Jap. Soc. of Civil Eng.*, 265, p. 73. (Japanese).

Hino, M., S. Yoshikawa and T. Kurihara (1977). Experience in Air Pollution Prediction by Statistical and Stochastic Prediction Techniques. *J. of Soc. of Civil Eng.*, 268, 47. (Japanese).

Ivakhnenko, A.G. (1970). Heuristic Self-Organization in Problems of Engineering Cybanetics. *Automatica*, 6, 207.

Schnaitman, C.A. and M.S. Kortzynski (1969). Kinetic Studies of Iron Oxidation by Whole Cells of Ferrobacillus Ferrooxidans. *Jr. Bacteriology*, 99, 552.

Tamura, H. (1978). Application of the Kalman Filter to Determination of Water Quality System. *J. of System Control*, 22, 20. (Japanese).

Tanaka, M. and H. Itagaki (1978). Application of GMDH to Determining Water Quality. *JSIDRE*, 74, 24.

Umita, T. (1978). Studies on the Treatment of Acid Drainage with Iron by Thiobacillus Ferroxidans. Masters Thesis, Tohoku University. (Japanese).

Practicable Models for Solid Waste Recycling

O. TABASARAN

Institut für Siedlungswasserbau, Wassergüte- und Abfallwirtschaft der Universität
Stuttgart, Federal Republic of Germany

ABSTRACT

In this paper three models for waste handling are discussed, with the objective of speeding the processing of wastes, including all steps from disposal to commercial management.

The three models are:

The Landfill Site offers a favourable location for carrying out simple material separation, where, in the first instance, suitable cover material can be recovered and then gradually, sorting of secondary materials for sale on the market can be developed.

The Federal Waste Recovery Model is aimed at the development of a technically sophisticated, economically viable and environmentally acceptable handling system, which will help to drive new sources of raw material and to reduce the amount of wastes for tipping. The technical concept envisages a series of carefully thought out, dry and wet separating devices in combination with a composting line, which will be tested for an initial period of seven years.

Used Tyres currently have a negative value in the majority of cases, whereas pyrolytic treatment leads to a range of usuable products and so achieves a positive socio-economic contribution. It is planned to try out at operational scale a system which has already been tested experimentally, the most important components of which are a pyrolysis trommel together with gas cleaning and condensing stages.

INTRODUCTION

The Council of European Communities states, in the Directive of 15 July 1975, that every regulation concerning waste management must have as an essential objective the protection of human health and the environment against adverse effects from the collection, transport, handling, storage and disposal of wastes, and that the processing of wastes and the utilization of recovered materials should be promoted in the interests of conserving natural raw material sources.

In this paper, some considerations regarding simple material separation are firstly commented upon, and then the concept of a technically sophisticated plant for the large-scale testing of raw materials sorting from municipal wastes and its industrial application is presented within the sense of recycling material back into the economic system. Then, a discussion of a model for pyrolysing used tyres follows as a further alternative for achieving waste recovery.

LANDFILL, COMBINED with a SIMPLE SEPARATION PLANT

What may be termed as a favourable compromise between straight landfill and the higher technologies, is the sorting out of valuable materials on the landfill site itself by the simplest means possible, thereby enabling both the recovery of suitable groups of material and the reduction of the quantity to be tipped. Proceeding in this way affords the further substantial advantage of being able to gradually build up the potential market and to accommodate, with calculable risk, fluctuations in sales. The plant can, be managed with minimum expenditure, in response to demand and the needs of the market, and the requirements of the customer can be determined without undue haste. Moreover, the knowledge gained about the quantities and qualities desired by the market can be easily and rapidly transformed to assist with decisions relating to the optimization of the system.

Materials Worthy of Separation in Municipal Refuse

The components of municipal refuse listed in Table 1, in general, merit separation.

Table 1: Materials Worthy of Separation in Municipal Refuse

Material Group	approximate price obtainable in the Federal Republic of Germany (DM/t)
Paper, Board	30-60
Ferrous metals	40-70
Plastics	0-1000
Glass	10-50
Organic Matter	0-10

The possibilities for utilizing these waste materials are diverse and vary from case to case. Waste constituents can be:

- re-used in accordance with its original specification, or;
- utilized as raw material for the manufacture of the primary product; or,
- employed for the production of a new product.

Planning Example for a Landfill Combined with a Simple Separation Plant

Since January 1978, the Landkreis (Country) Ludwigsburg in Baden-Württemberg has operated a central landfill site for household refuse and similar types of waste. The total site area is 46 hectares and is divided into sections.

The proposed concept of combining landfill with separation of valuable materials envisages that the vehicles, after leaving the main highway, first arrive at the vehicle weighbridge, where they can be weighed full and empty. They then drive to the internal transfer station, which is erected in front of the actual landfill and consists variously of staggered tipping ramps and special trenches; the arrangement can be gathered from Fig. 1. The intended separation of read vehicles and landfill site vehicles offers certain advantages such as, simpler site operation, no (or much smaller) tyre cleaning facilities, faster vehicle clearance, etc.

This type of refuse transfer facilitates selective waste sorting. Through the difference in height at the point of interchange, the feeding and evacuation of the separation plant by means of gravity becomes possible in which case, apart from a transfer bunker, no additional equipment is required.

In the first instance, it is intended to erect a test plant with the aim of optimizing the technical performance of the system and reducing the risk. Here, the wastes will be conveyed from the delivery bunker, which has a volume of about 15 m^3, via a chute, to the trommel screen, which has been selected as the appropriate device for this project, on the basis of existing experience. This screen has overall dimensions of 2 diameter and a length of 6 m, made up of a 1 long unperforated section, a 4 m long 220 mm coarse screen, and a further 1 m long section. The mounting is made up of two, twin-tyre truck axles arranged symmetrically to the axis of the trommel screen. Vertical adjustment of the mounting enables the trommel to be tilted from 0^0 to 10^0. The speed of rotation is infinitely variable between about 5 to 10 revolutions per minute. Slippage from the inclined position is prevented by a slide-wheel on which the outer ring of the trommel is supported. The crude refuse falls from about 1 m height into the trommel. An internal collar prevents a possible overrun during any short-term accumulation of refuse. 300 mm x 100 mm large, parallel plates, arranged at an acute angle to the axial direction, leads to good circulation of the material. The screw arrangement of the plates conveys the feed inside the trommel. Plastic bags are split open, aided by pointed and sharpedged members. The 1 m wide, 18 mm punched screen primarily divides out inert matter and glass as a fine product. The 4 m long, 220 mm coarse screen, which is made up of individually welded, polished tubular steel rings, is equipped in the first half with 30 cm long sliding scoops attached from the outside, which ensures the almost uncontaminated delivery of even the longer constituents, whereas in the second half, the textiles (in particular) trail around the collars and thus cannot be carried along by the refuse falling through. The largest portion of the middle-sized material is separated out in the first half of the coarse screen. The fine and coarse screen is shielded by vertical guard plates on the sides and a plate on the underside, which taper off towards the discharge conveyor. As plate cladding all round is not necessary, the trommel screen can be rapidly lifted clear without dismantling structural components.

Bulky items, which remain over in the coarse screen and are broken up by the plates, have only a short-term influence on the synchronisation of the trommel. After a few seconds, the normal state is achieved again through the damping effect of the tyres. The materials arising in the individual screen fractions are collected in containers.

The oversize fraction comes from the trommel screen onto a conveyor belt, over which a magnetic separator removes ferrous fragments from the material stream. The residue is subsequently hand-picked for paper, plastics and glass. The fine material arising through screening can be directly applied as cover material on the landfill site, as it consists predominantly of inert material. The medium fraction of up to 220 mm particle diameter is subjected to a decaying process. The material emerging after 4 to 6 weeks of decomposition can likewise be subsequently used as cover material on the landfill site, if no other possibilities for application are available. Good prospects exist for finding customers for the sorted glass, paper and plastics, as raw materials.

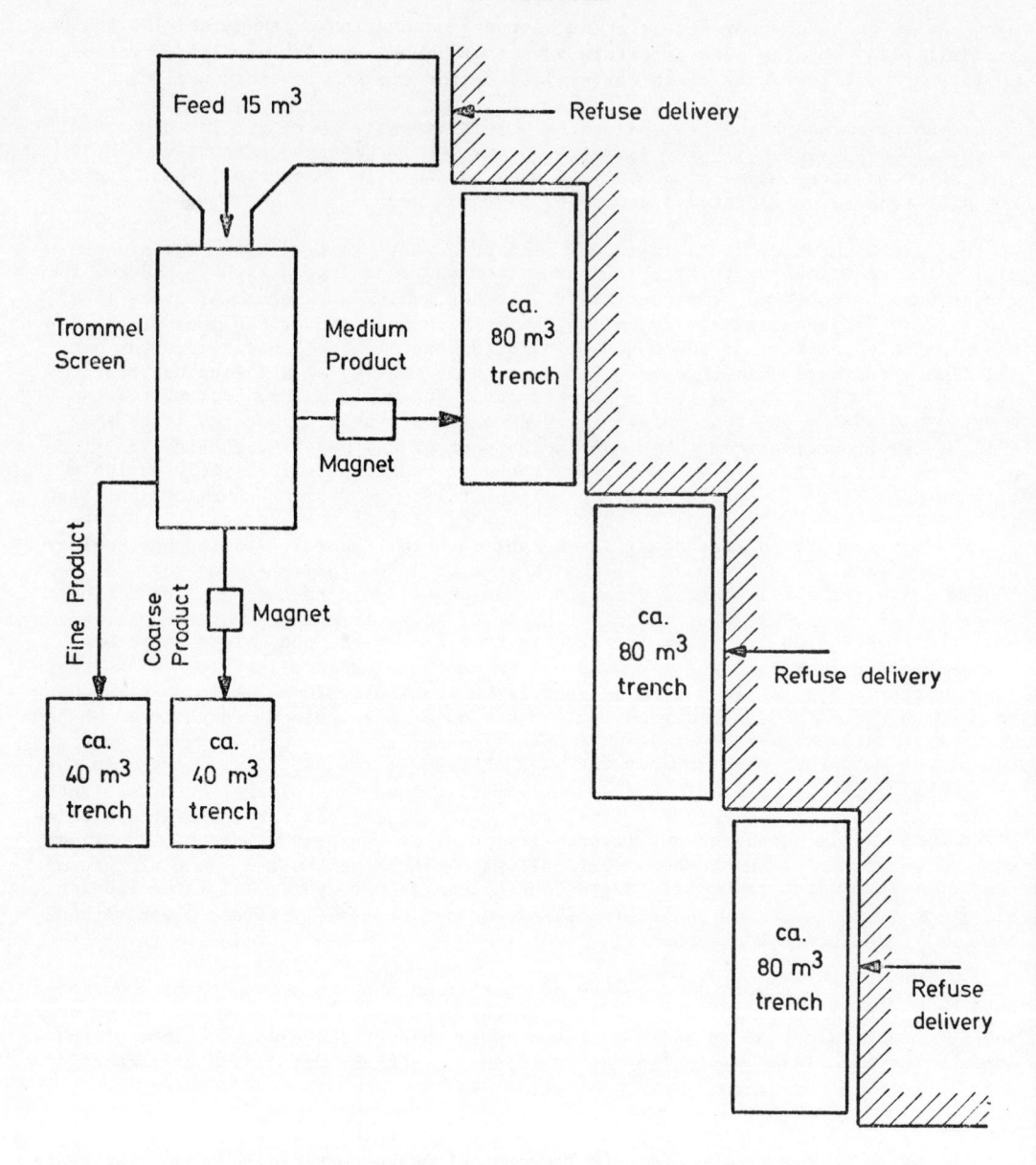

Fig. 1: Scheme for material sorting on the landfill site

First Stage of Concept Realization

Of the household refuse delivered at the landfill site, ca. 40 tonnes daily will be conveyed for separation in the first phase. The immediate objective is first of all the recovery of cover material. During this first stage, the aim will be to develop the markets for the raw materials such as paper, ferrous fragments and plastics, recovered from the coarse product, so that gradually, as comprehensive a recovery system as possible is built up. In this respect, it would appear important to secure powerful customers at the beginning of the series of experiments, the extent of the proceeds remaining of secondary importance.

Results obtained from investigations undertaken by the Institute for Water Engineering, Water Quality and Waste Management of the University of Stuttgart under contract from the Federal Environment Agency (Umweltbundesamt), in the absence of any other data, are taken as a basis for the present considerations. In these investigations the separating capacity of a trommel screen with 18 mm and 220 mm perforations and throughput of 10 to 14 t/h, were examined. From the results obtained the average material balance shown in Fig. 2 can be prepared. A somewhat different distribution of the material streams can occur, depending on refuse composition and the width of screen perforations selected.

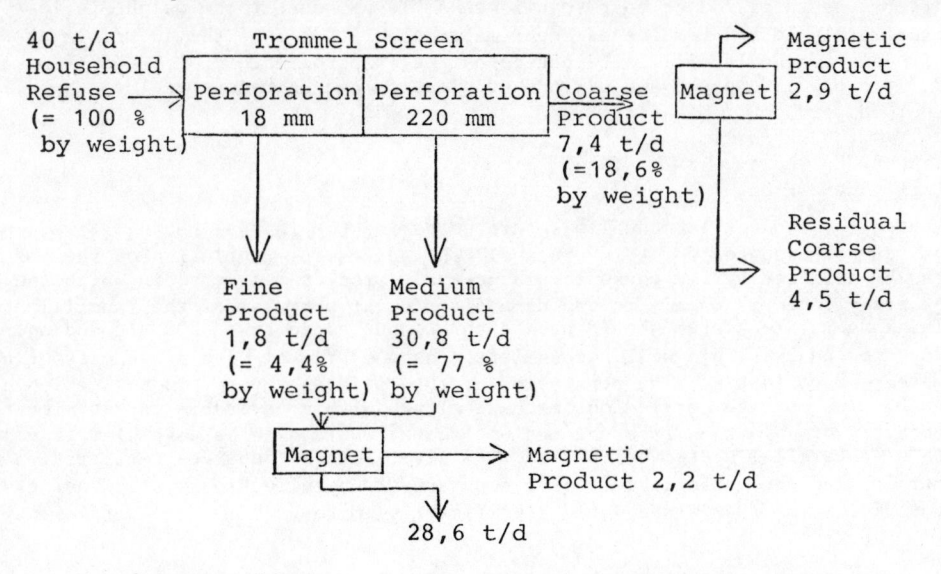

Fig. 2: Material Balance Diagram for a Trommel Screen

The composition of the medium product - from which the ferrous metals have been extracted - which represents the primary stream from the trommel, is given in Table 2 and the composition of the residual coarse product in Table 3.

Table 2: Composition of the Medium Products from the Trommel

Component	Generation Rate t/d	Percent by weight
Organic	21.4	74.7
Paper	4.5	15.6
Plastic	1.3	4.7
Glass	0.8	2.7
Other	0.6	2.3

As can be seen from Table 2, 90.3% by weight of the medium products are suitable for composting.

Table 3: Composition of Coarse Product

Component	Generation Rate t/d	Percent by Weight
Paper	3.24	72.0
Plastics	0.68	15.1
Textiles, Wood	0.46	10.2
Glass	0.10	2.2
Other	0.02	0.5

After a decay period of four to six weeks, 10 percent of the weight is lost, and approximately 26 t/d remains as cover material. 4.0 t/d of the 40 t/d input, is taken directly to the landfill site and deposited. Approximately, 1.5 t/d of metals, 3.2 t/d of waste paper and 0.7 t/d of plastics can be recovered from the 7.5 t/d of residual product.

Extension into a
Large Scale Plant

The coarse product is manually separated during the initial phase. Five men are required as personnel for this process. The successful completion of the initial phase can be followed by the phase for large-scale product recovery, in which the output quantity conforms mainly to the demand for cover material of the landfill operation. The large scale system should have a throughput capacity of 200 t/d and operate 250/ days per annum. This will necessitate a second trommel with a capacity of approximately 12 to 14 t/h. The separation of the residual course product, which amounts to 23 t/d, must be carried out mechanically. With a favorable market situation the recovery of, plastics from the medium product would also be possible. The setting-up of a parallel system on the landfill site, for the separate registration of commercial and the establishment of a separate glass collection system, using special containers in urban areas would significantly improve the overall efficiency.

THE FEDERAL WASTE
RECOVERY MODEL

The project which has become known under the term 'Federal Waste Recovery Model' has as its objective the mechanical sorting and possible processing of various fractions of household refuse, such as inert matter, glass, metals, paper, plastics and organic matter, for sale on the market, and thus ultimately to lay greater stress on the notion of materials management in municipal waste disposal. In this way, various separation techniques will be tested and outstanding questions concerning economic viability and forms of organisation will be answered.

The plant is being constructed on a site made available by the Municipality of Dusslingen in the District of Tuebingen and is managed by an Administrative Union and cater for a total of 400,000 inhabitants in the Districts of Reutlingen and Tuebingen. It will consist of three, technically-different, 15 t/h process lines arranged in parallel, and will be given a seven-year optimiation phase. The risk, associated with a developing system, of exceeding the level of costs usually regarded as reasonable by the waste disposal authority, is being absorbed through subsidies by the Federal Government and the State of Baden-Württemberg, up to a total of 47 million Deutschmarks.

The wastes, after being recorded and weighted, are conveyed into a deep bunker and then to the process line by a crane. There it is possible to extract further interesting materials, running in parallel, by systematic manual sorting.

While one of the lines constitutes conventional composting for the decomposition of organic components into a soil-improving agent, the other two include units for both dry and wet separation, such as magnetic separators, screens, air classifiers, upstream sorters, and ballistic separators in connection with mills. The prospect of also offering immediately saleable products on the market, besides secondary raw materials, will exist if enough suitable space is available. Residual matter and non-marketable products are tipped on a landfill site near the plant.

PYROLYSIS of USED TYRES

The pyrolysis of used tyres currently presents itself as a possibility for the recovery of secondary raw materials which is realizable without excessive difficulties. Its basic advantage lies in the spatial separation of the degassification process from the process of oxidation. The energy in the gas produced is sufficient for the self-sustainment of the pyrolysis process (carbonization). The liquid products with characteristics similar to mineral oil are storable. The gaseous emissions can be simply controlled.

The section which follows describes a system for pyrolysing used tyres, which until now has proved successful in test operation with rates up to 150 kg/h, and now stands almost ready for application as a treatment plant.

Concept for a Used Tyre Pyrolysis Plant

The processes employed in the used tyre pyrolysis plant, the flow diagram of which is shown in Fig. 3, consists of the following stages: (a) pyrolysis (carbonization); (b) condensing; (c) gas cleaning.

The continuous feed reactor consists of a rotary steel drum which is indirectly heated by specially designed internal heat-exchanger tubes.

Pyrolysis (carbonization), is followed by a two-stage fractional condensation, in which the high-boiling point components are liquified at approximately $200^\circ C$, in an oil-operated spray-cooler. All other fractions are condensed in a further cooler at $20^\circ C$. This dual arrangement has the advantage that a crude separation of the complex, liquid reaction products results and both fractions become more interesting as raw materials.

Sulphur compounds appear in the gas mainly as hydrogen sulphide, the splitting of which begins at relatively low temperatures of around $250^\circ C$. Through thermal dissociation of the hydrogen sulphide on hot or sugar-heated walls, elemental sulphur emerges which can react further with carbon and carbon monoxide to form carbonyl sulphide, amid formation of carbon bisulphide. Organic sulphur compounds such as carbon bisulphide, carbonyl sulphide, thiocyanide, thiophene and mercaptans, are as a rule only present in small quantities. The typical smell of pyrolysis oil and gases can be attributed primarily to the presence of sulphur compounds which are even in trace amounts.

Nitrogen compounds may be detected in the gas primarily as ammonia and hydrogen cya-

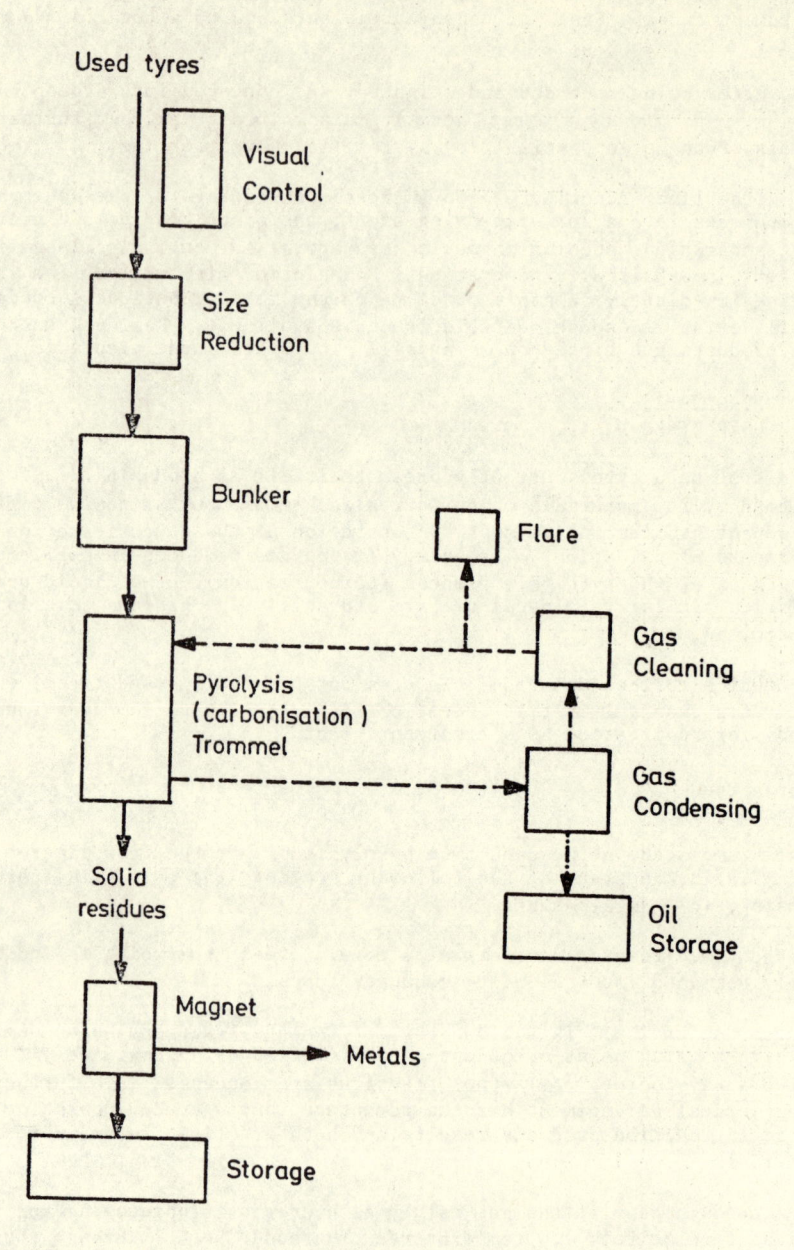

Fig. 3: Scheme of a plant for pyrolysing used tyres

Table 4: Chemical-Physical Characterisation of the High Calorific Value Gas Resulting from the Pyrolysis of Used Tyres After Condensation

Gas Components	Vol. %
H_2	32
CO	5
CO_2	8
O_2	(0 - 2)
N_2	5
CH_4	25
C_nH_m (C_2 to C_5)	21.5
C_nH_m (C_5 to C_6)	2.5
C_nH_m	1
H_u	42,000 kJ/m^3_n

Table 5: Chemical Characterization of the Distilled Pyrolysis Oil Recovered from User Tyres

Boiler Range	Specification of Fraction	% by weight	Presently identified classes of substance
(up to 200°C)	similar to benzine	15	ca.62% of single-nucleus aromatics, olefins, trace paraffins.
(200°C-360°C)	similar to gas oil	37	open-chain and cyclical olefins and diolefins, aromatics with olefinic side chains
(>360° C)	similar to lube oil	27	
	residue, decomposition product	21	

nide. The lower the temperature of carbonization, and the shorter the residence time of the pyrolysis gas in the higher temperature zones, the more intensive is the formation of ammonia.

In the gas cleaning stage, the hydrogen sulphide is led through a column filled with hydrated ferric oxide so that conversion to pyrite takes place. Hydrogen cyanide is also removed with the same cleaning agent.

Pyrolysis Products

The pyrolysis of 1 ton of used tyres yields on average

> 120 kg of gas
> 440 kg of oil
> 385 kg of carbon black
> 55 kg of steel

The energy content of used tyres, is to 32×10^6 kJ/ton, and it is distributed approximately as follows:

> 13% in the gas,
> 55% in the oil, and the remaining
> 32% in the char.

The cleaned gas consists mainly of hydrogen and lower hydrocarbons, the composition of which can be gathered from Table 4. This gas is more than sufficient to heat the pyrolysing trommel so that the plant can operate self-sustainingly in terms of heat. Excess gas is flared.

The condensing complex oily products, the characteristics of which are summarized in Table 4, can be used as a fuel in diesel engines, as a fuel for oil bunkers, as an additive in rubber manufacture, or find application in the chemical industry.

The solid residues are presented as a carbon-rich, free-running, practically odourless material with a heating value of around 33,500 kJ/kg. Because of that, it is possible to be used as fuel. It could also be used for example, after suitable treatment, as a filler, in the rubber industry as a filler and colouring agent in the chemical industry, or as an adsorbing agent, after first being granulated and treated with high temperature steam activation.

The steel enclosures, which are also belonging to the solid residues, can be easily extracted from the pyrolytic char by means of a magnet.

The recovery of zinc oxide is still at the development stage.

REFERENCES

Goosman, G., Willing, E. (1977). "The Federal Waste Recovery Model", *Der Landkreis*, 8/9, 387, Verlag W. Kohlhammer, Stuttgart.

Institute for Water Engineering of the University of Stuttgart, "Expert Opinion on the Question of Reorganization of Waste Management in the Districts of Reutlingen, Tuebingen and Zollernalb", prepared under contract from the Rural Districts of Reutlingen and Tuebingen (1975) unpublished.

Tabasaran, O. (1978). "Landfill site as Location for Waste Treatment Plants such as Simple Composting, Materials Separation and the Like". *Stuttgart Bulletins on Waste Management (Stuttgarter Berichte zur Abfallwirtschaft)*, 9.

Tabasaran, O., Besemer, G., Thomanetz, E. (1977). "Pyrolysis of Organic Wastes as a Method of Handling and as a Possibility for the Recovery of Energy and Raw Materials". *Mull und Abfall, 10.*, 293, E. Schmidt-Verlag, Berlin.

Experiences with a Two-Stage Refuse Composting System at the Refuse Composting Plant in Salzburg, Austria

H. MOOSS

Ruthner Industrieanlagen, Vienna, Austria

ABSTRACT

The composting plant in Salzburg was erected in the year 1976 after construction period of 16 months and was put into operation in September, 1977. The total costs amounted to 110,000,000 Austrian Shillings. At the moment, the plant is processing approximately 300 tons of refuse and 60-100 tons of sewage sludge per day.

The plant is operated in a two-stage process, consisting of a 24-hour treatment in a fermentation drum and a 4-week fermentation on an artificially aerated fermentation platform. On proper application of the process, the treatment costs for one ton of refuse amount to 244 Austrian Shillings.

The compost produced is tested in order to be used for several purposes, and can be considered to be a useful and valuable product.

HISTORICAL ASPECTS

The history of industrial composting of communal wastes goes back to the beginning of the 20th century. In this period several pilot composting plants were established in South Europe, e.g. Italy, France and Spain. The process used was an open window system, based on aerobic decomposting of organic matter, which can be symbolized by the following structural formula.

$$6 \, C_6H_{12}O_6 + 28 \, O_2 + 3 \, NH_3 = 26 \, CO_2 + 33 \, H_2O + HNO_3 + 2 \, C_5H_7O_2N$$

The first term stands for "dextrose", which can be taken as representative of "organic matter". The last term is one of the possible symbols for the cellular substance of micro-organisms, which are responsible for the composting process. Term No. 2 represents the oxygen demand, which has to be supplied necessarily to the material in order to obtain aerobic conditions. In fresh compost this demand varies according to the stage of decomposting and ranges roughly between 130 and 300 $mm^3/g/h$, during a period between the first up to approximately the seventh week.

Probably it was the difficulties in supplying sufficient oxygen which caused bad odour problems and finally led to the closing of most of these early pioneer plants. Thus, in order to improve the efficiency of this static system, and in order to avoid odour and compost-quality problems, tests with extended aeration systems were performed. One of the best known pilot plants was erected in 1953 in Baden Baden/Germany, where compost piles of 3.5 m in height with removable aeration tubes inside were used.

The next step was realized in the composting plants at Heidenheim in 1970 and at Landau, where stationary aeration devices were installed, consisting of aeration grooves at the bottom of a concrete platform, covered with a layer of 1.8 - 3 m of compost. But still there was the disadvantage of a big demand in area on account of the required detention time of 6-10 weeks. Also the insufficient homogeneity in the decaying material caused the problem of anaerobic zones, which consequently influenced the quality of the final compost products.

Thus, in a parallel development dynamic systems were established, which consequently led to the construction of the fermentation drum. It was in the year 1951 that in the composting plant at Gladsaxe the first experimental rotating drum was installed, which can be considered to be the prototype of the fermentation drum. From this time on, several types of fermentation drums were developed, working all on the same basic principle: communication, homogenization, sludge or water addition, opening of physical structure and prefermentation at temperatures between 30 and 50^0C.

It was obvious that the fermentation drum brought advantages, but still the cost was a big disadvantage. On account of the length of detention time (5-8 days) inside the drum in order to get good compost qualities, the capacity of fermentation drums had to be very low. This, in combination with relatively high investment costs, resulted in an unsatisfactory feasibility.

Therefore, in spite of a great number of manufactured fermentation drums, still the desire continued to find a system, which combines the advantages of both static and dynamic methods, and which also is able to avoid most of their problems.

THE TWO-STAGE SYSTEM

Based on the experiences of the past, and also initiated by a number of model tests, the Ruthner MSA-system was developed in 1975. This system consists of two main steps

1- dynamic prefermentation in a fermentation drum for 24 hours
2- static fermentation on an artificially aerated platform for 3-4 weeks

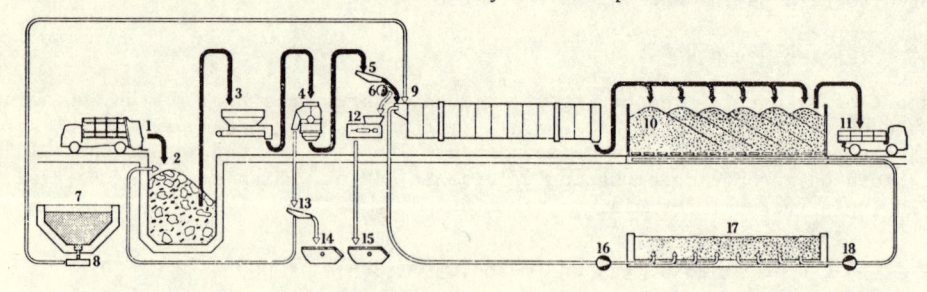

1	wastes supply	7	sludge bin	13	screen
2	wastes bin	8	sludge pump	14	uncomminutable
3	dosing equipment	9	fermentation drum	15	scrap bales
4	shredder	10	maturing hall	16	drum aeration
5	vibrating chute	11	compost transport	17	earth filter
6	iron separator	12	scrap baling press	18	compost aeration

Fig. 1: Scheme of the process

The aim of the system is to carry out the biological process under optimal conditions by keeping costs to the minimum and also creating as little environmental pollution as possible.

The process is briefly described below.

PRE-TREATMENT of SOLID WASTES

Before entering the proper process, the refuse has to be pretreated by a shredder, which comminutes large pieces and starts to open the physical structure of the material. Good results have been obtained when 95% of the particles are smaller than 40 mm. Due to this degree of comminution, an extended oxygen exchange and contact surface is guaranteed, and at the same time, due to the particle size not yet being exceedingly small, an optimal access of air into the material is still allowed. After shredding, the iron in the refuse is separated by magnetic separators, in order to recycle the iron and also to protect the machinery used in the following processes.

1st Step: Dynamic Fermentation

The prepared refuse is fed into the fermentation drum, which consists of a cylindric drum jacket, rotating between both a static front and rear wall, as indicated in Fig. 2.

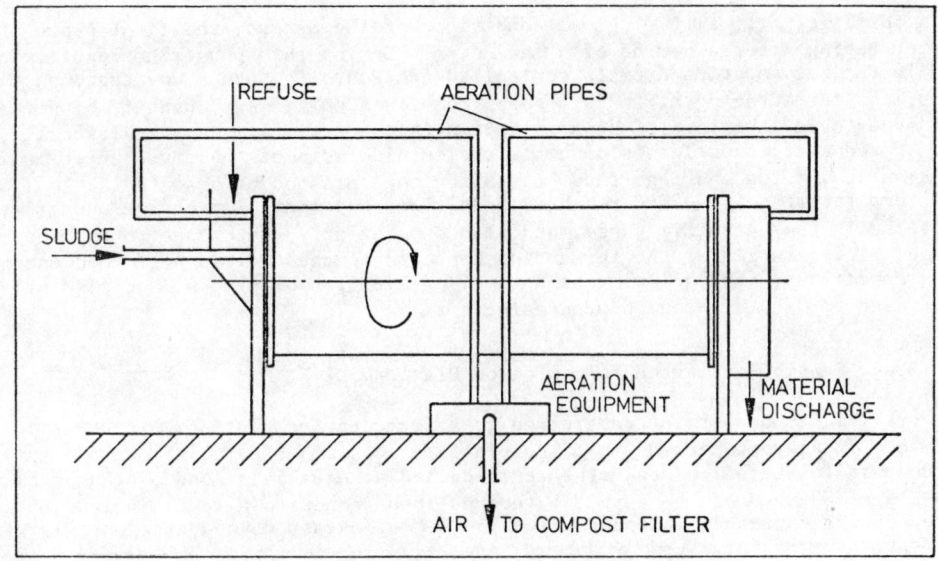

Fig. 2: Diagram of dynamic fermentation process

Feeding and discharge of refuse and sewage sludge is performed automatically. The supply of oxygen is maintained by means of a special aeration device, which enables the direction of the air to be changed and also part of the discharged air to be recycled into the fermentation drum. Thus, the generated heat can be utilized to the best advantage.

The main functions of the fermentation drum are as follows.

Addition of Sludge and Water

In order to reach the optimal water content of 45-50%, as well as the proper C/N ratio of 25:35, liquid sewage sludge containing up to 12% dry matter is pumped into the drum. For this purpose, it is possible to use either untreated or biologically pretreated sludge, as well as aerobically or anaerobically stabilized sludge coming from municipal sewage plants.

Homogenization and Biological Process

Due to the low rotating speed of the drum, it is possible to obtain a good mixture of the shredder refuse and the added sewage sludge. During the average detention time of 24 to 35 hours, the mixture produced is not only completely homogeneous, but also sufficiently well pretreated to undergo the biological process of opening the fiber structure of paper, textiles, kitchen wastes, etc.

Inoculation

During the 24-hour period of charging and discharging of the fermentation drum, there is always a certain amount of the previous day's drum content remaining in the fermentation drum. This quantity varies according to the exploitation of the capacity,and amounts to at least 5% of one day's charge. This amount, which is enriched with microorganisms and bacteria, produces an inoculation effect on refilling of the drum with the next day's charge. Due to that effect, the fermentation process is initiated immediately and uniformly.

Prefermentation

As a result of optimal biological conditions at the outset, the first phase of fermentation begins after a period of a few hours. During this phase the quantity of oxygen in the fermentation drum is controlled and dosed in such a way that, on the one hand, the fermentating mixture is supplied with a sufficient quantity of air and, on the other hand, the loss of heat is kept within certain limits. Additionally, in order to reduce further losses of heat, the outside skin of the drum jacket is furnished with an insulating coating, thus preventing the emission of heat to the environment. The temperatures of 40-60°C reached in this way especially encourage the growth of bacteria in the mesophile phase, and also subsequently the change to the thermophile phase. The aeration of the fermenting contents of the drum is performed under depression conditions, thus allowing an optimal utilization of oxygen, and also preventing the emission of odor-loaded processing air.

2nd Step: Static Rotting on Fermentation Platform

The preferment compost is transferred to a fermentation platform, where it is submitted to a final rotting process. The platform consists of a flat concrete surface in which parallel aeration conduits are installed at equally spaced distances of about 1.0-1.5 m. Oxygen can be supplied through these conduits alternatively with overpressure or with depression, depending on the actual oxygen demand at a certain stage of rotting.

The final rotting phase, which then commences, is marked by an intensive growth of fungi as well as by the development of actinomycetes and the network-like structures of mycelia. In the course of this process, the temperatures rise within a few days to a maximum of 70 - 75°C, thus allowing the final hygienization, together with the development of antibiotic inhibitors.

During the following phase of the development, the temperature remains stable, but wit increasing desiccation and decomposting of the material a slow decrease starts after a period of 2-3 weeks. Together with controls of temperature, which show relatively constant values of 65-75°C after a few days, the generation of CO_2 is a significant parameter for the intensity of the process. Fig. 3 shows the CO_2 production on the fermentation platform.

Starting with high amounts during the first few days, the CO_2 production decreases after one week to approximately 10 kg/h and stays constant at approximately 2-5 kg/h after a period of 3 weeks. After this time the main effect of the fermentation platform, namely, decomposting of light organic matter is achieved, which gives an indication of the most economical detention time on the platform.

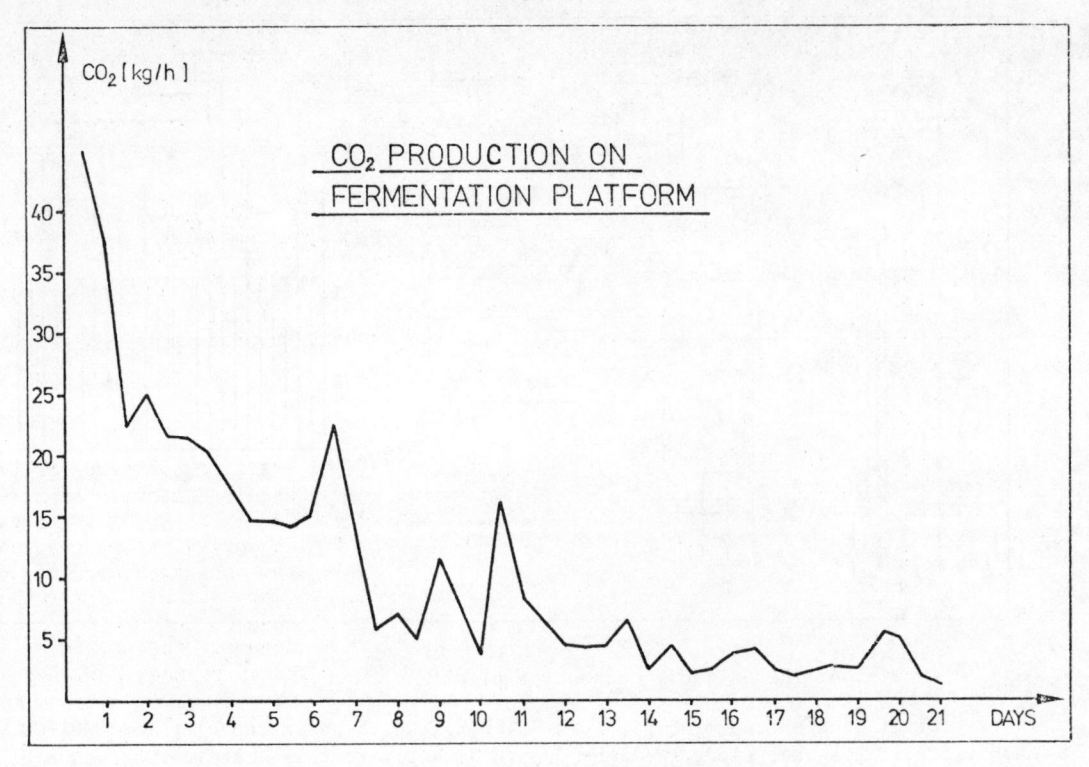

Fig. 3: CO_2 production on the fermentation platform

After a period of about 4 weeks, the compost, which still presents temperatures of 60-70°C, is discharged from the fermentation platform and may be used, after transitory stocking of approximately one month, for landscape-building and recultivation purposes. It also may be submitted to further processing in compost treatment lines in order to transform it into a high quality, matured compost for use in gardening and agriculture.

The Salzburg Plant

The lay-out of the composting plant is based on a capacity of 400 tons of refuse and 100 tons of liquid sewage sludge per day. In order to obtain a maximum reliability of operation, the mechanical arrangement is established in three parallel machine lines. A fourth line can be added in case of necessity, without any additional building construction, and will increase the capacity up to 600 tons per day of solid refuse (Fig. 4).

Household and industrial refuse, delivered by garbage, trucks, container vehicles and private vehicles, is dumped directly into a closed bin hall, which has access through 10 doors. In the case of a continuous refuse delivery, the vehicles can dump the material directly into the three feeding hoppers onto the feeding conveyors to the shredders. In the case of delivery of more than three vehicles at a time, the excess refuse is dumped into the storage bin, which has a capacity of 1500 m^3. A second bin with a capacity of 500 m^3 for industrial refuse is also situated in the bin-hall. From there the refuse is transported by a travelling crane, equipped with a spider-grab of 5 m^3 capacity. The crane cabin is situated next to the control center at the end of the storage bin. The operating procedure in the delivery hall, as well as the entire process, can be checked by 6 television cameras and monitors. Liquid sewage sludge is delivered in a storage container, which has a capacity of 200 m^3. After completion

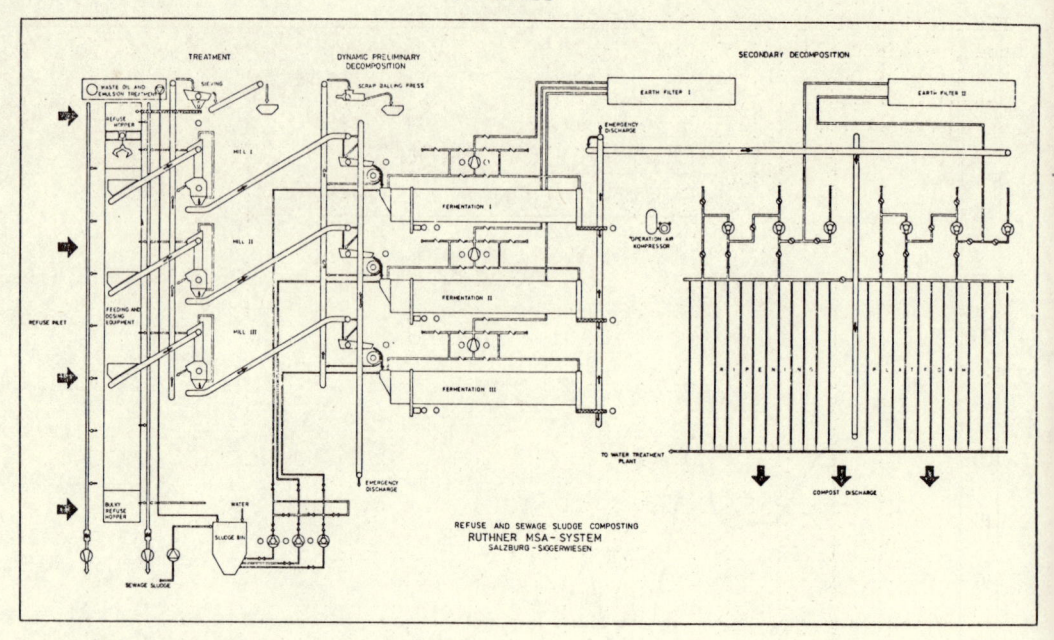

Fig. 4: The Salzburg composting plant

of the central sewage treatment plant in Salzburg, which has a capacity for 400 000 inhabitants, the sewage sludge will be pumped directly to this container by means of a pipe system. All delivery procedures for refuse and sewage sludge are handled in the closed hall, which enables dust and smell emissions, as well as the problem of loose flying papers, to be prevented. The dust, which is generated by the dumping vehicles in the hall, is suctioned by fans and cleaned by filters before exhausting.

The refuse is then continuously transported by steel plate conveyors to the three shredding units. The dosing is carried out using an adjustable conveyor speed, proper inclination of the conveyor and also the adjustable position of scrapers.

Through intake chutes, which are manufactured in a special way, in order to prevent the exit of pieces knocked back, the refuse is dropped directly into the shredders. Deflagrations and explosions, for example, from solvents and dust, may take place in this intake section, but because of expansion due to the wide construction, do not damage the shredder. The crushing equipment of the shredder consists of a rotor, with attached rotating hammers, moving through a knife-bridge construction. Non-crushable items, such as crank-shafts, railway rails, etc., are pushed out automatically through a spring-loaded clamp and are brought by belt conveyors to a container or a storage area. The quantity of non-crushable material amounts to less than 5%.

In order to reduce wear and tear costs, the hammers can be used on four cutting edges. For this purpose, the upper part of the shredder casing can be opened and removed by a special device. The entire working procedure for changing the hammers, and for opening and closing the shredder requires 1.5-2 hours.

The crushed refuse is conveyed by totally closed trough chain conveyors to the iron separation unit. Here, the material is continuously spead on vibration chutes and conveyed to magnetic drums where the iron parts are separated and conveyed to a scrape bailing press. The bars, with dimensions of 50 x 50 x 20 cm, are transported to containers, for dispatch and reuse in the steel industry.

Experiences with a Two-Stage Refuse Composting System

The pretreated refuse is then brought to the fermentation drums, where it undergoes the prefermentation process. The sludge is transported from the storage bin into the drum by means of eccentric spiral pumps, at a certain proportion to the refuse capacity. Fig. 5 shows the possible addition of sewage sludge or water, depending on the water content of the raw refuse.

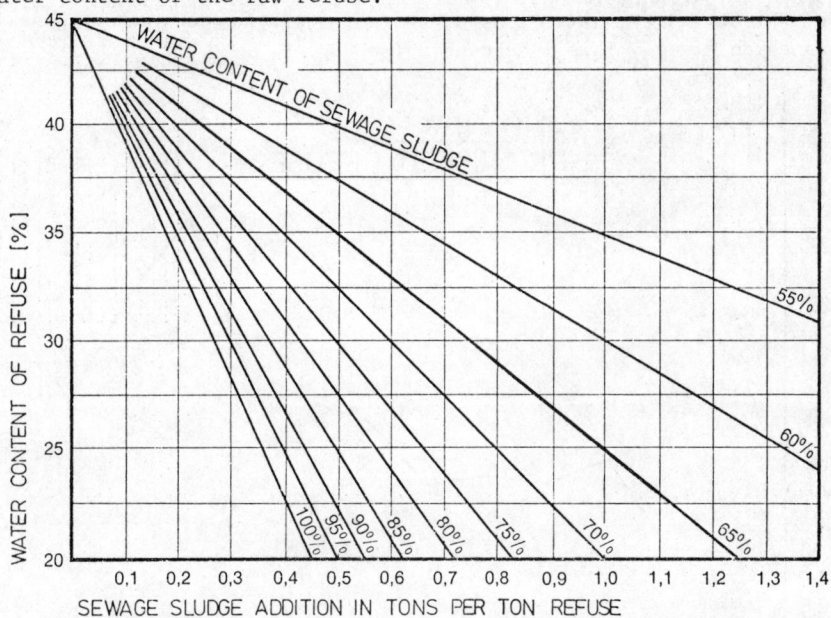

Fig. 5: Sewage sludge addition according to water content of raw refuse

Assuming a water content of 35% in the raw refuse, and a water content of sewage sludge of 95%, it can be seen from Fig. 5 that an addition of 200 kg sewage sludge/ton of refuse is possible.

The processing air, which is suctioned off from the fermentation drum, is conveyed by a pipe system to the compost filter, where it is rid of bad odours at a degree of efficiency of between 95 and 99%.

The fresh compost is now transported to the fermentation platform. By means of a feeding system, consisting of a longitudinal belt conveyor with a conveyable discharge van and a movable loading bridge with a drag chain conveyor, the fermentation platform is covered evenly at a height of 4.0 m. Now by means of the aeration system, air is suctioned or pressed through the fermenting material by means of aeration conduits. Each of these 77 conduits can be regulated separately: therefore, it is possible to obtain biologically optimal conditions in every section of the platform. The exhaust air from the suction operation is passed through the compost filter in a similar way as the processing air from the fermentation drums.

The fermentation process is then constantly checked by controls of temperature, CO_2 production and water content. Thus, if the water content of the material decreases below 40%, the moisture content can be adjusted by spray nozzles, which are attached in the roof construction in each section of the hall.

After a period of 3-4 weeks, the compost can be discharged by wheel-loaders and trucks and can be transported for storage or final compost-processing.

EXPERIENCES

Based on the decision of an international tender, the plant was erected on a turn-key basis during a period of 16 months by Ruthner Industrieanlagen A.G., Vienna, and put into operation in September, 1977. Since that time continuous treatment of the refuse of Salzburg City, as well as of an increasing number of surrounding communities, has taken place. This is shown in the Fig. 6.

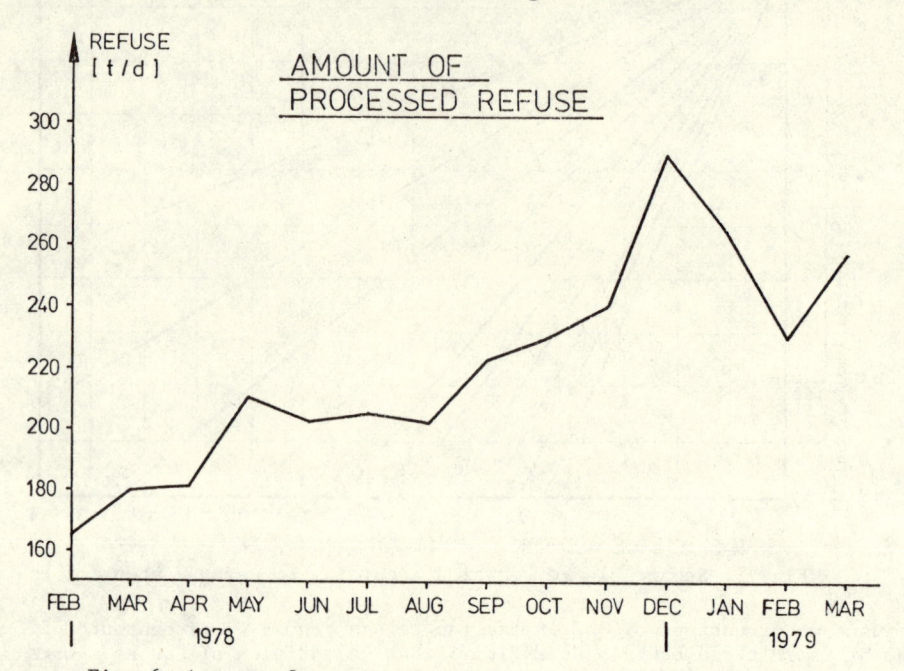

Fig. 6: Amount of processed refuse in the Salzburg area.

The value of December, '78 shows an irregular peak, on account of a seasonal delivery of dry sewage sludge, which was scheduled under "solid wastes". The following valley in February, 1979 was caused by a seasonal decrease in population in the Salzburg area during the winter holiday time. Fig. 7 shows the average water content of the raw refuse, as well as of the material in the fermentation drum and of the final compost on the fermentation platform.

The plot for raw refuse shows normal peaks, indicating high water contents in summer and low water contents in winter. The water content in the fermentation drum was regulated by sewage sludge and water addition, and was kept roughly at an average of 45%, which is considered to be the optimum. The middle plot shows the water content of compost after the period in the fermentation hall. Due to a lower capacity up to June, '78, the detention time on the platform was very long (approximately 10 weeks), which resulted in an extreme dessication of the material to a water content of below 30%.

The temperatures throughout the first phase in the fermentation drum are shown in Fig. 8. On account of the parallel direction of the air duct, the front end of the fermentation drum shows the lowest temperatures, ranging between 30 and 40^{0}C. Middle and rear end temperatures show amounts between 40 and 55^{0}C, which are generated by a full activation of the process in these zones. The low temperatures in January, 79 are not representative, as they were caused by a maintenance standstill of one fermentation drum, which resulted in a relative overload of the two other drums.

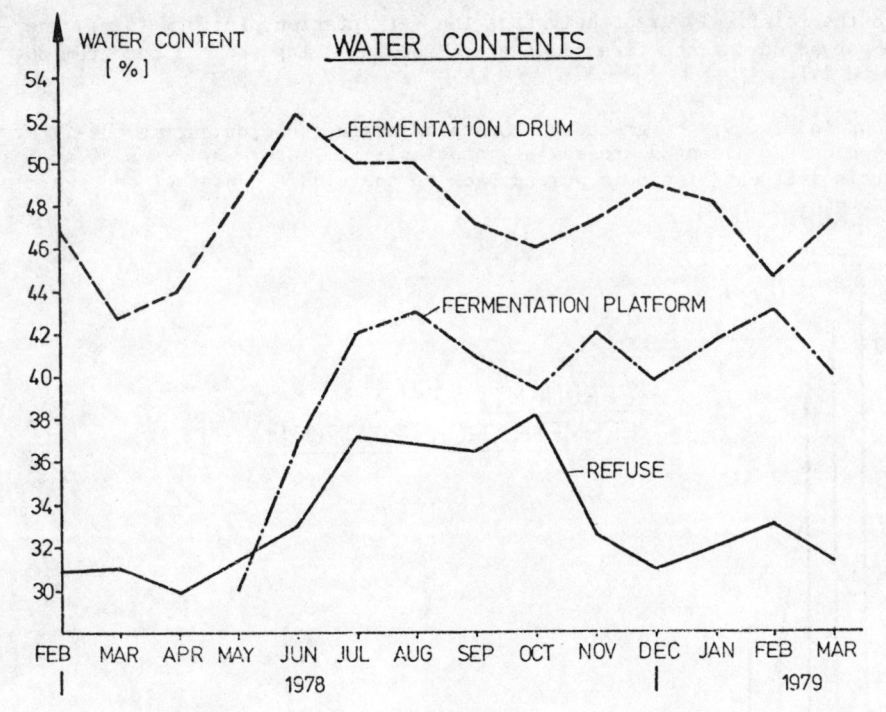

Experiences with a Two-Stage Refuse Composting System 421

Fig. 7: Average water contents.

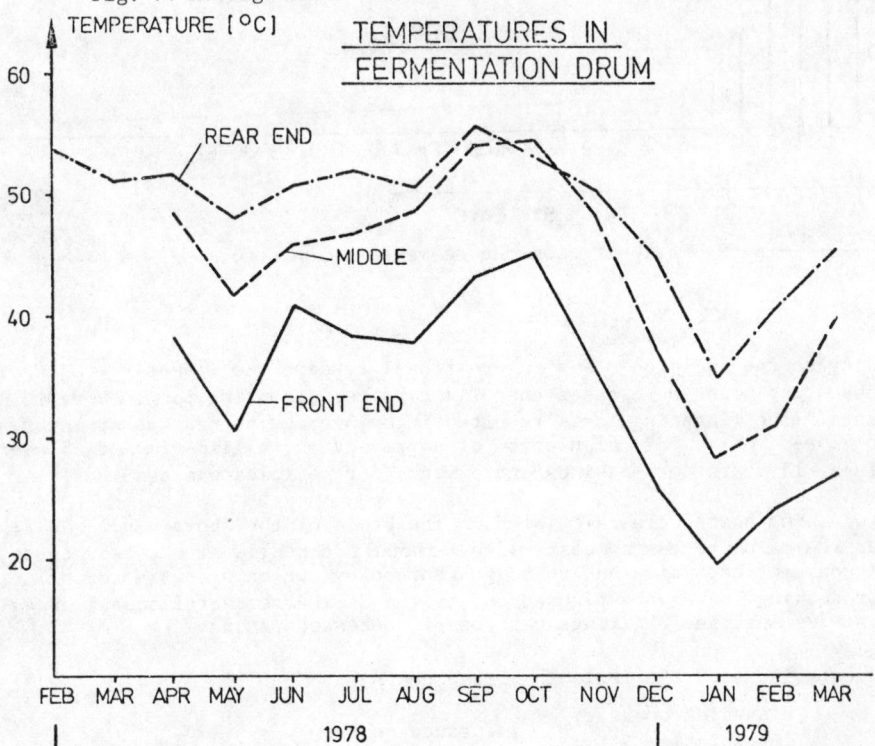

Fig. 8: Temperatures in the fermentation drum.

Due to the relatively great height of the fermentation platform (4 m), the effluent was observed during the first days after feeding being pressed out from the fermenting material.

As shown in Fig. 9, the greatest amounts of effluent occur during the first two days, after which the amount decreases asymptotically to zero. By means of a pipeline system, this effluent liquid is pumped back to the sludge container and recycled into the process.

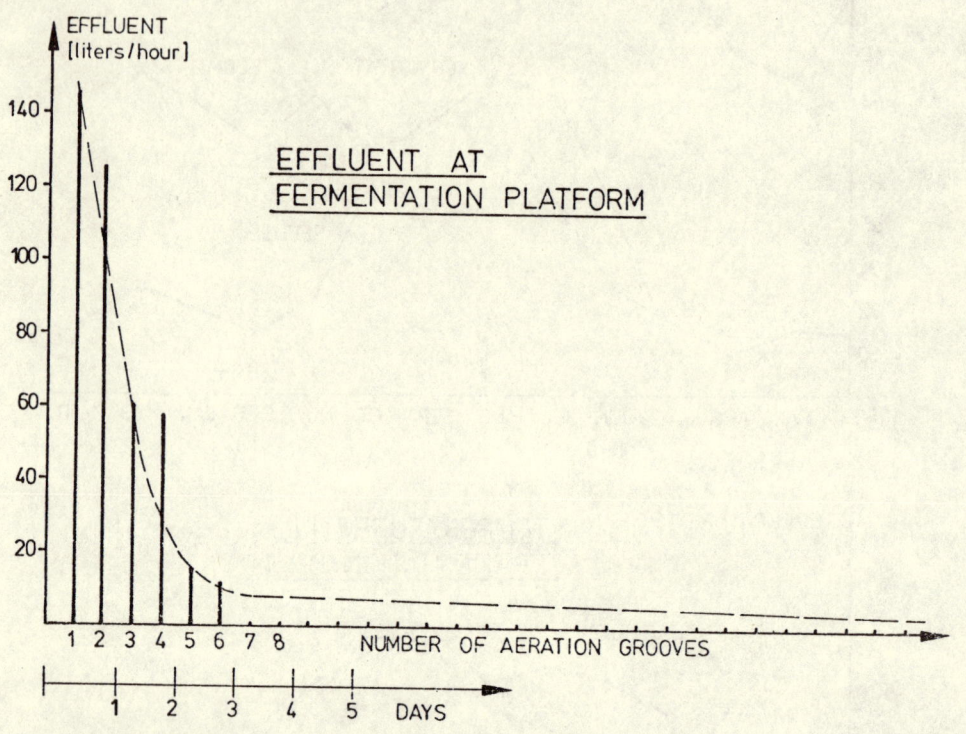

Fig. 9: Effluent from the fermentation platform.

FEASIBILITY

According to the decision of an international tender, the investing costs of the turnkey composting plant were based on a fixed price, amounting to 110,000,000 Austrian Shillings. The financing costs resulted in an average percentage of interest amounting to 7% per year. For calculation of depreciation, a life-time of 25 years for buildings, 15 years for machines and 5 years for vehicles was assumed.

The processing costs were calculated on the basis of the above-mentioned figures, as well as the main processing costs like personnel demand, wear and tear costs and electrical consumption, which had to be guaranteed by the main contractor. Thus, the actual processing costs were figured out by the Salzburger Abfallbeseitigungs GesmbH & Co KG at 244 Austrian Shillings per ton of processed refuse.

These costs included the following fixed costs:

        financing costs
        depreciation   insurance costs
        personnel   area rental costs

and also variable costs such as,

> energy costs
> wear and tear costs
> repair costs
> maintainence costs

The proportion of fixed costs to variable costs was approximately 2:1. In accordanc with refuse contracts with the authorities of Salzburg City as well as the surrounding communities; these costs were covered by communal refuse treatment fees. Thus, the charge for treatment of a locally applied container size of 90 liters amounted to 3.30 Austrian Shillings.

COMPOST

At present, a number of experiments to determine the compost quality at the composting plant are being conducted in cooperation with the University of Salzburg. An average analysis of the produced compost is given in the following table.

Analysis of Compost Quality at the Refuse Composting Plant in Salzburg

Compost Analysis		
Water content	%	32
Specific weight	g/ℓ	600
pH-value		7.3
Salt content	% i. TS	1.4
Annealing loss	% i. TS	51.5
Total organic matter	% i. TS	48.5
Total effective org. matter	% i. TS	23.7

Main Nutrient Matters		
Total nitrogen	% i. TS	1
Phosphate	% i. TS	0.5
Kali	% i. TS	0.3
Magnesium	% i. TS	0.9
Calcium	% i. TS	1.6
Humus material	% i. TS	7.8

Trace Nutrient Material in ppm (mg/kg)	
Total boron	30
Total manganese	630
Total copper	600
Total Zinc	930

Noxious material in ppm (mg/kg)	
Total Cadmium	11
Total led	250
Total Mercury	2.0
Total Arsenic	5.2
Total Chromium	350

The positive effect of compost is based on the content of nutrients, which represent a natural source of fertilizing compounds over an extended period of 1 to 3 years, and also on the physical structure, which improves the ability of water retention and also the resistance against wind and water erosion. Tests have shown that a mixture of top soil with compost of 1:1 to 1:3, depending on geological and climatic conditions, produces the best results in terms of growth and quality of plants and crops On the physical side, it should be mentioned, that heavy clay soils have especially shown good results in improving their abilities with respect to oxygen access and water retention.

The glass content of compost, which is often disliked when used for agricultural purposes, can be rendered innocuous by applying a new method of compost-milling. Thus, the particle-size of glass is reduced to dimensions of 0.2 to 0.5 mm, which is similar to the physical structure of sand.

DISCUSSIONS

Hamen, J. (Luxemburg) : Could you give more information about the compost filter in Salzburg for removing foul air? Do you use fresh compost and how often must the compost be renewed?

Mooss,H. (Austria) : Laboratory tests have shown a theoretical efficiency of 97-99%. We, however, had slight problems with the filter at first because its capacity was not large enough. At present 50 m^3 of air pass through 1 m^2 of filter, which appears to be adequate. Matured compost is used. Since the plant has been in operation for just 2 years, I can only say with regard to compost renewal that it should take place after a minimum of 2 years.

Hamen, J.(Austria) : A compost filter exists in one other plant in Austria, but I cannot give you details about its efficiency. In general, if the filter is large enough, the odour can be reduced to a minimum.

Patrick,P.K. (U.K.) : Could you give information about the design capacity of the Salzburg plant, its capital cost and its surface area?

Mooss,H.(Austria) : The design capacity of the plant is 400 t/d of refuse and 120 t/d of sludge. The capacity at the moment is 300 t/d and 60-100 t/d of refuse and sludge, respectively. The investment costs amounted to 110 million A.S. or ∅ 8.14 million. I cannot give you the exact surface area of the plant. The length and width of the buildings are 140 m and 70 m, respectively. The area of storage outside depends on casual demand.

Dunz, W. (Germany) : Have your experiments related to the growing of vegetables with compost yielded any results yet, and if so, when will they be published?

Mooss,H. (Austria) : At present the results of tests on the growth of the plants are available. Test results in connection with the contents of the vegetables are expected in the autumn of 1980.

Karlsson, L. (Sweeden) : No compost was sold initially because tests were being carried out on it. The selling of compost depends on the country concerned, the climate and soil conditions. In Austria compost is generally used in vineyards. However, since Salzburg is not a grape-growing area, most of the compost simply has to be disposed of. In Germany, however, the compost from the plant at Duisburg is sold for filter purposes to animal-incineration plants.

Bunning, W. (Holland) : What are your experiences with the compost filter during the winter?

Mooss, H. (Austria) : As the temperature of the air passed through the filter is between 40°C and 70°C, a stream of liquid can be observed in the winter, but this does not affect the functioning of the filter. Problems can, however, occur in the hot and dry summer periods when the filter material dries out, with the result that the function of the filter is reduced. This problem can be solved by spraying water on the filter.

Investigation and Reuse of Industrial Solid Waste from Stockholm

L. KARLSSON

VBB, Member of SWECO, Swedish Consulting Group, Stockholm, Sweden

ABSTRACT

An investigation of the amount and composition of industrial solid waste produced in Stockholm was carried out by VBB in 1976. This was followed in 1978 by a study of various potential methods of reusing the waste and an evaluation of their feasibility from the technical, economic, environmental and energy-saving point of view. The methods used in the recovery of material and energy are technically advanced. Emissions to the environment can be controlled. Recovered material is currently difficult to market and the benefits do not cover the costs. However, the costs of energy recovery from pulverized refuse-derived fuel could be met by a reduced demand for fuel oil.

INTRODUCTION

Increased utilization of science and technology has meant that the old society based on the primitive bartering economy has developed into an extremely industrialized society with a high production of goods. This change has in general led to an increased standard of living accompanied in many cases, however, by heavy exploitation of natural resources. Awareness of and anxiety for the limited assets has recently led to a greatly increased interest in economizing on raw material and recycling of wastes. The wide variety of material currently discarded should be regarded as a resource rather than merely as waste products.

In Sweden, interest in the recovery of solid waste is still growing, although most of it is still disposed of by controlled tipping. Frequent use is being made of domestic solid waste and waste from tree-felling and sawmills in connection with energy production and composting, and new technology for material recovery is being developed.

Although quite a lot is known both in Sweden and in other countries about the amount and composition of domestic solid waste, little is known about industrial solid waste. In 1976, VBB was commissioned by the Stockholm Public Health Committee to make an investigation of the industrial solid waste from Stockholm. The investigation was followed later on behalf of SKAFAB (Stockholm Municipality Waste Refinement Ltd.) by another investigation which took the form of a comparative survey of various potential methods of reusing the waste.

ORIGIN, AMOUNT AND COMPOSITION OF THE INDUSTRIAL SOLID WASTE

Industrial solid waste is in this context defined as those kinds of non-hazardous waste which cannot be referred to as common domestic waste, surplus material from earth works and sewage sludge excluded. The waste basically originates from four spheres of activity: industry, construction, commercial and dwelling management (bulky residential waste).

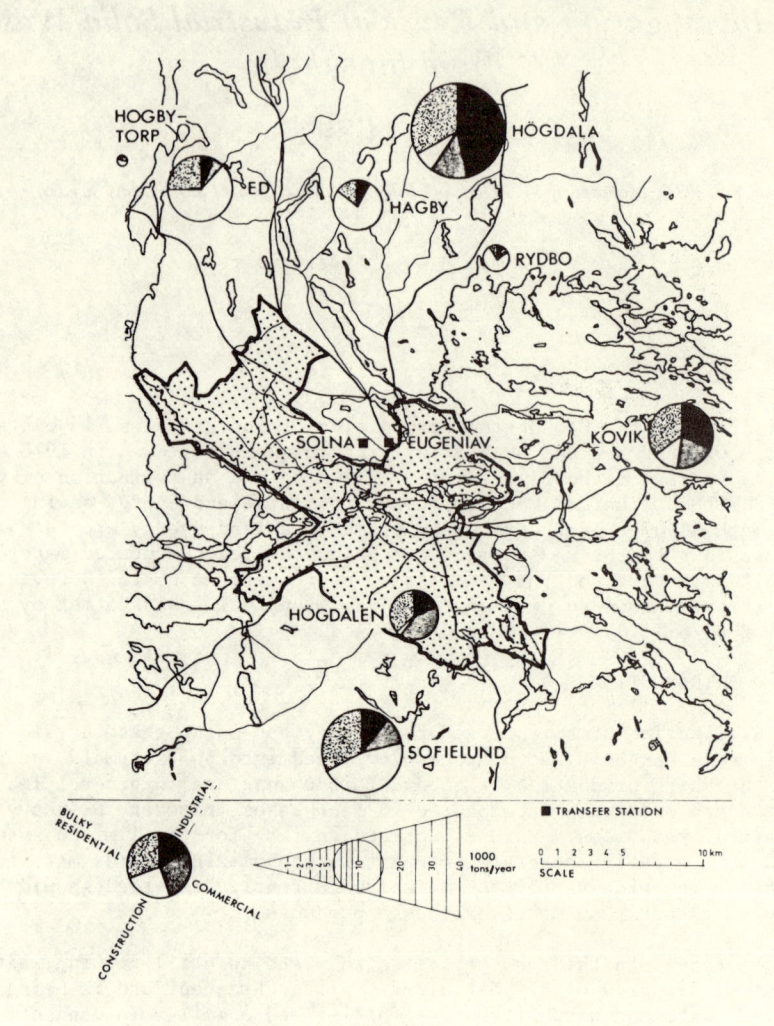

Fig. 1. Disposal Plants Receiving Industrial Solid Waste from the Municipality of Stockholm

Performance of the investigation

For a period of 1-2 weeks in the autumn of 1976, every trip-load arriving at all eight disposal plants or sites (Fig. 1) receiving industrial solid waste from Stockholm, as well as from other municipalities, was registered and visually defined with regard to origin, volume, type of waste and composition in terms of percentage of volume. At the four plants which received most of the waste, every tenth trip-load was manually sorted into components which were in turn measured for volume and weight. Construction waste was not measured since extensive knowledge was available from earlier investigations. The area of origin was recorded, but is not discussed any further on this paper because it is mainly of local interest only.

Available records of received amounts of waste were collected, although there tended to be a lack of uniformity between them from site to site. The figures were given in trip-loads per day, trip-loads per month, m^3 per day, m^3 per month and tons per month, and were used in order to calculate the total weight of industrial solid waste from Stockholm. When establishing the conversion factor the following assumptions were made:

- The amount of waste from Stockholm was proportional to the total amount received.

- The volume of waste was proportional to the number of trip-loads.

- The volume of waste was proportional to the weight.

Results of the investigation

The total registered volume of waste subdivided into the various origins and spheres of activity is shown in Table 1.

The seperated waste components were monitored for weight and volume and their proportions, expressed as percentages of the total volume, are given in Table 2. The seperated volume was found to be 40% more than the registered volume and since the average weight per m^3 of the components was 74 kg/m^3, unsorted registered waste could be expected to weigh 104 kg/m^3. Also there was a considerable difference between the volumes of the seperated components compared to the visually registered volumes. Visual judgement is thus a very rough method of defining the composition of the trip-loads.

Table 1 Registered Volume of Waste in m^3 per Two Weeks

Component	Industrial	Commercial	Construction	Residential	Total
Cardboard	1,009	859	1,694	1,400	4,961
Newspaper	157	28	74	337	596
Other paper	3,027	2,264	1,458	2,863	9,612
Wood	1,915	1,187	4,905	2,600	10,607
Glass	248	49	119	224	639
Metals	854	312	855	949	2,971
Plastics	945	482	661	730	2,816
Rubber	142	45	42	83	312
Textiles	192	52	110	377	731
Others	288	188	2,197	1,166	3,840
Unspecified	119	39	166	137	461
Total	8,894	5,503	12,274	10,862	37,533

Table 2 Proportion of Components in Seperated Waste

Component	Percent of total amount	
	Volume %	Weight %
Cardboard	14.6	4.3
Newspaper	1.2	3.3
Other paper	10.1	13.6
Wood	42.9	41.2
Glass	0.4	1.8
Metals	13.3	16.9
Plastics	8.2	4.5
Rubber	0.5	1.5
Textiles	4.6	6.4
Others	4.3	6.5

Table 3 Total Amount of Industrial Waste in Tons/Year

Component	Sphere of activity				
	Industrial	Commercial	Construction	Residential	Total
Cardboard	1,090	620	1,390	1,360	4,460
Newspaper	830	480	1,070	1,040	3,420
Other paper	3,440	1,960	4,390	4,300	14,090
Wood	10,420	5,930	13,310	13,020	42,680
Glass	460	260	580	570	1,870
Metals	4,270	2,430	5,460	5,340	17,500
Plastics	1,140	650	1,450	1,420	4,660
Rubber	380	220	490	470	1,560
Textiles	1,620	920	2,070	2,020	6,630
Others	1,640	940	2,100	2,060	6,740
Total	25,280	14,380	32,310	31,610	103,600

The total annual amount of industrial waste was calculated, based on the results of the seperation and the available records of loads to the different sites, as shown in Table 3. Almost all the metals were found to be magnetic, and of the component category headed "others", which included bricks, light-weight concrete, brush-wood, leaves, ashes etc., about 65% were combustible. The total figure for waste, i.e. 103,600 ton/year, represents a specific value of 156 kg per capita per year.

The average volume per trip-load was 8.3 m^3, not counting the loads received at Högdala and Kovik, which are normally delivered in 30 m^3 containers from the two transfer stations, where the waste is compacted three or four times.

FACTORS AFFECTING THE COMPOSITION

Possible schemes for the waste cycle in the present society are shown in Fig. 2. It should be noticed that some of the potential waste can be sorted out by the industries themselves and recycled to either internal or external production. The composition of waste is affected by reuse at the source, the state of the market, developments within industry, especially the packaging industry, and overruling Governmental decisions.

As a result of Governmental legislation, seperation of paper at the source is frequent in Sweden, especially that from households. Tests have been performed in connection the seperation of metals and glass, and within industry seperation is practised as long as it is an economically viable proposition for the industry in question. Depending mainly on the state of the market for recovered material, the proportion of paper and metals in the waste may change. The proportion of plastics has increased recently because of the increased use of plastics for packaging, and the trend may continue.

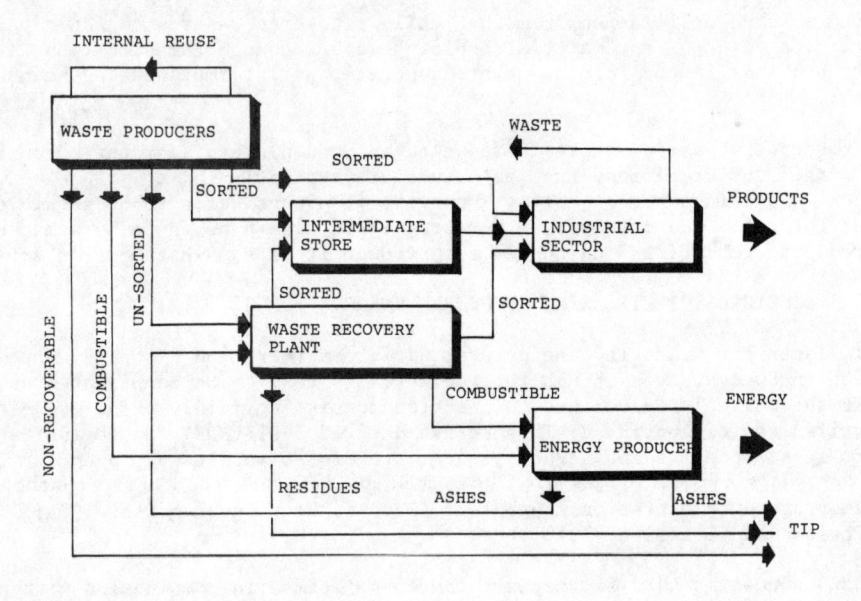

Fig. 2. Flow Chart Showing Possible Waste Cycle

METHODS OF RECOVERING MATERIAL

Recovery can be performed either at the source or at a central plant, although a variety of intermediate alternatives can also be found. The possible use of different containers - for paper, metals, glass etc., or for combustible wastes - depends on local conditions.

There are several different systems of centralized material recovery from domestic solid waste and technology is developing rapidly (Hagerty, 1977). In general, the methods used are manual seperation, dry mechanical seperation and wet mechanical seperation. Manual seperation of industrial solid waste is used at one plant in Sweden (Helsingborg) and pilot plants have been built at Högdalen, Stockholm for air classification of paper and plastics from domestic waste (Citron and Halén, 1976; Bahri, 1976) and at Kovik, Stockholm for ballistic seperation (Wannag, 1976).

Technology has been developed to seperate a light fraction consisting of paper, plastics, wood and textiles. It is also possible to seperate paper and plastics further from this fraction, but owing to impurities, the use of the products is very limited, at least in Sweden. Glass, magnetic metals and aluminium can also be seperated from domestic waste.

MARKET SITUATION FOR RECOVERED MATERIAL

About 500,000 tons of recovered paper was used in 1977 in 24 papermills in Sweden, five of which have plants de-inking. The waste paper, mainly consisting of cardboard and newspaper recovered at the source, is thoroughly classified, but since its market value is generally low, about 100 DM/ton excluding transportation, stockpiling is frequent. The paper industry is hesitant to use mechanically recovered paper, but a possible use could be for fluting cardboard and for newspaper although there is a risk of operational problems owing to impurities.

Today there is no utilization of mechanically recovered waste wood for paper or board production. A lack of raw material could, however, change the situation although impurity problems in the form of paint, concrete, nails, impregnating agents etc. may occur.

1.3 metric tons of waste metals is recycled in Sweden every year but because of the crisis within the steel industry the prices are currently low and range between 10-15 DM/ton, depending on the quality. However, it always seems to be possible to find a market for recovered metals. Furthermore, there also appears to be a market for recovered plastics, glass and rubber even though it is somewhat more uncertain.

METHODS OF UTILIZING RECOVERED ENERGY

Several plants for producing energy from direct incineration of solid waste are in operation in Sweden, even though the incinerators tend to be complicated and expensive. If the waste is pulverized, a fraction consisting mainly of paper, plastics and textiles can be seperated. This refuse-derived fuel (RDF) can be incinerated in simpler, less expensive incinerators, e.g. stoker-fed wooden chip ovens or fluidized bed ovens. Pyrolysis may also be a possibility even though the system has up to now been plagued by operational problems (Pavoni, Heer and Hagerty, 1975; Tchobanoglous, Theisen and Eliassen, 1977).

The heat produced is used in thermal plants, sometimes in combination with power production. In Sweden where the period during which heating is required is fairly long and the use of hot water even during the summer is high, the combination of waste incineration and district heating is often found to be a feasible proposition which permits the use of cheap hot water boilers with a maximum temperature of 120^0C

Based on specific calorific values (Lipták 1974), the energy content of the industrial solid waste in Stockholm has been calculated at about 12,000 kJ/kg for a water content of 20%. If metals are seperated, the remaining fractions will have a calorific value of about 14,000 kJ/kg, meaning that about 3 tons of waste correponds to 1 m^3 of fuel oil, the efficiency factor included.

COMPARISON BETWEEN ALTERNATIVE RECOVERY METHODS

Central reclaiming of solid waste implies advanced technology with a high level of operational skill and supervision. Landfill disposal should be available for emergency purposes as well as for the disposal of residues from the reclamation plant. Material recovery appears to be more sensitive to variations in the received waste.

Material recovery will in general save more energy than energy recovery from incineration. The difference can be calculated to correspond to about 200 l of fuel oil per ton of paper, 14,000 l per ton of aluminium and 530 l per ton of plastics. Recovery of glass and tinplate will save about 15% and 55%, respectively, of the energy needed for new production. However, the calculations above presuppose that the material recovered at the source is comparatively free from impurities.

Dust seperation is necessary if the waste is seperated using dry mechanical methods. Wet seperation gives heavy liquid pollution problems. Material recovery will also induce secondary emissions at the installations using the recovered material. The fumes from the incineration of solid waste contain more dust, heavy metals and hydrochloric acid but less sulphur dioxide than if fuel oil is used and scrubbers are considered necessary. In general, however, technology exist to keep the emissions to the environment within the desired limits with regard both to material recovery plants and to incineration plants.

An economic comparison between a material recovery plant and a thermal plant based on refuse-derived fuel from the industrial solid waste of Stockholm (100,000 ton/year) is given in Table 4. Both alternatives include pretreatment with hammer mills, but transportation costs are excluded. A capital cost of 13% of the investment and a single shift operation per working day are assumed. Residues which cannot be utilized are disposed of at a controlled tip. The benefit of refuse-derived fuel corresponds to a fuel oil price of 240 DM/m^3.

Energy recovery appears to be the most attractive proposition in the current economic situation. Further increases in the price of fuel oil may strengthen this trend and at the same time the market for recovered material may also improve rapidly. A waste reclamation plant should, therefore, be flexible enough to permit the later incorporation of material recovery facilities.

PROPOSED LAYOUT

The industrial solid waste will be received and pre-separated into a grindable fraction and a coarse fraction consisting mainly of bulky wood which will be chopped before grinding (Fig. 3). Other bulky waste such as, refrigerators, stoves etc., as well as tyres and uniform loads, will also be pre-separated when received. After grinding magnetic metals will be removed and the remaining refuse will be collected in containers for transport to a thermal plant, possibly after compaction.

In a plant of this type, close supervision and control are essential and thus the control room has been located centrally at a point where all operations can be observed and remotely controlled (Fig. 4). The grinding, however, should take place in a seperate locked room because of the risk of explosions and the need to control noise and dust.

Table 4 Material/Energy Recovery - Cost/Benefit

	Material recovery		Energy recovery	
	Amount 1000 t	Total cost M DM	Amount 1000 t	Total cost M DM
Costs				
Pretreatment, capital	100	1.1	100	1.1
operation	100	0.9	100	0.9
residue	15	0.3	15	0.3
Processing, capital	75	2.0	75	1.7
operation	75	1.4	75	1.0
residue	15	0.3	8	0.2
Total annual cost		6.0		5.2
Benefits				
Metals	10	0.2	10	0.2
Other	75	3.2	75	4.8
Total annual benefit		3.4		5.0

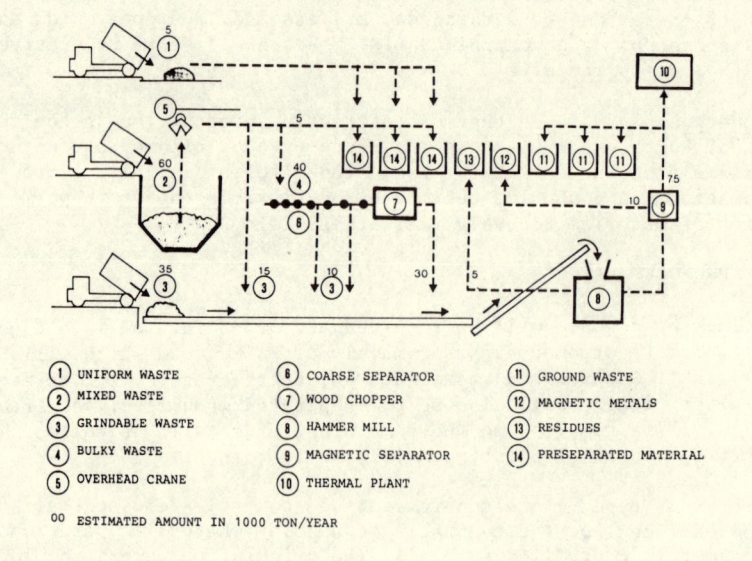

① UNIFORM WASTE
② MIXED WASTE
③ GRINDABLE WASTE
④ BULKY WASTE
⑤ OVERHEAD CRANE
⑥ COARSE SEPARATOR
⑦ WOOD CHOPPER
⑧ HAMMER MILL
⑨ MAGNETIC SEPARATOR
⑩ THERMAL PLANT
⑪ GROUND WASTE
⑫ MAGNETIC METALS
⑬ RESIDUES
⑭ PRESEPARATED MATERIAL

00 ESTIMATED AMOUNT IN 1000 TON/YEAR

Fig. 3. Flow Chart

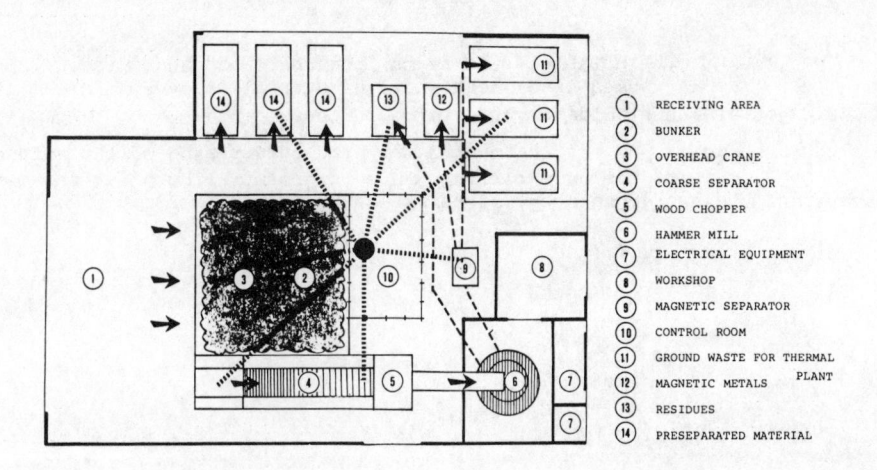

① RECEIVING AREA
② BUNKER
③ OVERHEAD CRANE
④ COARSE SEPARATOR
⑤ WOOD CHOPPER
⑥ HAMMER MILL
⑦ ELECTRICAL EQUIPMENT
⑧ WORKSHOP
⑨ MAGNETIC SEPARATOR
⑩ CONTROL ROOM
⑪ GROUND WASTE FOR THERMAL PLANT
⑫ MAGNETIC METALS
⑬ RESIDUES
⑭ PRESEPARATED MATERIAL

Fig. 4. Layout

REFERENCES

Bahri M. (1976). Automatic Seperation of Waste at Högdalen, (in Swedish). Återvinning. Teknik och ekonomi, Ingenjörsförlaget/Naturvårdsverket, Stockholm.

Citron B. and B. Halén (1976). *Mechanical Recovery from Domestic Solid Waste.* (in Swedish). Stvrelsen för Teknisk Utvecking, STV, Stockholm.

Hagerty, D. (1977). Current American Alternatives in Resource Recovery. *Solid Wastes,* June 1977.

Lipták, B.G. (1974) *Environmental Engineers Handbook.* Chilston Book Co. Radour Pennsylvania.

Pavoni J.L., J.E. Heer Jr. and D.J. Hagerty (1975). *Handbook of Solid Waste Disposal, Materials and Energy Recovery.* Van Nostrand Reinhold Co., New York.

Tchobanoglous G., H. Theisen and R. Eliassen (1977). *Solid Wastes.* Mc Graw-Hill Book Co., New York.

Wannag A. (1976). Plant for Seperation of Waste at Kovik. (In Swedish). Återvinning. *Teknik och Ekonomi,* Ingenjörsförlaget/Naturvårdsverked, Stockholm.

DISCUSSIONS

Patrick, P.K. (U.K.) : In Table 4 the terms "pretreatment" and "processing" are used in connection with material recovery and energy recovery. Could you please elucidate these terms further?

Karlsson, L. (Sweeden) : By pretreatment I mean the pulverizing of the refuse in the hammer mill. This is essential both for the material recovery and the energy recovery of refuse-derived fuels.

A Study of Selected Industrial Solid Wastes

E. ERDIN

*Environmental Engineering Department, Civil Engineering Faculty,
Ege University, Izmir, Turkey*

ABSTRACT

The results of an inquiry into the solid wastes from selected industrial plants in the Izmir Area are presented at the beginning of this study. The amount of solid wastes and their compositions have been investigated. The techniques for the disposal and possible reuse of wastes have been indicated for each industry that was studied.

Compost plants, which may be considered as industrial premises causing environmental pollution, are studied along with other industries in Izmir. Non-compostable portions of the municipal solid wastes are stored in piles formed on land. Leachates which seep from these piles pollute the surface waters and groundwater sources by carrying along the non-degradable wastes. Reusable materials in the piles, however, are generally picked up manually and recycled. Tin cans, glass, cardboard and paper are among these recycled materials.

INTRODUCTION

The amount of solid waste produced in the City of Izmir is about 1000 t/d. The constituents vary depending on the season, sources and the socio-economic situation. For example, the percentage of inorganics increases in winter because of the high ash content of the wastes.

The solid waste (600 t/d) is disposed of by storing it in the marshes of Bayrakli and Çiğli. A connection from the marshes carries the inorganic and organic pollution to Izmir Bay.

It is necessary that Izmir's solid waste be disposed without causing environmental pollution. It is also of importance to choose an economic method for the disposal.

Domestic solid wastes from the various industries are collected by the municipalities in the area. Industrial wastes are generally reused. Some of the investigations which were conducted in selected industrial facilities will be summarized here. As part of this study, the characteristics and quantities of various industrial solid wastes have been determined. The leachate from composting units has been analyzed physically and chemically. The solubility of leachate and compost with water (in a ratio of 1:20) have been determined as well. Color, odor, pH, EC_{25}, organic and inorganic composition and salinity are the properties analyzed.

MATERIAL AND METHOD

Material: Industrial wastes and leachate from composting units.

Method : Investigations on industrial plants have been conducted by questionnaires. The pH and E_h values have been measured by a Metrohm Herisau pH-meter (Pöpel, 1973). Organic contents have been determined by ashing the samples at 625°C. Water contents have been determined in the oven at 105°C (EAWAG, 1970).

EC_{25} and salinity have been measured with a Beckman Conductivity Bridge RC-19 (Bock, 1972).

Trace element concentrations have been measured with a Varian Techtron AA5 atomic absorption spectrophotometer (Grabner, 1977).

SOLID WASTES FROM SELECTED INDUSTRIES

Some examples from the chemical, dye, bicycle, wood, wine and petro-chemical industries, as well as from poultry farms lignite mines and raisin -processing plants are given below.

The Olgun Carpentry Shop produces 2 tons of wood shavings per month. This waste is used as fuel.

The Sanver Wine Distillery produces 70 tons of waste (deposits from wine tanks) annually. This cannot be used as fertilizer because of the grape seeds it contains. However, it is possible to produce oil from the grape seeds. The deposits in the wine tanks can be used to produce tartaric acid.

The Önder Raisin Processing and Export Company produces 3000-4000 tons of waste annually. This waste is used as animal feed and for the production of molasses and syrup. The wastes from the packing department are used as fuel for drying ovens.

The Industrial Turc Company, which manufactures Licorex and some chemicals, produces 3 tons of solid waste per day. Part of the waste is simply soil and used as fill material. The organics are used as fuel.

The wastes from the Yavuz Poultry Farm contain 20-30 tons of faeces and 4-5 tons of other solid wastes monthly. The disposal of the faeces is a nuisance, but other wastes are sold to factories producing animal food.

The Dewilux Dye, Resin and Polyester Industries Inc. processes its wastes for reuse. In this way, 49017 kg of solvent and 87219 kg of resin were obtained from 135236 kg of waste solvent between 8.11.1977 and 4.12.1978.

The DYO-Sadolin Factories, which produce dyes and brushes, burn their waste-washing sludge in an open area near Izmir. Solid wastes produced by the brush-manufacturing department are used as fuel. Packing wastes are either recycled or sold.

The Özyurt Chemical and Petrochemical Industry uses its tar-like waste (80 tons/yr) as fuel after mixing it with fuel-oil. Such wastes as tar and asphalt are hard to burn, and thus are buried in the ground. Fibrous wastes, such as cellulose and lignin-containing liquids, are mixed with liquid effluents from the plant.

Sediment deposits from the wastewater settling tanks of the Aliağa Petroleum Refinery were cleaned in 1978 for the first time since 1972. The depth of this deposit was 1.0-2.5 m. The total amount of settled sludge was 1500 m^3, with an 80% water content. The characteristics of the effluent taken from the settling unit on 15.2. 1979 were found to be as follows: pH=8.6, salinity= 130 mg/ℓ, total hardness=220 mg/ℓ oil=8.5 mg/ℓ, dissolved solids=750 mg/ℓ and suspended solids were less than 5 mg/ℓ.

These values were observed when the NaOH content was extremely high. 28.6% of the sludge taken out annually from the sedimentation tanks is organic and 71.4% is inorganic. The organic content of the tar-like waste from the Özyurt Chemical Industry is 61.1% and the inorganic content is 39.9%. The specific weight of the samples with a high organic content is low (approx. 1).

Solid waste from the sheet iron and piping industry in metal-processing plants, such as the Bisan Bicycle and Moped Factory, are sent to iron and steel mills, where they are recycled.

As can be seen from the examples above, reuse techniques are widely applied by the industries to solid wastes, except for sludge. Burning the wastes or disposing them into water-bodies causes other environmental problems in air, water and soil. To have a more comprehensive picture of the situation existing in Izmir it is necessary to enlarge the inventory of industries. This study is just an introduction to a much broader investigation of the industrial solid waste problems.

COMPOSTING OF MUNICIPAL SOLID WASTES

In Izmir, the municipal solid wastes produced amount to 400 t/d and are processed at composting units. Approximately, 250-300 t/d of organic fertilizer (compost) is produced from this heterogenous raw material (solid waste).

Leachate from Composting and Its Characteristics

One of the composting units is at Halkapınar, located near the municipal water supply facilities, and the other is at Çiğli, on a marshland which has connections with Izmir Bay. That is why it was deemed necessary to determine the characteristics, especially the chemical properties, of the leachate produced in the composting process and drained out from the piles. Leachate samples were taken from the units and from the drainage of the compost piles at different times.

Leachate from the Dano-Biodegradation unit was analyzed for organic and inorganic contents. At the time of the sampling, the solid wastes contained a great deal of water melon and its wastes. For example, the solid contents of the sample taken on 18.7.78 from the leachate which comes from the composting drum contained 50.5% organics and 49.5% inorganics. A sample taken on 18.7.1979 was found to contain 56% organics and 44% inorganics. The BOD_5, COD and total nitrogen values were 7,400 mg/ℓ, >9,280 mg/ℓ and 142 mg/ℓ, respectively. Very high heavy metal concentrations were determined in the ashes after incinerating the sample at 625°C.

Mn, As, Cr elements gave high values upon oxidation with HCl, but Pb, Zn, Cd, Fe and Cu were better digested with a $H_2SO_4+H_2O_2$ mixture. It is recommended that HCl be used as a digestion acid to enable better recovery of most trace elements (Techtron, 1977).

As it is clearly indicated in Table 1, the trace element concentrations in the leachate were higher than the concentrations obtained for the samples which were prepared by HCl extraction with 1/20 H_2O. Although it was impossible to detect any Pb, Cd, and Mn in the acid extracts it was possible to meet 4.5-18.6 ppm Pb, 0.02-0.12 ppm Cd, 5.2-15.1 ppm Cr, and 49.5-55 ppm Mn in the leachate. The low pH value of the leachate (pH=5) made it possible for the heavy metals to remain in solution (Table 1 & 2). However, pH values were in the order of 7-8 for the samples prepared by mixing the compost with water. It was also observed that decreasing the pH increased the solubility of heavy metals in water.

By taking samples from the piles before and after rainfall and analyzing them for heavy metals and salinity, it was proved that the rain had a dragging effect. The Pb concentration was 480 ppm in the sample taken before the rain and was reduced to

Table 1 Elements and Water-Soluble Trace Metals in Compost Leachate and Decomposition Water (as ppm)

Sampling Date and Character	Extracted in	Pb	Cd	Mn	Ni	As	Cr	Fe	Zn	Sn	Cu
11.5.1978 leachate	water	4.5	0.02	49.5	5.6	0.0	5.2	186.5	21.2	185	1.15
18.7.1978 leachate	water	18.6	0.12	55.0	4.75	0.0	15.1	750.0	43.4	–	0.2
Ashed dry solids in leachate	$H_2O_2+H_2SO_4$	525	3.5	1125.0	6.5	60	5.0	357.5	1810.0	–	45.0
	2N.HCl	127.2	0.1	1250.0	108.0	360	238.5	298.4	1685.4	–	13.3
18.7.1978 fresh compost	water (1 day) (1:20)	0.0	0.0	80.0	28.4	0.0	18.7	6.0	33.8	–	0.0
Tracing of the leaching effect of rainfall
11.10.1976 before rain	–	480	< 1	–	175	–	–	–	490	–	155
12.10.1976 after rain	–	80	< 1	–	175	–	–	–	420	–	85
Difference	–	400	–	–	–	–	–	–	70	–	70

Table 2   pH, $EC_{25} \times 10^6$ and % Salinity Values of Some Water* and Leachate Samples

Sampling Date and Character	pH	$EC_{25} \times 10^6$	%Salinity**	%Total Solids	Organic Matter%
Rain water:					
19.4.1978 from pluviometer	7.2	130	0.0083	0.021	-
30.5.1978 from pluviometer	7.3	255	0.0163	0.045	-
12.4.1978 from roof	7.3	180	0.0115	-	-
Surface Waters					
16.4.1978 Nif creek	8.1	275	0.0176	-	-
21.4.1978 Nif creek at source	7.4	501	0.0320	-	-
5.5.1978 Halkapinar water supply unit	7.3	527.5	0.0338	-	-
5.5.1978 Halkapinar Lake	7.4	565.8	0.0362	-	-
11.5.1978 Halkapinar Lake near compost plant	8.2	1,382.0	0.0884	-	-
Compost Pile Leakage					
11.5.1978	6.5	41,960	2.685	-	-
24.10.1978, 1 m away	7.1	21,133	1.353	4.26	2.45
24.10.1978, 10 m away	7.6	12,913	0.826	1.14	0.49
3.11.1978, 1 m away	6.8	25,113	1.607	8.41	4.85
3.11.1978, 10 m away	7.5	7,353	0.471	1.48	0.56
14.12.1978, 1 m away	6.9	31,660	2.026	6.88	3.71
30.12.1978, 1 m away	7.7	29,820	1.909	7.24	3.61
12.1. 1979, 1 m away	7.7	27,320	1.748	-	-
20.1. 1979, 1 m away	7.4	27,680	1.772	6.28	3.91
20.1. 1979, 10 m away	8.1	5,780	0.370	0.63	0.154
27.2. 1979, 1 m away	7.0	6,376	0.408	-	-
Drum Leachate					
18.7.1979	6.2	48,780	3.120	20.28	11.43
Landfill Leachate					
7.11.1978	8.8	9,204	0.589	0.662	0.100

* In collaboration with A. Müezzinoğlu (Water analysis only)

** Total Salinity is found by the formula, $\%S \dfrac{EC_{25} \times 10^6 \cdot .64}{100}$ (Kreeb, 1964).

400 ppm after the rain in the same pile. The initial and the reduced concentrations of the other elements were as follows:

$$Cr= 675-575 \text{ ppm}, \quad Zn= 490-420 \text{ ppm}, \quad Cu= 155-85 \text{ ppm}.$$

The composting leachate contained some elements like Pb, Cd and Hg, which are of importance from the environmental engineering point of view (Table 3). The leachate can be carried to surface waters or percolated through the soil into the ground water in the rainy seasons (spring, winter, fall). This must be prevented. It will therefore, be necessary to control the composting leachate and treat it at the Izmir composting plants. A more suitable method might be to collect this water by means of a drainage system. Another possibility would be to use ditches as collectors.

As it is clearly seen in Table 3, the trace metal concentrations and total nitrogen were higher in the samples taken from the bottom of the piles than the samples taken 10 meters away from the piles. The concentrations of various elements determined in the samples were much higher than the values which could be found in some natural ground waters, except for Zn and Cd. The Pb concentrations were especially high (varying between 2.8-28.0 ppm). Kick (1974) informs us that Pb concentrations vary between 0.1-11 ppm in soil, and the limit for accumulation is 100 ppm. The leachate from composting causes soil and water pollution (Tabasaran, 1979). In addition to that, the Halkapinar composting unit is near the Municipal Water Supply units of Izmir. That is why the measures mentioned above must be taken for the protection of the ground waters and water supply systems. Soil in the mentioned location is calcareous and this adds even further to the seriousness of the problem.

The Çiğli Compost Plant, on the other hand, is near a marshland which drains to Izmir Bay. Therefore, the leachate from this unit is responsible for carrying organic and inorganic pollutants into the already heavily polluted bay waters.

CONCLUSIONS

As has been seen from the section on selected industries, carpenter workshops, where wood is used as a raw material, produce solid wastes which do not create any disposal problems. Wastes from these workshops are disposed of by burning. Wine distillaries, which use grapes as raw material, produce wastes such as seed material, grape stones, etc., which are utilized as animal feed and syrup. Wastes from the licorex-producing industry are used as burning material owing to their high calorific value.

Wastes from the poultry industry are hard to remove and cause objectionable odors, flies and rodent problems. Leachate from poultry fertilizer piles causes water pollution problems in the region. The possibility of reusing these wastes is under investigation.

The disposal of sludge from the dye and petrochemical industry by burning, landfilling and random dumping on the ground should not be permitted. Control of these industrial wastes can be carried out by examination of their contents and characteristics. Evaluation of the reclamation and reuse possibilities of wastes from specific industries is of great importance from the pollution control standpoint. Open mine wastes, such as lignite excavation wastes and scraped earth, should not be dumped on the land, but used as a landfill in selected areas.

Metal industries, such as the bicycle-manufacturing industry, produce fewer wastes than the other industries investigated in this study.

In the composting process, textile, metals, paper, stones and ceramics etc., are by-products which should be stored for reuse, and those which are not usable should be

Table 3 Trace Elements (ppm) and N_{TOTAL} (mg/ℓt) in Compost Leachate after Rainfall

Sampling Date and Character	Zn	Sb	Ni	Fe	Mn	Pb	Cd	N
24.10.1978 1 m from pile	8.1	8.95	3.0	52	20	4.4	0.0	218.4
24.10.1978 10 m from pile	0.8	7.00	1.5	27	4.2	4.4	0.0	50.6
7.11.1978 1 m from pile	0.0	5.75	0.0	0.0	0.0	2.8	0.01	57.65
3.11.1978 1 m from pile	7.4	12.40	4.4	56.0	26.5	28.0	0.0	65.20
3.11.1978 10 m from pile	0.0	5.75	0.03	4.4	0.0	4.4	0.0	60.20
14.12.1978 1 m from pile	0.9	11.0	5.2	47.5	20.0	2.8	0.0	–
30.12.1978 1 m from pile	7.7	12.0	3.3	56.5	23.0	19.8	0.0	–
12.1.1979 1 m from pile	1.5	9.75	1.7	50.0	10.0	11.0	0.0	–

Natural waters rich and poor in heavy metals (Ernst, 1974):

	Zn	Ni	Fe	Mn	Pb	Cd	Cu	Co
Trace Metals Rich Waters	70	0.020	0.050	0.0	0.001	0.085	0.020	0.0
Waters with Low Trace Metal Concentrations	0.001	0.004	0.005	0.0	0.0	0.003	0.003	0.0

disposed of in a landfill. Furthermore, due to the fact that some leachate may be produced during the composting, an appropriate drainage system should be provided in these areas. Otherwise, the leachate which is produced may be mixed with the surface or underground waters.

With its very high pollutional strength, leachate may have detrimental effects on the environment. Because of that, an appropriate drainage network should be designed for the composting plants of Izmir and constructed as soon as possible.

REFERENCES

Bock, R. (1972). *Aufschlussmethoden der anorganischen, und organischen Chemie*. Verlag Chemie GmbH, Weinheim, Bergstr.

EAWAG (1970). *Methods of Sampling and Analysis of Solid Wastes* (Section on Solid Wastes).

Ernst, A. (1974). *Schwermetallvegetation der Erde*. Gustav Fischer Verlag, Stuttgart.

Grabner, E. (1977). Vergleich zweier Aufschlussverfahren zur Bestimmung der Metalle Fe, Cr, Zn, Cu, Cd, Mn in Klärschlämmen. *ISWA Informationsblatt, 23*, 25.

Kick, H. (1974). Die Problematik der anorganischen Schadstoffe bei der Kompostierung von Siedlungsabfällen. *Giessener Berichte zum Umweltschutz 4*, 51.

Kreeb, K. (1964). *Ökologische Grundlagen der Bewässerungskulturen in den Subtropen*. G. Fischer Verlag, Stuttgart.

Tabasaran, O. (1979). *Evsel ve Endüstriyel Katı Atıklar Kursu*. 25.6-6.7.1979, İTÜ, Istanbul.

Varian, T. (1977). *Analytical Methods for Flame Spectroscopy*, Australia.

Waste Management and Raw Materials Policy

W. SCHENKEL

Federal Environmental Agency, Berlin

ABSTRACT

As the third largest industrial producer, the Federal Republic of Germany currently consumes about 10% of the world output of mineral raw materials. Its own share of the world output amounts to less than 1%. This complete dependence of the Federal Republic of Germany on raw material imports, the world-wide growth in raw material consumption, the increasing uncertainty of supplies and the politically motivated raising of prices demand a Raw Material Safeguards Programme and a Waste Management Programme in which the special contribution which waste management can make, particularly with regard to consumer wastes, is described.

INTRODUCTION

Prof. Dr. M. Kürsten, on the occasion of his speech at the 1977 anniversary of the Association of German Engineers (VDI) on the theme, Better Raw Material Utilization - a Task for the Engineer, said: "The theme of raw materials is not only multi-faceted and complex but also fashionable and scintillating. Almost automatically the concepts of 'shortage' and 'crisis' come to mind, the fear of the free citizen in unheated houses, the spectre of economic collapse. But these doubts apparently beset only the pessimists. The optimist immediately reply that the situation is not nearly as bad as that. There are no grounds for acute alarm they claim and human ingenuity and technological progress will cope with these difficulties also."

In discussing the theme of this paper, Waste Management and Raw Materials Policy, the supply situation for a number of selected examples will be reviewed, and a description given of the specific contribution which waste management can make, particularly with regard to consumer wastes, towards reducing future bottlenecks.

The concepts of occurence and availability require a short explanation. The occurence of mineral raw materials relates to their incidence in nature and thus encompasses the type and origin as well as the quantity of deposits. Availability denotes our opportunities to extract the mineral resources and to channel these as raw materials into the economic system. Availability is therefore constrained by technological and economic limits as well as by political conditions.

RAW MATERIAL SUPPLY

Reference is made in this section to the contribution mentioned at the beginning, as it would appear to depict most competently the present situation. First of all, with regard to the supply situation, how long will raw material supplies last? In principle, there are sufficiently large quantities of all mineral raw materials contained in the earth's crust and in the oceans to meet all the conceivable requirements of humanity.

There are impressive and bewildering calculations which show, for example, exactly how large a cube of normal granitic crustal material would have to be to meet the world's annual demand for copper. Such a cube of rock would have to have edges at least 4 km. long. Besides this, it would also contain large amounts of iron, aluminium and other important elements.

Why then isn't granite, which is available in large quantities, used instead of the relatively rare ores? That is a question of energy. To extract copper from copper ores with concentrations of copper of over 0.5 %, around 8,000 kWh (up to 13,000 kWh) is required per ton of copper. The amount of copper contained in granite is only about 50 ppm. The energy necessary is estimated at about 600,000 kWh for the extraction of a single ton of copper from granite.

A typical hard coal power station of 150 MW with a daily input of 1,200 t coal, would deliver just enough energy to produce 6 t copper per day, corresponding to 1,200 t per annum. Current world production stands at 7.5 million tons of copper per year. Accordingly, for this production, the energy from 3,570 150 MW power stations would be necessary. The operation of these power stations would, for its part, be accompanied by a high consumption of primary energy materials. These facts indicate the limits of this optimistic approach. In order to be economically recoverable, mineral deposits must have a minimum concentration of extractable valuable and contain sufficient quantities so that exploitation is guaranteed for an economically defined minimum period. Only when these conditions are fulfilled, can reserves and supplies be talked about. At the margin of the broadly 'sub-economic' field, there is still an undetermined quantity resources which may prove economically recoverable with increasing prices or technological breakthroughs, so that they become reserves.

For the most important mineral raw materials such as petroleum, natural gas, copper, lead and zinc, there has been established since 1976 a growth rate in extraction which corresponds approximately to the annual increase in demand. Generally speaking, this curve shows a steady climb. In contrast, the increase in supplies takes place more spasmodically. Large new individual discoveries, for instance with copper, or the breakthrough in new technological processes, such as offshore technology in oil exploration, are reflected therein.

The static life-span of supplies at a given point in time specifies the range of supplies, assuming that no increase in consumption and no addition to supplies through new discoveries occurs. The dynamic life-span, in contrast, takes into consideration the average incremental growth rates of consumption and discoveries. It turns out that during the past thirty years the static life span of supplies has varied between 15 and 50 years. Superficially, that sounds reassuring. But the numerical data can also be portrayed differently. For example, the static life span of world oil supplies in 1950 amounted still to 20 years. These total supplies, however, were not exhausted in 20 years, but in fact in 12 years. In 1960, statically-viewed oil supplies were expected to last for 36 years. Today 75% of these have already been consumed. Their actual life span was thus placed not at 36 years, but rather at 20 years. The outlook appears similar for important metals. In 1946, copper supplies had a static life span of 50 years. But these supplies were already exhausted after 27 years. Under these premises, the figures for the static life span of important mineral raw materials, which are valid today, do not look so reassuring any more as can be seen from Table 1.

It any factual reckoning, however, account must be taken of the fact that new discoveries keep shifting the total dynamic balance in a positive direction. The costs of these new discoveries using, are rising more rapidly than even the rate of inflation. Canada, a country presumably rich in deposits of raw materials, there was for example a doubling in the metal value from mine production from 1.4 million dollars per year at the beginning of the '60's to 3.2 million dollars at the begining of the '70's, while expenditure on exploration almost tripled from 30 to 88 million dollars. At the same time, in spite of this higher expenditure the discovery rate sank, from

Table 1: Time Span for the Exhaustion of Important Mineral Raw Materials

Base Material	Growth Rate (% per annum)	Exponential Consumption (years)	Exponential Consumption with five-fold Supplies (years)	Consumption Remaining the same as in 1970 (years)
Iron	1.6	93	173	240
Aluminium	6.4	31	55	100
Lead	2.0	21	64	26
Copper	4.6	21	48	36
Zinc	2.9	18	50	23
Tin	1.1	15	61	17
Molybdnum	4.5	34	65	79
Chrome	2.6	95	154	420
Silver	2.7	13	42	16
Gold	4.1	9	29	11

3-4 to 2-3 deposits per year: in short, (diminishing) success in exploration has been with increasing expenditure.

The exploration costs are only the tip of the iceberg of investment costs. Hardly any modern mining operation exist today with investment costs under 100 million dollars, and one milli dollars is certainly not exceptional. According to a recent estimate, the international metal ore-mining industry in the western world must raise 28 milli dollars by the start of the '80's in order to install the technology to meet the required mining capacity. By way of comparison, the American 'Man on the Moon' programme cost around 10 million dollars.

The investment costs for oil and natural gas development are considerably higher. For example, up to 1985, investments of between 27 and 40 milliard dollars will be necessary for the exploration and development of the hydrocarbon resources of the Canadian Arctic and Alaska. The Federal Establishment for Geology and Raw Materials (BGR) has calculated exploration costs in the Near East at 3.5 milliard dollars up to 1985.

In connection with the staggeringly high investments, another factor appears ominously on the horizon: time. Today, it takes 6 to 10 years before a newly-discovered deposit can go into production. The interest changes during this amortisation period are high. Furthermore, this time span comparises a not insignificant part of the actual life span of the important raw materials detailed more precisely above. Under these assumptions, almost impossible demands are made on the vision and courage of the entrepreneur.

The conclude our considerations of the supply situation, the following evaluation appears to be justifiable.

Energy Raw Materials, Including Nuclear Fuels

Within this group, which economically is of the greatest importance, the hydrocarbons and particularly oil are in every respect in greatest jeopardy. Uranium ore is also threatened with shortages. Until now only coal, which is found in almost all industrial countries, was available for a longer period of time even though it is a more expensive alternative raw material. In addition, it has the advantage that its regional distribution corresponds extensively to the structure of demand. Indeed, coal has been a major factor in the industrialization of Europe.

Metallic Raw Materials

In this group, individual distinctions in the supply situation - as we saw earlier - and also differences in political and economic importance are sharply expressed. Copper, lead and zinc, together with uranium, tin, wolfram, molybdenum, antimony, silver, mercury and zirconium, belong to the particularly threatened materials. With respect to these, increased efforts in exploration as well as in technological development must be made in order to maintain the supply level for a considerable length of time.

With regard to natural supplies, nickel, iron, manganese, ilmenite-titanium, and platinum are less at risk; but in stating this, first and foremost the development of new technological processes for material extraction from low concentration ores must be assumed. There is greater latitude under the same assumption of new technological developments with regard to the raw materials for aluminium, vanadium, chrome and magnesium.

Non-metallic Raw Materials, Rocks and Solids

Very little has been said until now about this group, but it also involves risks and problems. With the import-dependent raw materials of this group, such as phosphate and also kaolin, the problems coincide extensively with those of the two groups mentioned above. A large number of indigenous raw materials also belong to this group, e.g. sand, gravel, cement-lime, clays, gypsum, salt and road-metal. Their socio-economic value is very high. In this group however, increasing bottlenecks are occuring on account of stricter quality requirements and environmental protection regulations High-grade materials are becoming scarce. Land-use planning and environmental protection limit the supplies available. Although today this area still lies completel within the responsibilities of the individual Federal States, overall Federal juridiction is desirable from the view-point of raw materials management.

Water

This raw material is indisputably one of the most immediate necessities of life. Among the remaining raw materials, it occupies an exceptional position insofar as, depending on climate, it is constantly renewing itself in a dynamic cyclical system. That the loading of this system has set limits is well-known. It is therefore all the more astounding that until now there has existed no uniform framework for the planning of water resources in the Federal Republic, despite the fact that the quantities of water available define simultaneously the feasible limit of economic growth and population size.

AVAILABILITY OF RAW MATERIALS

Regarding the availability of raw materials, the first limiting factor is their location. This is determined by natural geological factors. An unalterable fact is that 56% of the established reserves and 37% of the total world reserves of oil are concentrated in the countries of the Middle East. It should also be pointed out that the United Nations has not yet passed any resolution on the fact that one of the largest currently known deposits of uranium ore in the world lies in Southwest Africa (Namibia) It is upto the politicians to take these factors into account.

Until now, supply bottlenecks have been determined rather less by supply conditions than by events invoked by man himself, be it the current politically-motivated "Cartel `a la OPEC", or be it unforeseeable strikes, expropriations and other interferences in the market. Appart from these politically-motivated market interferences, ill-timed and insufficiently extensive provision of production capacity can lead to supply bottlenecks. It is predicted that the negative consequences of such a "shrinkage in exploration" will first become manifest in 10 to 15 years. Above all, those nations which are market-dependent have had to suffer, and now the Federal Republic of Germany has to a very large extent suddenly arrived at the same position. Table 2 shows this import-dependence for some selected examples.

As the third largest industrial producer, the Federal Republic of Germany currently consumes about 10% of the world output (in 1972, 640 million tons) of mineral raw materials, with a value of approximately 32 million D.M.

Table 2 : Import Dependence of the Federal Republic of Germany as a Percentage of Consumption

Base Material	1953 %	1963 %	1974 %
Iron	73.5	90.7	93.0
Aluminium	100.0	100.0	100.0
Lead	57.7	80.7	76.0
Copper	99.1	99.6	99.0
Zinc	31.4	63.1	70.0
Tin	100.0	100.0	100.0
Molybdenum	100.0	100.0	100.0
Chrome	100.0	100.0	100.0
Silver	100.0	100.0	100.0
Gold	100.0	100.0	100.0

Its own contribution to the world output amounts to less than 1%. Such problems are intensified still further against the background of the dialogue between the industrial and third-world contries on a new world economic order, by which mineral raw materials are acquiring a new political dimension as goods on the world market. In the long-term, the available quantities of raw materials are a growth-determining factor which, as with capital or labour-force potential, limits the production potential of an economic system. The future payment of higher raw material prices affects capital funds for investment and results in reduced consumption opportunities.

In the light of the above exposition, it is proposed that the following measures be taken:

1. Development of exploration methods with greater depths of penetration.

2. Methodical developments for rapid and broad prospecting of large areas

3. Greater inclusion of outlying areas in prospecting, e.g. permafrost regions, offshore-areas.

4. Further development of processes for secondary and tertiary extraction of petroleum.

5. Extensive mechanization and automation of mining techniques and processing, in order to make better use of bulk ores with a low metal concentration as well as of complex and finely disseminated ores.

6. Further development of hydro-metallurgical enrichment and extraction processes, which appear to be particularly well-suited for working deposits which are currently not economically recoverable.

7. Encouragement of diversification in the whole area of energy, metals and non-metals.

8. Intensified efforts to increase the recycle quota and raw materials substitution.

This complete dependence of the Federal Republic of Germany on raw material imports, the world-wide growth in raw material consumption, the increasing uncertainty of supplies and the politically motivated raising of prices demand new considerations and actions.

The conceptual aims of the Federal Government are summarised in a raw Material Safeguards Programme proposed in 1976:

1. Expansion of the raw material base for primary and secondary raw materials.

    a) Extending the reconnaissance areas for minarel deposits.
    b) Increasing the level of certainty of exploration methods.
    c) Reconnaissance of raw material deposits unused until now.
    d) Extending the workability of deposits.
    e) Increasing the efficiency of mining and processing techniques.

2. Conservation of raw materials.

    a) Substitution of scarce, raw and valuable materials.
    b) Using export substances and related production.
    c) Reducing specific material input.
    d) Cutting production losses.
    e) Improving material yields.
    f) Increasing durability.
    g) Improving reparability.
    h) Reducing wear and corrosion.
    i) Multiple usage of products.

3. Recycling of raw materials.

    a) Recovery and utilization of valuable materials from consumed products.
    b) Utilization of production wastes.
    c) Product design for recycling.

These requirements are also substantial conmponents of the 1975 Waste Management Programme of the Federal Government.

PREREQUISITES OF REUTILIZATION

As already stated, the disproportionate increase in the price levels of primary raw materials will, for various reasons, have to be taken into consideration. At the same time, the differential with regard to the high cost of recovery processes, which has existed until now, will be reduced; in other words, waste will become a potential secondary raw material and the cost of substitutes, including the new technology required, will become economically acceptable.

The development of technology for processing primary raw materials has continued for decades, in some cases even centuries. In contrast, the development of technologies for processing secondary raw materials, including in particular industrial and consumer wastes, must ensue in the shortest possible time. The interest of industry in the development of recovery/recycling processes and the development of low-waste and low-emission technologies will be stimulated, at least in the western industrial countries, by more stringent legislation with its associated, cost-intensive impositions. It has been established that, for example, the doubling of waste-disposal charges often results in the halving of the amount of waste. Recycling is frequently cheaper than disposing of the waste.

An important factor, which must be incorporated into economic considerations, is the collection and transport costs which arise from the need to convey the 'widely dispersed' wastes to a processing plant (increasing entropy). Stumm and Davis have portrayed this in the example of the recovery of copper wastes (Fig. 1).

The specific energy consumption for the extraction of raw copper from ore amounts to 13,000 kWh per ton of copper. If copper scrap is introduced in combination with the ore, the specific energy consumption with the introduction of 40% scrap falls to about 7,500 kWh per ton of copper. The total energy consumed for the production of a given amount of copper decreases accordingly with an increasing proportion of scrap. This applies when approximately 30% easily convertible scrab is used. However, 25-30% of copper is in the form of alloys, sludges and concentrates and 40% is in dilute solutions, pigments, dyes and effluents. If this dispersion were reversed, i.e. if the entropy, were reversed then the input of energy would increase immensely, along with costs and undesired environmental phenomena. Here, it has to be considered whether copper from such production cannot be replaced by other materials.

With correct appraisal of future technological development, energy consumption is a sensitive indicator as to whether extraction of a metal from ores or from waste materials is justified from an economic and ecological viewpoint.

Independent of this consideration, it can also be effectively demonstrated, using the example of copper, that recycling is not generally a suitable means for improving the raw material situation. If the recycling quota were increased from 40% to 100%, the growth in consumption maintained at 4.6% annually and an average product lifetime of 22 years assumed, then currently known supplies could be expected to last 3 years longer before exhaustion. If copper were to be substituted, the growth in consumption frozen at 0% and the mean lifetime of products extended from 22 to 30 years, then reserves would last up to 83 years longer with the recycling quota remaining at 40%; in other words, copper reserves cannot be preserved by recycling but only by freezing the in-

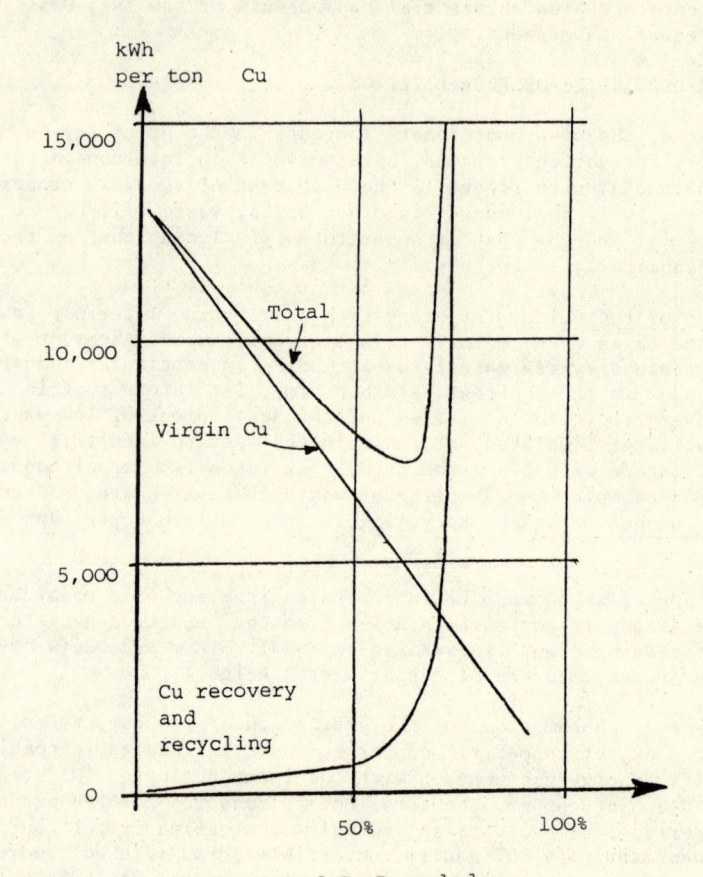

Fig. 1: Energy expenditure for the Manufacture of 1 ton of copper in relation to the proportion of recycled copper.

crease in consumption. Only if consumption remains stable, can reutilization bring an easing in primary raw material demand. That means raw material conserving and substituting techniques and consideration of present consumption habits.

It is possible that the real value of the current 'recycling euphoria' lies not so much in material recovery but rather more in the immense energy savings from the processing of secondary raw materials compared with that of primary raw materials. This has already been demonstrated with the example of copper. A further example is provided by the packaging industry. If the energy consumption for the manufacture of one-way and returnable bottles is examined then the following values shown in Table 3 can be given.

Table 3: Energy Consumption for the Packing of Liquids in Glass Bottles

		Energy Consumption	
Return of waste glass (%)	Manufacture of 1 kg of green glass (kcal)	Packing 1 l liquid in a 550 g one-way bottle (kcal)	Packing 1 l liquid in a 675 g returnable bottle with a trippage of 20 (kcal)
0	3640	2060	300
50	2880	1610	270
80	2340	1320	250

These figures include all energy expenditures from production to disposal of the glass bottles: transport of all raw materials to the glass works, waste glass collection, soda production, transport of bottles for filling, washing of returnable bottles and the disposal of wastes. The above table clearly shows the energy savings that can be expected with the utilization of waste glass and the advantages that can be obtained with the use of returnable containers.

In the following comparison, the energy savings from using secondary raw materials (waste) are also shown.

UTILIZATION (RECYCLING) OF CONSUMER WASTES

While in the industrial area waste recovery is already being carried out more intensively than the public realizes (with metal and plastic-processing operations, for example, reycling quotas of 80-95% are common), interest has recently been hesitatingly directed at municipal wastes as a raw material and energy source.

The difficulties which stand in the way of the development away from waste disposal to waste recovery in the municipal area are characterised particularly, with the materials presented here, by relatively low-value mixtures of varying materials and particle sizes, with strong regional and seasonal fluctuations in composition.

Table 4: Energy Consumption for the Processing of Primary and Secondary Raw Materials

Material Utilized	Product				
	Paper (Gcal/t)	Plastic (Gcal/t)	Glass (Gcal/t)	Iron (Gcal/t)	Aluminium (Gcal/t)
Primary raw material	1.5–2.5	7	2.8	6	14
Secondary raw material	0.17	0.1	(0.3)	0.6	2

The numerous processes and concepts developed and, in part, tested in the last few years can be classified as follows:

1. Separate collection (often referred to in the United States as "source separation"). This refers to the sepration of valuable materials from the waste at the source, prior to collection and transport.

2. Mechanical sorting. This refers to the separation of valuable materials by mechanical sorting processes after collection and transport of the wastes to central treatment plants.

3. New recovery-oriented thermal processes e.q. pyrolysis.

Separate collection is, to a large extent, material seperation by organizational measures. It has positive advantages in that a relatively low investment is required however, without too great a risk, it is also possible to introduce this type of system selectively in stages (based initially on one or two materials, e.g. paper, glass) which can respond flexibly to price fluctuations in the raw material market.

Separate collection can also be viewed advantageously with regard to the degree of cleanliness achievable, as seperation of valuable materials takes place before the contamination caused by collection. Difficulties exist in that this system is dependent upon the cooperation, persistence and conscientiousness of the participating population, i.e., upon unknown factors which are difficult to define and uncontrollable by technical means. Separate collection is, in fact, an old method of recovering valuable materials. It is currently experiencing a renaissance in numerous industrial countries

With mechanical sorting, as shown in Fig. 2, the following process steps are applied differing combinations for material separation:

1. Screening
2. Reduction
3. Sorting, based on the following:
    a) Aerodynamic characteristics (air classification).
    b) Magnetic properties (magnetic extraction).
    c) Density (density sorting in a wet medium).
    d) Optical sorting (colour sorting of the glass fraction).

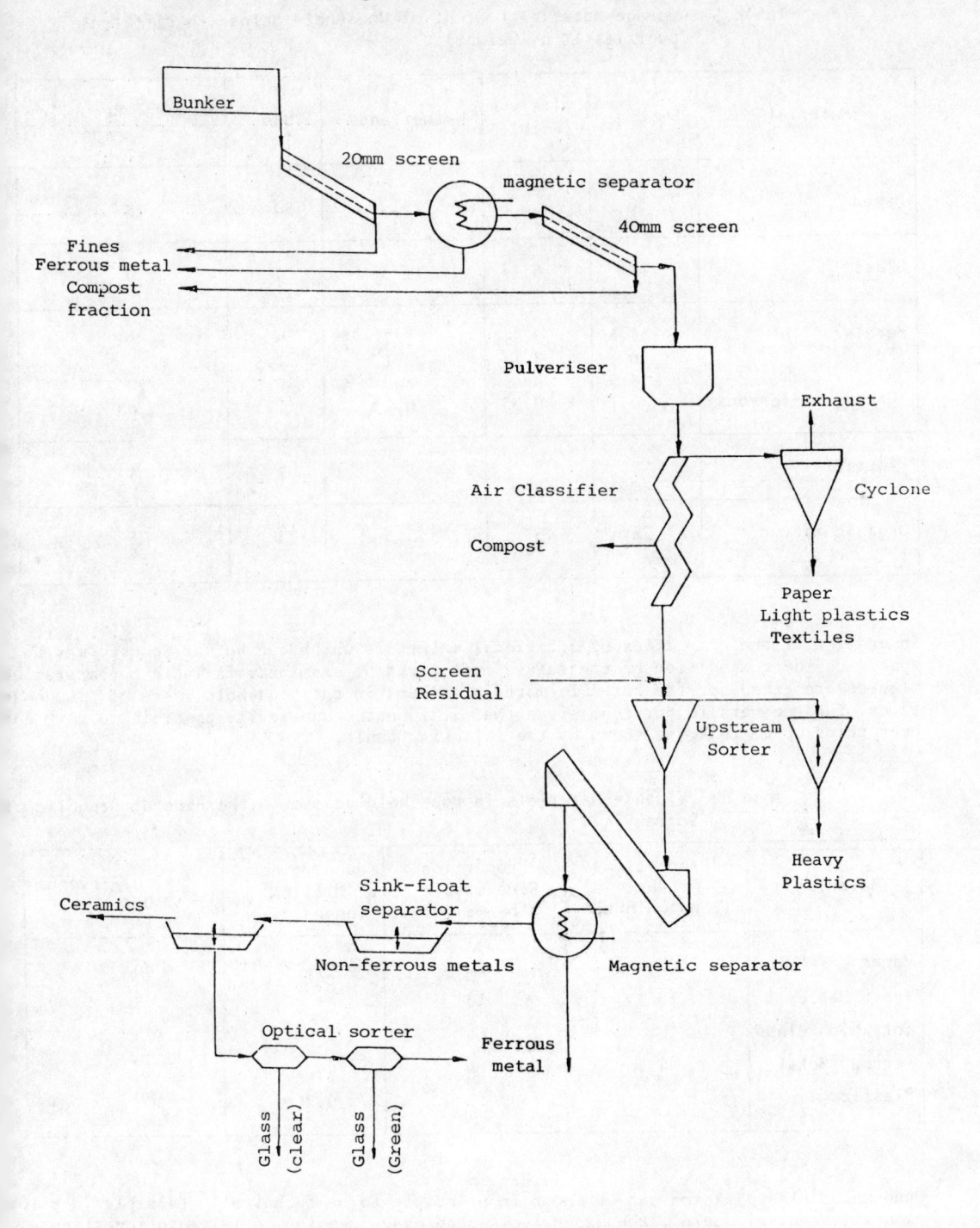

Fig. 2. Federal Waste Recovery Model (Process Line A).

Table 5: Average Material Content of Household Refuse in Different Countries (% by Weight)

Material	Sweden	Italy (Rome)	Netherlands	USA	Federal Republic of Germany
Paper	50	25	25	50	28
Glass	10	6	11	9	9
Metals: Fe Al Other non-ferrous	7 6 0.5 0.1	3.5 3.1 na 0.35	3.2 3.1 na 0.1	9 7.5 0.8 1.5	5 4.5 0.2 na
Plastics	9	8	5	1	4
Organic matter	22	45	45	23	35

In order that the quantities of individual materials which can be sorted out, may be specified the composition of the refuse must first be examined. In Table 5 comparative figures are given for the sortable materials found in the household waste of five countries. The recoverable portion, given that it is not economically possible to sort out everything up to 100%, is shown by the following table.

Table 6: Valuable Materials in Household Refuse in the Federal Republic of Germany

Material	Proportion Present (% by weight)	Proportion Recoverable (% by weight)	Quantities Million Tonnes p.a.	Maximum Revenue DM/t
Paper	28	20	3.6	60
Inert Matter	15	13	–	5
:of which Glass	9	6	1.08	70
Ferrous Metal	4.5	4	0.72	70
Plastics	4	2	0.36	200

Once material separation can be shown in principle to be technically feasible in pilot plants (technical scale), the expensive second stage remains, namely, to investigate the commercial and operational basis of automated material separation, especially the marketability of the separated products; this is only possible in a large-scale operational plant. The first demonstration plant is presently being planned in the

Federal Republic of Germany. This will be erected in the rural districts of Reutlingen and Tubingen and will serve to test two different sorting processes in combination with a composting works (thoughput of about 600 t/day). A range of pilot plants for materials recovery are in operation in other European countries (Stockholm/Sweden, Haarlem/ Netherlands, Stevenage/UK, Madrid/Spain). Large-scale plants have existed until now only in Italy (Rome and Perugia): however, the results achieved there are contingent upon local factors and cannot be readily transferred to other countries. The same applies to American developments.

A variation on the sorting process is represented by the processes for recovering a fuel from wastes. Instead of separating the mixed waste into a multitude of raw materials, division into a high calorific value, an organic fraction (25-40% by weight) and an inert residual fraction essentially results. Seperation of ferrous metal and glass is possible, as with the materials recovery processes mentioned above.

For the combustible fraction, the term 'Refuse-Derived Fuel' (RDF) has been adopted. Whereas in the case of refuse incineration, the treatment facility in which energy recovery takes place is designed to accept refuse as a raw material, RDF processes, by means of mechanical processing, are aimed at accommodating the raw material to firing systems developed for fossil fuels. The various existing RDF processes under development or, in part, already being tested on a large-scale (USA) are mainly based on combination of reduction, sorting and air classification similar to materials recovery processes. Depending on the process, the end-product comprising mainly paper, board, wood and plastics, is more or less finely processed, homogenized and, in certain cases according to the form of utilization, further refined (by drying, pelletizing, briquetting).

Pyrolysis processes rank among the new thermal processes; more precisely, degasification and gasification processes. These are processes for the thermal conversion of material in the absence of or with a deficiency of air. Depending on temperature range, gas and condensates such as oil, tar and pyrolytic coke, or with higher temperatures, gas and a solid slag residual are thus produced. In addition, there are the processes for high temperature incineration.

Given the large number of these processes (e.g. over 70 pyrolysis processes), it is not possible to go into great detail here. The majority are still at an early stage of development. A large-scale test plant based on the 'Andco-Torrax' system (high temperature gasification) has been operating in Luxembourg for a short time, and a similar unit is under construction in Frankfurt.

For these processes, the same objectives as for refuse incineration apply:

    1. Greatest possible reduction in volume and weight of the wastes being treated.

    2. Environmentally safe disposal and conversion of wastes.

In addition, the following advantages are anticipated:

    1. Recovery of saleable products to lower the costs (in the form of raw materials or storable energy such as gas and liquid hydrocarbons)

    2. Smaller quantities of flue-gas emissions (the order of magnitude for degasification being about 500 m^3/t) and thus, lower plant expenditure for gas-cleaning.

    3. Utilization of the solid residues (e.g. coke from pyrolsis), or tipping as the residues are completely sterile (with high temperature incineration).

Since these processes have not yet been proven on an operational scale, all the data until now are based more or less on theoretical speculation.

AN ATTEMPT AT ESTIMATING EFFECTIVENESS

Assuming that the future development of separation processes would, for reasons of cost, security of supply and operation, and sales distribution, proceed predominantly in densely-populated areas, than the following data would result, as developed by R.Turowski in a study by the Jülich nuclear research centre.

At the ministerial conference for regional planning, 24 areas in the Federal Republic of Germany were designated as densely populated regions in 1968. In 1970, about 46% of the population and 54% of the employees were living in these areas. With these data and the statistics on household refuse, it is possible to calculate the total amount of household refuse for 1971 in the densely populated areas at approximately 10.6×10^6 tons. This may be subdivided as follows:

Household refuse:

13.3×10^6 t - 54% in dense areas 7.18×10^6 t

Household-related commercial wastes:

5.2×10^6 t - 54% in dense areas 2.18×10^6 t

Bulky refuse:

1.9×10^6 t - 32% in dense areas 0.61×10^6 t

By considering the rate of increase in the total quantity of refuse and the growth in population in these dense areas, it may be assumed that the refuse obtained in these areas amounted to roughly 12×10^6 t in 1975.

SIGNIFICANCE FOR THE RAW MATERIAL BALANCE

Through the introduction of sorting processes, the quantities of valuable materials shown in the following table can be seperated out. Given a total of 12×10^6 tons of household refuse, the quantities of valuable materials shown in the following table can be separated out by the introduction of sorting processes.

PAPER

The material balance of the paper industry in 1973 showed a fibre pulp consumption of 6,541,689 t, of which just about a third was imported.

This was split as follows:

Woodpulp	:	2,006,647 t - 30.7%
Mechanical pulp	:	864,750 t - 13.2%
Waste paper	:	2,781,333 t - 42.5%
Other fibrous materials	:	888,961 t - 13.6%

With fibre losses of 2.85%, 6,355,366 t of paper and board products were manufactured from this. However, in order to meet demand, a further 2,054,634 t (balance of import and export statistics) had to be imported, the imports being confined essentially to bulk papers, newsprint, and paper and board for packaging purposes. It is precisely these types of paper which form the area in which paper from waste materials may be utilized.

Table 7: Raw Material Conservation through Recycling of Household Refuse

	Quantities of sortable material (t)	Quantities recoverable (t)	Total consumption of respective products 1974/75 (t)	Proportion of Consumption (%)
Paper	2,400,000	2,350,000	8,518,000	27.6
Glass	750,000	390,000*	3,210,000	12.2
Plastics	240,000	240,000	4,819,000	5.0
Iron & Steel.	490,000	490,000	39,041,000	1.3
Including tinplate	390,000			
Tin		1,600	13,000	12,3
Non-ferrous metals.	100,000			
Including aluminium	92,000	78,000	910,300	8.6

*Amount removable

Whether or not 2.4 million tons of paper, the approximate total which can be manufactured from refuse waste paper, can at any time be absorbed by the market and is capable of meeting roughly 28% of the paper and board consumption in the Federal Republic of Germany, depends mainly on external economic factors and on whether the graphic sanitary paper sector can be opened up for wastepaper.

GLASS

Quality requirements permit the use of refuse glass only in the manufacture of container glass, which nevertheless amounts to approximately 85% of hollow glass production. At any rate, even here the amount of foreign cullet which can be added has an upper limit, depending on colour.

Table 8 summarizes the present use of cullet and that which is regarded by the glass industry as feasible.

The production of the hollow glass industry had, up until the production recession of 1975, experienced growth rates of between 5 and 7% and had already exceeded the output of 3 million tons with a total input of cullet of 622,000 t, the proportion of cullet amounted to 22%. Only about one-fifth of this was foreign cullet. Taking the values regarded by the glass industry as feasible, the cullet input can be increased to just about 34%, of which 56% can be supplied by foreign cullet. In any case, these high inputs of foreign cullet are as yet untested in practice as cullet in this quantity is neither available nor has the glass industry had to suffer a shortage of raw materials, since these are sufficiently available domestically.

According to these specifications, the input of foreign cullet can only be increased to about 400,000 t annually, i.e. only slightly more than half of the cullet recoverable from refuse can be accomodated by the glass industry. In contradiction to this, the most recent statements of the glass industry are based on the assumption that the input or foreign cullet can in the medium-term be increased to 800,000 tons per annum. However, a prerequisite is that the cullet exist in colour-sorted form.

Table 8: Cullet Input in the Glass Industry

	Proportion of Production (%)	Input of Cullet 1974 (%)	Input of Cullet Technically Feasible (%)
Clear (flint) glass:	46	22.4	25
own cullet		19.3	15
foreign cullet		3.1	10
Brown glass:	26	19.2	50
own cullet		15.7	15
foreign cullet		3.5	35
Green glass:	28	24.3	35
own cullet		17.2	15
foreign cullet		7.1	20

PLASTICS

In the Federal Republic of Germany, 6-7 million tons of plastics are produced annual About 80% are thermoplastics. 240,000 t of plastics per year, the approximate total which can be extracted from household refuse, should be marketable despite their lower quality.

The primary material for the manufacture of thermoplastics is naphtha. If the production chain, with its associated quantity flows, is traced back to naphtha, then it is found that, per ton of mixed plastics such as would be found in household refuse, a naphtha input of 2,873 kg is required. In addition, this naphtha input is sufficient t provide for the further manufacture of 286 kg polypropylene as well as 201 kg of a mixture of pyrolysis gasoline and C_5-fractions for further utilization. Based on a total output of 1,286 kg, then 2,234 kg naphtha is required per ton of thermoplastic.

By recycling household refuse, around 536,000 tons of naphtha could be saved, which is slightly more than 5% of the annual consumption in the Federal Republic.

IRON AND STEEL

Just about 400,000 t of the 500,000 t of recoverable ferrous metal in household refuse is made up of tinplate. For metallurgical reasons, tinplate must, nevertheless, be de-tinned before being melted down again. In the event that the de-tinning works and this tinplate scrap has been de-tinned in the same manner as new tinplate scrap, then all of the ferrous metal contained in refuse can be returned to crude steel production. Up to 90% of this scrap is capable of being substituted for pig-iron as the primary mater ial in crude steel production. The amount of steel contained in refuse corresponds to about 1% of pig-iron production.

TIN

From the tinplate scrap in household refuse, 1,500 - 1,600 t tin is recoverable. This amounts to about 10 - 12 % of the annual demand in the Federal Republic. To recycle this amount would mean a considerable improvement in the raw material balance for tin.

ALUMINIUM

The recycling of aluminium in the Federal Republic of Germany currently absorbs only about 30% of the existing old scrap potential. The aluminium which arrives in private households in the form of domestic goods and packaging material is hardly touched. A large part of this is found again after use in household refuse, from which approximately 92,000 t annually may be sorted out and delivered to the smelting works. By using this potential, the recycling quota could be raised to 45 - 50 %. Around 92,000 t of aluminium scrap from household refuse, out of which just about 80,000 t of secondary aluminium can be smelted, is equivalent to 8 - 9 % of the annual consumption of primary pig and refined (secondary) aluminium in the Federal Republic of Germany.

REPERCUSSIONS ON THE PRIMARY ENERGY BALANCE

The net balancing of energy shows that the recycling of waste materials, when compared to the production of equivalent goods from primary raw materials, as a rule brings with it an energy saving of 30 - 70 %. The results are summarized in Table 9.

Table 9: Primary Energy Demand for the Production of Equivalent Products from Primary Raw Material Sources and from Household Refuse

	Energy Demand (kWh/t)			Energy saving (%)
	Primary Production	Production from household refuse	Difference	
Paper	5,712	4,198	1,514	27
Glass	5,060	2,868	2,192	43
Plastics	11,923	704	11,219	94
Crude steel	4,146	1,464	2,682	64
Tin	56,918	167,706	-110,788	-195
Aluminium	60,945	19,595	41,350	68

Plastics and tin can be regarded as exceptions. Plastics, as primary products, go through a long process chain from petroleum to the finished plastic. As waste material, they need only to be cleaned and plasticized before they can be worked into an extrudable granulate.

The recovery of tin, however, requires a three-fold greater primary energy demand than is necessary for the equivalent amount of imported tin on account of the particularly unfavourable pre-conditions of applying detinning techniques to tinplate from refuse cans.

From the data on the output of valuable materials, taking into consideration material losses in the production process, the opportunities for saving primary energy through the recycling of household refuse can be determined.

Table 10: Primary Energy Saving through Recycling of Household Refuse

	Production per ton of refuse (kg/t)	Primary energy saving per ton of refuse (kWh/t)	Production from refuse (t/a)	Primary energy saving (10^9 kWh/a)
Paper	198	300	2,350,000	3.56
Glass	31*	68	390,000*	0.85
Plastics	20	224	240,000	2.69
Crude steel	40	107	490,000	1.31
Tin	0.13	-14	1,600	-0.18
Aluminium	6.43	266	78,000	3.23
Total		951		11.46

*Quantity removable

Recycling of household refuse implies the following:

1. A specific primary energy saving of 951 kWh/t refuse equalling 3.42×10^9 j per ton refuse.

2. An improvement in the primary energy balance of about 11.46×10^9 kWh equalling 4.12×10^{16} J, if the recyclable amount of 12 million tons of household refuse is utilized.

The total possible annual primary energy saving constitutes about 0.4% of the primary energy consumption in 1975 and corresponds approximately to each of the following:

1. 1.41 million tons of hard coal.
2. 0.98 million tons of crude oil.
3. 1.29 million Nm^3 of natural gas.

These figures show very clearly that a substantial imrovement in the primary energy balance of the Federal Republic of Germany cannot be achieved by the recycling of household refuse. However, for the individual industrial branches considered, the primary energy saving through the manufacture of products from refuse can be really significant. If the energy consumption of individual industrial sectors is expressed in units of primary energy, then the values summarized in Table 11 for 1974 are obtained. With almost 9% and 6% respectively, the paper and board industry and the non-ferrous metals industry can achieve considerable primary energy savings by recycling household refuse.

Table 11: Primary Energy Savings through Recycling of Household Refuse, Broken down by the Industrial Sector and Measured against Consumption in 1974

	Primary energy consumption (10^9 kWh)	Primary energy saving (10^9 kWh)	%
Pulp, paper and board production	39.99	3.56	8.9
Glass and fine ceramics	30.92	0.85	2.8
Chemical industry	207.54	2.69	1.3
Iron & steel industry	328.68	1.29	0.4
Non-ferrous metals industry	55.08	3.23	5.9

In this context, a comparison with the incineration and pyrolysis processes, which utilize the heat value of refuse, is interesting. Nevertheless, the conversion efficiency of these processes remains low on account of the costly process control occasioned by the refuse characteristics. Allowing for the energy expenditure on collection, the improvement in the primary energy balance shown in Table 12, can alternatively be obtained from the 12 million tons of household refuse.

In comparing their effect on the primary energy balance, only incineration with the aim of producing heat performs better recycling. However, such plants are only seldom employed because refuse-recovery plants must also be continuously operated in view of the constant accumulation of refuse; moreover, the heat is not always capable of being removed. For this reason, plants with heat-power couplings are predominantly operated, whose contribution towards improving the primary energy balance is only slightly greater than through recycling. In any case, using the energy value of refuse tends to put a strain on the raw material balance.

The socio-economic gain in primary energy saving through recycling has not been considered in Table 12. This is not in itself measurable at the sorting plant; nevertheless, it should be included in any evaluation as a credit item in the same way as revenue from the recovered materials. The primary energy saving of 11.5 kWh per annum, for example, corresponds to 980,000 tons of crude oil. If this oil is valued at the average import price in 1975 of 222.96 DM/t, then from this one a credit of 18 DM/t refuse can be calculated. Increasing raw material and energy prices contributes to an improvement in the cost situation of a refuse recycling plant.

Refuse treatment through recycling is socially and economically sensible. However, it will not be introduced as long as the resulting costs to the operator of a plant are higher than the comparable costs of other systems which fulfill the same purpose. Here, the potential operators of such plants should be offered incentives to enable them to reduce their operating costs. The social benefits of such a step would be considerable.

Table 12: Substitution of Secondary Energy Carriers and the Improvement in the Primary Energy Balance thereby Achievable, through Using the Energy in Household Refuse

	Substitution of secondary energy carriers	Improvement in the primary energy balance
Process heat, Space heat	14.5×10^9 kWh	15.4×10^9 kWh
Power	4.3×10^9 kWh	11.3×10^9 kWh
Power and	1.4×10^9 kWh	11.8×10^9 kWh
Process heat, space heat in plants with heat-power coupling (power coefficient 31 kWh/GJ)	7.3×10^9 kWh	
Gas from refuse pyrolysis	8.0×10^9 kWh	8.0×10^9 kWh
Recycling		11.5×10^9 kWh

SUMMARY

The recycling of household refuse is a waste-handling process which has to compete with processes already practiced, such as landfill, incineration, composting as well as the projected process of pyrolysis. The extent to which this process can be carried out depends essentially on process costs. Other advantages are of only secondary importance to the operator of a recycling plant and will only gain greater weight if they are reflected in a reduction in operating costs.

The aim of the present work is to show the significance of household refuse recycling for the raw material and primary energy balance of the Federal Republic of Germany. The method of balancing net energy is an expedient by which the energy demand for the manufacture of products can be determined. This is composed of direct energy expenditures for the production process, as well as the so-called indirect energy expenditures for providing the production facilities and the raw materials and operating materials required. A comparison of the specific energy expenditures for obtaining equivalent products from virgin raw materials and from wastes shows clearly the effects on the primary energy balance of greater recycling.

Approximately 23 million ton of household and household-related wastes accumulate annually in the Federal Republic, of which 12 million tons can be recycled. The fractions of valuable materials capable of being separated out can be reworked into the six products: paper, glass, plastic, pig iron, tin and aluminium. If the potential 12 million tons of refuse is recycled, then the products so recovered can make a significant contribution towards meeting the demand for these products. With glass, tin and aluminium, about 10% of the demand could be met, and with paper, around 28%.

The net energy balances have shown that, as a rule, it is more advantageous in energy terms to manufacture products from secondary material than from virgin raw materials. Recycling of household refuse means the following:

1. A specific energy saving of 3.42×10^9 J/t of household refuse;

2. An improvement in the primary energy balance of about 4.12×10^{16} J per annum, with the recycling of 12 million tons of household refuse per year.

When compared to primary energy consumption in 1975, this brings a saving of 0.4%. Nevertheless, for the individual industrial sectors which process the waste material, the saving in primary energy attainable can be significant. In the non-ferrous metals industry, for example, it is around 6%, and in the paper industry, 9%.

Recycling also performs equally or even better, if compared to the improvement in the primary energy balance afforded by the incineration and pyrolysis processes, which utilize the heat value of the refuse. However, the conversion efficiency of these processes remains low because of the costly process control necessitated by the characteristics of the refuse. Besides which, this kind of energy use tends to put a strain on the raw materials balance.

The recycling of refuse is justified both in terms of conserving raw material reserves and promoting the rational use of energy. It fulfills both objectives in the same manner and is therefore best suited to application as a waste treatment process in a country with insufficient raw material and energy resources of its own.

CLOSING REMARKS

In the preceding discussion, it has been pointed out that an important source of our well-being, namely mineral raw materials, if not becoming more limited, will certainly become more expensive. The total import dependence of the Federal Republic of Germany makes the application of counter-measures imperative. In reviewing the diverse alternatives, the utilization of consumer wastes as secondary raw materials or as a potential raw material stock appears promising.

The limits and possible significance of recycling have been discussed both in terms of absolute numbers and considerations of principle. It has been emphasized that on the one hand, recycling is limited by the cost of energy, while on the other hand recycling is of interest to the user for reasons less to do with saving materials than saving energy. This is a very interesting field for teaching and research.

DISCUSSIONS

Patrick, P.K. (U.K.) : This paper is most valuable in epitomizing the problems and perspectives of waste-recycling. The benefits of recycling in national terms may be minimal, but they can be substantial for particular industries where the use of secondary materials can greatly reduce energy consumption. The problems of recycling are not mainly technical, but economic and institutional. Industry will not use secondary materials unless they are available at an economic price. The establishment of reasonably stable and long-term markets may require government intervention. One point to be borne in mind is that all recycling processes produce environmental pollution loads. On the other hand, the use of recycled materials by industry can reduce the pollution loads. The aim of waste management must be to minimize the environmental loads, while conserving natural resources.

Schenkel, W. (Germany) : I am very grateful to Mr. Patrick for his reference to the economic aspects of the problem. We in Germany are becoming increasingly aware that an economic basis must be provided for conserving and recycling of waste materials. For example, if the chemical industry is charged a high price for the treatment of its dangerous chemical wastes, it will regard this as a cost factor. It will try to lower such a cost by reducing the amount of waste produced. Also, tax legislation, regulations governing effluent discharge standards, the cost of waste disposal, as well as measures to stimulate capital investment related to the environment, provide abundant incentives. On the other hand, we observe to an increasing extent that ecological solutions are also economical solutions. The rising prices of petroleum and other raw materials are helping us to accomplish our goals.

Hernandez, J.W. (U.S.) : Isn't it possible that the application of advanced western technology, even in a limited form, to the problems of a country like Turkey will involve sacrificing some elements of public health and environmental protection built into it?

Schenkel, W. (Germany) : I am not sure. I believe that two different approaches can lead to the same end. For Turkey to reach the same goal as the West, it may be necessary to follow a policy more adopted to its needs. I agree about the importance of public health. However, our country will perhaps become more unhealthy because we are striving too hard for the sake of hygiene.

Samsunlu, A. (Turkey) : May I ask Prof. Hernandez what he means by "sacrificing some elements of public health?"

Hernandez, J.W. (U.S.) : Let me give a brief example. For the disposal of municipal wastewater in Istanbul, I would select the following procedures: collection, primary sedimentation and irrigation on selected fields, where only limited grazing would be permitted. This is a very cheap and profitable method of wastewater disposal; however, it involves a public health hazard, since the secondary and tertiary treatment processes, both obligatory in the U.S., have been omitted. Thus, in the adapting of western disposal methods to Turkey a certain sacrifice is made in terms of public health.

Schenkel, W. (Germany) : Let me give a counter-example from our country. The tendency in wastewater treatment in Germany is to centralize the treatment plants. We then observe that although the sludge from small individual plants handling only domestic wastewater was usable in agriculture, the sludge from the combined plants is no longer so. Now we need an incineration plant. We have made many mistakes in Germany. If you in Turkey learn from them, you can arrive at new and better solutions to your particular problems.

Samsunlu, A. (Turkey) : The suggestions of Prof. Hernandez for the treatment of the municipal wastewater of Istanbul surprise me, because in the mid-60's such ideas were rejected by a team of foreign experts, who instead proposed and supported with financial aid the construction of sea outfalls. Now Prof. Hernandez, who comes from the same country as these experts, recommends, from the point of view of public health, the disposal of the treated municipal wastewater by irrigation instead of discharging it into the sea. This apparent contradiction in opinion should be discussed.

Schenkel, W. (Germany) : We have had a similar experience in our country. For years we believed that our increasing energy needs could be met by the introduction of nuclear energy. Now, as a result of the efforts of a powerful and active group of intellectuals, the efficacy of following a nuclear energy policy has been called into question. It is, I believe, vital for an engineer constantly to re-examine his solutions. The same solution may not be applicable to several problems, even if similar. My question is whether Turkish engineers, with their specific knowledge gained from abroad, accept the challenge of searching for original and independent solutions appropriate to Turkey.

Samsunlu, A. (Turkey) : We are capable of finding solutions to our own problems, but we do not have the means. Ours is the fate of all underdeveloped countries unable to provide their own technology. We are, therefore, at the mercy of foreign organizations, which in supporting a given project with money and technology, impose their own ideas and conditions. To fulfill these conditions, further assistance from foreign experts is required; thus, our dependence on the developed countries is increased. How do we escape from this vicious circle?

Schenkel, W. (Germany) : You have opened a great wound, which is the result of wrong policies with respect to the development of the poorer countries. Up to now the richer countries have regarded aid to the developing nations as a means of promoting and exporting their own technology and expertise. We must now awaken our colleaques to the mistakes of the past and to the real needs of the developing countries.

Erdin, E. (Turkey) : How do you view the use of solar energy in Turkey?

Schenkel, W. (Germany) : I wish to impress on Turkish engineers that they should not just copy the methods used by western industrial nations. They must adapt them to the actual local conditions. I can make no concrete proposals for the solution of Turkey's energy problems, since I do not have sufficient knowledge of the country. We in Germany, to help counteract the effect of increasing energy crisis and to reduce energy costs, have been forced to re-examine our wastewater treatment technology with the aim of making our filter plants economically self-sufficient with respect to energy. This will involve, for example, optimizing the decomposition of the sludge for energy purposes. With regard to Turkey, which unlike Germany has long periods of sunshine, I can only surmise that solar energy could be used for operating the purification technology.

Costs of and Innovative Solutions for Industrial Waste Treatment

N. L. NEMEROW

University of Miami, Coral Gables, Florida, USA

ABSTRACT

Some fundamental differences in industrial and municipal waste are presented. Methods of expressing costs of industrial waste treatment are given. Four modern and creative solutions to the industrial waste disposal problems of today are presented and described in some detail. These include recovery and/or reuse, tying discharge levels to stream resources, marketing of available stream resources, and construction of enviromentally-optimized industrial complexes.

BACKGROUND OF INDUSTRIAL WASTE TREATMENT

The pollutants contained in municipal wastewaters are quite well known to most workers in the field of environmental engineering. They include suspended solids, BOD, microorganisms, and the more recently recognized dissolved minerals. Liquid wastes of industries may contain these and many more contaminants. Possibilities include color, toxic organic chemicals, heavy metals, oil, detergents, acids and alkalis, odors, heat, radioactivity, as well as many types of colloidal solids. In addition, industrial effluents are much more unpredictable both in quantity and quality at any given time. The potential combination of various unpredictable contaminants occurring at any time of the day, week, or season of the year makes proper treatment very difficult.

Industry has at least 12 alternatives available to it to solve its wastewater treatment problem. These are shown in the author's recent text (Nemerow, 1978). The major paths open to industry include treatment by the municipality of all its wastes either before or after treatment by the industry or treatment to some degree solely by the industry. The challenge to us once again is to determine "which path industry should follow." The challenge can only be solved by investigating the parameters involved. After the facts are revealed, they must be evaluated and integrated into appropriate models to yield the optimum solutions. While technical parameters, such as quality and quantity of wastewater and type of municipal treatment available, are the primary ones, economical, political, social and psychological factors are also extremely important.

Traditionally, two systems have been used for controlling the amount of industrial contaminants which can be discharged safely into our watercourses. One is manifested by establishing and maintaining receiving stream water quality standards. This prevents industry from putting more contaminants in the receiving waters than they can tolerate according to stream quality criteria. The standard selected will depend primarily upon the stream use. The total quantity of contaminants allowed will also depend upon the quantity of stream flow available. Because of the variability in both streamflow and industrial contaminants and the increasing number of pollutant contributors, this control method is difficult to apply. The other system uses an industrial effluent concentration or quantity to limit excessive contaminants. Basically each type of industry is expected to limit its common pollutants to a reasonable level. This method is much easier to apply, and more equitable, but also bears little correlation to the conditions in the receiving water.

Because the stream standards proved so difficult to administer and the effluent standards easier to monitor and control, more consideration and some modification of the latter were called for. For these reasons we are now trying a system of effluent guidelines for each of about 30 prime wet industries. Each of these industries is allowed to discharge only a reasonable percentage - say 15 to 50 - of their major and most common raw wastewater pollutants. It's too early to say whether this system will prevent excessive pollutants from entering our streams. However, the individual states still retain the right to demand additional treatment by industry in situation where the guidelines alone do not protect adequately the quality of specific receiving waters.

COST OF INDUSTRIAL WASTE TREATMENT

The capital and operating costs of industrial wastewater treatment must be computed, clearly expressed, and presented to management in a readily understandable way. Once the mysterious and imaginary, so-called excessive costs are identified, acceptance by industry becomes much easier. I recommend that all industrial waste treatment costs be calculated on an annual basis to include both capital and operating expenses. This makes them comparable to other industrial manufacturing costs such as labor, transportation, utilities, rent, marketing, and even administration charges. Open comparison of waste treatment with these other production costs will result in enlightened and rational decisions.

In the past, industry used two methods to express the cost of waste treatment; one for public relations to discourage implementation and the other for internal propaganda, also to discourage implementation. In the first the cost is presented simply as a lump sum, present-day charge. When heard by society as a single, unrelated, and presumed unnecessary cost, it sounds absolutely unreasonable and preposterous. For example, $5 million for power plant cooling towers. In the second, it is expressed as a cost of per share of stock to be subtracted from low or already diminished corporate earnings. When industrial managers and corporate board members consider these expenses and their consequences, naturally they react in a negative manner to implementing waste treatment.

It is much more realistic, and fair to express the cost of wastewater treatment as a charge per unit of production cost. The latter cost is often referred to as "value added." Therefore an appropriate expression could be cents per cents of value added for the end industrial product - or percentage of value added. Such information would provide industry with a direct comparison of its cost of waste treatment with other direct costs. It would also provide clues as to how to pay for these costs once a decision was made to proceed. In many instances, the cost per unit of production is so small that it can be absorbed readily by industry without raising the product market price. It might also force the industry to seek methods of lowering production costs by ingenious innovations in order to maintain a constant profit level. Another method suggested by the writer (Nemerow, 1972) is the sales index; i.e., waste treatment cost as percentage of an industry's sales volume. Sales records are usually readily available and generally public information whereas production costs may be difficult to obtain and assess.

Whether costs of waste treatment are related to production costs or sales the resulting percentages are best compared on an industry by industry basis. For example, in the pulp and paper industry (Nemerow, 1972), the average cost of waste treatment

(85% BOD removal) was found to be 4.34 cents per sales dollar. Specific paper mills requiring waste treatment costs greater than this value may be extremely hard-pressed to comply with EPA guidelines. In another unrelated type of industry, however, a sales index of even 4.34 cents may be too high for any plant within the industry to afford. In the nitrogen fertilizer industry, moreover, the ratio is 0.5% for ammonia, ammonium nitrate and ammonium sulfate. The figure for urea is about 2.0% (Charmichael, 1978). For phosphate fertilizer plants the ratios vary between 0.5 and 0.3% (Charmichael, 1978). During the last decade the writer has been collecting reliable economic data from certain industrial plants. In these cases an attempt has been made to relate waste treatment cost to value - added or production cost. Some of the percentages obtained are included in the table on the following page.

The writer (Nemerow, 1978) has also expressed the cost of waste treatment as dollars per million gallons of wastewater treated. This is a conventional method for domestic sewage, but because of the dissimilarities described in the first paragraph, it is not highly recommended for industry.

We hear a great deal these days about not being able to afford to pay for pollution control. We hear that we can't have both jobs and capital spending for growth and at the same time clean air and water. For the equivalent of secondary-type treatment, these statements are simply untrue for industry. As you can see from the relatively low cost of industrial waste treatment shown in the examples in the preceding section, these arguments are not valid. All but the unnessential, borderline, and unprofitable industrial plants not only can afford waste treatment costs, but also should not attempt to by-pass them with subterfuges of one type or another. When industry is forced to remove more pollutants than can be accomplished with secondary treatment, other systems for solution may be preferable to the more costly tertiary treatment.

SOME MODERN AND INNOVATIVE SOLUTIONS TO INDUSTRIAL WASTE TREATMENT

Four modern and innovative solutions to industrial waste treatment are (or will become) available to us. They can be summarized briefly as (1) recovery and/or reuse; (2) tying effluent levels to stream resources; (3) marketing of stream resources; and (4) construction of environmentally optimized industrial complexes. In this paper I am emphasizing these solutions - their meanings, problems of implementation, and likely results.

Reuse directly within a plant or indirectly by other industrial plants or municipalities offers a ray of hope for solving some of industry's problems. There are several obstacles to this ideal solution. First, a strong program in education must precede a sucessful reuse program. We should acquaint plant personnel with the potential value contained in wastewater. Second, a rather detailed qualitative analysis of the waste must be made available over a relatively long period of time. This information will fortify plant managers' convictions to reuse these wastes. Third, and probably most important, there must be an acceptable user for the wastewater located nearby. The recipient should have a water-use need closely correlated to the quantity and quality of wastewater available from the donor industrial plant.

Industrial wastewaters vary in quantity and quality from time to time. As the type and amount of product changes, so does industry's wastewater. The same is true for the pollution-carrying capacity of receiving streams. Most streams exhibit both high and low-flow periods corresponding to the time of year, rainfall and topography. One possibility to avoid excessive waste treatment costs is to gear production at the industrial plant to the stream resources available at the time. When little stream flow (stream resource) is available, industrial production could be reduced; and likewise when high stream flow or stream resource exists, plant production can be increased again. Long-term industrial production probably would be unchanged by

Table 1 Industrial Waste Treatment Costs

Type Industry	Type Production	Waste Treatment Cost Dollar Value Added (%)	Ref.
1. Cannery	Beans	1.11	Nemerow, 1972
2. Cannery	Tomatoes, Peaches	1.08	Nemerow, 1972
3. Poultry Processing	Chickens	1.096	Nemerow, 1970
4. Tannery	Upper Leather	0.42	Nemerow, 1972
5. Iron & Steel	Finished Steel	1.8	Nemerow, 1976
6. Petro-chemical	Ethane-Propane Cracking	0.5	Charmichael, 1977
	Naphtha Cracking	0.9	Charmichael, 1977
	Gas-Oil Cracking	1.4	Charmichael, 1977
	Ammonia Via Natural Gas	1.1	Charmichael, 1977
	Via Coal Gas	7.2	Charmichael, 1977
	Via Heavy Oil Gasification	3.1	Charmichael, 1977
7. Fertilizer	Nitrogen Fertilizer	Max. $0.78/ton Product	Charmichael, 1978
	NPK Plant	$2.15/ton Product	Charmichael, 1978

this technique. This method of controlled production is commonly used under prevailing conditions of raw material supply, market conditions and labor available and could likewise be applied to stream resources.

In practice, a telemetering system from the stream to the plant production could be installed to control production. However, both industry and water pollution control agencies would need to be convinced to use this method. Industry is often wary of another constraint placed on its production. Regulatory agencies, on the other hand, doubt the complying honesty and integrity of industry in compliance with production requirements.

One sure way to persuade industry to seek acceptable solutions to wastewater treatment problems is to penalize them for violating the laws of nature. These laws are often different from the laws of man. The law of nature dealing with wastewater states that you shall not discharge into a watercourse more contaminant than can be assimilated without an adverse environmental impact. A potential method for assuring society that industry will not violate this law of nature is to charge it a fee for using the existing resource. A unit price for the stream resource roughly equivalent to the total benefit to society of that resource can be established. Each consumer (industry, people, or agriculture) could purchase given amounts or units of these resources for that price. As the stream resources diminish, higher unit prices would be established to discourage over-consumption leading to a state of pollution. As the unit price becomes too high for industrial consumers, they would seek alternative solutions to avoid payment and pollution. The author (Nemerow, 1974) recently proposed this system of optimization of receiving water resources. The system has not been implemented so far because of the relatively inexpensive cost of waste treatment and the relatively large effort which must be expended to initiate the system.

The ultimate in solutions to industrial wastewater problems is one utilizing environmentally optimized industrial complexes. These complexes are designed to contain a number of compatible industrial plants collocated and practising endless reuse and recirculation of all residues, liquid, solid and air. The raw materials for each plant are intended to be the waste materials from other plants of the complex. In this way the environment suffers a minimum or no adverse impact, and production costs are minimized. The key to the success of these complexes lies in our ability to select the appropriate combination of plants and production capacities for each complex and to obtain industry's cooperation in the whole concept. The author (Nemerow, 1979) described this novel solution in the case of the pulp and paper industry. Other wet industries which may benefit from industrial complexing include the steel, fertilizer, tanning, textile, oil-refining, petro-chemical, and certain food processing industries. Many typical plants in these industries are so large that they seldom construct new plants. This would delay carrying out the complexing principal solution. Another obvious deterrent in this method of solution is determining the optimum combination of plants for specific complexes. And, further, the assistance of some industrial development agency would be necessary to obtain agreement for locating plants of these types and sizes in complexes. Despite these hurdles, the benefits to be gained from such solutions far outweigh the efforts which must be made beforehand. In fact, industrial complexing offers one of the most promising long-range solutions to not only environmental pollution problems, but also to all industrial economic problems of the future.

REFERENCES

Carmichael, B.(1977) Environmental Problems of the Petrochemical Industry. UNIDO, Vienna, Austria, 53.

Carmichael, B. (1978) Environmental Problems in the Fertilizer Industry. UNIDO, Vienna, Austria, 12.

Nemerow, L. (1972) Benefit-Related Expenditures for Industrial Waste Treatment, *Cost of Water Pollution Control National Symposium*, Raleigh, North Carolina.

Nemerow, L. (1974) *Scientific Stream Pollution Analysis*. McGraw-Hill Company, New York City.

Nemerow, L. (1976) Iron and Steel Environmental Management. Prepared for UNIDO, Vienna, Austria, 26.

Nemerow, L. (1978) *Industrial Water Pollution-Origins, Characteristics, and Treatment*. Addison-Wesley Co., Reading, Mass., U.S.A.

Nemerow, N.L., Farooq, S. Sengupta, S. (1979) Environmentally Optimized Industrial Complexes for the Pulp and Papermill Industries. Presented at Industrial Waste Treatment Conference for Pulp and Papermills in Calcutta, India, December 1977. To be published in 1979 in *Environmental International*, Washington, D.C.

Economic Design of Industrial Wastewater Treatment Systems in Brazil

S. A. S. ALMEIDA* and R. G. LUDWIG**

*Catholic University of Rio de Janeiro
**Encibra S.A., Consulting Engineers, Brazil

ABSTRACT

Innovative methods for increasing the cost-effectiveness of industrial waste treatment facilities have been introduced in Brazil particularly through the use of fine-screening operations in place of conventional flotation systems for pre-treatment and in place of primary settling tanks. Rotary, static and vibrating screens have been installed and excellent results are anticipated, particularly for wastes such as from meat-processing and latex production where a high content of large-size solids are found. Costs and area requirements are significantly less for such plants and solid-handling problems are reduced.

INTRODUCTION

Brazil is currently experiencing rapid industrial development, being the most important industrial center of Latin America. At the same time, the cuontry is fighting poverty, inflation and other problems common to developing countries. Although funds are limited, the desire to protect the environment against irreversible damage is fervent in both the people and government and has led to the establishment of industrial pollution emission standards.

Legislation for the control of pollution resulting from the discharge of industrial wastewaters was initiated in Brazil in the early nineteen-seventies. Similar to regulations affecting water supply and sanitation, this control is essentially under State jurisdiction or supervision with very little input from cities or from the federal government.

The pioneer state in industrial pollution control legislation was São Paulo, the richest and most industrialized state in Brazil. Soon after, the state of Rio de Janeiro started action to control industrial pollution, with the creation of CECA (State of Rio de Janeiro Commission for Environmental Control), a legislative committee consisting of members of the staff of the State Secretary of Public Works. FEEMA (State of Rio de Janeiro Foundation for Environmental Engineering) serves as the technical agent for CECA. This foundation, which currently has over 1500 employees, has the responsibility of advising CECA on technical matters concerning pollution control for all industries within the State boundaries. Inspection is made of each industry and the needs for waste treatment are stipulated, where necessary. If action is not taken within reasonable time limits, warnings are issued and followed by fines in cases of non-compliance. All industries must appoint a technical representative, either an individual engineer or a consulting firm, to act on their behalf regarding FEEMA regulations, as well as to design an adequate wastewater treatment system. Each new industry must receive approval of its proposed pollution control system by FEEMA, before the issuance of the permit to start operation, and before it can apply for special development agency loans.

EFFLUENT STANDARDS FOR THE STATE OF RIO DE JANEIRO

Although the authors have designed industrial control systems all over Brazil, the standards presented in Table 1 are those issued by the State of Rio de Janeiro, where most of the authors' work has been located.

Table 1 Industrial Effluent Standards for the State of Rio de Janeiro

Parameter	Maximum Concentration mg/ℓ (except where indicated)
Chromium +6	0.5
Chromium +3	1.0
Copper	0.5
Cadmium	0.1
Mercury	0.01
Nickel	1.0
Lead	0.5
Zinc	1.0
Arsenium	0.1
Silver	0.1
Barium	5.0
Selenium	0.05
Cyanides	0.2
Phenols	0.2
Sulfides	1.0
Fluorides	10.0
Pesticides (Organophosphates)	0.1
Chlorinated organics not listed above (pesticides, solvents)	0.05
Total Phosphorus[1]	1.0
Total Nitrogen[1]	10.0
pH	between 5.0 and 9.0 units
Temperature	<40°C
Settleable solids (1 hr Imhoff Cone)	1.0 mg/ℓ
Oil & Grease (mineral)	20 mg/ℓ
Oil & Grease (Vegetal and animal)	30 mg/ℓ
BOD_5	Secondary treatment for new industries. For existing industries, degree of treatment is not to be set by FEEMA, in accordance with the receiving water body, but never less than primary treatment.

[1] Applicable to effluents discharged to lakes, only.

PRELIMINARY OR PRIMARY TREATMENT ALTERNATIVES PRECEDING BIOLOGICAL TREATMENT SYSTEMS

Probably the most successful innovation introduced with the aim of reducing costs of pollution control plants in Brazil is the use of fine screens to replace more sophisticated and/or expensive units traditionally used for preliminary or primary treatment operations. Two types of fine screens have been tested and used, namely, rotary screens (for the larger flows) and static screens (for the smaller flows). Both types provide for self-cleaning. Figure 1 presents sketches of the two types mentioned.

Preliminary Treatment

Food-processing industries discharge wastes which contain significant quantities of relatively large-sized solids which pass through conventional bar-screening devices, but which settle in conveyance systems and/or clog pumping units. Such wastes also contain solids which float and therefore reach biological reactors, especially where the extended aeration process is used. This latter problem is very common in systems handling slaughterhouse wastes. Traditionally, simple but not very highly effective grease traps or sophisticated and expensive flotation units have been utilized to overcome this problem.

Pre-treatment of such wastes by fine screening was first proposed in Brazil in 1973. This first plant, as well as many others incorporating fine screens, is at present in operation, providing greater simplicity and comparable efficiency. Where flotation processes are used, the material removed is difficult to collect and entraps relatively large amounts of water. In contrast, screenings are self-separating from the liquid stream and, in some cases, have a moisture content as low as 60% (for large openings).

Sizing. The sizing of fine screens for pre-treatment is a function of hydraulic flow and the diameter of the larger solids to be removed. Tables 2 and 3 furnish the sizing data for rotary and static screens, respectively. The rotary screens mentioned in Table 2 are of the "slot-type"; the capacity for screens of the "hole-type" is significantly reduced.

Cost. The costs of flotation and the fine-screening operation are presented in figure 2, as a function of flow for flotation and of flow-and-screen opening for fine-screening.

Table 2 Sizing of Rotary Screens for Pre-Treatment
Cylinder Ø 0.96 m x 3.00 m Based on SS = 300 mg/ℓ

Slot Opening (mm)	Flow (ℓ/s)
0.50	160
0.75	220
1.00	280
1.25	325
1.50	380
2.50	490

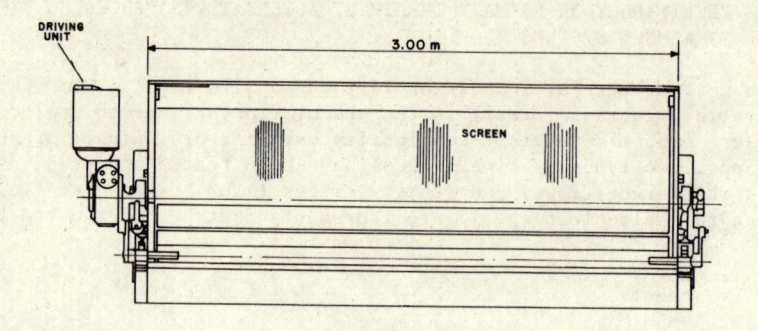

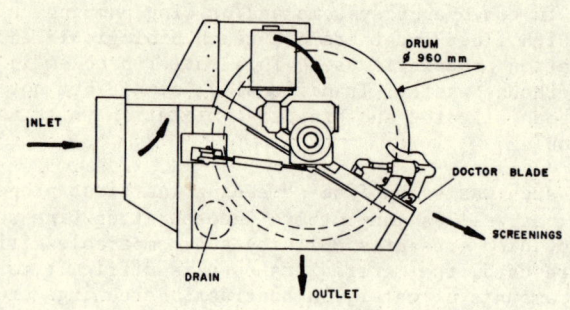

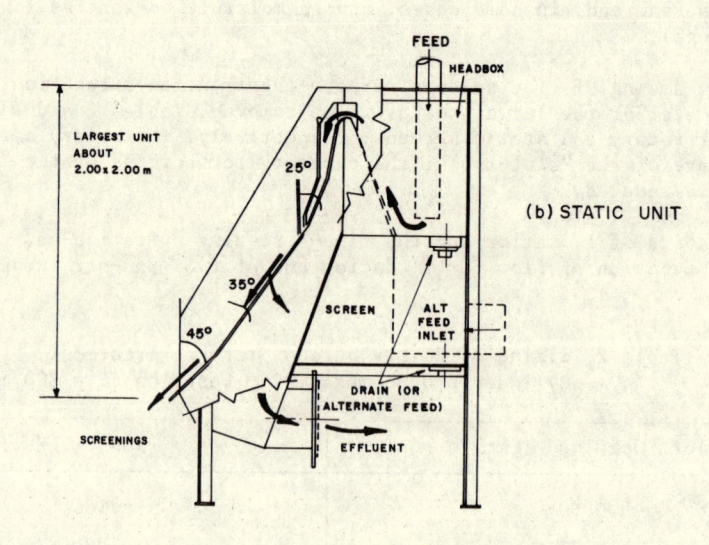

Fig. 1. Screening unit details

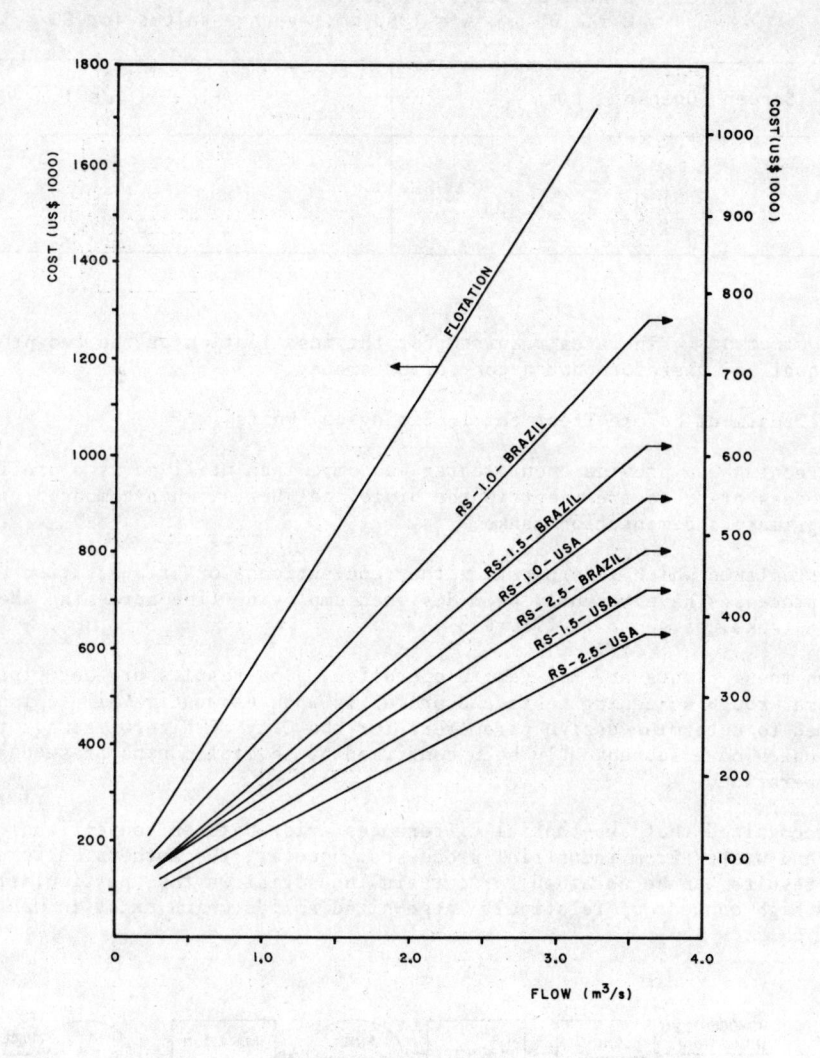

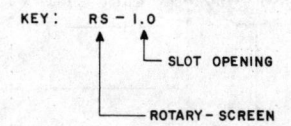

Fig. 2. Capital costs for preliminary treatment using flotation and rotary screens

Table 3 Sizing of Static Screens
H = 2100 mm; W = 1950 mm; Average Values for SS = 300 mg/ℓ

Screen Opening (mm)	Flow (ℓ/s)
0.5	25
1.0	35
1.50	50

Area requirements. The area required for the installation of the two processes is about equal and therefore not a control parameter.

Primary Treatment before Treatment in Biological Units

Fine-screening can provide even greater economy, when utilized as a preliminary treatment process prior to treatment in the biological units, when compared with conventional primary sedimentation tanks.

Treatment plants which incorporate either conventional or step-aeration activated sludge processes have recently been designed employing fine-screening ahead of the aeration tanks, as shown in figure 3.

Although these plants are not yet in operation, good results are anticipated, based upon some rough screening tests and primarily upon exhaustive pilot-plant tests performed to determine design parameters for the City of Niteroi sewage treatment plant, which have subsequently been confirmed by several months of successful full-scale operation.

It is recognized that substantial differences exist between conventional municipal sewage and wastes from industrial processes. However, the authors believe that even better results can be obtained for certain industrial wastes, particularly those with a high content of relatively large-sized solids which exist in wastes resulting

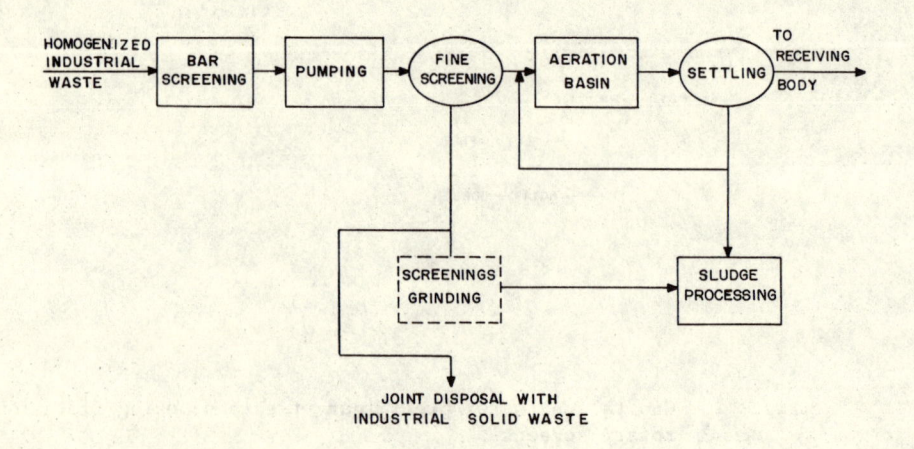

Fig. 3. Diagram for Conventional or Step-Aeration Activated Sludge Using Fine-Screening

from meat-processing and latex production, than can be obtained for municipal sewage where most of the solids are of dimensions which would require extremely fine-meshed screens to achieve results comparable with those from conventional primary settling.

The average results of screening operations at a plant where fine screens have been used in place of primary settling for municipal sewage are summarized below. This particular plant employs five rotary screens (Rotostrainers), with a maximum capacity of approximately 800 l/s (each unit with 160 l/s capacity), with one additional stand-by unit and slot openings of 0.5 mm.

To achieve 95% efficiency in the overall treatment process, the design must consider the lower BOD removal resulting from fine-screening, as compared with conventional primary settling tanks, and size the aeration units accordingly. At the above-mentioned municipal plant, where fine screens (0.5 mm openings) were used in place of primary tanks and no allowance was made in the aeration tanks for the lower BOD removal, performance tests have shown an overall reduction in BOD removal of only about 5%.

Table 4 Average Efficiencies for Rotary Screens Employed for Municipal Sewage

Slot Opening	Removals (%) BOD	SS
0.5 mm (pilot & full-scale)	25	40
0.75 mm (pilot only)	20	25

Cost. The costs of primary treatment using conventional settling tank units and of fine-screening are shown in Figure 4. Curve 1 indicates the cost of primary settling as a function of flow (retention of 2 hours); curve 2 indicates the cost of screening in Brazil using screens with openings of 0.5 mm, and curve 3 indicates the costs of screening in the United States. Differentiation is made between Brazil and the USA because screening equipment costs are much higher in Brazil, due to the imposition of heavy duties on certain pieces of equipment which must be imported. The screening alternative would thus be more attractive where such duties are less severe. For conventional settling, the costs in the two countries are quite similar, because concrete costs are about the same and this type of equipment is made entirely in Brazil at a comparable cost. The cost curves include all public works, mechanical and electrical equipment, and piping.

Area requirements. This is the main parameter in favor of the fine-screening alternative. Area requirements are reduced by a considerable factor, as can be noted from Figure 5. For example, for a flow of 1000 l/s, the area required by primary settling is 2700 m^2 and for 0.5 mm fine screening 160 m^2, indicating a factor of 17 to 1. Calculations of the required area for the primary sedimentation alternative are based on the use of circular tanks incorporating a surface loading rate of 40 m^3/m^2-d. As a rule of thumb, the area required for screening is 24 m^2 per unit.

Sludge production. Sludge produced in the fine-screening process has considerably less volume than that produced from the primary settling alternative. This is due not only to the reduction in removal but primarily to the lower water content of the materials removed.

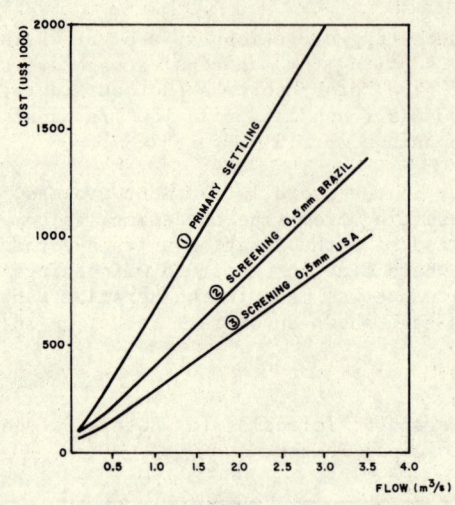

FIGURE 4 – CAPITAL COSTS FOR PRIMARY TREATMENT USING SETTLING TANKS AND ROTARY SCREENS 0,5mm

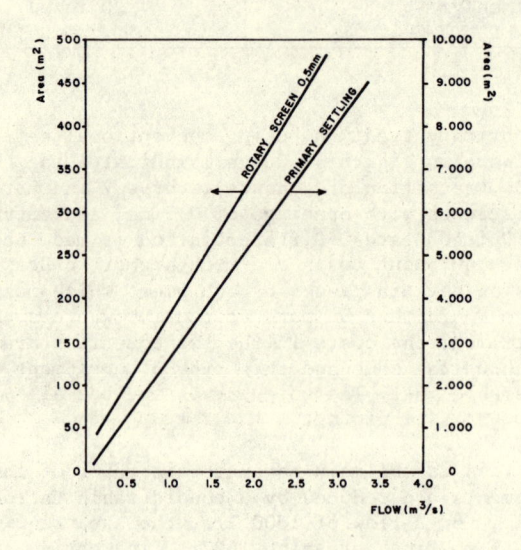

Fig. 5. Area Requirements for Primary Treatment Using Settling Tanks and Rotary Screens 0.5 mm

While conventional primary sludge contains about 95% moisture (the exact percentage depending on waste characteristics and settling tank design operation), the sludge removed on any type of fine screen is usually in the range of 10-15% solids and can reach up to 20%. This fact means that for a flow of 1 m^3/s and a suspended solids content of 600 mg/l, the primary settling alternative would produce about 674 m^3 of sludge per day and the fine-screening process 140 m^3/d, the latter being much easier to handle and not requiring further dewatering as for most final disposal methods.

On the other hand, in further dewatering is necessary for the disposal of screening solids, experience in Brazil has demonstrated the need for grinding such solids to finer particles to avoid damage to the filtration dewatering equipment.

Other devices have been tested for the removal of excess water from screening solids with the "Roto-Press", produced in Sweden, found to be the most cost-effective unit.

VIBRATING SCREENS

The use of vibrating screens in place of rotary or static screens can be an even more economical solution for pre-treatment operations.

Such screens were first utilized for this purpose in Brazil at Mouran Slaughterhouse Industries, State of São Paulo. The original design specified the use of a rotary screen to remove large grease pieces and blood clots prior to the extended aeration activated sludge process. However, for economical reasons, the rotary screen was replaced by a set of four vibrating screens, installed at an angle, permitting the screenings to drop into a container. This modification reduced the screening equipment cost from US$ 18,000 to US$ 8,000.

The difference in price between the two types of screening equipment is due both to duties on certain parts of the rotary screening equipment and to the fact that production is much greater for vibrating screens, which are used for many other purposes.

Although the plant is still operating experimentally, the results obtained there are favorable.

OTHER SIMPLIFICATION PROCEDURES

Except for a few sophisticated installations, all industrial pollution control plants in Brazil are manually operated, incorporating only a few pieces of instrumentation that are absolutely necessary. Because of the low cost of labor in Brazil, manual operation methods actually result in lower costs as compared with the utilization of control instruments, which are expensive due to limited domestic production of such equipment. On the other hand, costs of pollution control equipment manufactured in Brazil compare favorably with costs in developed countries.

For physico-chemical plants, all chemical dosages are determined by the plant operators at intervals during the day from laboratory analyses and flow data, thus avoiding automatic transmission of flow readings to the chemical feeders as well as the need for zeta potential meters for automatic correction of the chemical dosage. Aeration is also controlled by the operator, without the use of automatic dissolved oxygen probes. In most cases, flow is measured using weirs or Parshall Flumes with scale readings converted to flow, using prepared charts. Experience has shown that Brazilian plant operators have the capacity to adapt themselves to this type of operation inasmuch as most of the plants are achieving high levels of efficiency.

CONCLUSIONS

Legislation for the control of pollution resulting from industrial waste discharges was initiated in Brazil during the early 1970's. In the State of Rio de Janeiro, the controlling commission (CECA) is under the authority of the State Secretary of Public Works, with technical control and monitoring carried out by FEEMA, the State Foundation for Environmental Engineering.

Industries, now faced with the cost of providing expensive waste treatment facilities, are searching for cost-effective solutions. Some innovative processes have been introduced by the authors in plant designs, particularly for the food-processing industries.

Preliminary treatment through the use of fine-screening devices in place of flotation system has resulted in considerable cost savings and ease of handling the solids removed, without any significant effect on the following biological treatment processes.

The use of fine screening as an alternative to primary sedimentation tanks can provide even greater economy. On the basis of favorable results are anticipated for plants designed using this concept. Although the Niteroi plant is handling municipal sewage, it is expected that equal or greater efficiency will be obtained in similar plants handling wastes, such as from meat-processing and latex production, where a high conter of relatively large-sized solids is found. The cost data show significant savings in Brazil, where rotary screening equipment has replaced primary tanks; these savings are shown to be greater for the USA due to the elimination of customs duties. Area requirements for the screening alternative are shown to be only 6% of that required for primary tanks. Screening solid have a low volume and a low moisture content, and usually require no further dewatering for disposal.

The use of vibrating screens in place of rotary or static screens may, on the basis of preliminary results at a slaughterhouse, be an even more economical solution for waste pre-treatment operations.

Because of low labor costs, manual operation methods result in lower costs when compared with the use of sophisticated instrumentation. Experience has shown that Brazilian plant operators have the capacity to provide excellent process control using manual procedures judging from the high degree of efficiency achieved at most plants.

DISCUSSIONS

Hanisch, B. (Germany)	:	I have never seen these rotary screens you have mentioned.
Almeida, A.S. (Brazil)	:	They first started in the U.S., then Sweden, Australia and Brazil too started to make them.
Hanisch, B. (Germany)	:	How does the construction cost of such screens compare with normal bar screens?
Almeida, A.S. (Brazil)	:	They are a much more expensive piece of equipment. Bar screens can only be produced with large openings, otherwise they clog easily. Rotary screens do not replace bar screens; they are used in addition to them. Since they do a different job, there is really no basis for comparison.

Wastewater Treatment by Using Bamboo Bladed Rotors

S. MUTHUSWAMY

NEERI, Madras, India

ABSTRACT

Simple wastewater treatment systems are essential for treating industrial liquid wastes effectively. The major cost involved in operating such systems is the power required for aeration. Introduction of light-weight bamboo-bladed rotors is a recent improvement which reduces power consumption as well as installation and maintenance costs. The aeration capacity and power consumption varies mainly with the water spraying capacity in the space around the rotor periphery and with the speed of the rotor. The available data on such characteristics has been reviewed. Much emphasis is placed in tropical countries on the operation of the oxidation ditch and on making further improvements in the cage rotor. It is necessary to establish the optimum submergence for the cage rotor to get maximum BOD reduction per unit of electrical energy input. The performance of the oxidation ditch, when the bamboo-bladed rotor and the batch system are used, has been evaluated with respect to the reduction of BOD, suspended solids, nitrogen and phosphorus.

Suitable design criteria can be established for the bamboo-bladed rotor under varying loading conditions. Its performance has been studied under field conditions using the batch process in the sewage treatment demonstration plant at The Centre for Environmental Engineering, College of Engineering, Madras. The results of the experiment are indicated in terms of power consumption in kwh per meter of rotor length. The removal of BOD in kg/unit energy input for different submergences was very good. The optimum submergence was 13.5 cm. It was observed that the steel bladed got corroded within a period of 6 months whereas the noncorrosive bamboo blades lasted 14 months. On the basis of this data, an attempt can be made to design an oxidation ditch to treat industrial liquid wastes.

INTRODUCTION

Treatment of wastewater in rural areas is a problem because of lack of adequate financing and technical know-how. In India even the corporations and the big municipalities are considering low-cost wastewater treatment units. They are still improving their technical skills to reduce the cost of even the low-cost wastewater treatment. Thus the light-weight bamboo-bladed rotor has been developed. The cost of a treatment unit depends mainly on (i) the materials from which the ditch has been constructed (ii) the material from which the cage rotor is built (iii) the weight of the cage rotor, (iv) the oxygenation capacity and the power consumption.

The bamboo-blades, because of its concave shape, can hold more water and produce more bubbles during aeration; hence the oxygenation capacity of a system with bamboo-blades is greater. By conducting a field study, a bamboo-bladed rotor was established to be more economic with a lower power consumption for the same BOD removal and other MLSS (Mixed Liquor Suspended Solids) buildup.

EXPERIMENTAL SET-UP

At the sewage treatment demonstration plant of The Centre for Environmental Engineering, College of Engineering, Guindy, Madras, detailed experimental studies were conducted in an oxidation ditch of 68100 ℓ capacity. The sewage can be treated using either the continuous or the batch processes. The flow channel is open-ended, with a trapezoidal cross-section of 0.8 m at the bottom and 2.03 m at the top. The channel has a depth of 1.27 m and a side slope of 0.8:1. The embankment top is 0.91 m wide and 0.45 m of the top is lined with plain concrete. The lining has been extended to the inner side slopes and bottom to control corrosion and weed growth. The cage rotor has been mounted on brick masonry walls with a thickness of 0.45 m. The operational capacity of the ditch is 46400 ℓ at a liquid depth of 1.00 m.

A 90° V notch of height 0.16 m has been embedded in the inlet channel to measure the flow rate. A cast iron pipe, 125 mm in diameter, has been laid to convey the sewage from the inlet to the ditch. The pipe has been equipped with a downward bend to avoid turbulence in front of the rotor during pumping. A hose outlet, 10 cm in diameter, has been placed exactly opposite of the ditch inlet to discharge the supernatant (Fig. 1).

DETAILS OF THE CAGE ROTOR

A recent development in reducing the weight of the cage rotor is achieved by the use of bamboo-blades. The rotor has been built with a hollow shaft 6.5 cm in diameter. Two end discs consist of 0.8 mm thick steel plates. There are 12 radial arms, each with a radius of 0.35 m and made up of 25 x 25 mm M.S. angles projecting from the central shaft. The twelve radial angles have been connected by 12 longitudinal arms, which are made also of the same M.S. angles (Fig. 2).

Details of the Bamboo Blades

A uniform bamboo post having a girth of 17 cm has been taken and suitably prepared (Fig. 3). The rotor has a gross length and diameter of 1 m and 0.7 m, respectively. The bamboo blades have been attached to the longitudinal arms with a clear spacing of 5 cm between each. Thus the first longitudinal arm accommodated 10 blades while the next one had 9 blades in the gaps left as spacing in the first arm. The total number of blades used is 114 (6 x 10 + 6 x 9 = 114). All parts of the rotor, inclusive of bamboo blades, are coated with anticorrosive paints.

FIELD STUDY

During the field study the rotor was kept at a constant level by fixing the shaft at a height of 1.3 m above the bottom level of the ditch. The liquid depth in the ditch was varied to get the various submergence levels. The observations were carried out for submergence of 11 cm, 13.5 cm, 16 cm, and 18.5 cm. The corresponding liquid levels from the bottom were 95 cm, 97.5 cm 100 cm and 102.5 cm, respectively.

Ditch Operation

The MLSS level varied in the range of 5000 to 6000 mg/ℓ. Based on the submergence of the blades, aeration was allowed for a specific time period. A reduction gear with a ratio of 1:20 was coupled to the 2 H.P. electrical motor to reduce the rotor speed from 1440 to 72 rpm. The ditch was operated using a batch system. The operational cost, based on power consumption for a flow of 68100 ℓ/d using the bam-

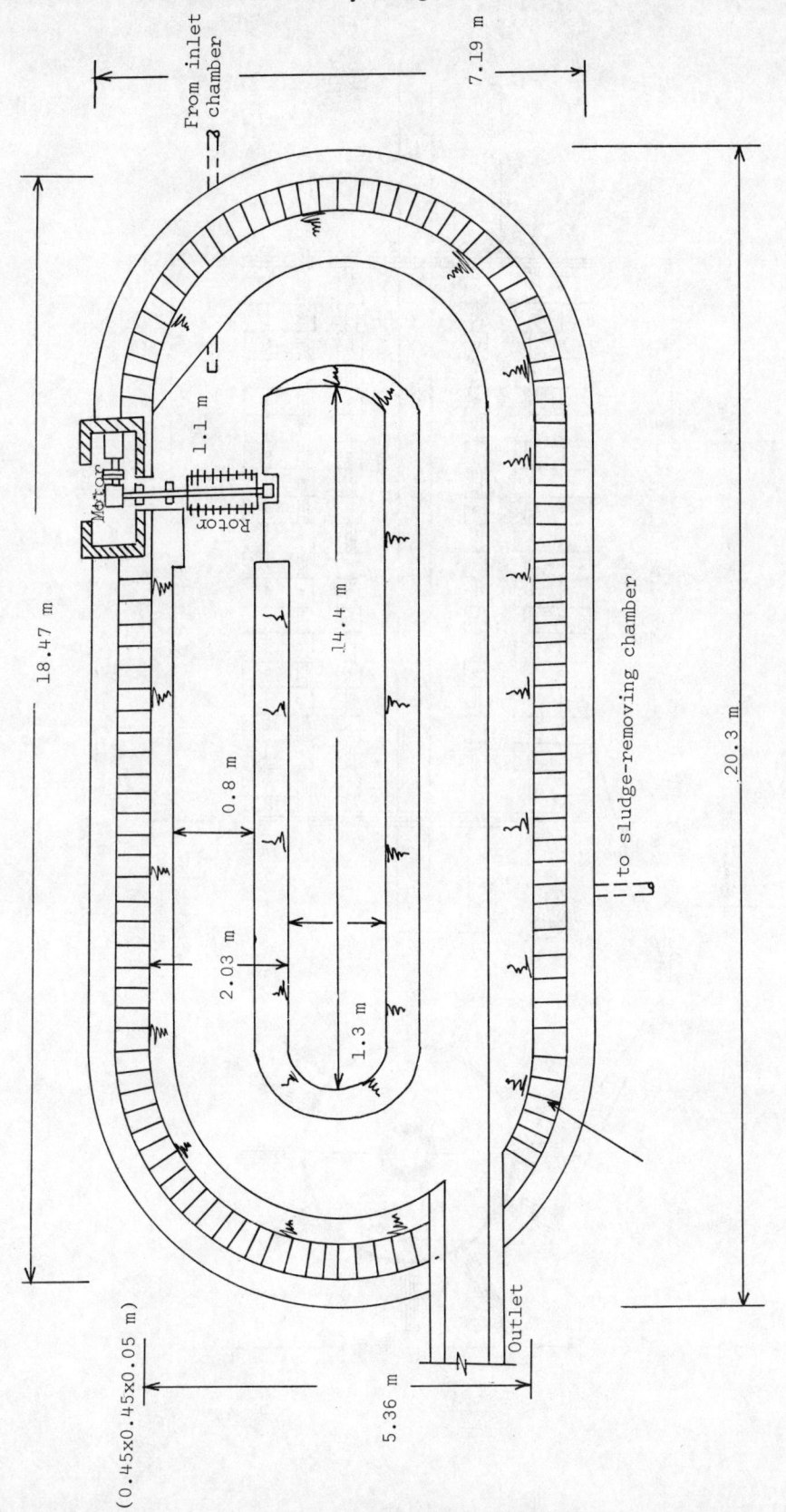

Fig. 1: Plan of the Oxidation Ditch

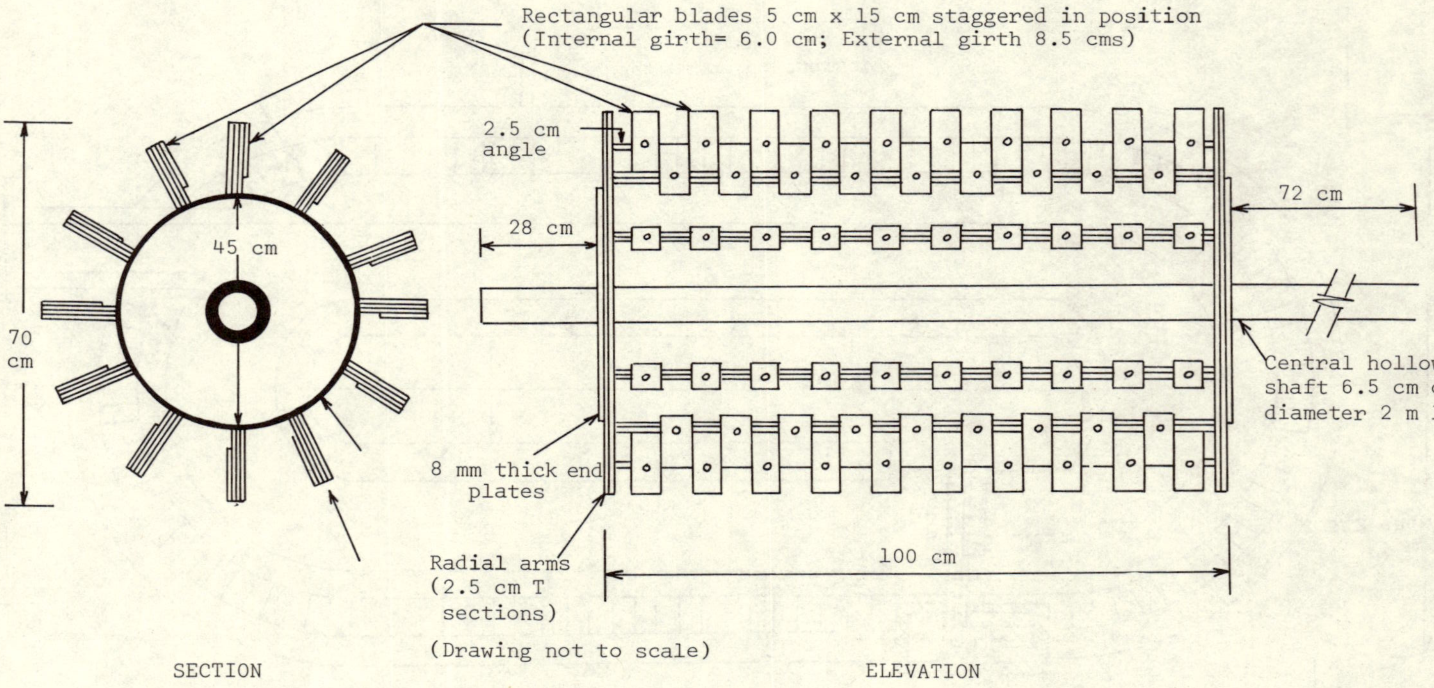

Fig. 2: Details of cage rotor with bamboo blades.

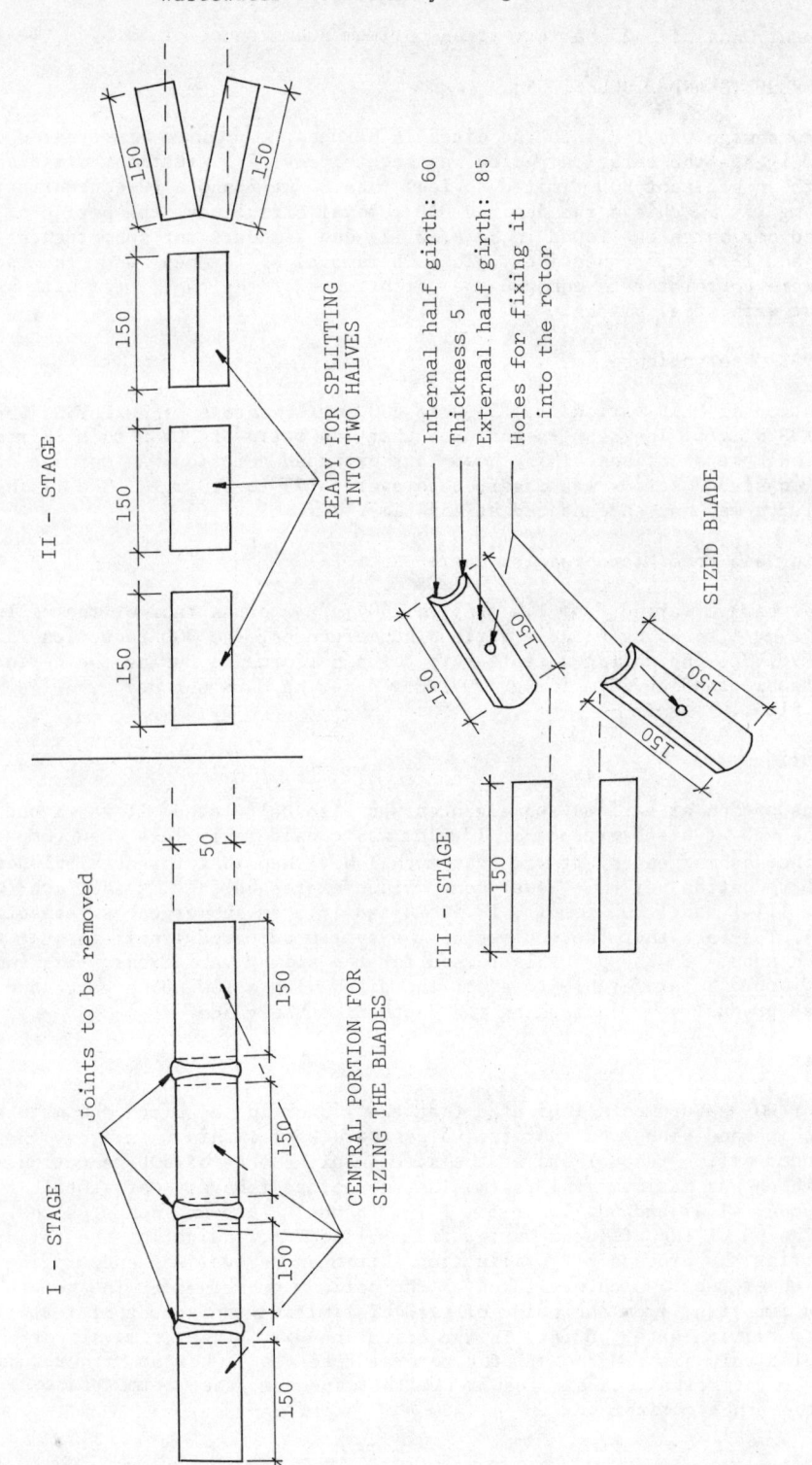

Fig. 3: Size of the bamboo blades.

boo-bladed rotor, was Rs 3.25 per day at an optimum submergence of 13.5 cm (Table 1).

RESULTS AND DISCUSSIONS

22700 ℓ of raw sewage was fed into the ditch in batches. 3 batches were treated daily. For each submergence the aeration period was set to provide a treatment efficiency over 90%, with an effluent BOD limited to less than 20 mg/ℓ and a DO concentration of about 1.0 mg/ℓ. To achieve the desired BOD removal efficiency, the period of aeration required per batch was found to be 4, 3, 2½ and 2½ hours for submergences of 11, 13.5, 16 and 18.5 cm, respectively. The BOD removal efficiency and the power consumption were correlated to compare the performance of the cage rotor with bamboo blades to that with steel blades.

MLSS and MLVSS Observations

The concentration of MLSS varied from 5000 to 6000 mg/ℓ whereas the MLVSS varied from 1696 to 3326 mg/ℓ. The results indicate that the ratio of MLVSS to MLSS remained below 0.6 on most occasions. This shows that extended aeration was carried out. The increase in sludge solids was observed to average 75 to 210 mg/ℓ. The rising or bulking of sludge was not encountered at any time.

Organic Loading and Food/Microorganism Ratio

The BOD of raw sewage varied from 140 mg/ℓ to 240 mg/ℓ whereas that of the effluent varied from 2 mg/ℓ to 25 mg/ℓ under various submergences. The BOD reduction varied from 84.5 to 98.83%, indicating satisfactory ditch performance during the period of study. The organic loading rate in kg BOD removed per kg MLSS per day varied from 0.02025 to 0.5125.

Power Consumption

The power consumption at various submergences was also calculated. It was found that the power consumed at a submergence of 13.5 cm was considerably less when compared to that at other submergences. It was also much lower than that for steel-bladed rotors. The BOD reduction per unit power consumption expressed in kg BOD/kwh was 1.369, 1.517, 1.141 and 1.203 for 11, 13.5, 16 and 18.5 cm submergences respectively (Table 2). The fact that the ecology of the system adjusted itself to suit the rate of oxygen supply is the probable reason for the slight variations. This indicates that it would be preferable to start the ditch with a low submergence and improve the load gradually by increasing the depth of submergence.

Other Observations

The results of DO measurements indicated that the liquor in the ditch was maintained under aerobic conditions and that the values were kept within a very narrow range. Suspended solids removal was at the same level as that of BOD reduction per unit power, which was maximum at 13.5 cm. The percentage removal of total solids was 53.13, 56.64, 41.47 and 53.71 for 11, 13.5, 16 and 18.5 cm submergences, respectively. The pH of the effluent varied from 7.1 to 7.8, indicating the formation of buffers during the process of purification. Nitrogen removal was encouraging at the 13.5 cm submergence, which was taken as the optimum submergence. The removal of phosphates is important from the point of view of limiting the amount of fertilizers reaching the receiving water bodies. In the oxidation ditch the removal of phosphates was relatively high. Moreover, the removal rate was higher at higher submergences. Further investigations are required with respect to the time factor, the biomass and the other variables.

Table 1 Operational Cost at the Optimum Submergence of 13.5 cm (Based on power consumption for a flow of 68.100 ℓ/d using a bamboo-bladed rotor)

Date	5-day 20°C BOD Raw (mg/ℓ)	5-day 20°C BOD Effluent (mg/ℓ)	BOD reduction mg/ℓ	% removal	BOD reduction in mg/d	Power consumption per hr in kwh	Power consumption per day 3 batch (9h)	Cost per day at 40NP per kwh Rs np
4.4.77	140	6	134	95.75	9.14	0.833	7.5	3.00
7.4.77	175	7	162	96.00	11.45	0.822	7.4	2.96
8.4.77	195	10	185	94.90	12.60	0.833	7.5	3.00
9.4.77	210	9	201	95.75	14.70	0.833	7.5	3.00
10.4.77	200	12	188	94.00	12.80	0.833	7.5	3.00
11.4.77	190	15	175	92.11	12.61	0.833	7.5	3.00
12.4.77	170	14	156	92.25	10.62	0.855	7.7	3.08
13.4.77	180	12	168	93.25	11.45	0.833	7.5	3.00
14.4.77	160	10	150	93.95	10.20	0.833	7.5	3.00
15.4.77	170	12	158	93.00	10.24	0.833	7.5	3.00
16.4.77	190	10	180	94.75	10.28	0.833	7.5	3.00
17.4.77	195	8	187	95.90	12.75	0.833	7.5	3.00
18.4.77	160	7	153	95.60	10.41	0.833	7.5	3.00
19.4.77	165	12	153	92.60	10.41	0.833	7.5	3.00
20.4.77	170	13	157	92.40	10.70	0.833	7.5	3.00
							$\bar{x}$ =	3.01

Maintenance cost per day

Cost for power = Rs 3.01
Oil, grease and cotton waste etc. = Rs 0.24
TOTAL = Rs 3.25/d

Table 2   BOD Reduction and Energy Consumption for Various Submergences

Date	BOD reduction kg BOD/d	Gross energy consumption kwh	BOD reduction per unit power kg BOD/kwh/d
\multicolumn{4}{c}{11 cm submergence}
23.3.77	9.55	9.00	1.060
24.3.77	10.70	9.00	1.190
25.3.77	10.61	9.00	1.810
26.3.77	12.31	9.00	1.369
27.3.77	12.95	9.00	1.415
\multicolumn{4}{c}{13.5 cm submergence}
6.4.77	9.14	7.5	1.219
7.4.77	11.45	7.4	1.547
8.4.77	12.60	7.5	1.680
9.4.77	13.70	7.5	1.829
10.4.77	12.20	7.5	1.709
\multicolumn{4}{c}{16 cm submergence}
20.2.77	10.80	9.84	1.100
21.2.77	7.55	9.09	0.830
22.2.77	8.30	7.80	1.065
23.2.77	11.83	7.80	1.518
24.2.77	8.92	7.80	1.190
\multicolumn{4}{c}{18.5 cm submergence}
8.3.77	12.75	9.8	1.301
9.3.77	11.59	9.8	1.183
10.3.77	9.95	9.8	1.015
11.3.77	12.20	9.5	1.250
13.3.77	12.40	9.8	1.265

CONCLUSIONS

1. Batch processing was preferred for reducing the recirculation and operational costs. The ditch adjusted itself according to oxygen supply, aeration period and the required BOD reduction.

2. Steel-bladed rotors installed consumed 0.91, 1.00, 1.43 and 1.56 kws per meter when run at 11, 13.5, 16 and 18.5 cm submergences, while the bamboo-bladed rotor consumed 0.7525, 0.8344, 1.071 and 1.337 kw per meter at the same submergence levels. Thus considerable savings could be obtained by introducing the bamboo-bladed rotor.

3. The BOD reduction was also better when compared to the unit energy input. The BOD reductions per unit energy expressed in kg BOD removed per kwh were 1.111, 1.212, 1.129 and 1.100 for steel and 1.369, 1.517, 1.141 and 1.203 for bamboo for 11, 13.5, 16 and 18.5 cm submergences, respectively.

4. For an optimum submergence of 13.5 cm, the maximum removal of ammonia nitrogen was 63.65%. Nitrite nitrogen appeared in the range of 0.1 mg/ℓ to 0.75 mg/ℓ and nitrate nitrogen in the range of 7.0 mg/ℓ to 13.5 mg/ℓ.

5. Bamboo-blades performed well without any corrosion or damage during the study. The bamboo-blades required replacement only after a period of 14 months, whereas, due to corrosion, the steel required replacement within a period of 6 months. Therefore, bamboo blades are more economical.

ACKNOWLEDGEMENT

My sincere thanks are due to my student Mr. E. Lakshmanan, who carried out this work under my guidance. My thanks are also due to Dr. B. B. Sundaresan, who constantly provided assistance in conducting this study.

REFERENCES

Arceivala, S.J. and S.R. Alagarsamy (1972). Design and Construction of Oxidation Ditches under Indian Conditions. *Symposium on Low Cost Waste Treatment,* p. 173, CPHERI, Nagpur, India.

Handa, B.K., B.V.R. Subramaniyam and G.J. Mohanrao (1972). Preliminary Studies on the Performance of Oxidation Ditch. *Symposium on Low Cost Waste Treatment,* p. 194, CPHERI, Nagpur, India.

Haridass, G. (1973). Field Studies on Load Variation in an Oxidation Ditch. M.Sc. Degree Thesis, University of Madras.

Jaiprakash Narain, G.B. (1971). Performance Studies on an Oxidation Ditch. M.Sc. Degree Thesis, University of Madras.

Kshirsagar, S.R. (1968). Oxidation Ditches in the Netherlands. *Environmental Health,* *10,* No. 2, 97.

Pasveer, A. (1960). Waste Treatment. *Proc. of the Second Symposium on the Treatment of Wastewater.* p. 126, Pergamon Press, England.

Pasveer, A. (1972). The Oxidation Ditch - Principle, Results and Applications. *Symposium on Low Cost Waste Treatment,* p. 167, CPHERI, Nagpur, India.

A Mathematical Model for the Nutrient Cycle in Reservoirs

M. KARPUZCU

Environmental Engineering Division, Istanbul Technical University, Istanbul, Turkey

ABSTRACT

In this study, a mathematical model developed for the determination of nitrogen and phosphorus compounds in reservoirs is considered and then applied to the Indian Creek Reservoir. In the model, four nitrogen compounds, NH_3, $NO_2 + NO_3$, particulate nitrogen in biomass (PN) and dissolved organic nitrogen (DON), as well as three phoshporus compounds, phosphate, particulate organic phosphorus and dissolved organic phosphorus, are taken into account. The model is tested using the measurements made by EPA in the reservoir. Observations and predictions are shown on the same graph and compared.

INTRODUCTION

Population growth and industrialization have led to the need for limited the beneficial and economy in the use of water resources. Recently, the increasing mechanization of agriculture and the use of ground-water for irrigation have caused an increase in the consumption of fertilizers. In order to meet this need for fertilizers, it has been necessary to open new factories for the production of phosphates and nitrates. Fertilizer factories, like the others, cause environmental problems, especially water pollution in the region. Since industrial wastewater contains phosphorus and nitrogen compounds, it is more harmful for reservoirs, lakes and estuaries. These compounds of phosphorus and nitrogen, which are essential nutrients for the production of algae in natural bodies of water, cause eutrophication and change the trophic status of the water.

In general, it has not been possible to remove the nutrients by the classical methods of treatment; indeed, for the removal of these elements it is necessary to apply tertiary treatment. Tertiary treatment is expensive and the effect of the effluent on water bodies is still not well understood. Because of this, it is necessary to evaluate predictive nutrient models before the planning and construction of treatment systems. Such models can be used to predict the water quality with respect to increased or decreased phosphorus and nitrogen loads caused by the different activities.

NUTRIENT MODEL

Nutrients Involved

This study is concerned with the following compounds as macro-nutrients:

- Ammonium N (NH_3)
- Nitrite and Nitrate N ($NO_2 + NO_3$)
- Particulate N
- Dissolved Organic N
- Phosphate P
- Particulate Organic P
- Dissolved Organic P

The cycle of nitrogen and phosphorus compounds in the model is assumed to be as shown in Fig. 1. In the presence of solar energy, NH_3, NO_3 and phosphate are taken up to form algae (containing particulate N and P). When the algae decompose, they release NH_3 and phosphate back into the water column. Most of the algae (particulate N and P) settle to the bottom of the waterbody, where a certain amount of NH_3 and PO_4 is released and returned to the water column. In this study, nutrients released from sediment are not considered. First order reactions are used for all transformations.

Formulation of the Model

Based on the representation in Fig. 1 and the principle of mass conservation, the mathematical expressions for well-mixed reservoirs are as follows:

$$V \frac{dN_1}{dt} = Q_{in} C_{N_1} - Q_{out} N_1 - VR_{12} N_1 - VR_{13} N_1 + VR_{41} N_4 + W_{N_1} \qquad (1)$$

$$V \frac{dN_2}{dt} = Q_{in} C_{N_2} - Q_{out} N_2 - VR_{23} N_2 + VR_{12} N_1 + W_{N_2} \qquad (2)$$

$$V \frac{dN_3}{dt} = VR_{13} N_1 + VR_{23} N_2 - VGN_3 - VR_{34} N_3 - Q_{out} N_3 \qquad (3)$$

$$V \frac{dN_4}{dt} = Q_{in} C_{N_4} - Q_{out} N_4 + VR_{34} VN_3 - VR_{41} N_4 + W_{N_4} \qquad (4)$$

$$V \frac{dP_1}{dt} = Q_{in} C_{P_1} - Q_{out} P_1 - VK_{12} P_1 + VK_{31} P_3 + W_{P_1} \qquad (5)$$

$$V \frac{dP_2}{dt} = VK_{12} P_1 - Q_{out} P_2 - VK_{23} P_2 - VGP_2 \qquad (6)$$

$$V \frac{dP_3}{dt} = Q_{in} C_{P_3} - Q_{out} P_3 + K_{23} P_2 - K_{31} P_3 + W_{P_3} \qquad (7)$$

Where Q_{in} (L^3/T) is the adjective inflow; Q_{out} (L^3/T) is the adjective outflow; $V(L^3)$ is the volume of the reservoir; $W(M/T)$ is the nutrient load; C_{N_1}, C_{N_2}, ... and C_{P_1}, C_{P_2} are the concentrations of nitrogen and phosphorus compounds in the tributary inputs; N_1, N_2, N_3, N_4, P_1, P_2, P_3 are the concentrations of nitrogen and phosphorus compounds in the water column; R and K are reaction rates; and G is the sedimentation rate (O'Melia, 1977).

The reaction rates are dependent on temperature, solar radiation, and the available nutrients of N and P. In this study the daily average reaction rates are assumed to be as follows:

$$R_{12} = 0.03 \, (1.08)^{T-20}$$

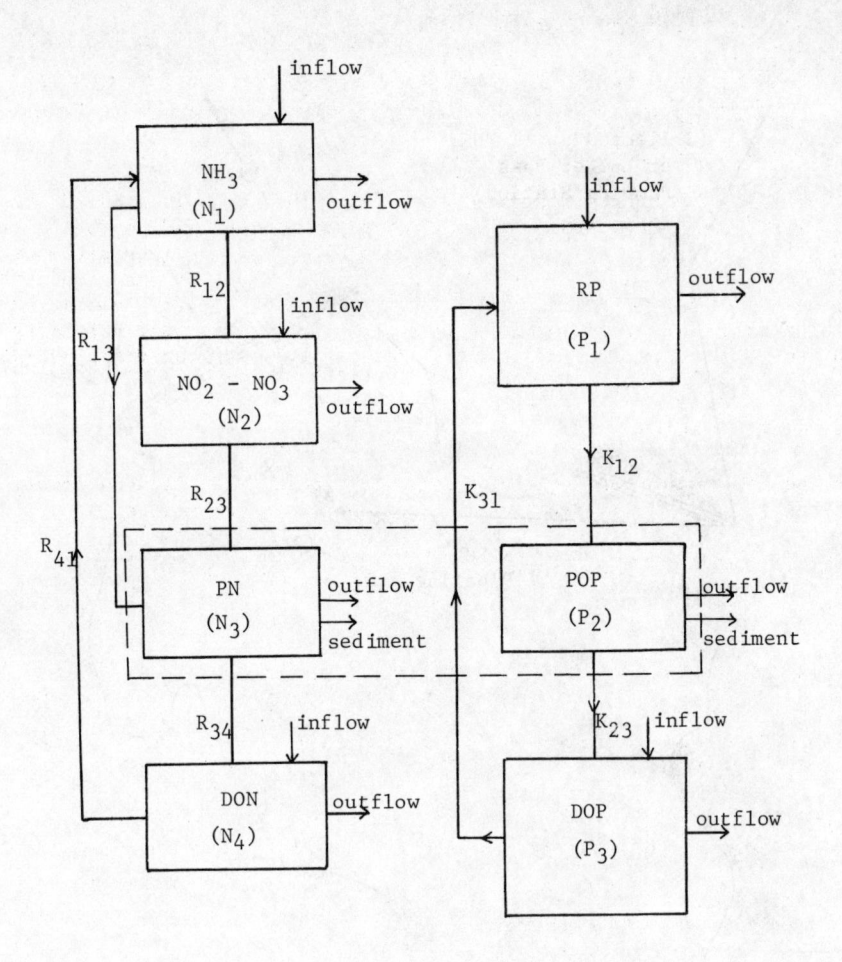

Figure 1. Nutrient Model for Reservoirs.

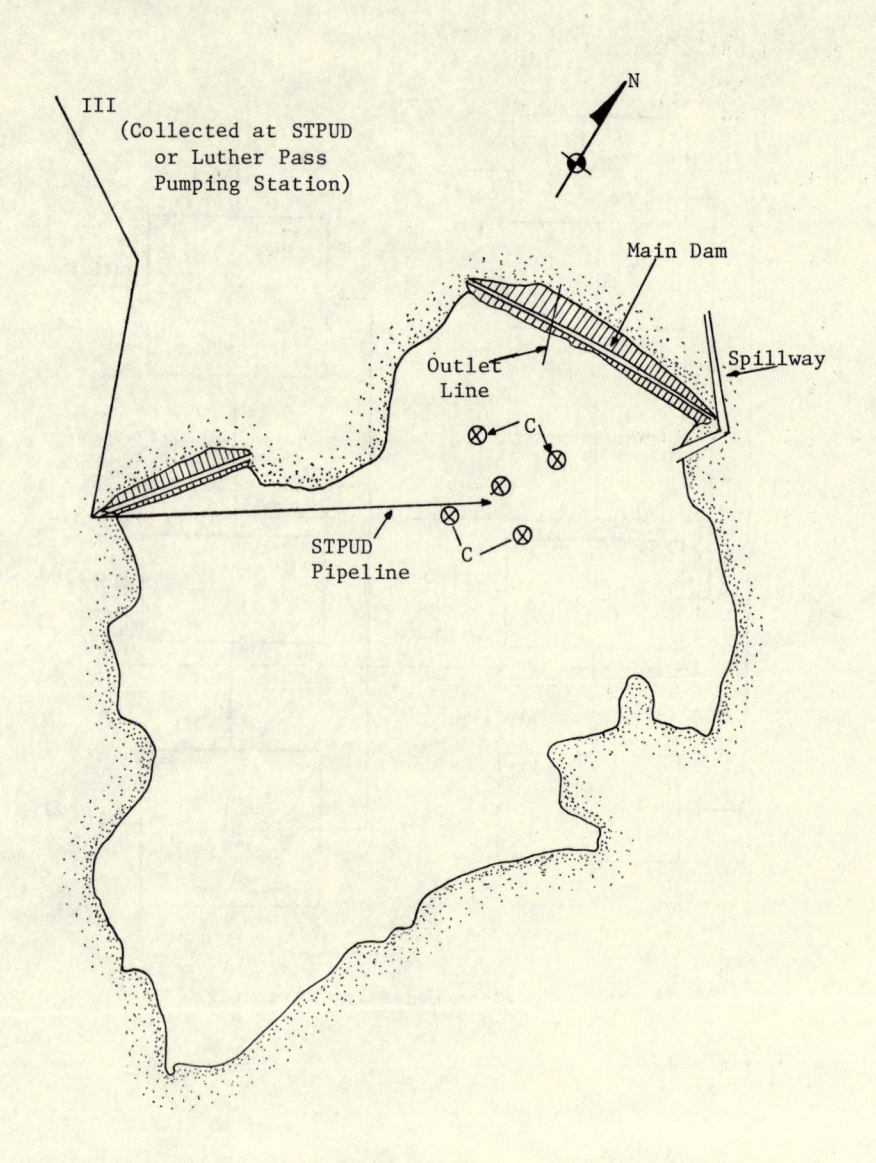

Figure 2. Location of Sampling Stations at Indian Creek Reservoir.

$R_{13} = 0.15 \ (1.08)^{T-30} \cdot \{(N_1 + N_2)/(0.30 + N_1 + N_2)\} \ \{P_1/(0.01 + P_1)\}$

$R_{23} = 0.06 \ (1.08)^{T-30} \cdot \{(N_1 + N_2)/(0.6 + N_1 + N_2)\} \ \{P_1/(0.01 + P_1)\}$

$R_{34} = 0.04 \ (1.03)^{T-20}$

$R_{41} = 0.01 \cdot T$

$K_{12} = R_{23}$

$K_{23} = R_{34}$

$K_{31} = R_{41}$

Where T (°C) is the temperature of the water column (Chang, 1977).

When Eqs. (1) through (7) are rewritten in terms of nutrient concentrations, the following set of ordinary differential equations can be derived:

$$\frac{dN_1}{dt} = A_{11} N_1 + A_{14} N_4 + W_1 \qquad (8)$$

$$\frac{dN_2}{dt} = A_{21} N_1 + A_{22} N_2 + W_2 \qquad (9)$$

$$\frac{dN_3}{dt} = A_{31} N_1 + A_{32} N_2 + A_{33} N_3 \qquad (10)$$

$$\frac{dN_4}{dt} = A_{43} N_3 + A_{44} N_4 + W_4 \qquad (11)$$

$$\frac{dP_1}{dt} = B_{11} P_1 + B_{13} P_3 + W'_1 \qquad (12)$$

$$\frac{dP_2}{dt} = B_{21} P_1 + B_{22} P_2 \qquad (13)$$

$$\frac{dP_3}{dt} = B_{32} P_2 + B_{33} P_3 + W'_3 \qquad (14)$$

Where

$A_{11} = -R_{12} - R_{13} - Q_{out}/V$

$A_{14} = R_{41}$

$A_{21} = R_{12}$

$$A_{22} = -R_{23} - Q_{out}/V$$

$$A_{31} = R_{13}$$

$$A_{32} = R_{23}$$

$$A_{33} = -R_{33} - G - Q_{out}/V$$

$$A_{43} = R_{34}$$

$$A_{44} = -R_{41} - Q_{out}/V$$

$$B_{11} = -R_{13} - R_{23} - Q_{out}/V$$

$$B_{13} = R_{41}$$

$$B_{21} = R_{13}$$

$$B_{22} = -R_{34} - G - Q_{out}/V$$

$$B_{32} = R_{34}$$

$$B_{33} = -R_{41} - Q_{out}/V$$

Solution of the Equations

Due to the analytical difficulty in solving Eqs. (8) through (14), a numerical integration scheme is needed. There are various numerical methods which can be used; in this study, the second-order two-step Runge method using the Euler Predictor step together with a half-step integrator has been employed (Di Toro and others, 1971).

Let $y = f(t,y)$ be the vector differential equation and let Δt be the step size. The half-step integrator is initially calculated as

$$yy = y_o + \Delta t/2 \cdot f(t_o, y_o)$$

Where y_o is the initial condition vector at t_o. An estimate of the value of y at the end of the time interval $(t_o + \Delta t)$, denoted by y_1, can be calculated as follows:

$$y_1 = y_o + \Delta t \cdot f(t_o + \Delta t/2, yy)$$

When these two steps are applied to Eq. (8), the following equations can be obtained:

$$NN_1 = N_{1_o} + (A_{11} N_{1_o} + A_{14} N_{4_o} + W_1) \cdot \Delta t/2 \tag{15}$$

and

$$N_{1_{i+i}} = N_{1_o} + (A_{11} NN_1 + A_{14} NN_4 + W_1) \cdot \Delta t \qquad (16)$$

Similar equations have been written and solved by computer.

APPLICATION OF THE MODEL

Because of the lack of available data in Turkey, the model has been applied to the Indian Creek Reservoir, U.S.A.

Characteristics of the Indian Creek Reservoir

The Indian Creek Reservoir was constructed in 1968 as a rockfill dam, 20.7 m in height. The reservoir has a maximum depth of 17 m and maximum mean depth of 6 m. At maximum water surface elevation, the surface area of the reservoir and the drainage area above the dam are 64.8 ha and 688 ha, respectively. The maximum volume of impounded water is about 3,860,000 m^3 (Mc Gauhey and others, 1971).

The average annual precipitation at the reservoir is reported to be about 51 cm, with 70% of this total occurring during the winter season of the year (Hill and others, 1968). The anticipated composition of impounded water on the basis of annual influent is as follows:

Meteorological water   30%
Reclaimed water   70%

In contrast to meteorological water, inputs to the reservoir through the export pipeline of the STPUD (South Tahoe Public Utility District) are continuous the year around, being greatest during the summer season when the transient population of the Lake Tahoe Basin is greatest. Because of the relatively small percentage of surface water in comparison with reclaimed water and an operating schedule which calls for the discharge of water for irrigation, the water impounded in the Indian Creek Reservoir is considered as predominantly reclaimed water.

Water Quality Prediction

The predictive model has been used to forecast water quality in the Indian Creek Reservoir and resultant predictions have been compared with observations. Beginning in April 1969, influent and composite samples have been collected from the reservoir and chemical analyses of the water samples have been made according to Standard Methods (1965) with respect to BOD, pH, organic nitrogen, ammonia, total phosphorus etc. by EPA (1975). Sampling stations have been chosen at the locations indicated in Fig. 2.

The variables used as inputs for the model include temperature, the reclaimed water influent, discharge from the reservoir, the nutrient concentration in the influent water and the depth of the reservoir. The volume variation of the reservoir is described as a function of depth, H, as follows:

$V = -3505 + 117.5 \cdot H, \qquad 36 < H < 56$
(correlation coefficient, $r^2 = 0.99$)

The reaction rates are computed as functions of temperature and nutrient concentrations. The measurements made in April, 1969 are used as the initial concentration of the nutrients considered. The increment of the integration procedure is 0.25 days.

Figure 3.

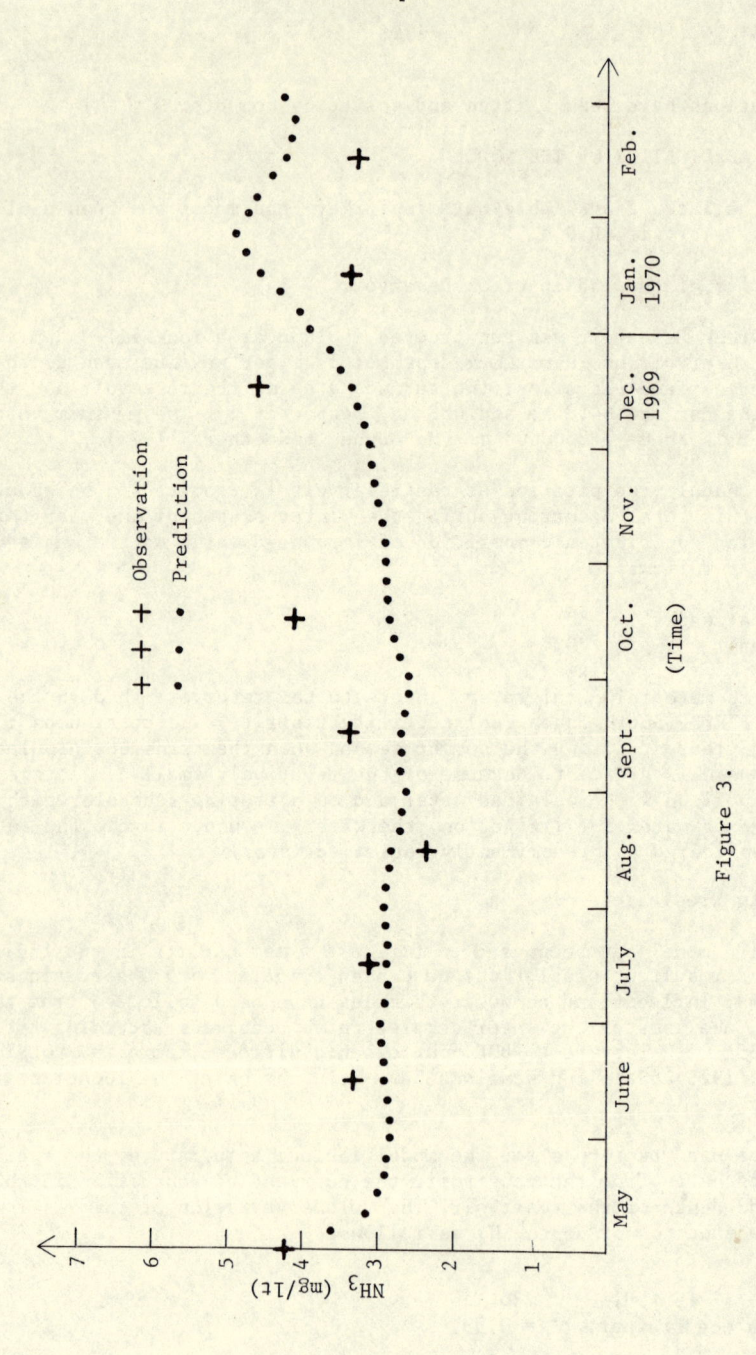

Figure 4.

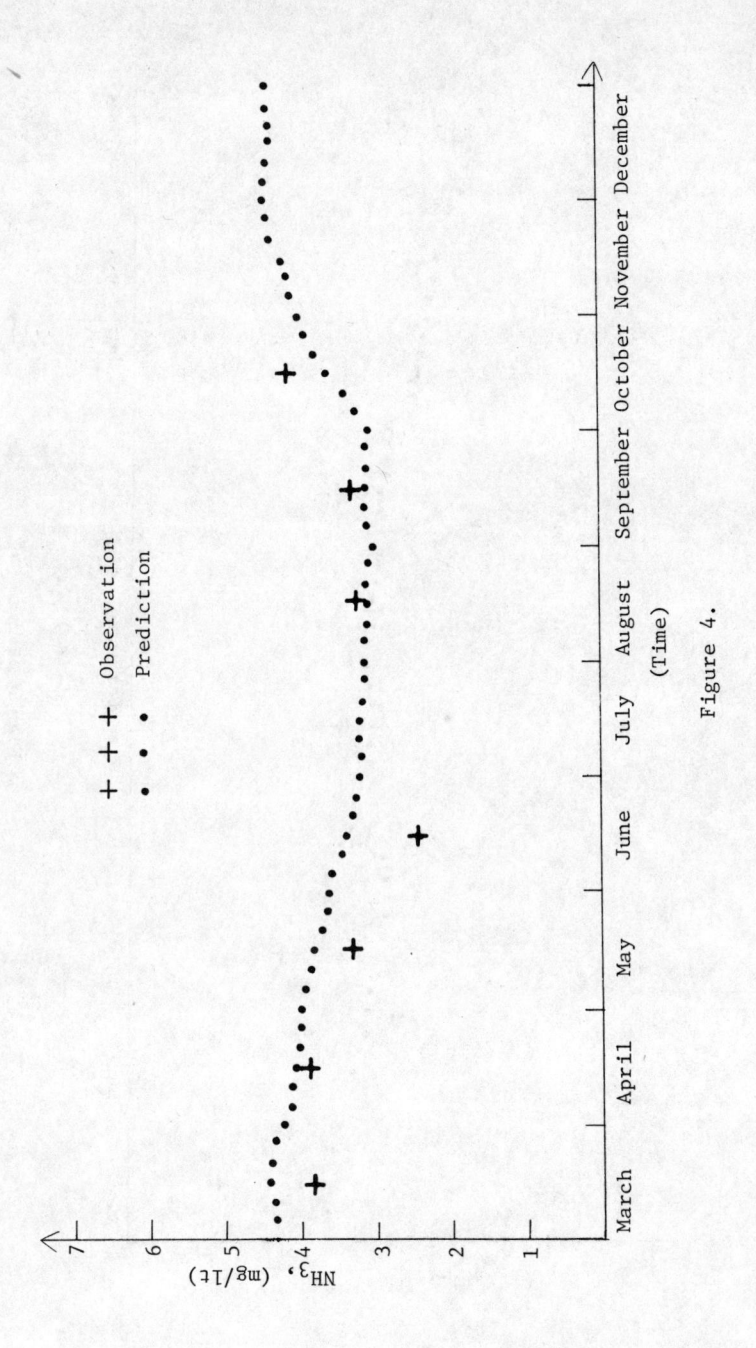

Figure 5.

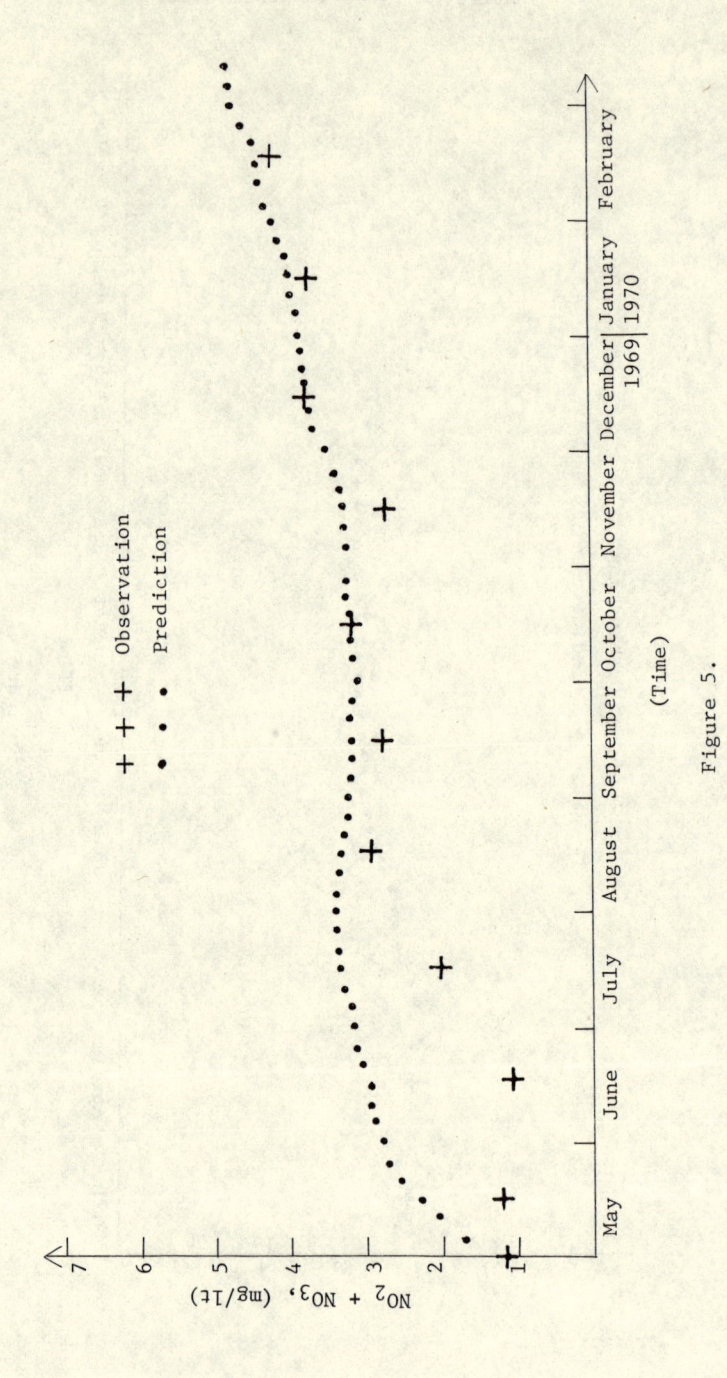

Figure 6.

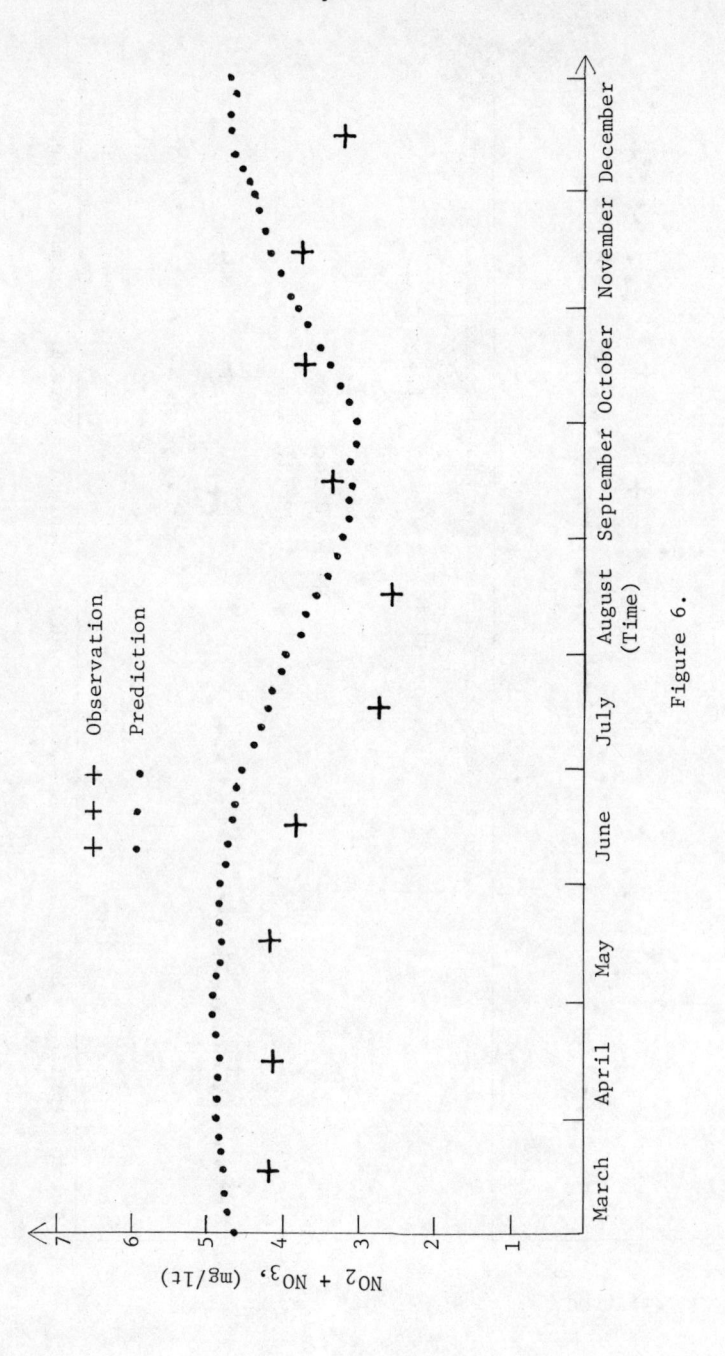

Figure 7.

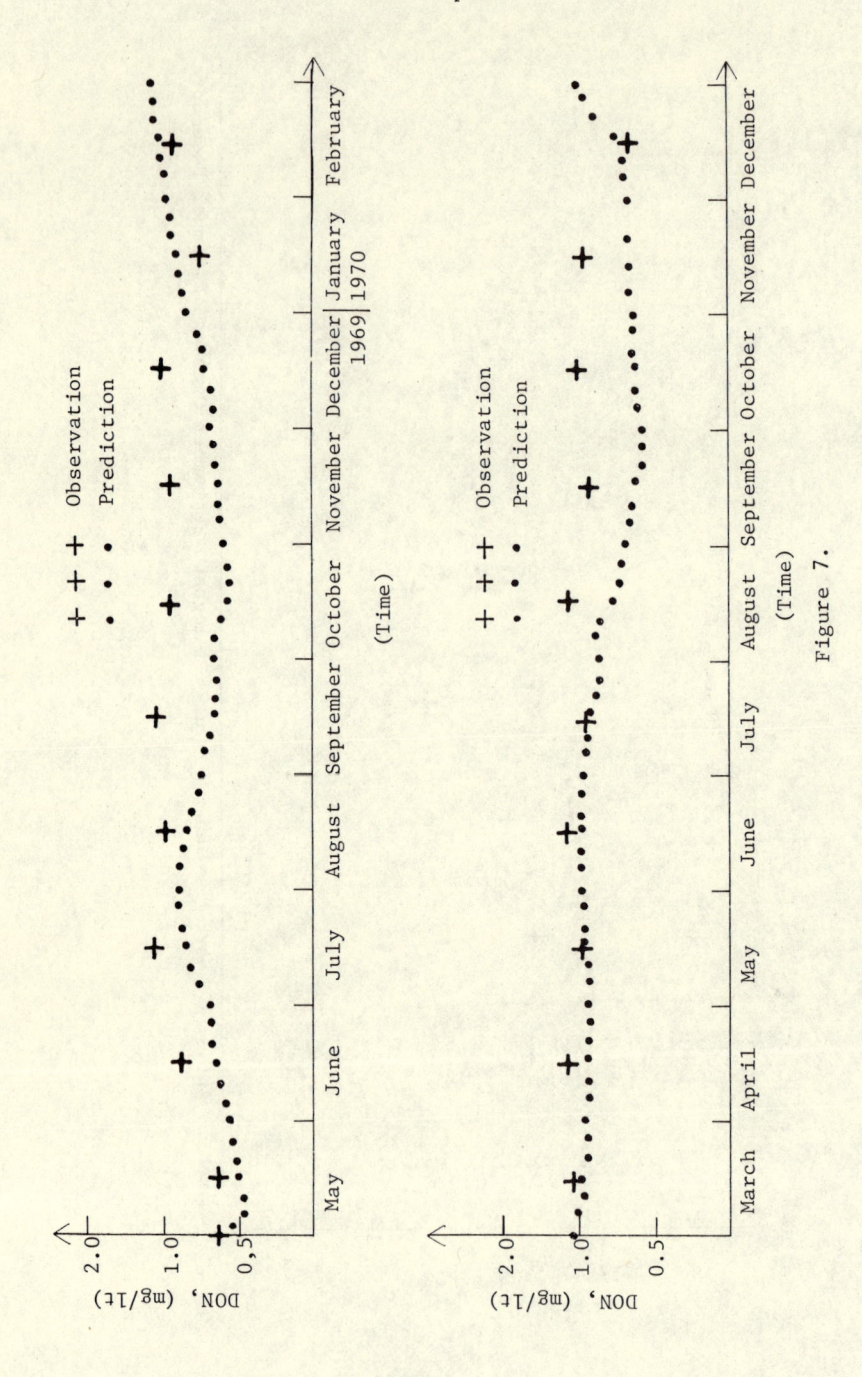

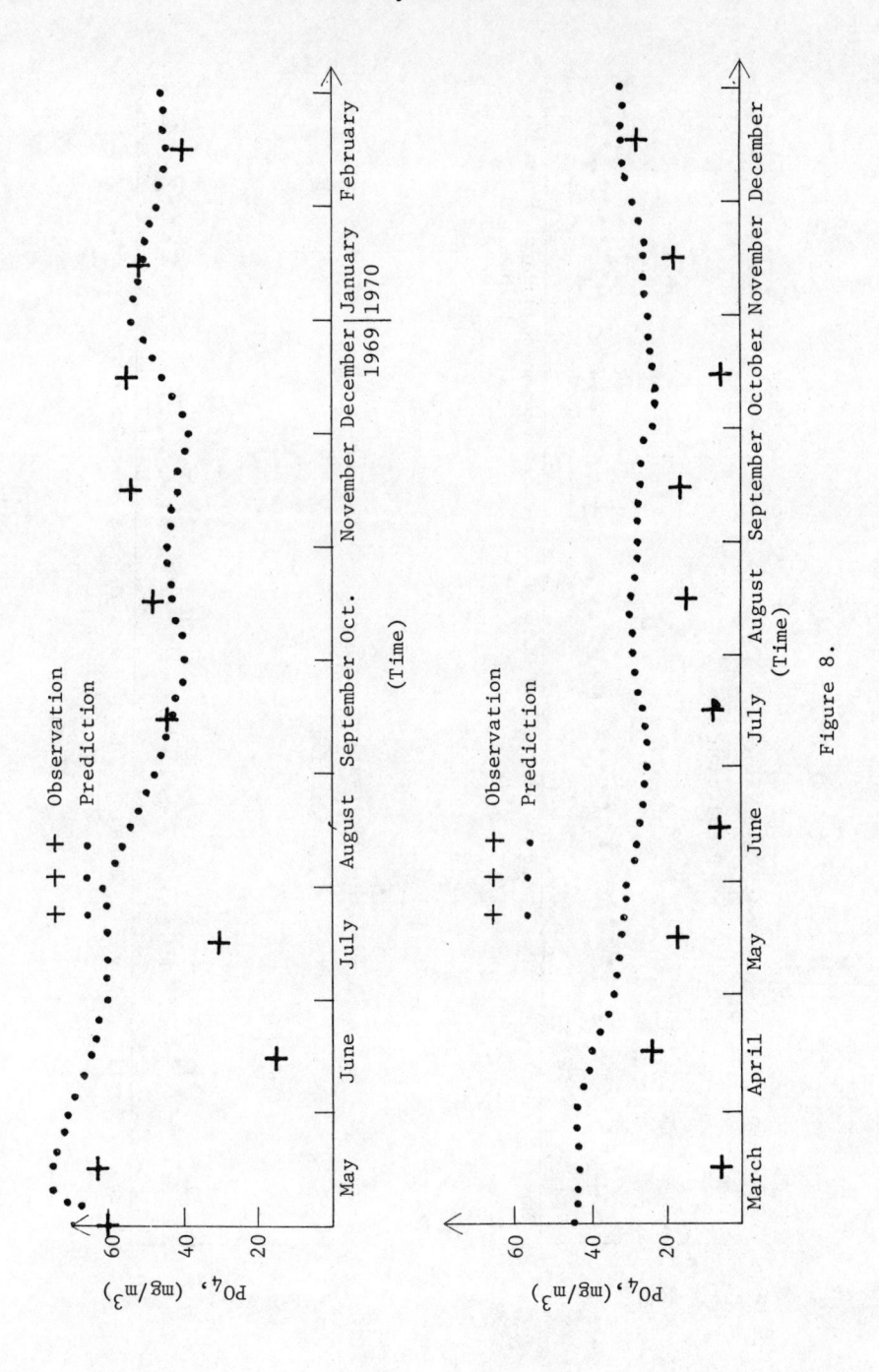

Nutrient Cycle in Reservoirs

Figure 8.

Figure 9.

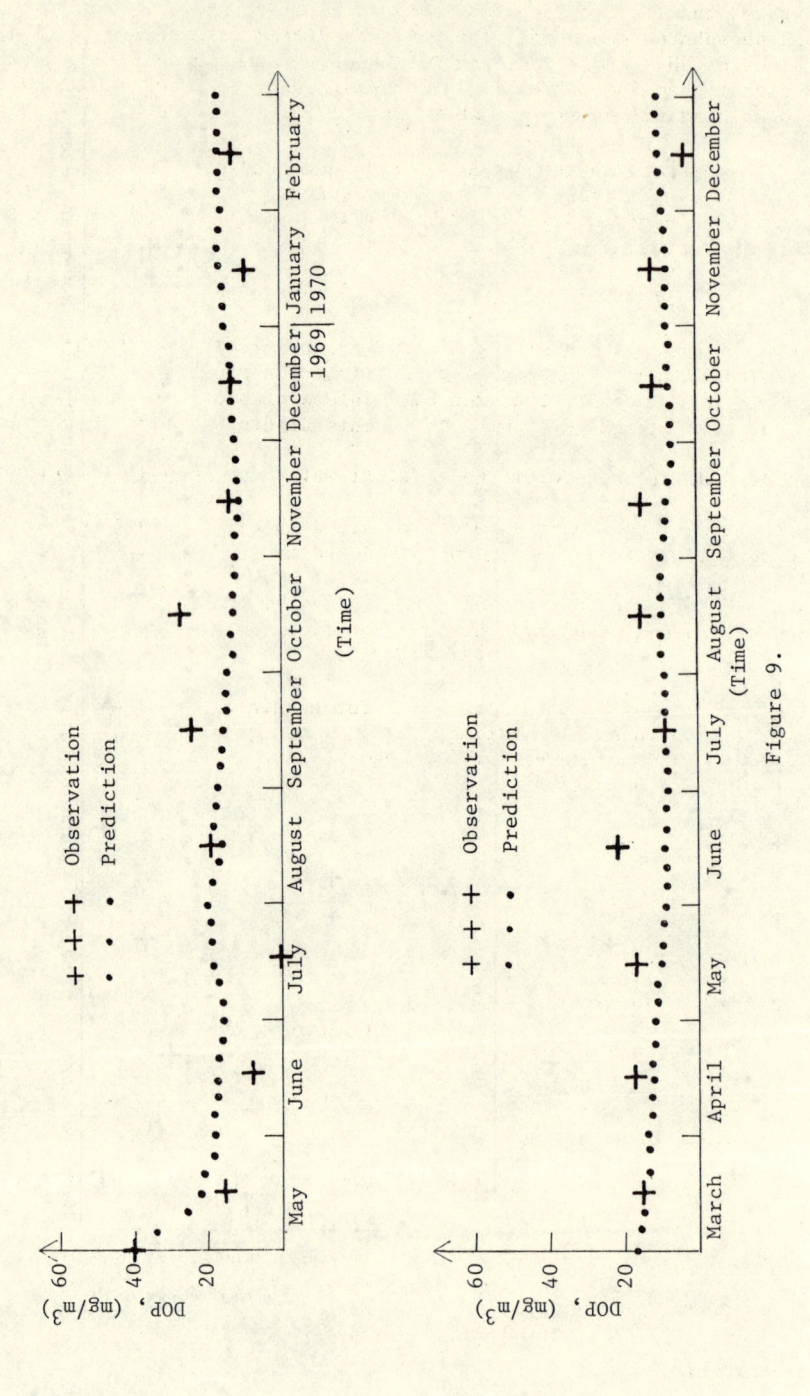

The calibrated results recorded from 1969 to 1971 are shown in Fig. 3 to 9 for nitrogen and phosphorus compounds. The model-predicted values agree well with the observed values of NH_3, $NO_2 + NO_3$ and DOP. However, the phosphate predictions are somewhat higher than the corresponding observed values. There are no PN or POP data, so their predicted values are not shown here.

Although the agreement between observations and predictions is not excellent for every compound, overall there appears to be quite a good match, and thus the model considered can be used to predict the water as a result of the various activities in the catchment area.

CONCLUSIONS

A predictive phosphorus and nitrogen model for reservoirs has been constructed and tested. The comparison of predicted values from the model with observed data is quite encouraging. However, the predicted values of phosphate are higher than the observed values. Possible explanations of this discrepancy are as follows:

(i) The continuous leakage of meteorological water due to poor seating of the valve,

(ii) Particulate inorganic phosphorus, existing in clay as apatite, and its accumulation in the sediment have not been considered in the model.

If those parameters had been included in the system inputs, then the agreement between observation and prediction would probably have been improved for this compound.

The nutrient model may be used as a basis for examining and predicting the effects on water quality of possible nutrient loadings. If the municipalities in the reservoir basins provide increasing nutrient removal, the effects of N and P loads on the waterbodies can be estimated.

REFERENCES

APHA - AWWA - WPCF (1965). *Standart Methods for the Examination of Water and Wastewater*. 12th Ed. American Public Health Association, New York.

Di Toro, D.M. and others (1971). A Dynamic Model of the Phytoplankton Population in the Sacramento-San Joaquin Delta. In *Advances in Chemistry Series*, No. 106, Nonequilibrium Systems in Natural Water Chemistry.

Hill, C.A. and others (1968). *Feasibility Report on Indian Creek Reservoir*, South Tahoe Public Utility District, California.

Lake Tahoe Area Council (1975). *Eutraophication of Surface Waters-Lake Tahoe's Indian Creek Reservoir*. EPA-660/3-75-003, February 1975.

McGauhey, P.H. and others (1971). *Eutrophication of Surface Waters-Lake Tahoe: Indian Creek Reservoir*. EPA-16010 DNY 07/71.

O'Melia, R.C. (1977). *Personal Communication*.

AUTHOR INDEX

Akçın, G.	260, 295	Fleckseder, H.	269 - 282
Ali, H. I.	23, 47		
Al-Khatib, E.	369 - 382		
Almeida, S.A.S.	475 - 484	Göçer, I.	186
Anderson, G.K.	131,141,142	Günay, J.	21
Atımtay, A.	99,114,160,204	Gür, A.	260,269
Baysal, B.	283 - 296	Hamen, J.	424
Bewtra, J.K.	23 - 47	Hamza, A.	249 - 260
Buning, W.G.W.	260,303,316 424	Hanisch, B.	1-7, 21, 46
		Harremoës, P.	49 - 63
Curi, K.	86, 87, 186, 189 - 205	Hernandez, J.W.	7, 15-20,21, 282,295,466
Çetin, C.I.	260	Howell, J.A.	207 - 213
Diyamandoğlu, V.	189 - 205	Ishihara, J.E.	345 - 355
Donelly, T.	131 - 141		
Droste, R.L.	23 - 47	Karlsson, L.	187, 424, 427 - 436
Dunn, I.J.	73 - 85	Karpuzcu, M.	495 - 510
Dunz, W.	424	Khan, M.Z.A.	261 - 268
		Koppers, H.M.M.	303 - 316
Eckenfelder, W.W.Jr.	65 - 72		
Erdin, E.	315, 437-444, 467	Krauth, K.	297 - 302
Erenler, E.	9		21,46,47,226,
Esen, İ.İ.	369 - 382	Leentvaar, J.	259, 296, 303-316

Author Index

Letten, D.J.	131 - 141	Staab, K.F.	297 - 302
Lo, S.N.	316	Suerth, M.	215 - 220
Ludwig, R.G.	475 - 484	Şendökmen, N.	317 - 328
		Şengül, F.	115 - 129, 283 - 296
Matsumoto, J.	345 - 355	Şentürk, H.	46, 179
Mitwally, H.H.	315, 383-390		
Mooss, H.	413 - 425		
Müezzinoğlu, A.	115-129, 260, 283-296, 344	Tabasaran, O.	302, 366, 403 - 412
Muthuswamy, S.	485 - 493	Taygun, N.	141, 179, 205, 317-328, 366
Muttamara, S.	227 - 240	Thanh, N.C.	227 - 240
		Tokuz, R.Y.	101 - 114
Nemerow, N.L.	469 - 474	Tünay, O.	161 - 168
Okubo, T.	345 - 355	Uslu, M.	99
Omura, T.	391 - 402	Uzman, S.	73 - 85
Onuma, M.	391 - 402	Ülkü, G.	317 - 328
Orhon, D.	161 - 168	Ürün, H.	89 - 99
Patrick, P.K.	424, 436, 465	Velicangil, Ö.	207 - 213
Rüffer, H.M.	129, 221-226, 344, 355	Velioğlu, S.G.	114, 189-205, 282, 296, 367
		Yılmaz, C.	204
Saad, S.G.	329 - 344	Yılmaz, M.	315
Saatçı, A.C.	241 - 247	Yongaçoğlu, S.	73-85, 86, 87
Samsunlu, A.	46, 86, 87, 141, 179, 220, 283-296, 302, 355, 357-368, 390, 466		
		Walker, B.	73 - 85
Sarıkaya, H.Z.	114, 143-159, 366		
Schenkel, W.	445 - 467		
Sekoulov, I.	169 - 179		
Sestini, U.	181 - 187, 204, 315		

SUBJECT INDEX

ACTIVATED SLUDGE
  Brewery   231
  Effect of Phenol   73-85
  Effect of Recycling Ratio   161-168
  Models   27
  Response to High Salinity   101-113
  Treatment of
    Fish-Processing
    Wastewaters   345-355
  Treatment of Metal
    Containing Wastes   161-168

ADVANCED TREATMENT   49

ANAEROBIC TREATMENT   131-141

BACTERIAL RESPIRATION   357

BACTERIAL GROWTH
  Effect of Stirring   31

BIODEGRADABILITY   67, 115-129
  of Brewery Wastes   360

BIOLOGICAL FILTER
  Inclined Plane   89-99

BIOLOGICAL TREATMENT   49, 297-302

BOD
  Removal   52

BREWERY WASTEWATER
  Degree of Biodegradability   360

CELLULOSE INDUSTRY
  (*see also* Pulp & Paper)   241-247
  Biological Wastewater
    Treatment   243
  Dewatering   244
  Incineration Plant   244
  Mechanical-Chemical
    Treatment   242

CHEESE INDUSTRY   215-220
  Whey Waste   140
  Treatment by
    Ultrafiltration   207

CHEMICAL INDUSTRY   5

COAGULATION-
FLOCCULATION   303-316

COMPOSTING   413-425

DAIRY
  Wastewater
    Characteristics   134

DENITRIFICATION   57

EDIBLE OIL INDUSTRY
  Wastewater
    Characteristics   134

EFFLUENT CHARGE ACT   6

EFFLUENT QUALITY
STANDARDS
  Italy   184
  Rio de Janeiro   476

ENERGY SAVING
  by Refuse Recycling   462

FERROUS IRON OXIDATION   143

FISH-PROCESSING
INDUSTRY   345-355

FOOD PRODUCTION
INDUSTRY   138, 215-220

HEAVY METALS
REMOVAL   69

HIDE GLUE FACTORIES   297-302

Subject Index

INDUSTRIAL
WASTEWATER
- Characteristics — 134
- High-Strength — 131-141
- Problems in Alexandria — 383-390
- Quantity — 2-3
- Treatability — 357-366
- Treatment — 29-46
- Treatment Economics — 475-484

IRRIGATION WITH
INDUSTRIAL WASTES — 15-21

KITAKAMI RIVER
- Water Quality in — 391-402

MATHEMATICAL MODEL
- for the Nutrient Cycle — 501-515

METAL PROCESSING
WASTEWATER
- Degree of Biodegradability — 360

NITRIFICATION — 56

NITROGEN REMOVAL — 57

NUTRIENT CYCLE — 501-515

OLIVE OIL INDUSTRY
- Treatment of Wastes — 189-205
- Waste Characteristics — 138, 190

OXIDATION DITCH
- Bamboo-Bladed Rotors — 485-493

OXYGEN UPTAKE
- Correlation with Phenol Uptake — 73-85

OZONATION — 70

PAPER INDUSTRY
(see also Pulp & Paper) — 249-260
- A Case Study — 283-296
- Sludge Disposal — 261-268

PHENOL UPTAKE — 73-85

PHOSPHORUS REMOVAL — 59

POLLUTION CONTROL — 369-382

POWER INPUT
- in Biological Treatment — 23-46

PULP INDUSTRY
- Reduction of Effluents — 269-282
- Sludge Disposal — 261-268
- Wastewater — 5

PYROLYSIS
- of Used Tyres — 409

RECYCLING
- Example from the Industry — 11
- Industrial Solid Wastes from Stockholm — 427-436
- Paper Industry — 249-260
- Solid Wastes — 403-412

REFUSE COMPOSITION — 456

REUSE
- of Industrial Solid Waste from Stockholm — 427-436
- Textile Industry — 11
- Wastewater in Paper reprocessing — 249-260

ROTARY SCREENS — 481

ROTORS
- Bamboo Bladed — 485-493

SHUAIBA INDUSTRIAL
AREA — 369-382

SLUDGE DISPOSAL — 261-268

SLURRY FREEZING
PROCESS — 261-268

SMALL INDUSTRIES — 181-187

SOLID WASTES
- Glass — 459
- Industrial — 437-444
- Iron & Steel — 460
- Models for Recycling — 403-412
- Paper — 458
- Plastics — 460
- Recycling — 462
- Refuse Composting — 413-425
- Reuse — 427-436

STARCH INDUSTRY
- Waste — 137
- Waste Treatment — 329-344

Subject Index

SUGAR INDUSTRY	303-316, 317-328	WASTEWATER	
		Characteristics	
TANNERY WASTEWATER		Brewery	227
Degree of		Hide-Glue Industry	299
Biodegradability	360	Olive-Oil Industry	190
TEXTILE INDUSTRY		Yeast Industry	223
Recycling	11	Wine Distillery	134
		Disposal	1-7
TOXIC POLLUTANTS		Irrigation	15-21
CONTROL	65-72	Renovation	249-260
		Reuse	9-14
TRICKLING FILTER			
Brewery, Pilot Scale	237	WASTEWATER TREATMENT	
		Brazil	475-484
ULTRAFILTRATION		Cost	469-471
MEMBRANES	207-213	Innovative	
		Solution	471-473
UNOX	175	Plant Model	169-179
		Using Bamboo-Rotors	485-493
YEAST FACTORY			
WASTES		WINE DISTILLERY	134
Characteristics	134, 223		
Treatment	221-226		
WASTE MANAGEMENT			
POLICY	445-467		